Building Regulations in Brief

Ray Tricker

Third edition

ELSEVIER
BUTTERWORTH
HEINEMANN

AMSTERDAM • BOSTON • HEIDELBERG • LONDON • NEW YORK • OXFORD
PARIS • SAN DIEGO • SAN FRANCISCO • SINGAPORE • SYDNEY • TOKYO

Elsevier Butterworth-Heinemann
Linacre House, Jordan Hill, Oxford OX2 8DP
30 Corporate Drive, Burlington, MA 01803

First published 2003
Reprinted (twice) 2003
Second edition 2004
Reprinted 2004
Third edition 2005

British Library Cataloguing in Publication Data
A catalogue record for this book is available from the British Library

Library of Congress Cataloguing in Publication Data
A catalogue record for this book is available from the Library of Congress

ISBN 0 7506 6703 6

For information on all Elsevier Butterworth-Heinemann publications
visit our website at http://books.elsevier.com

Typeset by Charon Tec Pvt. Ltd, India 0853291   Ref 692.3 T
www.charontec.com
Printed and bound in Great Britain by Biddles Ltd, King's Lynn, Norfolk

Contents

Foreword   xi

Preface   xiii

1   The Building Act 1984   **1**

1.1   Aim of the Building Act 1984   1
1.2   What happens if I contravene any of these requirements?   3
1.3   Who polices the Act?   3
1.4   Are there any exemptions from Building Regulations?   3
1.5   What about civil liability?   5
1.6   What does the Building Act 1984 contain?   5
1.7   What are the Supplementary Regulations?   5
1.8   What are 'Approved Documents'?   12
1.9   What is the 'Building Regulations Advisory Committee'?   12
1.10   What is 'type approval'?   13
1.11   Does the Fire Authority have any say in
        Building Regulations?   13
1.12   How are buildings classified?   13
1.13   What are the duties of the local authority?   13
1.14   What are the powers of the local authority?   15
1.15   Who are approved inspectors?   15
1.16   What causes some plans for building work to be rejected?   18
1.17   Can I apply for a relaxation in certain circumstances?   18
1.18   Can I change a plan of work once it has been approved?   19
1.19   Must I complete the approved work in a certain time?   19
1.20   How is my building work evaluated for conformance
        with the Building Regulations?   19
1.21   What about dangerous buildings?   20
1.22   What about defective buildings?   22
1.23   What are the rights of the owner or occupier of the premises?   23
1.24   Can I appeal against a local authority's ruling?   23
        Appendix A   Contents of the Building Act 1984   24

2   The Building Regulations 2000   **28**

2.1   What is the purpose of the Building Regulations?   28
2.2   Why do we need the Building Regulations?   28
2.3   What building work is covered by the Building Regulations?   29

2.4	What are the requirements associated with the Building Regulations?	30
2.5	What are the Approved Documents?	30
2.6	Are there any exemptions?	41
2.7	What happens if I do not comply with an Approved Document?	42
2.8	Do I need Building Regulations approval?	42
2.9	How do I obtain Building Regulations approval?	46
2.10	What are building control bodies?	46
2.11	How do I apply for building control?	48
2.12	Full plans application	49
2.13	Building notice procedure	50
2.14	How long is a building notice valid?	53
2.15	What can I do if my plans are rejected?	53
2.16	What happens if I wish to seek a determination but the work in question has started?	55
2.17	When can I start work?	55
2.18	Planning officers	56
2.19	Building inspectors	56
2.20	Notice of commencement and completion of certain stages of work	56
2.21	What are the requirements relating to building work?	58
2.22	Do I need to employ a professional builder?	58
2.23	Unauthorized building work	59
2.24	Why do I need a completion certificate?	60
2.25	How do I get a completion certificate when the work is finished?	60
2.26	Where can I find out more?	60
Appendix 2A	Example application form	62
Appendix 2B	Example planning permission form	64
Appendix 2C	Example of an application for listed building consent	66
Appendix 2D	Typical application for agricultural/forestry determination	68
Appendix 2E	Example of an application for consent to display advertisements	70
Appendix 2F	Regularization (Example)	72
3	**The requirements of the Building Regulations**	**74**
3.1	Part A – Structure	75
3.2	Part B – Fire safety	76
3.3	Part C – Site preparation and resistance to contaminants and moisture	79
3.4	Part D – Toxic substances	80
3.5	Part E – Resistance to the passage of sound	81
3.6	Part F – Ventilation	83

3.7	Part G – Hygiene	84
3.8	Part H – Drainage and waste disposal	86
3.9	Part J – Combustion appliances and fuel storage systems	91
3.10	Part K – Protection from falling, collision and impact	93
3.11	Part L1 – Conservation of fuel and power in dwellings	95
3.12	Part L2 – Conservation of fuel and power in buildings other than dwellings	96
3.13	Part M – Access to and use of buildings	97
3.14	Part N – Glazing – safety in relation to impact, opening and cleaning	99
3.15	Part P – Electrical safety	100

4 Planning permission   **101**

4.1	Planning controls	102
4.2	Who requires planning permission?	102
4.3	Who controls planning permission?	103
4.4	What is planning permission?	105
4.5	What types of planning permission are available?	106
4.6	How do I apply for planning permission?	106
4.7	Do I really need planning permission?	107
4.8	How should I set about gaining planning permission?	107
4.9	What sort of plans will I have to submit?	112
4.10	What is meant by 'building works'?	113
4.11	What important areas should I take into consideration?	113
4.12	What are the government's restrictions on planning applications?	114
4.13	How do I apply for planning permission?	115
4.14	What is the planning permission process?	116
4.15	Can I appeal if my application is refused?	122
4.16	Before you start work	123
4.17	What could happen if you don't bother to obtain planning permission?	130
4.18	How much does it cost?	131

5 Requirements for planning permission and Building Regulations approval   **136**

5.1	Decoration and repairs inside and outside a building	137
5.2	Structural alterations inside	138
5.3	Replacing windows and doors	139
5.4	Electrical work	140
5.5	Plumbing	141
5.6	Central heating	142
5.7	Oil-storage tank	142
5.8	Planting a hedge	142
5.9	Building a garden wall or fence	143
5.10	Felling or lopping trees	143

5.11   Laying a path or a driveway   144
5.12   Building a hardstanding for a car, caravan or boat   145
5.13   Installing a swimming pool   146
5.14   Erecting aerials, satellite dishes and flagpoles   146
5.15   Advertising   147
5.16   Building a porch   148
5.17   Outbuildings   149
5.18   Garages   151
5.19   Building a conservatory   152
5.20   Loft conversions, roof extensions and dormer windows   153
5.21   Building an extension   155
5.22   Conversions   158
5.23   Change of use   159
5.24   Building a new house   164
5.25   Infilling   165
5.26   Demolition   166

6   Meeting the requirements of the Building Regulations   169

6.1   Foundations   173
6.2   Buildings – size   198
6.3   Drainage   203
6.4   Water supplies   235
6.5   Cellars   236
6.6   Floors and ceilings   238
6.7   Walls   286
6.8   Ceilings   365
6.9   Roofs   368
6.10   Chimneys   386
6.11   Stairs   416
6.12   Windows   434
6.13   Doors   441
6.14   Vertical circulation within the building   446
6.15   Corridors and passageways   453
6.16   Facilities in buildings other than dwellings   456
6.17   Water (and earth) closets   469
6.18   Electrical safety   492
6.19   Combustion appliances   500
6.20   Hot water storage   509
6.21   Liquid fuel   511
6.22   Cavities and concealed spaces   515
6.23   Kitchens and utility rooms   518
6.24   Storage of food   518
6.25   Refuse facilities   519
6.26   Fire resistance   520
6.27   Means of escape   524

6.28	Bathrooms	534
6.29	Loft conversions	537
6.30	Entrance and access	541
6.31	Extensions to buildings	566
6.32	External balconies	568
6.33	Garages	569
6.34	Conservatories	570
6.35	Rooms for residential purposes	572
6.36	Rooms for residential purposes resulting from a material change of use	575
6.37	Reverberation in the common internal parts of buildings containing flats or rooms for residential purposes	577
6.38	Internal walls and floors (new buildings)	579
6.39	Regulation 7 – Materials and workmanship	583
6.40	Work on existing constructions	587

Appendix A	*Access and facilities for disabled people*	596
Appendix B	*Conservation of fuel and power*	625
Appendix C	*Sound insulation*	640
Appendix D	*Guidance to the requirements of Part P – Electrical safety*	646
Bibliography		662
Useful contact names and addresses		683
Index		689

Foreword

Subject to specified exemptions, all building work in England and Wales (a separate system of building control applies to Scotland and Northern Ireland) is governed by Building Regulations. This is a statutory instrument, which sets out the minimum requirements and performance standards for the design and construction of buildings, and extensions to buildings.

The current regulations are the Building Regulations 2000. These take into consideration some major changes in technical requirement (such as conservation of fuel and power) and some procedural changes allowing local authorities to regularize unauthorized development.

Although the 2000 regulations are comparatively short, they rely on their technical detail being available in a series of Approved Documents and a vast number of British, European and international standards, codes of practice, drafts for development, published documents and other non-statuary guidance documents.

The main problem, from the point of view of the average builder and DIY enthusiast, is that the Building Regulations are too professional for their purposes. They cover every aspect of building, are far too detailed and contain too many options. All the builder or DIY person really requires is sufficient information to enable them to comply with the regulations in the simplest and most cost-effective manner possible.

Building inspectors, acting on behalf of local authorities, are primarily concerned with whether a building complies with the requirements of the Building Regulations and to do this, they need to 'see the calculations'. But how do the DIY enthusiast and/or builder obtain these calculations? Where can they find, for instance, the policy and requirements for load bearing elements of a structure?!

Builders, through experience, are normally aware of the overall requirements for foundations, drains, walls, central heating, air conditioning, safety, security, glazing, electricity, plumbing, roofing, floors, etc., but they still need a reminder when they come across a different situation for the first time (e.g. what if they are going to construct a building on soft soil, how deep should the foundations have to be?).

On the other hand, the DIY enthusiast, keen on building his own extension, conservatory, garage or workshop etc. usually has no past experience and needs the relevant information – but in a form that he can easily understand without having had the advantage of many years experience. In fact, what he really needs is a rule of thumb guide to the basic requirements.

From a number of surveys it has emerged that the majority of builders and virtually all DIY enthusiasts are self taught and most of their knowledge is gained through experience. When they hit a problem, it is usually discussed over a pint in the local pub with friends in the building trade as opposed to seeking professional help. What they really need is a reference book to enable them to understand (or remind themselves of) the official requirements.

The aim of my book, therefore, is to provide the reader with an in-brief guide that can act as an *aide-mémoire* to the current requirements of the Building Regulations. Intended readers are primarily builders and the DIY fraternity (who need to know the regulations but do not require the detail), but the book, with its ready reference and no-nonsense approach, will be equally useful to students, architects, designers, building surveyors and inspectors, etc.

This edition of the book, as well as including the requirements from the new Part P (Electrical safety) and the new versions of Part A (Structure) and Part C (Site preparation and resistance to contaminants and moisture) also include outline details of the new (i.e. proposed) Part Q (Electronic communications services) – whose draft is currently out for consultation and comments. Possible changes that are likely to be made in the near future to Parts F and L have also been noted in the relevant text.

 Note: If any reader has any thoughts about the contents of this book (such as areas where perhaps they feel I have not given sufficient coverage, omissions and/or mistakes, etc.) then please let me know by e-mailing me at ray@herne. org.uk and I will make suitable amendments in the next edition of this book.

Preface

The Great Fire of London in 1666 was probably the single most significant event to shape today's legislation! The rapid growth of fire through co-joined timber buildings highlighted the need to consider the possible spread of fire between properties and this consideration resulted in the publication of the first building construction legislation in 1667 requiring all buildings to have some form of fire resistance.

Two hundred years later, the Industrial Revolution had meant poor living and working conditions in ever expanding, densely populated urban areas. Outbreaks of cholera and other serious diseases, through poor sanitation, damp conditions and lack of ventilation, forced the government to take action and building control took on the greater role of health and safety through the first Public Health Act of 1875. This Act had two major revisions in 1936 and 1961, leading to the first set of national building standards (i.e. the Building Regulations 1965). Over the years these regulations have been amended and updated and the current document is the Building Regulations 2000.

The Building Regulations are approved by the Secretary of State and are intended to provide guidance to some of the more common building situations as well as providing a practical guide to meeting the requirements of Regulation 7 of the Building Act 1984, which states:

Materials and workmanship
7.   Building work shall be carried out –

(a) with adequate and proper materials which –
     (i)   are appropriate for the circumstances in which they are used,
     (ii) are adequately mixed or prepared, and
     (iii) are applied, used or fixed so as adequately to perform the functions for which they are designed; and

(b) in a workmanlike manner.

What are the current regulations?

The current legislation is the Building Regulations 2000 (Statutory Instrument No 2531) which is made by the Secretary of State for the Environment under powers delegated by parliament under the Building Act 1984. Since then,

the Building Regulations have received a number of Building Amendment
Regulations as shown below.

Table P.1 Statutory instruments currently in place

The Building Regulations 2000 (SI 2000 No 2531)

Made	*13 September 2000*
Laid before Parliament	*22 September 2000*
Came into force	*1 January 2001*

Statutory Instrument	Made	Laid before Parliament	Coming into force
SI 2001 No 3335	4 Oct 2001	11 Oct 2001	1 Apr 2002
SI 2002 No 440	28 Feb 2002	5 Mar 2002	1 Apr 2002
SI 2002 No 2871	16 Nov 2002	25 Nov 2002	1 Jul 2003 (less sound insulation) 1 Jan 2004 (sound insulation)
SI 2003 No 2692	17 Oct 2003	27 Oct 2003	1 Dec 2003 (Regulations 1, 2(1) & (8) plus 3(5)) 1 May 2003 (remainder)
SI 2004 No 1465	28 May 2004	8 Jun 2004	1 Jul 2004 Regulations 1(1), (2), (4) and (5) 1 Dec 2004 (remainder)
SI 2004 No 3210	6 Dec 2004	10 Dec 2004	31 Dec 2004

 Note: Copies of the above documents are available from TSO (☎ 0870 600 5522) and through booksellers. They can also be viewed on the OPDM website at www.safety.odpm.gov.uk/bregs/brpub/01.htm

The Building Act 1984

By Act of Parliament, the Secretary of State is responsible for ensuring that
the health, welfare and convenience of persons living in or working in
(or nearby) buildings is secured. This Act is called the Building Act 1984
and one of its prime purposes is to assist in the conservation of fuel and
power, prevent waste, undue consumption, and the misuse and contamination
of water.

It imposes on owners and occupiers of buildings a set of requirements concerning the design and construction of buildings and the provision of services,
fittings and equipment used in (or in connection with) buildings.

The Building Act 1984 consists of five parts:

Part 1   The Building Regulations
Part 2   Supervision of Building Work etc. other than by a Local Authority
Part 3   Other provisions about buildings

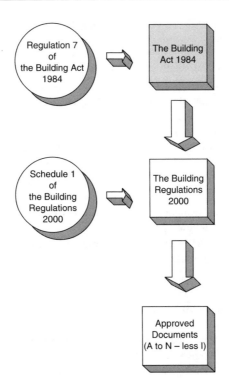

Figure P.1 Implementing the Building Act

Part 4   General
Part 5   Supplementary

Part 5 then contains seven schedules whose prime function is to list the principal areas requiring regulation and to show how the Building Regulations are to be controlled by local authorities. These schedules are:

Schedule 1 – Building Regulations;
Schedule 2 – Relaxation of building regulations;
Schedule 3 – Inner London;
Schedule 4 – Provisions consequential upon public body's notice;
Schedule 5 – Transitional provisions;
Schedule 6 – Consequential amendments;
Schedule 7 – Repeals.

Schedule 1 is the most important (from the point of view of builders) as it shows, in general terms, how the Building Regulations are to be administered by local authorities, the approved methods of construction and the approved types of materials that are to be used in (or in connection with) buildings.

 The Building Act 1984 does not apply to Scotland or to Northern Ireland.

The Building Regulations describe the mandatory requirements for completing **all** building work including:

- accommodation for specific purposes (e.g. for disabled persons);
- air pressure plants;
- cesspools (and other methods for treating and disposing of foul matter);
- dimensions of rooms and other spaces (inside buildings);
- drainage (including waste disposal units);
- emission of smoke, gases, fumes, grit or dust (or other noxious or offensive substances);
- fire precautions (services, fittings and equipment, means of escape);
- lifts (escalators, hoists, conveyors and moving footways);
- materials and components (suitability, durability and use);
- means of access to and egress from;
- natural lighting and ventilation of buildings;
- open spaces around buildings;
- prevention of infestation;
- provision of power outlets;
- resistance to moisture and decay;
- site preparation;
- solid fuel, oil, gas, electricity installations (including appliances, storage tanks, heat exchangers, ducts, fans and other equipment);
- standards of heating, artificial lighting, mechanical ventilation and air-conditioning;
- structural strength and stability (overloading, impact and explosion, under-pinning, safeguarding of adjacent buildings);
- telecommunications services (wiring installations for telephones, radio and television);
- third party liability (danger and obstruction to persons working or passing by building work);
- transmission of heat;
- transmission of sound;
- waste (storage, treatment and removal);
- water services, fittings and fixed equipment (including wells and bore-holes for supplying water); and
- matters connected with (or ancillary to) any of the foregoing matters.

The Building Regulations

Building Regulations 2000 (Statutory Instrument No 2531) has been made by the Secretary of State for the Environment under powers delegated by Parliament under the Building Act 1984. They are a set of minimum require-ments and basic performance standards designed to secure the health, safety and welfare of people in and around buildings and to conserve fuel and energy in England and Wales.

They are legal requirements laid down by parliament and based on the Building Act 1984. The Building Regulations:

- are approved by parliament;
- deal with the minimum standards of design and building work for the construction of domestic, commercial and industrial buildings;
- set out the procedure for ensuring that building work meets the standards laid down;
- are designed to ensure structural stability;
- promote the use of suitable materials to provide adequate durability, fire and weather resistance, and the prevention of damp;
- stipulate the minimum amount of ventilation and natural light to be provided for habitable rooms;
- ensure the health and safety of people in and around buildings (by providing functional requirements for building design and construction);
- promote energy efficiency in buildings;
- contribute to meeting the needs of disabled people.

The level of safety and standards acceptable are set out as guidance in the approved documents. Compliance with the detailed guidance of the Approved Documents is usually considered as evidence that the Building Regulations themselves have been complied with.

Approved Documents

The Building Regulations are supported by separate documents which correspond to the different areas covered by the regulations. These are called 'Approved Documents' and they contain practical and technical guidance on ways in which the requirements of Schedule 1 and Regulation 7 of the Building Act 1984 can be met.

Each Approved Document reproduces the actual *requirements* contained in the Building Regulations relevant to the subject area. This is then followed by *practical and technical guidance* (together with examples) showing how the requirements can be met in some of the more common building situations. There may, however, be alternative ways of complying with the requirements to those shown in the Approved Documents and you are, therefore, under no obligation to adopt any particular solution in an Approved Document if you prefer to meet the requirement(s) in some other way.

The current set of approved documents are in 13 parts, A to N (less 'I') and consist of:

A   Structural
B   Fire safety
C   Site preparation and resistance to moisture
D   Toxic substances
E   Resistance to the passage of sound

F   Ventilation
G   Hygiene
H   Drainage and waste disposal
J   Combustion appliances and fuel storage systems
K   Protection from falling, collision and impact
L   Conservation of fuel and power
M   Access and facilities for disabled people
N   Glazing – safety in relation to impact, opening and cleaning
P   Electrical safety

Parts A to D, F to K (except for paragraphs H2 and J6), N and P of Schedule 1 do not require anything to be done except for the purpose of securing reasonable standards of health and safety for persons in (or about) buildings and for any others who may be affected by buildings, or matters connected with buildings.

 At the time of writing this edition of the book, there is also another Approved Document that is in the process of being written, namely Part Q (Electronic communication services) which is currently at the Consultation Stage.

Amendments are also being made to the following documents which, although only at Consultation Draft, will be issued during the next 2 years.

- Part F (Ventilation)
- Part L (Conservation of fuel and power)

 Notes:

(1)   Paragraphs H2 and J6 are excluded from Regulation 8 because they deal directly with the prevention of contamination of water.
(2)   Parts E and M (which deal, respectively, with resistance to the passage of sound, and access to and use of buildings) are excluded from Regulation 8 because they address the welfare and convenience of building users.
(3)   Part L is excluded from Regulation 8 because it addresses the conservation of fuel and power.

Planning permission

Planning permission is the single biggest hurdle for anyone who has acquired land on which to build a house, or wants to extend or carry out other building work on property. There is never a guarantee that permission will be given and without it no project can start. Yet the system is not at all user-friendly.

There is a bewildering array of formalities to go through and ever more stringent requirements to satisfy. Planning permission has never been more difficult to get, nor so sought after. Every year over half-a-million applications are made and the number is rising.

The purpose of the planning system is to protect the environment as well as public amenities and facilities. It is **not** designed to protect the interests of one person over another. Within the framework of legislation approved by parliament, councils are tasked to ensure that development is allowed where it is needed, while ensuring that the character and amenity of the area are not adversely affected by new buildings or changes in the use of existing buildings and/or land.

Provided, that the work you are completing does not affect the external appearance of the building, you are allowed to make certain changes to your home without having to apply to the local council for permission. These are called 'Permitted Development Rights', but the majority of building work, that you are likely to complete will, however, probably require you to have planning permission – so be warned!

The actual details of planning requirements are complex but for most domestic developments, the planning authority is only really concerned with construction work such as an extension to the house (e.g. a conservatory) or the provision of a new garage or new outbuildings. Structures like walls and fences also need to be considered because their height or siting might well infringe the rights of their neighbours and other members of the community. The planning authority will also want to approve any change of use, such as converting a house into flats or running a business from premises previously occupied as a dwelling only.

Aim of this book

The prime aim of this book is to provide builders and DIY people with an aide memoire and a quick reference to the requirements of the Building Regulations. This book provides a user-friendly background to the Building Act 1984 and its associated Building Regulations. It explains the meaning of the Building Regulations, their current status, requirements, associated documentation and how local authorities and councils view their importance. It goes on to describe the content of the guidance documents (i.e. the 'Approved Documents') published by the Secretary of State and, in a series of 'what ifs', provides answers to the most common questions that DIY enthusiasts and builders might ask concerning building projects.

The book is structured as follows:

Chapter 1 – The Building Act 1984
Chapter 2 – The Building Regulations 2000
Chapter 3 – The requirements of the Building Regulations
Chapter 4 – Planning permission
Chapter 5 – How to comply with the requirements of the Building Regulations
Chapter 6 – Meeting the requirements of the Building Regulations

These chapters are then supported by the following appendices:

Appendix A   Access and facilities for disabled people
Appendix B   Conservation of fuel and power
Appendix C   Sound insulation
Appendix D   Electrical safety

and concludes with a bibliography, useful names and addresses, and a full index.

The following symbols will help you get the most out of this book:

 an important requirement or point

 a good idea or suggestion

 further amplification or information

Main changes in the third edition of *The Building Regulations in Brief*

The third edition of this book has been produced and rewritten around the new versions and amendments to:

- Part P (Electrical safety) which came into effect on 1 Jan 2005.
- Part A (Structure) which came into effect on 1 Jan 2005.
- Part C (Site preparation and resistance to contaminants and moisture) which finally came into effect on 1 Dec 2004.
- Part E (Resistance to passage sound) which is an amendment of the previous 2003 edition and which came into effect on 1 Jan 2005.

Consideration has also been given to the revised proposals for Parts L and F (which are currently being circulated for consultation and comment) as well as the new part covering electronic communication services (Part Q). Although these documents are at the time of writing still 'unofficial', where appropriate I have drawn attention to the possible changes that might happen in the future with the following marker:

Possible future amendment	

This book will be further amended (and published as new editions) to reflect these (and other) ongoing changes as soon as these documents are formalized. In the meantime you should treat any information contained in these boxes as 'draft proposals'.

Part P – Electrical safety

This is a new part to the Building Regulations which alters the way in which fixed electrical installations (which are intended to operate at low voltage or extra-low voltage and which are not controlled by the Electricity Supply Regulations 1988 as amended, or the Electricity at Work Regulations 1989 as amended) may be installed in future.

The prime purpose of these changes is to increase the safety of householders by improving the design, installation, inspection and testing of electrical installations in dwellings when they (i.e. the installations) are being newly built, extended or altered, and requires that:

- all new electrical wiring or electrical components for domestic premises (or small commercial premises linked to domestic accommodation) will have to be designed and installed in accordance with the Building Regulations;
- all fixed electrical installations (i.e. wiring and appliances fixed to the building fabric such as socket outlets, switches, consumer units and ceiling fittings) shall be designed, installed, inspected, tested and certified to BS 7671.

Part P also introduces a requirement for new cable core colours for AC power circuits.

 It is understood that the Government is also intending to introduce a scheme whereby domestic installations are checked at regular intervals (as well as when they are sold and/or purchased) to make sure that they comply with the Building Regulations. This would mean, of course, that if you had an installation which was **not** correctly certified, then your house insurance may well **not** be valid.

Part A – Structure

Principal changes include the removal of the much loved timber tables (now available as a TRADA document) and the return of disproportionate collapse requirements for certain types of buildings. Different methods of calculating wind speed and loadings have also been introduced but the primary changes concern:

A1 and A2

Traditional dwellings:

- the sizing of timber floors and roofs for traditional house construction has been removed as the timber tables are now published by TRADA;
- a revised map of basic wind speeds in accordance with BS 6399: Part 2 is now included;
- stainless steel cavity wall ties have been specified for all houses regardless of their location;

- the guidance on masonry walls to dwellings has been extended to enable the rules to be applicable when using either the appropriate British Standards or the emerging BS EN CEN Standards;
- concrete foundations to houses have been revised to align with the recommendations given in the British Standards and other authoritative guidance. Recommendations on minimum foundation depths have also been included to counter the impact of predicted climate changes;
- the design and construction of domestic garages has been extensively updated to reflect modern practice.

A3

Disproportionate collapse:

- the *Application Limit to the Requirement* (i.e. the five storey limit) has been removed so as to bring **all** buildings under the control of the A3 Requirement.

 Note: Although Part A is primarily concerned with the traditional masonry construction of buildings, it is recognized that other forms of construction – such as timber framed, prefabricated timber, light steel and pre-cast concrete – are already in use in the housing sector and are (if correctly installed) equally compliant with the recommendations of this Part.

Part C – Site preparation and resistance to contaminants and moisture

The principal changes are a much greater emphasis on control of contaminated sites and the introduction of requirements for ventilation of roof voids (formerly F2). The chief modifications concern:

Site preparation

- Site investigation is now recommended as the method for determining how much unsuitable material should be removed.

Resistance to contaminants

- Remedial measures for dealing with land affected by contaminants have been expanded to include biological, chemical and physical treatment processes.
- The area of land that is subject to measures to deal with contaminants now includes the land around the building.

Resistance to moisture

- More information concerning the interface between walls and doors and windows and the need to check rebates (in the most exposed parts of the country).
- Former Requirement F2: Condensation in roofs, has been transferred to Part C as it deals with effects on the building fabric rather than ventilation for the health of occupants.

1

The Building Act 1984

1.1 Aim of the Building Act 1984 (Building Act 1984 Section 1)

By Act of Parliament, the Secretary of State is responsible for ensuring that the health, welfare and convenience of persons living in or working in (or nearby) buildings is secured.

This Act is called the Building Act 1984 and one of its prime purposes is to assist in the conservation of fuel and power, to prevent waste, undue consumption, misuse and contamination of water. It imposes on owners and occupiers of buildings a set of requirements concerning the design and construction of buildings and the provision of services, fittings and equipment used in (or in connection with) buildings. These involve, and cover:

- a method of controlling (inspecting and reporting) buildings;
- how services, fittings and equipment may be used;
- the inspection and maintenance of any service, fitting or equipment used.

1.1.1 What about the rest of the United Kingdom?

As shown in Table 1.1, the Building Act 1984 does **not** apply to Scotland or Northern Ireland.

Scotland

Within Scotland, the requirements for buildings are controlled by the *Building (Scotland) Act 2003*. The *Building (Scotland) Regulations 2004* then set the

Table 1.1 Building legislation

	Act	Regulations	Implementation
England and Wales	Building Act 1984	Building Regulations 2000	Approved Documents
Scotland	Building (Scotland) Act 2003	Building (Scotland) Regulations 2004	Technical Handbooks
Northern Ireland	Building Regulations (Northern Ireland) Order 1979	Building Regulations (Northern Ireland) 2000	'Deemed to satisfy' by meeting supporting publications

functional standards under this Act. The methods for implementing these requirements are similar to England and Wales, except that the guidance documents (i.e. for achieving compliance) are contained in two *Technical Handbooks*, one for domestic work and one for non-domestic. Each handbook has a general section and 6 technical sections.

The main procedural difference between the Scottish system and the others is that a building warrant is **still** required before work can start in Scotland.

	England and Wales		Scotland		Northern Ireland
Part A	Structure	Section 1	Structure	Technical Booklet D	Structure
Part B	Fire safety	Section 2	Fire	Technical Booklet E	Fire safety
Part C	Site preparation and resistance to contaminants and water	Section 3	Environment	Technical Booklet C	Preparation of site and resistance to moisture
Part D	Toxic substances	Section 3	Environment	Technical Booklet B	Materials and workmanship
Part E	Resistance to the passage of sound	Section 5	Noise	Technical Booklet G	Sound insulation of dwellings
Part F	Ventilation	Section 3	Environment	Technical Booklet K	Ventilation
Part G	Hygiene	Section 3	Environment	Technical Booklet P	Sanitary appliances and unvented hot water storage systems
Part H	Drainage and waste disposal	Section 3	Environment	Technical Booklet J	Solid waste in buildings
				Technical Booklet N	Drainage
Part J	Combustion appliances and fuel storage systems	Section 3	Environment	Technical Booklet L	Heat-producing appliances and liquefied petroleum gas installations
		Section 4	Safety		
Part K	Protection from falling, collision and impact	Section 4	Safety	Technical Booklet H	Stairs, ramps and protection from impact
Part L	Conservation of fuel and power	Section 6	Energy	Technical Booklet F	Conservation of fuel and power
Part M	Access and facilities for disabled people	Section 4	Safety	Technical Booklet R	Access for facilities and disabled people
Part N	Glazing	Section 6	Energy	Technical Booklet V	Glazing
Part P	Electrical safety	Section 4	Safety		

Northern Ireland

On the other hand, the *Building Regulations (Northern Ireland) Order 1979* (as amended by the *Planning and Building Regulations (Amendment) (NI) Order 1990*) is the main legislation for Northern Ireland and the *Building Regulations (Northern Ireland) 2000* then details the requirements for meeting this legislation.

Supporting publications (such as British Standards, BRE publications and/or Technical Booklets published by the Department) are then used to ensure that the requirements are implemented (*deemed to satisfy*).

1.2 What happens if I contravene any of these requirements? (Building Act 1984 Sections 2, 7, 35, 36 and 112)

If you contravene the Building Regulations or wilfully obstruct a person acting *in the execution of the Building Act 1984 or of its associated Building Regulations*, then on summary conviction, you could be liable to a fine or, in exceptional circumstances, even a short holiday in HM Prisons!

1.3 Who polices the Act?

Under the terms of the Building Act 1984, local authorities are responsible for ensuring that any building work being completed conforms to the require-ments of the associated Building Regulations. They have the authority to:

- make you take down and remove or rebuild anything that contravenes a regulation;
- make you complete alterations so that your work complies with the Building Regulations;
- employ a third party (and then send you the bill!) to take down and rebuild non-conforming buildings or parts of buildings.

They can, in certain circumstances, even take you to court and have you fined – especially if you fail to complete the removal or rebuilding of the non-conforming work.

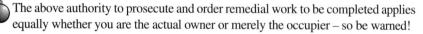

The above authority to prosecute and order remedial work to be completed applies equally whether you are the actual owner or merely the occupier – so be warned!

1.4 Are there any exemptions from Building Regulations? (Building Act 1984 Sections 3, 4 and 5)

The following are exempt from the Building Regulations:

- A '*public body*' (i.e. local authorities, county councils and any other body '*that acts under an enactment for public purposes and not for its own profit*').

This can be rather a grey area and it is best to seek advice if you think that you come under this category;

- Buildings belonging to '*statutory undertakers*' (e.g. a water board).

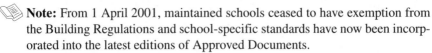 **Note:** From 1 April 2001, maintained schools ceased to have exemption from the Building Regulations and school-specific standards have now been incorporated into the latest editions of Approved Documents.

Purpose-built student living accommodation (including flats) should thus be treated as hotel/motel accommodation in respect of space requirements and internal facilities.

1.4.1 What about Crown buildings? (Building Act 1984 Sections 44a, d and 87)

Although the majority of the requirements of the Building Regulations are applicable to Crown buildings (i.e. a building in which there is a Crown or Duchy of Lancaster or Duchy of Cornwall interest) or government buildings (held in trust for Her Majesty) there are occasional deviations and before submitting plans for work on a Crown building you should seek the advice of the Treasury.

1.4.2 What about buildings in Inner London? (Building Act 1984 Sections 44, 46 and 88)

You will find that the majority of the requirements found in the Building Regulations are also applicable to buildings in Inner London boroughs (i.e. Inner Temple and Middle Temple). There are, however, some important deviations (see Section 1.7.3) and before submitting plans you should seek the advice of the local authority concerned.

1.4.3 What about the UK Atomic Energy Authority? (Building Act 1984 Section 45)

The Building Regulations do **not** apply to buildings belonging to or occupied by the United Kingdom Atomic Energy Authority (UKAEA) unless they are dwelling houses and offices.

1.4.4 What about the British Airports Authority? (Building Act 1984 Section 45)

The Building Regulations do **not** apply to buildings belonging to or occupied by the British Airports Authority, unless it is a house, hotel or building used as offices or showrooms.

1.4.5 What about the Civil Aviation Authority? (Building Act 1984 Section 45)

The Building Regulations do **not** apply to buildings belonging to or occupied by the Civil Aviation Authority, unless it is a house, hotel or building used as offices or showrooms.

1.5 What about civil liability? (Building Act 1984 Section 38)

It is an aim of the Building Act 1984 that all building work is completed safely and without risk to people employed on the site or visiting the site etc. Any contravention of the Building Regulations that causes injury (or death) to any person is liable to prosecution in the normal way.

1.6 What does the Building Act 1984 contain?

The Building Act 1984 consists of five parts:

Part 1   The Building Regulations
Part 2   Supervision of building work etc. other than by a local authority
Part 3   Other provisions about buildings
Part 4   General
Part 5   Supplementary

These parts are then broken down into a number of sections and subsections as shown in Appendix A to this chapter.

1.7 What are the Supplementary Regulations?

Part 5 of the Building Act contains seven schedules whose function is to list the principal areas requiring regulation and to show how the Building Regulations are to be controlled by the local authority. These schedules are:

- Schedule 1 – Building Regulations;
- Schedule 2 – Relaxation of Building Regulations;
- Schedule 3 – Inner London;
- Schedule 4 – Provisions consequential upon public body's notice;
- Schedule 5 – Transitional provisions;
- Schedule 6 – Consequential amendments;
- Schedule 7 – Repeals.

1.7.1 What is Schedule 1 of Part 5 of the Building Act 1984?

The Building Regulations are a statutory instrument, authorized by parliament, which details how the generic requirements of the Building Act are to be met. Compliance with Building Regulations is required for **all**:

- alterations and extensions of buildings (including services, fixtures and fittings);
- provision of new services, fittings or equipment;

unless (in most circumstances) **the increased area of the alteration or extension is less than 30 m^2** ($35.9 y^2$) in which case the Building Regulations provide the generic and specific requirements for this work.

Building Regulations **also** apply to alterations and extensions being completed on buildings erected **before** the date on which the regulations came into force.

Schedule 1 of the Building Act 1984 shows, in general terms, how the Building Regulations are to be administered by local authorities, the approved methods of construction and the approved types of materials that are to be used in (or in connection with) buildings.

How are the Building Regulations controlled?

To assist local authorities, Section 1 shows:

- how notices are given;
- how plans of proposed work (or work already executed) are deposited;
- how copies of deposited plans are administered and retained;
- how documents are to be controlled;
- how work is tested;
- how samples are taken;
- how local authorities can seek external expertise to assist them in their duties;
- how certificates signifying compliance with the Building Regulations are to be issued;
- how local authorities can accept certificates from a person (or persons) nominated to act on their behalf;
- how proposed work can be prohibited;
- when a dispute arises, how local authorities can refer the matter to the Secretary of State;
- what fees (and what level of fees) local authorities can charge.

What are the requirements of the Building Regulations?

Schedule 1 describes the mandatory requirements for completing **all** building work. These include:

- accommodation for specific purposes (e.g. for disabled persons);
- air pressure plants;
- cesspools (and other methods for treating and disposing of foul matter);
- emission of smoke, gases, fumes, grit or dust (or other noxious and/or offensive substances);
- dimensions of rooms and other spaces (inside buildings);
- drainage (including waste disposal units);
- electrical safety;
- fire precautions (services, fittings and equipment, means of escape);
- lifts (escalators, hoists, conveyors and moving footways);
- materials and components (suitability, durability and use);
- means of access to and egress from;
- natural lighting and ventilation of buildings;
- open spaces around buildings;
- prevention of infestation;
- provision of power outlets;

- resistance to moisture and decay;
- site preparation;
- solid fuel, oil, gas and electricity installations (including appliances, storage tanks, heat exchangers, ducts, fans and other equipment);
- standards of heating, artificial lighting, mechanical ventilation and air-conditioning;
- structural strength and stability (overloading, impact and explosion, under-pinning, safeguarding of adjacent buildings);
- third party liability (danger and obstruction to persons working or passing by building work);
- transmission of heat;
- transmission of sound;
- waste (storage, treatment and removal);
- water services, fittings and fixed equipment (including wells and bore-holes for supplying water);

and matters connected with (or ancillary to) any of the foregoing matters.

1.7.2 What is Schedule 2 of the Building Act 1984?

This Schedule provides guidance in connection with work that has been carried out prior to a local authority (under the Building Act 1984 Section 36) dispensing with or relaxing some of the requirements contained in the Building Regulations.

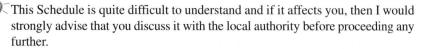 This Schedule is quite difficult to understand and if it affects you, then I would strongly advise that you discuss it with the local authority before proceeding any further.

1.7.3 What is Schedule 3 of the Building Act 1984?

Schedule 3 applies to how Building Regulations are to be used in Inner London and, as well as ruling which sections of the Act may be omitted, also details the requirements for drainage to Inner London buildings and shows how by-laws con-cerning the relation to the demolition of buildings (in Inner London) may be made.

What sections of the Building Act 1984 are not applied to Inner London?

In Inner London, because of its existing and changed circumstances (compared to other cities in England and Wales), certain sections of the Building Act are inappropriate (see Tables 1.2 and 1.3) and additional requirements – which are applicable to Inner London **only** – have been approved instead. These primarily cover drainage and demolition of buildings.

What about the buildings and drainage to buildings in Inner London?

Under the terms of the Building Act 1984, it is not lawful in an Inner London borough to erect a house/other building, or to rebuild a house/other building

Table 1.2 Sections inapplicable to Inner London

Section	Sub-section
Buildings	• Provision of food storage accommodation in house. • Entrances, exits etc. to be required in certain cases. • Means of escape from fire. • Raising of chimney. • Cellars and rooms below subsoil water level. • Consents under Section 74.
Defective premises, demolition etc.	• Dangerous building. • Dangerous building – emergency measures. • Ruinous and dilapidated buildings and neglected sites. • Notice to local authority of intended demolition. • Local authority's power to serve notice about demolition. • Notices under Section 81. • Appeal against notice under Section 81.

Table 1.3 Sections inapplicable to temples

Section	Sub-section
Drainage	• Drainage of building. • Use and ventilation of soil pipes. • Repair etc. of drain.
Buildings	• Provision of food storage accommodation in house. • Entrances, exits etc. to be required in certain cases. • Means of escape from fire. • Raising of chimney. • Cellars and rooms below subsoil water level. • Consents under Section 74.
Defective premises, demolition etc.	• Dangerous building. • Dangerous building – emergency measures. • Ruinous and dilapidated buildings and neglected sites. • Notice to local authority of intended demolition. • Local authority's power to serve notice about demolition. • Notices under Section 81. • Appeal against notice under Section 81.

that has been pulled down to (or below) floor level, **unless** that house/building is provided with drains in conformance with the borough council's requirements. These drains must be suitable for the drainage of the whole building and all works, apparatus and materials used in connection with these drains must satisfy the council's requirements.

It is not lawful to occupy a house or other building in Inner London that has been erected or rebuilt in contravention of the above restriction.

The basic requirements of all Inner London borough councils are that:

- the drains must be connected into a sewer that is (or is intended to be constructed) nearby;
- if a suitable sewer is not available then a covered cesspool or other place should be used, provided that it is not under any house or other building;
- the drains must provide efficient gravitational drainage at all times and under all circumstances and conditions.

If it is impossible or unfeasible to provide gravitational drainage to all parts of the building, then (but depending on the circumstances) the council may allow pumping and/or some other form of lifting apparatus to be used.

In **all** circumstances the council have the authority (under this Schedule of the Act) to order the owner/occupier:

- to construct a covered drain from the house or building into the sewer;
- to provide proper paved or water-resistant sloping surfaces for carrying surface water into the drain;
- to provide proper sinks, inlets and outlets (siphoned or otherwise trapped), for preventing the emission of effluvia from the drain – or any connection to it;
- to provide a proper water supply and water-supplying pipes, cisterns and apparatus for scouring the drain;
- to provide proper sand traps, expanding inlets and other apparatus for preventing the entry of improper substances into the drain.

You are not allowed to commence any work on drains, dig out the foundations of a house or to rebuild a house in Inner London **unless**, at least seven days previously, you have provided a notice of intent to the borough council.

If a house or building in an Inner London borough (regardless of when it was first erected), has insufficient drainage and there is no proper sewer within 200 feet of any part of the house or building, the borough council may serve on the owner written notice requiring that person:

- to construct a covered watertight cesspool or tank or other suitable receptacle (provided that it is not under the house); and
- to construct and lay a covered drain leading from the house or building into that cesspool, tank or receptacle.

The Inner London borough council have the authority to carry out irregular inspections of drains and cesspools constructed by the owner and, if they prove to be unsuitable, they have the authority to make the owner alter, repair or abandon them if they contravene council regulations.

What about Inner London's by-laws?

By authority of the Building Act 1984, the Greater London Authority (GLA) may make by-laws in relation to the demolition of buildings in the Inner London boroughs and regulate and (in certain circumstances) mandate, concerning:

- the fixing of floor level fans on buildings undergoing demolition;
- the hoarding up of windows in a building where all the sashes and glass have been removed;
- the demolition of internal parts of buildings before any external walls are taken down;
- using screens and mats as a precaution against dust;
- the hours during which ceilings may be broken down and mortar may be shot, or be allowed to fall, into any lower floor.

The GLA may also make by-laws with respect to closets, sanitary conveniences, ashpits, cesspools and receptacles for dung (and their accessories) for buildings being erected or altered in Inner London.

1.7.4 What is Schedule 4 of the Building Act 1984?

Schedule 4 of the Building Act 1984 concerns the authority and ruling of public bodies' notices and certificates.

What is a public body's plans certificate?

When a public body (i.e. local authorities, county councils and any other body '*that acts under an enactment for public purposes and not for its own profit*') is satisfied that the work specified in their (as well as another) public body's notices has been completed as detailed (and in full accordance with the Building Regulations) then that public body will give that local authority a certificate of completion.

This certificate is called a 'Public Bodies Plans Certificate' and can relate either to the whole, or to part of, the work specified in the public body's notice. Acceptance by the local authority signifies satisfactory completion of the planned work and the public body's notice ceases to apply to that work.

What is a public body's final certificate?

When a public body is satisfied that all work specified in their (or another's) public body's notice has been completed in compliance with the Building Regulations, then that public body will give the local authority a certificate of completion. This is referred to as a 'Final Certificate'.

How long is the duration of a public body's notice?

A public body's notice comes into force when it is accepted by the local authority and continues in force until the expiry of an agreed period of time.

 Local authorities are authorized by the Building Regulations to extend the notice in certain circumstances.

1.7.5 What is Schedule 5 of the Building Act 1984?

Schedule 5 lists the transitional effect of the Building Act 1984 concerning:

- The Public Health Act 1936;
- The Clean Air Act 1956;
- The Housing Act 1957;
- The Public Health Act 1961;
- The London Government Act 1963;
- The Local Government Act 1972;
- The Health and Safety at Work etc. Act 1974;
- The Local Government (Miscellaneous Provisions) Act 1982.

1.7.6 What is Schedule 6 of the Building Act 1984?

Schedule 6 lists the consequential amendments that will have to be made to existing Acts of Parliament owing to the acceptance of the Building Act 1984. These amendments concern:

- The Restriction of Ribbon Development Act 1935;
- The Public Health Act 1936;
- The Atomic Energy Authority Act 1954;
- The Clean Air Act 1956;
- The Housing Act 1957;
- The Radioactive Substances Act 1960;
- The Public Health Act 1961;
- The London Government Act 1963;
- The Offices, Shops and Railway Premises Act 1963;
- The Faculty Jurisdiction Measure 1964;
- The Fire Precautions Act 1971;
- The Local Government Act 1972;
- The Safety of Sports Grounds Act 1975;
- The Local Land Charges Act 1975;
- The Development of Rural Wales Act 1976;
- The Local Government (Miscellaneous Provisions) 1976;
- The Interpretation Act 1978;
- The Highways Act 1980;
- New Towns Act 1981;
- The Local Government (Miscellaneous Provisions) 1982;
- The Public Health (Control of Disease) Act 1984.

1.7.7 What is Schedule 7 of the Building Act 1984?

Schedule 7 lists the cancellation (repeal) of some sections of existing Acts of Parliament, owing to acceptance of the Building Act 1984. These cancellations concern:

- The Public Health Act 1936;
- The Education Act 1944;
- The Water Act 1945;
- The Town and Country Planning Act 1947;
- The Atomic Energy Authority Act 1954;
- The Radioactive Substances Act 1960;
- The Public Health Act 1961;
- The London Government Act 1963;
- The Greater London Council (General Powers) Act 1967;
- The Fire Precautions Act 1971;
- The Local Government Act 1972;
- The Water Act 1973;
- The Health and Safety at Work etc. Act 1974;
- The Control of Pollution Act 1974;
- The Airports Authority Act 1975;
- The Local Government (Miscellaneous Provisions) Act 1976;
- The Criminal Law Act 1977;
- The City of London (Various Powers) Act 1977;
- The Education Act 1980;
- The Highways Act 1980;
- The Water Act 1981;
- The Civil Aviation Act 1982;
- The Local Government (Miscellaneous Provisions) Act 1982.

1.8 What are 'Approved Documents'? (Building Act 1984 Section 6)

The Secretary of State makes available a series of documents (called 'Approved Documents') which are intended to provide practical guidance with respect to the requirements of the Building Regulations (for details see Chapter 3).

1.9 What is the 'Building Regulations Advisory Committee'? (Building Act 1984 Section 14)

The Building Act allows the Secretary of State to appoint a committee (known as the Building Regulations Advisory Committee) to review, amend, improve and produce new Building Regulations and associated documentation (e.g. such as Approved Documents – see above).

1.10 What is 'type approval'? (Building Act 1984 Sections 12 and 13)

Type approval is where the Secretary of State is empowered to approve a particular type of building matter as complying, either generally or specifically, with a particular requirement of the Building Regulations. This power of approval is normally delegated by the Secretary of State to the local council or other nominated public body.

1.11 Does the Fire Authority have any say in Building Regulations? (Building Act 1984 Section 15)

When a requirement 'encroaches' on something that is normally handled by the Fire Authority under the Fire Precautions Act 1971 (e.g. provision of means of escape, structural fire precautions etc.) then the local authority **must** consult the fire authority before making any decision.

1.12 How are buildings classified? (Building Act 1984 Section 35)

For the purpose of the Building Act, buildings are normally classified:

- by reference to size;
- by description;
- by design;
- by purpose;
- by location.

or (to quote the Building Act of 1984) '*any other suitable characteristic*'!

1.13 What are the duties of the local authority? (Building Act 1984 Section 91)

It is the duty of local authorities to ensure that requirements of the Building Act 1984 are carried out (and that the appropriate associated Building Regulations are enforced) subject to:

- the provisions of Part I of the Public Health Act 1936 (relating to united districts and joint boards);

- Section 151 of the Local Government, Planning and Land Act 1980 (relating to urban development areas);
- Section 1(3) of the Public Health (Control of Disease) Act 1984 (relating to port health authorities).

1.13.1 What document controls must local authorities have in place? (Building Act 1984 Sections 92 and 93)

All notices, applications, orders, consents, demands and other documents, authorized, required by or given to, that are required by this Act or by a local authority (or an officer of a local authority), need to be in writing and in the format laid down by the Secretary of State.

All documents that a local authority is required to provide under the Building Act 1984 shall be signed by:

- the proper officer for this authority;
- the district surveyor (for documents relating to matters within his province);
- an officer authorized by the authority to sign documents (of a particular kind).

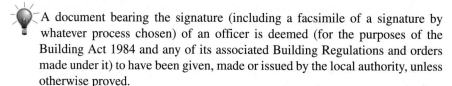A document bearing the signature (including a facsimile of a signature by whatever process chosen) of an officer is deemed (for the purposes of the Building Act 1984 and any of its associated Building Regulations and orders made under it) to have been given, made or issued by the local authority, unless otherwise proved.

1.13.2 How do local authorities 'serve' notices and documents? (Building Act 1984 Section 94)

Any notice, order, consent, demand or other document that is authorized or required by the Building Act 1984 can be given or served to a person:

- by delivering it to the person concerned;
- by leaving it, or sending it in a prepaid letter addressed to him, at his usual or last known residence.

Or if it is not possible to ascertain the name and address of the person to or on whom it should be given or served (or if the premises are unoccupied) then the notice, order, consent, demand or other document can be addressed to the 'owner' or 'occupier' of the premises (naming them) and delivering it to '*some person on the premises*' or, if there isn't anyone at the premises to whom it can be delivered, then a copy of the document can be fixed to a conspicuous part of the premises.

1.14 What are the powers of the local authority? (Building Act 1984 Sections 97–101)

The powers of the local authority, as given by the Building Act 1984 and its associated Building Regulations, include:

- overall responsibility for the construction and maintenance of sewers and drains and the laying and maintenance of water mains and pipes;
- the authority to make the owner or occupier of any premises complete essential and remedial work in connection with the Building Act 1984 (particularly with respect to the construction, laying, alteration or repair of a sewer or drain);
- the authority to complete remedial and essential work themselves (on repayment of expenses) if the owner or occupier refuses to do this work himself;
- the ability to sell any materials that have been removed, by them, from any premises when executing works under this Act (paying all proceeds, less expenses, from this sale to the owner or occupier).

 This does not apply to any refuse that is, or has been, removed by the local authority.

1.14.1 Have the local authority any power to enter premises? (Building Act 1984 Section 95)

An authorized officer of a local authority has a right to enter any premises, at all 'reasonable hours' (except for a factory or workplace in which 24 hours' notice has to be given) for the purpose of:

- ascertaining whether there is (or has been) a contravention of this Act (or of any Building Regulations) that it is the duty of the local authority to enforce;
- ascertaining whether or not any circumstances exist that would require local authority action or for them having to complete any work;
- taking any action, or executing any work, authorized or required by this Act, or by Building Regulations;
- carrying out their functions as a local authority.

If the local authority is refused admission to any premises (or the premises are unoccupied) then the local authority can apply to a Justice of the Peace for a warrant authorizing entry.

1.15 Who are approved inspectors? (Building Act 1984 Section 49)

An approved inspector is a person who is approved by the Secretary of State (or a body such as a local authority or county council designated by the Secretary of State) to inspect, supervise and to authorize building work. Lists of approved inspectors are available from all local authorities.

Building Act 1984 Section 57

If an approved inspector gives a notice or certificate that falsely claims to comply with the Building Regulations and/or the Building Act of 1984, then he is liable to prosecution.

1.15.1 What is an initial notice? (Building Act 1984 Section 47)

An approved inspector will have to present an initial notice and plan of work to the local authority. Once accepted, the approved inspector is authorized to inspect and supervise all work being completed and to provide certificates and notices. Acceptance of an initial notice by a local authority is treated (for the purposes of conformance with Section 13 of the Fire Precautions Act 1971 regarding suitable means of escape) as '*depositing plans of work*'.

Under Section 47 of the Building Act, the local authority is required to accept all certificates and notices, unless the initial notice and plans contravene a local ruling. Whilst the initial notice continues in force, the local authority are not allowed to give a notice in relation to any of the work being carried out or take any action for a contravention of Building Regulations.

If the local authority rejects the initial notice for any reason, then the approved inspector can appeal to a magistrates' court for a ruling. If still dissatisfied, he can appeal to the crown court.

Cancellation of initial notice (Building Act 1984 Sections 52 and 53)

If work has not commenced within three years (beginning the date on which the certificate was accepted), the local authority can cancel the initial notice.

If an approved inspector is unable to carry out or complete his functions, or is of the opinion that there is a contravention of the Building Regulations, then he can cancel the initial notice lodged with the local authority.

Equally, if the person carrying out the work has good reason to consider that the approved inspector is unable (or unwilling) to carry out his functions, then that person can cancel the initial notice given to the local authority.

The fact that the initial notice has ceased to be in force does not affect the right of an approved inspector, however, to give a new initial notice relating to any of the work that was previously specified in the original notice.

1.15.2 What are plans certificates? (Building Act 1984 Section 50)

When an approved inspector has inspected and is satisfied himself that the plans of work specified in the initial notice do not contravene the Building Regulations in any way, he will provide a certificate (referred to as a 'plans certificate') to the local authority. This plans certificate:

- can relate to the whole or part of the work specified in the initial notice;
- does not have any effect unless the local authority accepts it;
- may only be rejected by the local authority *on prescribed grounds*.

 If, however, work has not commenced within three years (beginning the date on which the certificate was accepted), the local authority may rescind their acceptance, by notice to the approved inspector or the person shown on the initial notice, giving their grounds for cancellation.

1.15.3 What are final certificates? (Building Act 1984 Section 51)

Once the approved inspector is satisfied that all work has been completed in accordance with the work specified in the initial notice, he will provide a certificate (referred to as a 'Final Certificate') to the local authority and the person who carried out the work. This certificate will detail his acceptance of the work and, once acknowledged by the local authority, the approved inspector's job will have been completed and (from the point of view of local authority) he will have been considered *to have discharged his duties*.

1.15.4 Who retains all these records? (Building Act 1984 Section 56)

Local authorities are required to keep a register of all initial notices and certificates given by approved inspectors and to retain all relevant and associated documents concerning those notices and certificates. The local authority is further required to make this register available for public inspection during normal working hours.

1.15.5 Can public bodies supervise their own work? (Building Act 1984 Sections 54 and 55)

If a public body (e.g. local authority or county council) is of the opinion that building work that is to be completed on one of its own buildings can be adequately supervised by one of its employees and/or agents, then they can provide the local authority with a notice (referred to as a 'public bodies notice') together with their plan of work.

Once accepted by the local authority, the public body is authorized to inspect and supervise all work being completed and to provide certificates and notices. Acceptance by a local authority of public bodies notice is treated (for the purposes of conformance with Section 13 of the Fire Precautions Act 1971 regarding suitable means of escape) as 'depositing plans of work'.

If the local authority rejects the public bodies notice for any reason, then they can appeal to a magistrates' court for a ruling. If still dissatisfied, they can appeal to the crown court.

1.16 What causes some plans for building work to be rejected? (Building Act 1984 Sections 16 and 17)

The local authority will reject all plans for building work that:

- are defective;
- contravene any of the Building Regulations.

In all cases the local authority will advise the person putting forward the plans why they have been rejected (giving details of the relevant regulation or section) and, where possible, indicate what amendments and/or modifications will have to be made in order to get them approved. The person who initially put forward the plans is then responsible for making amendments/alterations and resubmitting them for approval.

If a plan for proposed building work is accompanied by a certificate from a person or persons approved by the Secretary of State (or someone designated by him), then only in extreme circumstances can the local authority reject the plans.

1.17 Can I apply for a relaxation in certain circumstances? (Building Act 1984 Sections 7–11, 30 and 39)

The Building Act allows the local authority to dispense with, or relax, a Building Regulation if they believe that that requirement is unreasonable in relation to a particular type of work being carried out.

Schedule 2 of the Building Act 1984 provides guidance and rules for the application of Building Regulations to work that has been carried out **prior to** the local authority (under the Building Act 1984 Section 36) dispensing with, or relaxing, some of the requirements contained in the Building Regulations. This schedule is quite difficult to understand and if it affects you, then I would strongly advise that you discuss it with the local authority before proceeding any further.

For the majority of cases, applications for dispensing with or relaxing Building Regulations can be settled locally. In more complicated cases, however, the local authority can seek guidance from the Secretary of State who will give a direction as to whether the requirement may be relaxed or dispensed with (unconditionally or subject to certain conditions).

If a question arises between the local authority and the person who has executed (or has proposed to execute any) work regarding:

- the application of Building Regulations;
- whether the plans are in conformity with the Building Regulations;
- whether the work has been executed in conformance with these plans;

then the question can be referred to the Secretary of State for determination. In these cases, the Secretary of State's decision will be deemed final.

The Building Act allows the local authority to charge a fee for reviewing and deciding on these matters with different fees for different cases.

1.18 Can I change a plan of work once it has been approved? (Building Act 1984 Section 31)

If the person intending to carry out building work has had their plan (or plans) passed by the local authority, but then wants to change them, that person will have to submit (to the local authority) a set of revised plans showing precisely how they want to deviate from the approved plan and ask for their approval. If the deviation or change is a small one this can usually be achieved by talking to the local planning officer, but if it is a major change, then it could result in the resubmission of a complete plan of the revised building work.

1.19 Must I complete the approved work in a certain time? (Building Act 1984 Section 32)

Once a building plan has been passed by the local authority, then '**work must commence**' within three years from the date that it was approved. Failure to do so could result in the local authority cancelling the approved plans and you will have to resubmit them if you want to carry on with your project.

The phrase 'work must commence' can vary from local authority to local authority. Normally this will mean physically laying the foundations of the building but in other cases it could mean that far more work has to be completed in the three year time span. It is always best to check with the local authority and ask for clarification about this restriction when your plans are first approved.

1.20 How is my building work evaluated for conformance with the Building Regulations? (Building Act 1984 Section 32)

Part of the local authority's duty is to make regular checks that all building work being completed is in conformance with the approved plan and the Building

Regulations. These checks would normally be completed at certain stages of the work (e.g. the excavation of foundations) and tests will include:

- tests of the soil or subsoil of the site of the building;
- tests of any material, component or combination of components that has been, is being, or is proposed to be used in the construction of a building;
- tests of any service, fitting or equipment that has been, is being, or is proposed to be provided in or in connection with a building.

The cost of carrying out these tests will normally be charged to the owner or occupier of the building.

 The local authority has the power to ask the person responsible for the building work to complete some of these tests on their behalf.

1.20.1 Can I build on a site that contains offensive material?
(Building Act 1984 Section 29)

If the site you are intending to erect a building or extension on is:

- ground that has been filled up with material impregnated with faecal or offensive animal or offensive vegetable mater;
- ground upon which any such material has been deposited;

then that material must be removed or rendered innocuous before work can commence.

This requirement normally rests with the current owner/occupier of the building, but in certain circumstances (for example, if the site was previously used as a chicken farm or similar) then the previous owner might be held responsible. If the requirements of this particular section are applicable to you, then it is recommended that you seek guidance from the local authority before committing yourself!

1.21 What about dangerous buildings?
(Building Act 1984 Sections 77 and 78)

If a building, or part of a building or structure, is in such a dangerous condition (or is used to carry loads which would make it dangerous) then the local authority may apply to a magistrates' court to make an order requiring the owner:

- to carry out work to avert the danger;
- to demolish the building or structure, or any dangerous part of it, and remove any rubbish resulting from the demolition.

The local authority can also make an order restricting its use until such time as a magistrates' court is satisfied that all necessary works have been completed.

1.21.1 Emergency measures

In emergencies, the local authority can make the owner take immediate action to remove the danger or they can complete the necessary action themselves. In these cases, the local authority is entitled to recover from the owner such expenses reasonably incurred by them. For example:

- fencing off the building or structure;
- arranging for the building/structure to be monitored.

1.21.2 Can I demolish a dangerous building? (Building Act 1984 Section 80)

You must have good reasons for knocking down a building, such as making way for rebuilding or improvement (which in most cases would be incorporated in the same planning application).

Be careful, penalties can be very severe for demolishing something illegally!

You are not allowed to begin any demolition work (even on a dangerous building) unless you have given the local authority notice of your intention and this has either been acknowledged by the local authority or the relevant notification period has expired. In this notice you will have to:

- specify the building to be demolished;
- state the reason(s) for wanting to demolish it;
- show how you intend to demolish it.

Copies of this notice will have to be sent to:

- the local authority;
- the occupier of any building adjacent to the building in question;
- British Gas;
- the area electricity board in whose area the building is situated.

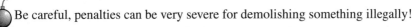This regulation does not apply to the demolition of an internal part of an occupied building, or a greenhouse, conservatory, shed or prefabricated garage (that forms part of that building) or an agricultural building defined in Section 26 of the General Rate Act 1967.

1.21.3 Can I be made to demolish a dangerous building? (Building Act 1984 Sections 81, 82 and 83)

If the local authority considers that a building is so dangerous that it should be demolished, they are also entitled to issue a notice to the owner requiring him:

- to shore up any building adjacent to the building to which the notice relates;
- to weatherproof any surfaces of an adjacent building that are exposed by the demolition;

- to repair and make good any damage to an adjacent building caused by the demolition or by the negligent act or omission of any person engaged in it;
- to remove material or rubbish resulting from the demolition and clear the site;
- to disconnect, seal and remove any sewer or drain in or under the building;
- to make good the surface of the ground that has been disturbed in connection with this removal of drains etc.;
- (in accordance with the Water Act 1945 (interference with valves and other apparatus) and the Gas Act 1972 (public safety)), to arrange with the relevant statutory undertakers (e.g. the water authority, British Gas and the electricity supplier) for the disconnection of gas, electricity and water supplies to the building;
- to leave the site in a satisfactory condition following completion of all demolition work.

 Before complying with this notice, the owner must give the local authority 48 hours' notice of commencement.

 In certain circumstances, the owner of an adjacent building may be liable to assist in the cost of shoring up their part of the building and waterproofing the surfaces. It could be worthwhile checking this point with the local authority!

1.22 What about defective buildings?
(Building Act 1984 Sections 76, 79 and 80)

If a building or structure is, because of its ruinous or dilapidated condition, liable to cause damage to (or be a nuisance to) the amenities of the neighbourhood, then the local authority can require the owner:

- to carry out necessary repairs and/or restoration; or
- to demolish the building or structure (or any part of it) and to remove all of the rubbish or other material resulting from this demolition.

If, however, the building or structure is in a defective state and remedial action (envisaged under Sections 93 to 96 of the Public Health Act) would cause an unreasonable delay, then the local authority can serve an abatement notice stating that within nine days **they** intend to complete such works as they deem necessary to remedy the defective state and recover the '*expenses reasonably incurred in so doing*' from the person on whom the notice was sent.

If appropriate, the owner can (within seven days) after the local authority's notice has been served, serve a counter-notice stating that he intends to remedy the defects specified in the first-mentioned notice himself.

 A local authority is **not** entitled to serve a notice, or commence any work in accordance with a notice that they have served, if the execution of the works would (to their knowledge) be in contravention of a building preservation order that has been made under Section 29 of the Town and Country Planning Act.

1.23 What are the rights of the owner or occupier of the premises? (Building Act 1984 Sections 102–107)

When a person has been given a notice by a local authority to complete work, he has the right to appeal to a magistrates' court on any of the following grounds:

- that the notice or requirement is not justified by the terms of the provision under which it purports to have been given;
- that there has been some informality, defect or error in (or in connection with) the notice;
- that the authority have refused (unreasonably) to approve completion of alternative works, or that the works required by the notice to be executed are unreasonable or unnecessary;
- that the time limit set to complete the work is insufficient;
- that the notice should lawfully have been served on the occupier of the premises in question instead of on the owner (or vice versa);
- that some other person (who is likely to benefit from completion of the work) should share in the expense of the works.

1.24 Can I appeal against a local authority's ruling? (Building Act 1984 Sections 40 and 41)

If you have grounds for disagreeing with the local authority's ruling to remove or renew 'offending work', then you are entitled to appeal to the local magistrates' court and they will rule whether the local authority were correct and entitled to give you this ruling, or whether they should withdraw the notice.

If you then disagree with the magistrates' ruling, you have the right to appeal to the crown court.

Where the Secretary of State has given a ruling, however, this ruling shall be considered as being final.

1.24.1 What about compensation? (Building Act 1984 Sections 103–110)

If an owner or occupier considers that a ruling obtained from the local authority is incorrect, he can appeal (in the first case) to the local magistrates' court. If, on appeal, the magistrates rule against the local authority, then the owner/occupier of the building concerned is entitled to compensation from the local authority. If, on the other hand, the magistrates rule in favour of the local authority, then the local authority is entitled to recover any expenses that they have incurred.

Be sure of your facts before you ask a magistrates' court for a ruling!

1.24.2 What happens if the plans mean building over an existing sewer etc.? (Building Act 1984 Section 18)

Before the local authority can approve a plan for building work which means having to first erect a building or extension over an existing sewer or drain, they must notify and seek the advice of the water authority.

 As part of the Public Health Act 1936 and the Control of Pollution Act 1974, local authorities are required to keep maps of all sewers etc.

Appendix A Contents of the Building Act 1984

Part 1 The Building Regulations

Section	Sub-section
Power to make building regulations	• Power to make building regulations. • Continuing requirements.
Exemption from building regulations	• Exemption of particular classes of buildings etc. • Exemption of educational buildings and buildings of statutory undertakers. • Exemption of public bodies from procedural requirements of building regulations.
Approved Documents	• Approval of documents for purposes of building regulations. • Compliance or non-compliance with Approved Documents.
Relaxation of building regulations	• Relaxation of building regulations. • Application for relaxation. • Advertisement of proposal for relaxation of building regulations. • Type relaxation of building regulations.
Type approval of building matter	• Power of Secretary of State to approve type of building matter. • Delegation of power to approve.
Consultation	• Consultation with Building Regulations Advisory Committee and other bodies. • Consultation with fire authority.
Passing of plans	• Passing or rejection of plans. • Approval of persons to give certificates etc. • Building over sewer etc. • Use of short-lived materials. • Use of materials unsuitable for permanent building. • Provision of drainage. • Drainage of buildings in combination. • Provision of facilities for refuse. • Provision of exits etc. • Provision of water supply. • Provision of closets. • Provision of bathrooms. • Provision for food storage. • Site containing offensive material.
Determination of questions	

Section	Sub-section
Proposed departure from plans	
Lapse of deposit of plans	
Tests for conformity with building regulations	
Classification of buildings	
Breach of building regulations	• Penalty for contravening building regulations. • Removal or alteration of offending work. • Obtaining of report where Section 36 notice given. • Civil liability.
Appeals in certain cases	• Appeal against refusal etc. to relax building regulations. • Appeal against Section 36 notice. • Appeal to Crown Court. • Appeal and statement of case to High Court in certain cases. • Procedure on appeal to Secretary of State on certain matters.
Application of building regulations to Crown etc.	• Application to Crown. • Application to United Kingdom Atomic Energy Authority.
Inner London	

Part 2 Supervision of Building Work etc. otherwise than by a local authority

Section	Sub-section
Supervision of plans and work by approved inspectors	• Giving and acceptance of initial notice. • Effect of initial notice. • Approved inspectors. • Plans certificates. • Final certificates. • Cancellation of initial notice. • Effect of initial notice ceasing to be in force.
Supervision of their own work by public bodies	
Supplementary	• Appeals. • Recording and furnishing of information. • Offences. • Construction of Part 11.

Part 3 Other provisions about buildings

Section	Sub-section
Drainage	• Drainage of building. • Use and ventilation of soil pipes.

(Continued)

Appendix A   (*Continued*)

Section	Sub-section
	• Repair etc. of drain.
	• Disconnection of drain.
	• Improper construction or repair of water closet or drain.
Provision of sanitary conveniences	• Provision of closets in building.
	• Provision of sanitary conveniences in workplace.
	• Replacement of earth closets etc.
	• Loan of temporary sanitary conveniences.
	• Erection of public conveniences.
Buildings	• Provision of water supply in occupied house.
	• Provision of food storage accommodation in house.
	• Entrances, exits etc. to be required in certain cases.
	• Means of escape from fire.
	• Raising of chimney.
	• Cellars and rooms below subsoil water level.
	• Consents under Section 74.
Defective premises, demolition etc.	• Defective premises.
	• Dangerous building.
	• Dangerous building – emergency measures.
	• Ruinous and dilapidated buildings and neglected sites.
	• Notice to local authority of intended demolition.
	• Local authority's power to serve notice about demolition.
	• Notices under Section 81.
	• Appeal against notice under Section 81.
Yards and passages	• Paving and drainage of yards and passages.
	• Maintenance of entrances to courtyards.
Appeal to Crown Court	
Application of provisions to Crown property	
Inner London	
Miscellaneous	• References in Acts to building byelaws.
	• Facilities for inspecting local Acts.

Part 4 General

Section	Sub-section
Duties of local authorities	
Documents	• Form of documents.
	• Authentication of documents.
	• Service of documents.
Entry on premises	• Power to enter premises.
	• Supplementary provisions as to entry.
Execution of works	• Power to require occupier to permit work.
	• Content and enforcement of notice requiring works.
	• Sale of materials.
	• Breaking open of streets.
Appeal against notice requiring works	

Section	Sub-section
General provisions about appeals and applications	• Procedure on appeal or application to magistrates' court. • Local authority to give effect to appeal. • Judge not disqualified by liability to rates.
Compensation and recovery of sums	• Compensation for damage. • Recovery of expenses etc. • Payments by instalments. • Inclusion of several sums in one complaint. • Liability of agent or trustee. • Arbitration.
Obstruction	
Prosecutions	• Prosecution of offences. • Continuing offences.
Protection of members etc. of authorities	
Default powers	• Default powers of Secretary of State. • Expenses of Secretary of State. • Variation or revocation of order transferring powers.
Local inquiries	
Orders	
Interpretation	• Meaning of 'building'. • Meaning of 'building regulations'. • Meaning of 'construct' and 'erect'. • Meaning of deposit of plans. • Construction and availability of sewers. • General interpretation. • Construction of certain references concerning temples.
Savings	• Protection for dock and railway undertakings. • Saving for Local Land Charges Act 1975. • Saving for other laws. • Restriction of application of Part IV to Schedule 3.

Part 5 Supplementary

Section	Sub-section
Supplementary	• Transitional provisions. • Consequential amendments and repeals. • Commencement. • Short title and extent.
Schedule 1 – Building regulations.	
Schedule 2 – Relaxation of building regulations.	
Schedule 3 – Inner London.	
Schedule 4 – Provisions consequential upon public body's notice.	
Schedule 5 – Transitional provisions.	
Schedule 6 – Consequential amendments.	
Schedule 7 – Repeals.	

2

The Building Regulations 2000

Even when planning permission is not required, most building works, including alterations to existing structures, are subject to minimum standards of construction to safeguard public health and safety.

2.1 What is the purpose of the Building Regulations?

The Building Regulations are legal requirements laid down by parliament, based on the Building Act 1984. They are approved by parliament and deal with the **minimum** standards of design and building work for the construction of domestic, commercial and industrial buildings.

Building Regulations ensure that new developments or alterations and/or extensions to buildings are all carried out to an agreed standard that protects the health and safety of people in and around the building.

 Building standards are enforced by your local building control officer, but for matters concerning drainage or sanitary installations, you will need to consult their technical services department.

Builders and developers are required by law to obtain building control approval, which is an independent check that the Building Regulations have been complied with. There are two types of building control providers – the local authority and approved inspectors.

2.2 Why do we need the Building Regulations?

As mentioned in the Preface, the Great Fire of London in 1666 was the single most significant event to have shaped today's legislation. The rapid growth of the fire through timber buildings built next to each other highlighted the need for builders to consider the possible spread of fire between properties when rebuilding work commenced. This resulted in the first building construction legislation that required all buildings to have some form of fire resistance.

During the Industrial Revolution (200 years later) poor living and working conditions in ever expanding, densely populated urban areas caused outbreaks of cholera and other serious diseases. Poor sanitation, damp conditions and lack of ventilation forced the government to take action and building control took on the greater role of health and safety through the first Public Health Act of 1875. This Act had two major revisions in 1936 and 1961 and led to the first set of national building standards – the Building Regulations 1965.

The current legislation is the Building Regulations 2000 (Statutory Instrument No 2531) which is made by the Secretary of State for the Environment under powers delegated by parliament under the Building Act of 1984.

The Building Regulations are a set of minimum requirements designed to secure the health, safety and welfare of people in and around buildings and to conserve fuel and energy in England and Wales. They are basic performance standards and the level of safety and acceptable standards are set out as guidance in the Approved Documents (which are quite frequently referred to as 'Parts' of the Building Regulations). Compliance with the detailed guidance of the Approved Documents is usually considered as evidence that the Regulations themselves have been complied with.

 Alternate ways of achieving the same level of safety, or accessibility, are also acceptable.

2.3 What building work is covered by the Building Regulations?

The Building Regulations cover all new building work. This means that if you want to put up a new building, extend or alter an existing one, or provide new and/or additional fittings in a building such as drains or heat-producing appliances, washing and sanitary facilities and hot water storage (particularly unvented hot water systems), the Building Regulations will probably apply. They may also apply to certain changes of use of an existing building (even though construction work may not be intended) as the 'change of use' could involve the building having to meet different requirements of the Regulations.

It should be remembered, however, that although it may appear that the Regulations do not apply to some of the work you wish to undertake, the end result of doing that work could well lead to you contravening some of the Regulations. You should also recognize that some work – whether or not controlled – could have implications for an adjacent property. In such cases it would be advisable to take professional advice and consult the local authority or an approved inspector. Some examples are:

- removing a buttressed support to a party wall;
- underpinning part of a building;
- removing a tree close to a wall of an adjoining property;
- adding floor screed to a balcony which may reduce the height of a safety barrier;

- building parapets which may increase snow accumulation and lead to an excessive increase in loading on roofs.

2.4 What are the requirements associated with the Building Regulations?

The Building Regulations contain a list of requirements (referred to as 'Schedule 1') that are designed to ensure the health and safety of people in and around buildings; to promote energy conservation; and to provide access and facilities for disabled people. In total there are 14 parts (A–P less I) to these requirements and these cover subjects such as structure, fire and electrical safety, ventilation, drainage etc.

The requirements are expressed in broad, functional terms in order to give designers and builders the maximum flexibility in preparing their plans.

2.5 What are the Approved Documents?

Approved Documents contain practical and technical guidance on ways in which the requirements of each part of the Building Regulations can be met.

Each Approved Document reproduces the *requirements* contained in the Building Regulations relevant to the subject area. This is then followed by *practical and technical guidance*, with examples, on how the requirements can be met in some of the more common building situations. There may, however, be alternative ways of complying with the requirements to those shown in the Approved Documents and you are, therefore, under no obligation to adopt any particular solution in an Approved Document if you prefer to meet the relevant requirement(s) in some other way.

Just because an Approved Document has not been complied with, however, does not necessarily mean that the work is wrong. The circumstances of each particular case should be considered when an application is made to make sure that adequate levels of safety will be achieved.

Note: The Building Regulations are constantly reviewed to meet the growing demand for better, safer and more accessible buildings as well as the need to reflect emerging harmonized European Standards. Building Regulations were last consolidated in SI 2000:2531, since then several of the Approved Documents have been republished as new editions and others are now under active review. Where there are any issues common to one or more parts (such as the guidance on airtightness in Part L corresponding to the requirements for ventilation in Part F) these have been taken into consideration.

Any changes necessary are brought into operation after consultation with all interested parties. This has meant several amendments since the publication of the Building Regulations in 2000 with the emphasis in more recent years being on:

- increased thermal insulation to conserve energy and reduce global warming;
- providing better access and facilities for disabled people;
- a more comprehensive, one stop approach to fire safety requirements;
- the need for protection against sound from within a dwelling-house, other parts of the building and/or adjoining buildings;
- improvement of acoustic conditions in schools.

Draft proposals for amending some of the existing Approved Documents are also now well under way and currently consist of:

- Part F – Ventilation – with the aim of minimizing the amount of uncontrolled air leaking through the building fabric;
- Part L – Conservation of fuel and power – and the need to reduce global warming by improving energy performance and efficiency of buildings.

A consultation draft proposal is also being considered for an additional Approved Document for:

- Electronic communication services (Part Q).

The Approved Documents are in 14 parts (A to P less I) and consist of:

A   Structure
B   Fire safety
C   Site preparation and resistance to contaminants and water moisture
D   Toxic substances
E   Resistance to the passage of sound
F   Ventilation
G   Hygiene
H   Drainage and waste disposal
J   Combustion appliances and fuel storage systems
K   Protection from falling, collision and impact
L   Conservation of fuel and power
M   Access and facilities for disabled people
N   Glazing – safety in relation to impact, opening and cleaning
P   Electrical safety

 Parts A to D, F to K and N (except for paragraphs H2 and J6 which are excluded from regulation 8 because they deal directly with preventing contamination of water) of Schedule 1 do not require anything to be done except for the purpose of securing reasonable standards of health and safety for persons in or about buildings (and any others who may be affected by buildings) or matters connected with buildings.

Parts E and M (which deal, respectively, with resistance to the passage of sound, and access and facilities for disabled people) are excluded from regulation 8 because they address the welfare and convenience of building users.

Part L is excluded from regulation 8 because it addresses the conservation of fuel and power.

 You can buy a copy of the Approved Documents (and the Building Act 1984 if you wish) from the Stationery Office (TSO), PO Box 29, Duke Street, Norwich, NR3 1GN (Tel: 0870 600 5522, Fax: 0870 600 5533, www.tso. co.uk), or some book shops. Occasionally they are available from libraries. Alternatively, you can download pdf copies of the Approved Documents from www.hsmo.gov.uk (then legislation/uk/acts) or www.safety.odpm.gov.uk.

2.5.1 Part A Structure

So that buildings do not collapse, requirements ensure that:

- all structural elements of a building can safely carry the loads expected to be placed on them;
- foundations are adequate for any movement of the ground (for example, caused by landslip or subsidence);
- large buildings are strong enough to withstand (for example) an explosion without collapsing.

2.5.2 Part B Fire safety

The Regulations consider six aspects of fire safety in the construction of buildings. These are:

(1) that the design of a building enables occupants to escape to a place of safety, by their own efforts, in the event of a fire;
(2) that the internal linings of a building do not support a rapid spread of fire;
(3) that the structure of the building does not collapse prematurely;
(4) that the slow spread of fire through the building (as well as in unseen cavities and voids) is prevented by providing fire resisting walls and/or partitions where necessary;
(5) that the spread of fire between buildings is limited by spacing them apart and controlling the number and size of openings on boundaries;
(6) that the building is designed to enable the fire brigade to fight a fire and rescue any persons caught in a fire.

2.5.3 Part C Site preparation and resistance to contaminants and moisture

There are four requirements to this part:

(1) that before any building works commence, all vegetation and topsoil are removed;
(2) that any contaminated ground is either treated, neutralized or removed before a building is erected;
(3) that subsoil drainage is provided to waterlogged sites;

(4) that **all** floors, walls and roof of a building should not be adversely affected by interstitial condensation.

2.5.4 Part D Toxic substances

This part requires walls to be constructed in such a way that any fumes filling a cavity are prevented from penetrating the building.

2.5.5 Part E Resistance to passage of sound

This part has four main requirements:

(1) that dwellings shall provide reasonable resistance to sound from other parts of the same building and/or from adjoining buildings;
(2) that internal walls and floors of dwellings shall provide reasonable resistance to sound;
(3) that the common internal parts of buildings (containing flats or rooms for residential purposes) shall prevent unreasonable reverberation;
(4) that school rooms shall be acoustically insulated against noise.

2.5.6 Part F Ventilation

There are two aspects considered by this part:

(1) adequate ventilation must be provided to kitchens, bath and shower rooms, sanitary accommodation and other habitable rooms (both domestic and non-domestic);
(2) roofs need to be well vented (or designed) to prevent moist air causing condensation damage.

Proposed amendment to the Building Regulations	**Part F – Ventilation** The main changes proposed for Part F are that: • Part F should move away from its current prescriptive approach (e.g. specifying the required background ventilation area for each room) to a performance-based approach that recommends what level of ventilation should be sufficient and allows the designers to cater for this; • the old Part F1 (which covered condensation in roofs of dwellings) will now be moved to Part C (which covers the control of interstitial and surface condensation in floors, walls and roofs); • the section on ventilation of non-domestic buildings needs to be revised;

- the guidance given in Part F will need to be improved – particularly concerning:
  - the types of ventilation achieved in buildings (i.e. infiltration and purpose-provided ventilation);
  - the ventilation of rooms – through other rooms and spaces;
  - ventilating car parks to reflect the change in the requirement with regard to predicted CO levels;
- the need to provide continuous air extraction in rooms containing printers and photocopiers in substantial use (so as to limit the emission of VOCs and ozone which can affect health) will now be included;
- it is now recommended that there should be **no** smoking in office buildings but if a smoking room is provided the aim should be to protect **non**-smokers outside the room from passive smoking;
- one other proposed (major) change is that in future background ventilators will need to be installed when one or more windows (fitted with trickle ventilators) are replaced or an equivalent ventilation opening will have to be provided in the same room;
- the need to make provisions to help historic buildings 'breathe' in order to control moisture and potential long-term decay problems is emphasized;
- climate change (which is likely to lead to higher temperatures and wind speeds) now needs to be taken into account when considering ventilation rates;
- to assist designers example calculations of ventilator sizing for dwellings are given in one of the appendices (as well as tables of ventilator areas for many common situations). Another appendix provides guidance on minimizing the ingress of external pollution into buildings in urban areas.

Note: The main reason for reviewing Part F was not (as has been mentioned in the media) to move towards having completely airtight buildings, but to minimize the amount of uncontrolled air leakage through the building fabric. Adequate ventilation in buildings is essential for controlling moisture and other indoor pollutants so it is important that any improved level of airtightness does not compromise indoor air quality. It is for these reasons the review of Part F closely followed the ongoing review of Part L.

2.5.7 Part G Hygiene

There are three aspects included in this part:

(1) buildings are required to have satisfactory sanitary conveniences and washing facilities;
(2) all dwellings are required to have a fixed bath or shower with hot and cold water;
(3) unvented hot water systems over a certain size are required to have safety provisions to prevent explosion.

2.5.8 Part H Drainage and waste disposal

There are four aspects of this part:

(1) new drains taking foul water from buildings are required to discharge into a foul water sewer (or other suitable outfall), be watertight and be accessible for cleaning;
(2) where no public sewer is available, holding tanks or sewage treatment plants should be made available;
(3) new drains taking rainwater from roofs of buildings need to be watertight, accessible for cleaning and (if there is no sewer available) discharge to a suitable surface water sewer or ditch, soakaway, or watercourse;
(4) storage facilities, reasonably close to the building, need to be provided for refuse collection.

2.5.9 Part J Combustion appliances and fuel storage systems

There are three main aspects to this part:

(1) heat producing appliances must be provided with a supply of fresh air to prevent carbon monoxide poisoning to the building's occupants;
(2) chimneys and flues need to be adequately designed so that smoke and other products of combustion are safely discharged to the outside air;
(3) fireplaces and heat producing appliances should be designed and positioned so as to avoid the building's structure from igniting.

2.5.10 Part K Protection from falling, collision and impact

There are five main aspects to this part:

(1) to avoid accidents on stairs, ladders and ramps; the physical dimensions need to be suitable for the use of the building;
(2) to avoid persons falling off stairwells, balconies, floors, some roofs; light wells and basement areas (or similar sunken areas) connected to a building need to be suitably guarded according to the building's use;
(3) to avoid vehicles falling off buildings; car park floors, ramps and other raised areas need to be provided with vehicle barriers;

(4)   to avoid danger to people from colliding with an open window, skylight, or ventilator; some form of guarding may be needed;

(5)   measures need to be taken to avoid the opening and closing of powered sliding or open-upwards doors and gates falling onto any person and/or trapping them.

Possible future amendment	Approved Document K contains general guidance on stair and ramp design. The guidance in Approved Document M (2004) reflects more recent ergonomic research conducted to support BS 8300 and takes precedence over Approved Document K in conflicting areas.
	Further research on stairs is currently being undertaken and will be reflected in future revisions of Approved Document K.

2.5.11 Part L Conservation of fuel and power

There are four main aspects to this part:

(1)   roofs, walls, windows, doors and floors need to have resistance to loss of heat (the amount will vary according to the size and use of building);
(2)   controls need to be available to enable occupants to turn off electric lighting;
(3)   controls need to be available to enable low energy lights to be used;
(4)   controls need to be provided for boilers so as to avoid inefficient usage and waste.

Proposed amendment to the Building Regulations	**Part L – Conservation of fuel and power**
	Since publication of Part L in 2002, the Government have acknowledged that global warming has become a reality and as a consequence (in 2003) published an Energy White Paper proposing significant cuts in the UK's carbon dioxide emissions – the main contributor to global warming – by at least 60%, by 2050.
	Currently, at least half of all carbon dioxide emissions come from buildings and it is envisaged that the proposals for amending Building Regulations would, if implemented, not only lead to an improvement in the energy efficiency of new buildings but also reduce the amount of gas, oil, coal and wood that would have to be

burnt in order to provide electricity, heating and hot water for these buildings.

Significant improvements to the Building Regulation's energy efficiency provisions are, therefore, seen as a major contributor towards achieving the target of an initial 20% reduction in carbon emissions by 2010 and the proposed changes to Part L will include:

- setting performance standards for buildings as a whole (rather than for construction and service element's component parts);

This proposal has been the source of a lot of interest from component manufacturers who have spent a lot of money retooling their works in order to meet Part L and Part M – and now realize that there is a possibility of this becoming outdated and having to do it all over again!

- carrying out pre-completion testing of the building's airtightness;
- laying down requirements for calculating energy performance in accordance with a national standard;
- making the results of such calculations known to prospective purchasers and tenants whenever buildings are constructed, sold or rented out (and displaying these results in certain larger buildings that are occupied by public authorities);
- promoting improvements in building energy performance (as opposed to setting specific targets).

Note: It is noted that improvements on a lesser scale would also be obtained whenever people carry out work on existing buildings.

2.5.12 Part M Access and facilities for disabled people

There are six main aspects to this part:

(1) dwellings – people, including disabled people, should be able:
   - to reach the principal, or suitable alternative, entrance to the dwelling from the point of access;
   - to gain access into and within the principal storey of the dwelling;
   - to gain access to sanitary conveniences at no higher storey than the principal storey.

(2)   regardless of disability, age or gender it should be possible for people in buildings (other than dwellings):
- to reach the principal entrance to the building from the site boundary, from car parking (within the site) and from other buildings on the same site (e.g. such as a university campus, school or hospital);
- to have access into and within, any storey of the building;
- to have access and use of the building's facilities.

(3)   the structure and amenities of a building should not constitute a hazard to users (especially people with impaired sight);

(4)   suitable accommodation should be made available for people in wheelchairs (or people with other disabilities) in audience or spectator seating;

(5)   people with a hearing or sight impairment should be provided with some form of aid to communication in auditoria, meeting rooms, reception areas, ticket offices and at information points;

(6)   sanitary accommodation should be available for **all** users of the building.

2.5.13 Part N Glazing – safety in relation to impact, opening and cleaning

There are four main aspects in this part:

(1)   glazing in locations where people might collide with the glass should either be robust enough not to break, or be constructed of safety glass, or be provided with suitable guarding;

(2)   large sheets of glazing need to be made obvious so that people do not collide with that glazing;

(3)   where non-dwelling windows, skylights and ventilators are openable by people, controls and/or limiters need to be provided to ensure safe operation and prevent persons falling through a window;

(4)   safe access for cleaning both sides of non-dwelling windows, skylights etc. over 2.0 m above ground needs to be available.

Possible future amendment	Approved Document N contains guidance on the use of symbols and markings on glazed doors and screens. The guidance now given in Approved Document M (2003) is as a result of more recent experience of 'door manifestation' and takes precedence over the guidance currently provided in Approved Document N in conflicting areas until such time as Approved Document N is revised.

2.5.14 Part P – Electrical safety

There are two aspects to this part:

(1)   the design, installation, inspection and testing of electrical installations should be planned so as to protect persons from fire or injury;

(2)   sufficient information shall be provided so that persons operating, maintaining or altering an electrical installation can do so with reasonable safety.

2.5.15 Future Approved Documents

At the time of writing this book, a consultation draft proposal is currently under consideration for the following **additional** Approved Document:

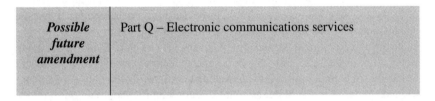

Possible future amendment	Part Q – Electronic communications services

Electronic communications services, in their widest sense, convey a range of communications by a range of media. These can include speech, music (and other sounds), visual images, broadband and/or signals communicated by electric, magnetic, electromagnetic, collector-chemical or electromechanical media.

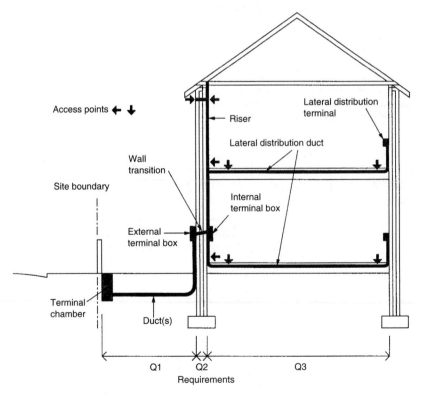

Figure 2.1 Outline requirements for the distribution of electronic communications services in buildings

Note: The generic term 'broadband' is used to describe the technology that delivers higher capacity two-way communication to private homes or business premises. Broadband can be delivered by several technologies, including copper and fibre optic cables, radio or satellite services.

The increased use of these electronic communications services in buildings has resulted in them being supplied/routed in quite a number of different ways. The problems of trying to route cables carrying these services into and around existing buildings has become quite a difficulty to both owners and occupiers of these buildings and may, as a result, have caused a certain amount of damage and disruption to the building fabric and surrounding ground.

The three Requirements (i.e. Q1, Q2 and Q3) of Part Q of the Building Regulations aim to ensure that future electronic communications services should be capable of being installed into an existing building with the minimum amount of inconvenience to the building owner/occupier and without any unnecessary disruption of the building fabric and/or the surrounding ground.

 Note: Although Requirements Q1, Q2 and Q3 apply to buildings of all purpose groups, the guidance in Approved Document Q is limited to the provisions considered necessary for dwellings, i.e. dwelling-houses, flats and maisonettes.

Q1 – Means of supply to the building

Requirement	Limits on application
Reasonable provision shall be made to enable the ready installation and removal of cable-based electronic communications services from an appropriate boundary of the site of the building to the building.	Not yet fully established

Electronic communications services should be supplied (and be capable of being removed at some time in the future) through existing ducts between the boundary of the building site and a point of entry into the building, **without** having to excavate the ground within the curtilage of the site of the building.

Sufficient terminal chambers and associated ducts capable of serving all of accommodation units (e.g. flats) in the building should be provided so as to facilitate future installation and/or removal of electronic communications services.

Q2 – Means of supply into the building

Requirement	Limits on application
Reasonable provision shall be made to enable electronic communications services from cable or wireless networks to be readily supplied from the exterior of the building to the interior (of the building).	Not yet fully established

Electronic communications services should be supplied to the inside of a dwelling (without having to unnecessarily disturb the fabric of the building) via an external terminal box, through a suitable wall transition, to an internal terminal box.

Q3 – Means of supply around the building

Requirement	Limits on application
There shall be reasonable provision of [cable ducts] within the building to facilitate the future installation and supply of electronic communications services to each floor of the building and to at least one suitable room on each floor of the building.	Not yet fully established

Electronic communications services should be capable of being distributed around the building, currently or at some time in the future, inside ducts to at least:

- each floor of each dwelling (including the loft space and basement where these exist);
- one habitable room at each level of each dwelling.

 The requirements in Part Q of Schedule 1 to the Building Regulations do **not** require anything to be done except for the purposes of '*securing reasonable levels of convenience for building users*'.

2.6 Are there any exemptions?

The Building Regulations do not apply to:

- a building belonging to statutory undertakers;
- a building belonging to the United Kingdom Atomic Energy Authority;
- a building belonging to the British Airports Authority;
- a building belonging to the Civil Aviation Authority;

unless it is a house, hotel or a building used as offices or showrooms not forming part of any of the above premises.

 Note: From 1 April 2001, maintained schools ceased to have exemption from the Building Regulations and school-specific standards have now been incorporated into the latest editions of Approved Documents.

Purpose-built student living accommodation (including flats) should thus be treated as hotel/motel accommodation in respect of space requirements and internal facilities.

2.7 What happens if I do not comply with an Approved Document?

Not actually complying with an Approved Document (which is after all only meant as a guidance document) doesn't mean that you are liable to any civil or criminal prosecution. If, however, you have contravened a Building Regulation then not having complied with the recommendations contained in the Approved Documents may be held against you.

2.8 Do I need Building Regulations approval?

If you are considering carrying out building work to your property then you may need to apply to your local authority for *Building Regulations approval*.

For most types of building work (e.g. extensions, alterations, conversions and drainage works), you will be required to submit a Building Regulations application prior to commencing any work.

Certain types of extensions and small detached buildings are exempt from Building Regulations control – particularly if the increased area of the alteration or extension is less than $30\,m^2$. However, you may **still** be required to apply for planning permission.

 If you are in any doubt about whether you need to apply for permission, you should contact your local authority building control department before commencing any work to your property (in all cases, you may require planning permission).

2.8.1 Building work needing formal approval

The Building Regulations apply to any building that involves:

- the erection of a new building or re-erection of an existing building;
- the extension of a building;
- the '*material alteration*' of a building;
- the '*material change of use*' of a building;
- the installation, alteration or extension of a controlled service or fitting to a building.

2.8.2 Typical examples of work needing approval

- Altered openings for new windows in roofs or walls;
- Cellars (particularly in London);
- Electrical installations;
- Erection of new buildings that are not exempt;

- Home extensions such as for a kitchen, bedroom, lounge, etc.;
- Installation of baths, showers, WCs which involve new drainage or waste plumbing;
- Installation of cavity insulation;
- Installation of new heating appliances;
- Internal structural alterations, such as the removal of a load-bearing wall or partition;
- Loft conversions;
- New chimneys or flues;
- Replacing roof coverings (unless exactly like for like repair);
- Underpinning of foundations.

2.8.3 Exempt buildings

There are certain buildings and work that are exempt from control. This is generally because they are buildings controlled by other legislation or because it would not be reasonable to control.

The following list, although not extensive, provides an indication of the main exemptions. These Regulations do not apply to:

- local authorities;
- county councils;
- public bodies;
- the Metropolitan Police Authority;
- any building constructed in accordance with the Explosives Acts 1875 and 1923;
- any building erected under the Nuclear Installations Act 1965;
- any building included in the schedule of monuments maintained under Section 1 of the Ancient Monuments and Archaeological Areas Act 1979;
- buildings not frequented by people;
- greenhouses and agricultural buildings – unless they are being used for retailing, packing or exhibiting;
- temporary buildings, i.e. a building which is not intended to remain erected for more than 28 days;
- ancillary buildings, e.g. an office on a building site;
- a detached building with a floor area less than $15\,m^2$ and containing no sleeping accommodation;
- a small detached building with a floor area less than $30\,m^2$, which contains no sleeping accommodation, is less than 1 m from the boundary and is constructed substantially of non-combustible material;
- a detached building designed and intended to shelter people from the effects of nuclear, chemical or conventional weapons;
- a conservatory whose floor area is less than $30\,m^2$ provided that it is wholly or partly glazed;
- a porch whose floor area is less than $3\,m^2$;

- a covered yard or covered way whose floor area is less than 30 m²;
- a carport open on at least two sides whose floor area is less than 30 m².

The power to dispense with or relax any requirement contained in these Regulations rests with the local authority. It is therefore advisable to contact your local authority building control officer with details of your particular exemption claim so that you obtain a written reply agreeing the exemption. This will aid any future sale of the property!

Are there any other exemptions from the requirement to give building notice or deposit full plans?

The installations listed in Table 2.1 are exempt from having to give building notice or deposit full plans, **provided** that the person carrying out the work is as indicated in the second column.

In addition, provided any associated building work required to ensure that the appliance (service or fitting detailed above) complies with the applicable requirements contained in Schedule 1 – unless it is a heat producing gas appliance) which

(a)   has a net rated heat input of 70 kilowatts or less; and
(b)   is installed in a building with no more than three storeys (excluding any basement).

'*appliance*' includes any fittings or services, other than a hot water storage vessel that does not incorporate a vent pipe to the atmosphere, which form part of the space heating or hot water system served by the combustion appliance; and

'*building work*' does **not** include the provision of a masonry chimney.

2.8.4 Where can I obtain assistance in understanding the requirements?

Local councils can provide assistance with:

- advice about how to incorporate the most efficient energy safety measures into your scheme;
- advice about the use of materials;
- advice on electrical safety;
- advice on fire safety measures (including safe evacuation of buildings in the event of an emergency);
- at what stages local councils need to inspect your work;
- deciding what type of application is most appropriate for your proposal;
- how to apply for Building Regulations approval;
- how to prepare your application (and what information is required);
- how to provide adequate access for disabled people;
- what your Building Regulation Completion Certificate means to you.

Table 2.1 Exemptions from giving building notice or depositing full plans

Type of work	Person carrying out work
Installation of a **heat-producing gas appliance**.	A person, or an employee of a person approved in accordance with Regulation 3 of the Gas Safety (Installation and Use) Regulations 1998.
Installation of an **oil-fired combustion appliance** with a rated heat output of 45 kilowatts or less, installed in a building with no more than three storeys*.	An individual registered under the Oil Firing Registration Scheme by the Oil Firing Technical Association for the Petroleum Industry Ltd.
Installation of **oil storage tanks** and the pipes connecting them to combustion appliances.	An individual registered under the Oil Firing Registration Scheme by the Oil Firing Technical Association for the Petroleum Industry.
Installation of a **solid fuel burning combustion appliance** which has a rated heat output of 50 kilowatts or less installed in a building with no more than three storeys*.	An individual registered under the Registration Scheme for Companies and Engineers involved in the Installation and Maintenance of Domestic Solid Fuel Fired Equipment by HETAS Ltd.
Installation of a **service or fitting** which is installed in or in connection with a building with no more than three storeys* and which does not involve connection to a drainage system at a depth greater than 750 mm from the surface.	An individual registered under the Approved Contractor Person Scheme (Building Regulations) by the Institute of Plumbing.
Installation of a **foul water drainage system** which is installed in or in connection with a building with no more than three storeys* and which does not involve connection to a drainage system at a depth greater than 750 mm from the surface.	An individual registered under the Approved Contractor Person Scheme (Building Regulations) by the Institute of Plumbing.
Installation of a **rainwater drainage system** in relation to which paragraph H3 of Schedule 1 imposes a requirement, which is installed in or in connection with a building with no more than three storeys* and which does not involve connection to a drainage system at a depth greater than 750 mm from the surface.	An individual registered under the Approved Contractor Person Scheme (Building Regulations) by the Institute of Plumbing.
Installation of a **hot water vessel** which is installed in or in connection with a building with no more than three storeys* and which does not involve connection to a drainage system at a depth greater than 750 mm from the surface.	An individual registered under the Approved Contractor Person Scheme (Building Regulations) by the Institute of Plumbing.
Installation, as a replacement, of a **window, rooflight, roof window or door** in an existing building.	A person registered under the Fenestration Self-Assessment Scheme by Fensa Ltd.
Electrical work.	A competent person registered with one of the Part P self-certification schemes.

*Excluding any basement.

2.9 How do I obtain Building Regulations approval?

You, as the owner or builder, are required to fill in an application form and return it, along with basic drawings and relevant information, to the building control office at least two days before work commences. Alternatively, you may submit full detailed plans for approval. Whatever method you adopt, it may save time and trouble if you make an appointment to discuss your scheme with the building control officer well before you intend carrying out any work.

The building control officer will be happy to discuss your intentions, including proposed structural details and dimensions together with any lists of the materials you intend to use, so that he can point out any obvious contravention of the Building Regulations before you make an official application for approval. At the same time he can suggest whether it is necessary to approach other authorities to discuss planning, sanitation, fire escapes and so on.

The building control officer will ask you to inform the office when crucial stages of the work are ready for inspection (by a surveyor) in order to make sure the work is carried out according to your original specification. Should the surveyor be dissatisfied with any aspect of the work, he may suggest ways to remedy the situation.

 When the building is finished you must notify the council.

It would be to your advantage to ask for written confirmation that the work was satisfactory as this will help to reassure a prospective buyer when you come to sell the property.

2.10 What are building control bodies?

Your local authority has a general duty to see that all building work complies with the Building Regulations. To ensure that your particular building work complies with the Building Regulations you must use one of the two services available to check and approve plans and to inspect your work as appropriate. The two services are the local authority building control service or the service provided by the private sector in the form of an approved inspector. Both building control bodies will charge for their services. Both may offer advice before work is started.

2.10.1 What will the local authority do?

This rather depends on whether you are submitting:

(1)   full plans application submission; or
(2)   building notice application.

In both cases the building control office will carry out site inspections at various stages.

 The total fee is the same whichever method is chosen.

Full plans

If you use the full plans procedure, the local authority will check your plans and consult any appropriate authorities (such as fire and water authorities). If your plans comply with the Building Regulations you will receive a notice that they have been approved. If the local authority are not satisfied, then you may be asked to make amendments or provide more details. Alternatively, a conditional approval may be issued which will either specify modifications that must be made to the plans, or will specify further plans that must be deposited. A local authority may only apply conditions if you have either requested them to do so or have consented to them doing so. A request or consent must be made in writing. If your plans are rejected the reasons will be stated in the notice.

Building notice

If you use the building notice procedure, as with full plans applications, the work will normally be inspected as it proceeds; but you will not receive any notice indicating whether your proposal has been passed or rejected. Instead, you will be advised where the work itself is found (by the building control officer) not to comply with the Regulations.

Where a building notice has been given, the person carrying out building work or making a material change of use is required to provide plans showing how they intend conforming with the requirements of the Building Regulations. The local authority may also require further information such as structural design calculations of plans.

2.10.2 What will the approved inspector do?

If you use an approved inspector they will give you advice, check plans, issue a plans certificate, inspect the work etc. as agreed between you both. You and the inspector will jointly notify the local authority on what is termed an initial notice. Once that has been accepted by the local authority, the approved inspector will then be responsible for the supervision of building work. Although the local authority will have no further involvement, you may still have to supply them with limited information to enable them to be satisfied about certain aspects linked to Building Regulations (e.g. about the point of connection to an existing sewer).

If the approved inspector is not satisfied with your proposals you may alter your plans according to his advice; or you may seek a ruling from the Secretary of State regarding any disagreement between you. The approved inspector might also suggest an alternative form of construction, and, provided that the work has not been started, you can apply to the local authority for a relaxation or a dispensation from one (or more) of the Regulations' requirements and, in the event of a refusal by the local authority, appeal to the Secretary of State.

If, however, you do not exercise these options and you do not do what the approved inspector has advised to achieve compliance, the inspector will not be able to issue a final certificate. The inspector will also be obliged to notify the local authority so that they can consider whether to use their powers of enforcement.

2.10.3 What is the difference between a full plans application and the building notice procedure?

A person who intends carrying out any building work or making a material change of use to a building, shall:

- either provide the local authority with a building notice or
- deposit full plans with the local authority

subject to the following exclusions listed in Section 2.11.1 below.

For a full plans application, plans need to be produced showing all constructional details, preferably well in advance of your intended commencement on site. For the building notice procedure less detailed plans are required. In both cases, your application or notice should be submitted to the local authority and should be accompanied by any relevant calculations, to demonstrate compliance with safety requirements concerning the structure of the building.

If the use of the building is a 'designated use' under the Fire Precautions Act 1971, the application method **must** be a 'full plans' submission. This is to allow the local building control office to consult the fire brigade to see if they have any comments on the adequacy of the building's proposed means of escape in the event of fire.

Approved plans are valid for at least three years.

2.11 How do I apply for building control?

If your prospective work will involve any form of structure, you could need building control approval.

Some types of work may need both planning permission and building control approval; others may need only one or the other. The process of assessing a proposed building project is carried out through an evaluation of submitted information and plans and the inspection of work as the building progresses.

Take advantage of the free advice that local authorities offer, and discuss your ideas well in advance.

2.11.1 What applications do not require submission plans?

The following building works do **not** require the submission of plans:

- in respect of any work specified in an initial notice, an amendment notice or a public body's notice, which is in force;

- where a person intends to have electrical installation work completed by a competent firm registered under the NICIEC Approved Contractor scheme;
- where a person intends to have installed (by a person, or an employee of a person approved in accordance with Regulation 3 of the Gas Safety (Installation and Use) Regulations 1998) a heat-producing gas appliance;
- where Regulation 20 of the Building (Approved Inspectors etc.) Regulations 2000 (local authority powers in relation to partly completed work) applies.

2.11.2 Other considerations

Depending on the type of work involved, you may need to get approval from several sources before starting. The list below provides a few examples:

- There may be legal objections to alterations being made to your property.
- A solicitor might need to be consulted to see if any covenants or other forms of restriction are listed in the title deeds to your property and if any other person or party needs to be consulted before you carry out your work.
- You may need planning permission for a particular type of development work.
- If a building is listed or is within a Conservation Area or an Area of Outstanding Natural Beauty, special rules apply.

2.12 Full plans application

This type of application can be used for any type of building work, but it **must** be used where the proposed premises are to be used as a factory, office, shop, hotel, boarding house or railway premises.

A full plans application requires the submission of fully detailed plans, specifications, calculations and other supporting details to enable the building control officer to ascertain compliance with the Building Regulations. The amount of detail depends on the size and type of building works proposed, but as a minimum will have to consist of:

- a description of the proposed building work or material change of use;
- plan(s) showing what work will be completed; plus
- a location plan showing where the building is located relative to neighbouring streets.

 The full plans application may be accompanied by a request (from the person carrying out such building work) that on completion of the work, he wishes the local authority to issue a completion certificate.

Two copies of the full plans application need to be sent to the local authority except in cases where the proposed building work relates to the erection, extension or material alteration of a building (other than a dwelling-house or flat) and where fire safety imposes an additional requirement, in which case five copies are required.

A full plans application will be thoroughly checked by the local authority who are required to pass or reject your plans within a certain time limit (usually eight weeks); or they may add conditions to an approval (with your written agreement). If they are satisfied that the work shown on the plans complies with the Regulations, you will be issued with an approval notice (within a period of five weeks or up to two months) showing that your plans were approved as complying with the Building Regulations.

If your plans are rejected, and you do not consider it is necessary to alter them, you will have two options available to you:

- you may seek a 'determination' from the Secretary of State if you believe your work complies with the Regulations (but you must apply before work starts);
- if you acknowledge that your proposals do not necessarily comply with a particular requirement in the Regulations and feel that it is too onerous in your particular circumstances, you may apply for a relaxation or dispensation of that particular requirement from the local authority. You can make this sort of application at any time you like but it is obviously sensible to do so as soon as possible and preferably before work starts. If the local authority refuses your application, you may then appeal to the Secretary of State within a month of the date of receipt of the rejection notice.

2.12.1 Consultation with sewerage undertaker

Where applicable, the local authority shall consult the sewerage undertaker as soon as practicable after the plans have been deposited, and before issuing any completion certificate in relation to the building work.

2.12.2 Advantages of submitting full plans application

The advantages of the full plans method are that:

- a (free) completion certificate will be issued on satisfactory completion of the work;
- a formal notice of approval or rejection will be issued within five weeks (unless the applicant agrees to extend this to two months);
- only when work starts on site (and the building control officer has completed his initial visit) is the remaining part of the fee invoiced;
- the plans can be examined and approved in advance (for an advance payment of (typically) 25% of the total fee).

2.13 Building notice procedure

Under the building notice procedure no approval notice is given. There is also no procedure to seek a determination from the Secretary of State if there is a disagreement between you and the local authority – unless plans are subsequently deposited. However, the advantage of the building notice procedure is that it will allow you to carry out **minor works** without the need to prepare full plans. You must, however, feel confident that the work will comply with the Regulations or you risk having to correct any work you carry out at the request of the local authority.

A building notice is particularly suited to minor works (for example, a householder wishing to install another WC). For such building work, detailed plans are unnecessary and most matters can be agreed when the building control officer visits your property. You do not need to have detailed plans prepared, but in some cases you may be asked to supply extra information.

This method is **not** allowed for any work on listed buildings or buildings in a Conservation Area.

As no formal approval is given, good liaison between the builder and the building control officer is essential to ensure that work does not have to be re-done.

The submission of a marked-up sketch showing the location of the building, although not mandatory, is recommended.

This type of application may be used for all types of building work, so long as no part of the premises is used for any of the purposes mentioned above under the full plans application.

2.13.1 What do I have to include in a building notice?

A building notice shall:

- state the name and address of the person intending to carry out the work;
- be signed by that person or on that person's behalf;
- contain, or be accompanied by:
  - a description of the proposed building work or material change of use;
  - particulars of the location of the building;
  - the use or intended use of that building.

Extension of a building

When planning a building extension, the building notice needs to be accompanied by:

- a plan to a scale of not less than 1:1250 showing:
  - the size and position of the building, or the building as extended and its relationship to adjoining boundaries;
  - the boundaries of the curtilage of the building, or the building as extended, and the size, position and use of every other building or proposed building within that curtilage;
  - the width and position of any street on or within the boundaries of the curtilage of the building or the building as extended;
- a statement specifying the number of storeys (each basement level being counted as one storey), in the building to which the proposal relates;

- particulars of:
  - the provisions to be made for the drainage of the building or extension;
  - the steps to be taken to comply with any local enactment which applies.

Insertion of insulating material into the cavity walls of a building

For cavity wall insulations, the building notice needs to be accompanied by a statement which specifies:

- the name and type of insulating material to be used;
- the name of any European Technical Approval issuing body that has approved the insulating material;
- the requirements of Schedule 1 in relation to which the issuing body has approved the insulating material;
- any European Economic Area (EEA) national standard with which the insulating material conforms;
- the name of any body that has issued any current approval to the installer of the insulating material.

Provision of a hot water storage system

A building notice in respect of a proposed hot water system shall be accompanied by a statement which specifies:

- the name, make, model and type of hot water storage system to be installed;
- the name of the body (if any) that has approved or certified the system;
- the name of the body (if any) that has issued any current registered operative identity card to the installer or proposed installer of the system.

Electrical installations

All proposals to carry out electrical installation work **must** be notified to the local authority's building control body before work begins, **unless** the proposed installation work is undertaken by a person who is a competent person registered with an electrical self-certification scheme and does not include the provision of a new circuit.

Non-notifiable work (such as replacing an electrical fitting) can be completed by a DIY enthusiast (family member or friends) but needs to be installed in accordance with manufacturers' instructions and done in such a way that they do not present a safety hazard. This work does **not** need to be notified to a local authority building control body (unless it is installed in an area of high risk such as a kitchen or a bathroom, etc.) **but** all DIY electrical work (unless completed by a qualified professional – who is responsible for issuing a Minor Electrical Installation Certificate) will still need to be checked, certified and tested by a competent electrician.

Any work that involves adding a new circuit to a dwelling will need to be either notified to the building control body (who will then inspect the work) or needs to be carried out by a competent person who is registered under a Government Approved Part P Self-Certification Scheme.

Work involving any of the following will also have to be notified:

- locations containing a bath tub or shower basin;
- swimming pools or paddling pools;
- hot air saunas;
- electric floor or ceiling heating systems;
- garden lighting or power installations;
- solar photovoltaic (PV) power supply systems;
- small-scale generators such as microCHP units;
- extra-low voltage lighting installations, other than pre-assembled, CE-marked lighting sets.

 Note: Where a person who is **not** registered to self-certify, intends to carry out the electrical installation, then a Building Regulation (i.e. a building notice or full plans) application will need to be submitted together with the appropriate fee, based on the estimated cost of the electrical installation. The building control body will then arrange to have the electrical installation inspected at first fix stage and tested upon completion.

2.14 How long is a building notice valid?

A building notice shall cease to have effect three years from the date when that notice was given to the local authority, unless, before the expiry of that period:

- the building work to which the notice related has commenced; or
- the material change of use described in the notice was made.

 The approved plans may be used (i.e. built to) for at least three years, **even if the Building Regulations change during this time**.

2.15 What can I do if my plans are rejected?

If your plans were initially rejected, you can start work **provided** you give the necessary notice of commencement required under Regulation 14 of the Building Regulations and are satisfied that the building work itself now complies with the Regulations. However, it would **not** be advisable to follow this course if you are in any doubt and have not taken professional advice. Instead:

- you should resubmit your full plans application with amendments to ensure that they comply with Building Regulations; or
- if you think your plans comply (and that the decision to reject is, therefore, unjustified) you can refer the matter to the Secretary of State for the Environment, Transport and the Regions, or the Secretary of State for Wales (as appropriate) for their determination, but usually only before the work has started; or

- you could (in particular cases) ask the local authority to relax or dispense with their rejection. If the local authority refuse your application you could then appeal to the appropriate Secretary of State within one month of the refusal.

In the first two cases, the address to write to is the Department of the Environment, Transport and the Regions (DETR), 3/C1, Eland House, Bressenden Place, London SW1E 5DU. In Wales, you should refer the matter to the Secretary of State for Wales, Welsh Office, Crown Buildings, Cathays Park, Cardiff CF1 3NQ.

A fee is payable for determinations but not for appeals. The fee is half the plan fee (excluding VAT) subject to a minimum of £50 and a maximum of £500. The DETR or the Welsh Office will then seek comments from the local authority on your application (or appeal) which will be copied to you. You will then have a further opportunity to comment before a decision is issued by the Secretary of State.

2.15.1 Do my neighbours have the right to object to what is proposed in my Building Regulations application?

Basically – *no*! But whilst there is no requirement in the Building Regulations to consult neighbours, it would be prudent to do so. In any event, you should be careful that the work does not encroach on their property since this could well lead to bad feeling and possibly an application for an injunction for the removal of the work.

Objections may be raised under other legislation, particularly if your proposal is subject to approval under the Town and Country Planning legislation or the Party Wall etc. Act of 1996.

The Party Wall Act 1996 came into force on 1 July 1997 and is largely based on Part VI of the London Building Acts (Amendment) Act 1939 – which started life as a Private Members Bill sponsored by the Earl of Lytton.

In a nutshell, this Act says that if you intend to carry out building work which involves:

- work on an existing wall shared with another property;
- building on the boundary with a neighbouring property;
- excavating near an adjoining building;

you **must** find out whether that work falls within the scope of the Act. If it does, then you must serve the statutory notice on all those defined by the Act as 'adjoining owners'. You should, however, remember that reaching agreement with adjoining owners on a project that falls within the scope of the Act does **not** remove the possible need for planning permission or Building Regulations approval.

Note: If you are not sure whether the Act applies to the work that you are planning, you should seek professional advice (see 'Useful contact names and addresses' at the end of the book).

A free explanatory booklet on the Party Wall etc. Act 1996 (*The Party Wall Act 1996 – Explanatory Booklet* (product code 02 BR 00862)) is available from the DETR Free Literature, PO Box No 236, Wetherby, LS23 7NB (Tel: 0870 1226 236, Fax: 0870 1226 237).

2.16 What happens if I wish to seek a determination but the work in question has started?

You will only need to seek a determination if you believe the proposals in your full plans application comply with the Regulations but the local authority disagrees. You may apply for a determination either before or after the local authority has formally rejected your full plans application. The legal procedure is intended to deal with compliance of '*proposed*' work only and, in general, applications relating to work which is substantially completed cannot be accepted. Exceptionally, however, applications for '*late*' determinations may be accepted – but it is in your best interest to always ensure that you apply for a determination well before you start work.

2.17 When can I start work?

Again, it depends on whether you are using the local authority or the approved inspector.

2.17.1 Using the local authority

Once you have given a building notice or submitted a full plans application, you can start work at any time. However, you must give the local authority a commencement notice at least two clear days (not including the day on which you give notice and any Saturday, Sunday, bank or public holiday) before you start.

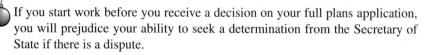 If you start work before you receive a decision on your full plans application, you will prejudice your ability to seek a determination from the Secretary of State if there is a dispute.

2.17.2 Using an approved inspector

If you use an approved inspector you may, subject to any arrangements you may have agreed with the inspector, start work as soon as the initial notice is accepted by the local authority (or is deemed to have been accepted if nothing is heard from the local authority within five working days of the notice being given). Work may not start if the initial notice is rejected, however.

2.18 Planning officers

Before construction begins, planning officers determine whether the plans for the building or other structure comply with the Building Regulations and if they are suited to the engineering and environmental demands of the building site. Building inspectors are then responsible for inspecting the structural quality and general safety of buildings.

2.19 Building inspectors

Building inspectors examine the construction, alteration, or repair of buildings, highways and streets, sewer and water systems, dams, bridges, and other structures to ensure compliance with building codes and ordinances, zoning regulations, and contract specifications.

Building codes and standards are the primary means by which building construction is regulated in the UK to assure the health and safety of the general public. Inspectors make an initial inspection during the first phase of construction and then complete follow-up inspections throughout the construction project in order to monitor compliance with regulations.

The inspectors will visit the worksite before the foundation is poured to inspect the soil condition and positioning and depth of the footings. Later, they return to the site to inspect the foundation after it has been completed. The size and type of structure, as well as the rate of completion, determine the number of other site visits they must make. Upon completion of the project, they make a final comprehensive inspection.

2.20 Notice of commencement and completion of certain stages of work

A person who proposes carrying out building work shall not start work unless:

- he has given the local authority notice that he intends to commence work; and
- at least two days have elapsed since the end of the day on which he gave the notice.

2.20.1 Notice of completion of certain stages of work

The person responsible for completing the building work is also responsible for notifying the local authority a minimum of five days **prior** to commencing any work involving excavations for foundations, foundations themselves, any damp-proof course any concrete or other material to be laid over a site and drains or sewers.

Upon completion of this work (especially work that will eventually be covered up by later work) the person responsible for the building work shall give five days' notice of intention to backfill.

A person who has laid, haunched or covered any drain or sewer shall (not more than five days after that work has been completed) give the local authority notice to that effect.

Where a building is being erected and that building (or any part of it) is to be occupied before completion, the person carrying out that work shall give the local authority at least five days' notice before the building, or any part of it is, occupied.

The person carrying out the building work shall **not**:

- cover up any foundation (or excavation for a foundation), any damp-proof course or any concrete or other material laid over a site; or
- cover up (in any way) any drains or sewers **unless** he has given the local authority notice that he intends to commence that work and at least one day has elapsed since the end of the day on which he gave the notice.

 Where a person fails to comply with the above, then the local authority can insist that he shall cut into, lay open or pull down '*so much of the work as to enable the authority to ascertain whether these Regulations have been complied with, or not*'. If the local authority then notifies the owner/builder that certain work contravenes the requirements in these Regulations, then the owner/builder shall, after completing the remedial work, notify the local authority of its completion.

This requirement does not apply in respect of any work specified in an initial notice, an amendment notice or a public body's notice that is in force.

2.20.2 What kind of tests are the local authorities likely to make?

To establish whether building work has been carried out in conformance with the Building Regulations, local authorities will test to ensure that all work has been carried out:

- in a workmanlike manner;
- with adequate and proper materials which:
  - are appropriate for the circumstances in which they are used,
  - are adequately mixed or prepared and
  - are applied, used or fixed so as to adequately perform the functions for which they are designed;
- complies with the requirements of Part H of Schedule 1 (drainage and waste disposal);
- so as to enable them to ascertain whether the materials used comply with the provisions of these Regulations.

2.20.3 Energy rating

Where a new dwelling is being created, the person carrying out the building work shall calculate (and inform the local authority of) the dwelling's energy rating not later than five days after the work has been completed and, where a new dwelling is created, at least five days before intended occupation of the dwelling.

If the building is not to be immediately occupied, then the person carrying out the building work shall affix (not later than five days after the work has been completed) in a conspicuous place in the dwelling, a notice stating the energy rating of the dwelling.

 Details of the correct procedures for calculating the energy rating are available from local authorities.

2.21 What are the requirements relating to building work?

In all cases, building work shall be carried out so that it:

(a)   *it complies with the applicable requirements contained in Schedule 1; and*
(b)   *in complying with any such requirement there is no failure to comply with any other such requirement.*

*Building work shall be carried out so that, **after** it has been completed:*

(a)   *any building which is extended or to which a material alteration is made; or*
(b)   *any building in, or in connection with which, a controlled service or fitting is provided, extended or materially altered; or*
(c)   *any controlled service or fitting, complies with the applicable require-ments of Schedule 1 or, where it did not comply with any such require-ment, is no more unsatisfactory in relation to that requirement than before the work was carried out.*

2.22 Do I need to employ a professional builder?

Unless you have a reasonable working knowledge of building construction it would be advisable before you start work to get some professional advice (e.g. from an architect, or a structural engineer, or a building surveyor) and/or choose a recognized builder to carry out the work. It is also advisable to con-sult the local authority building control officer or an approved inspector in advance.

2.23 Unauthorized building work

If, for any reason, building work has been done without a building notice or full plans of the work being deposited with the local authority; or a notice of commencement of work being given, then the applicant may apply, in writing, to the local authority for a regularization certificate.

This application will need to include:

* a description of the unauthorized work;
* a plan of the unauthorized work; and
* a plan showing any additional work that is required for compliance with the requirements relating to building work in the Building Regulations.

Local authorities may then '*require the applicant to take such reasonable steps, including laying open the unauthorised work for inspection by the authority, making tests and taking samples, as the authority think appropriate to ascertain what work, if any, is required to secure that the relevant requirements are met*'.

When the applicant has taken any such steps required by the local authority, the local authority will notify the applicant:

* if no work is required to secure compliance with the relevant requirements;
* of the work which is required to comply with the relevant requirements;
* of the requirements which can be dispensed with or relaxed.

2.23.1 What happens if I do work without approval?

The local authority has a general duty to see that all building work complies with the Regulations – except where it is formally under the control of an approved inspector. Where a local authority is controlling the work and finds after its completion that it does not comply, then the local authority may require you to alter or remove it. If you fail to do this the local authority may serve a notice requiring you to do so and you will be liable for the costs.

2.23.2 What are the penalties for contravening the Building Regulations?

If you contravene the Building Regulations by building without notifying the local authority or by carrying out work which does not comply, the local authority can prosecute. If you are convicted, you are liable to a penalty not exceeding £5000 (at the date of publication of this book) plus £50 for each day on which each individual contravention is not put right after you have been convicted. If you do not put the work right when asked to do so, the local authority have power to do it themselves and recover costs from you.

2.24 Why do I need a completion certificate?

A completion certificate certifies that the local authority are satisfied that the work complies with the relevant requirements of Schedule 1 of the Building Regulations, '*in so far as they have been able to ascertain after taking all reasonable steps*'.

 A completion certificate is a valuable document that should be kept in a safe place!

2.25 How do I get a completion certificate when the work is finished?

The local authority shall give a completion certificate only when they have received the completion notice and have been able to ascertain that the relevant requirements of Schedule 1 (specified in the certificate) have been satisfied.

Where full plans are submitted for work that is also subject to the Fire Precautions Act 1971, the local authority must issue you with a completion certificate concerning compliance with the fire safety requirements of the Building Regulations once work has finished. In other circumstances, you may ask to be given one when the work is finished, but you must make your request **when you first submit your plans**.

 If you use an approved inspector, they must issue a final certificate to the local authority when the work is completed.

2.26 Where can I find out more?

You can find out more from:

- the local authority's building control department;
- an approved inspector; or
- other sources.

2.26.1 Local authority

Each local authority in England and Wales (i.e. unitary, district and London boroughs in England and county and county borough councils in Wales) has a building control section whose general duty is to see that work complies with the Building Regulations – except where it is formally under the control of an approved inspector. Most local authorities have their own website and these usually contain a wealth of useful information, the majority of which is downloadable as read-only pdf files.

Individual local authorities co-ordinate their services regionally and nationally (and provide a range of national approval schemes) via LABC (Local

Authority Building Control) Services. You can find out more about LABC Services through its website at www.labc-services.co.uk but your local authority building control department will be pleased to give you information and advice. They may offer to let you see their copies of the Building Act 1984, the Building Regulations 2000 and their associated Approved Documents that provide additional guidance.

The *Fire and Building Regulations Procedural Guide* which deals with procedures for building work to which the Fire Precautions Act 1971 applies, and the Department of the Environment, Transport and the Regions (DETR) leaflet on safety of garden walls, are amongst the documentation and advice that is available, free of charge, from your local authority.

The DETR's (and the Welsh Office's) separate booklets on planning permission for small businesses and householders are also available free of charge from your local authorities.

2.26.2 Approved inspectors

Approved inspectors are companies or individuals authorized under the Building Act 1984 to carry out building control work in England and Wales.

The Construction Industry Council (CIC) is responsible for deciding all applications for approved inspector status. You can find out more about the CIC's role (including how to apply to become an approved inspector) through its website at www.cic.org.uk.

A list of approved inspectors can be viewed at the Association of Corporate Approved Inspectors (ACAI) website at www.acai.org.uk.

2.26.3 Other sources

Most of the documents can be purchased from The Stationery Office (TSO – formally HMSO), 29 Duke Street, Norwich, NR3 1GN or from any main bookshop. Orders to TSO can be telephoned to 0870 600 5522 or faxed to 0870 600 5533 and their website is www.tso.co.uk. Copies should also be available in public reference libraries.

Appendix 2A   Example application form

Building Control Service

RIDDIFORD DISTRICT COUNCIL

Building Act 1984 – Building Regulations 2000

DEPOSIT OF BUILDING NOTICE

(Do not use where Plan Approval is required, if Building over or within 3 metres of a Public Sewer/drain, if Building is Designated/Workplace or if the Building will front onto a Private street.)

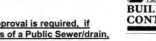

Local Authority
BUILDING CONTROL

Submit **Two** Copies of **Forms** To:	Complete in BLOCK Capitals and BLACK Ink.
Director of Planning and Technical Services Riddiford District Council Riddiford House Riddiford, Devon EX19 8DW	Application No: .... FEE: .... Cheque/PO/Cash: .... VAT: .... Total Fee: .... Accepted: .... *For Office Use Only*

Please Read The Notes Before Completing This Form

1. **Applicants Details (See Note 1):**   Owners Details (if different)
   Name:
   Address:

   Post Code: Tel. No:
   Fax No: E-Mail:

2. **Agents Details** (if any) to whom correspondence should be sent
   Name:
   Address:
   Post Code: Tel. No:
   Fax No: E-Mail:

3. **Location** of building to which work relates
   Address:

   Post Code: Tel. No:

4. **Proposed Work**
   Description:
   Date of commencement (See Note 6):

5. **Use of building**
   a.   If new building or extension state proposed use:
   b.   If existing building state present use:

6. **Method of Drainage**
   a.   Foul water:
   b.   Surface water:

7. **Means of water supply:**

8. Have you received Planning Consent for this work?   YES/NO
   If yes, please give Consent No. and Date of Decision:

9. **Conservation of Fuel & Power** (See Note 3.2) Please indicate the method used to show compliance with Part L
   For Dwellings   For Other Buildings
   ☐ Elemental Method   ☐ Elemental Method
   ☐ Target U-Value Method   ☐ Whole Building Method
   ☐ Carbon Index Method (Calculations to be enclosed) ☐ Carbon Emission Calculation Method
   ☐ Material Alteration / Change of Use   ☐ Material Alteration / Change of Use

10. **Fees** (See Guidance Note on Fees for information)
    a.   If Table A work - please state number of small domestic buildings:
                        - please state number of different types of buildings:
    b.   If Table B work - please state floor area m²:
    c.   If Table C work - please state estimated cost of work excluding VAT £:

11. **Statement**
    This notice is given in relation to the building work as described, and is submitted in accordance with Regulation 11(1)(a) and is accompanied by the appropriate fee

    Name: Signature: Date:

PLEASE TURN OVER	**JANUARY 2005**

Notes:

1 The applicant is the person proposing to carry out building work, e.g. the building's owner.

2 Two copies of this notice should be completed and submitted. If a sewer connection is proposed, please submit an additional set of plans.

3 Where the proposed work includes the erection of a new building, material change of use or extension this notice shall be accompanied by the following:

3.1 a block plan to a scale of not less than 1:1250 showing:

3.1.1 the size and the position of the building, or the building as extended, and its relationship to adjoining boundaries;

3.1.2 the boundaries of the curtilage of the building, or the building as extended, and the size, position and use of every other building or proposed building within that curtilage;

3.1.3 the width and position of any street or within the boundaries of the curtilage of the building or the building as extended;

3.1.4 the provision to be made for the drainage of the building or extension.

3.2 the requirements of Part L (Conservation of Fuel and Power) must be considered for all dwellings and all other buildings. The requirements also apply to material alterations and changes of use.

3.2.1 extensions to dwellings less than 6 square metres would only need to be as energy efficient as the existing dwelling. Therefore question 9 is not applicable.

3.2.2 for other buildings with a floor area exceeding $100 \, m^2$ floor area, details of artificial lighting systems will be required with the application.

3.2.3 Regulation 16 requires the provision of Energy Ratings calculated by the Government's Standard Assessment Procedure (SAP) for new dwellings and dwellings created as the result of material changes of use. The SAP calculation must be provided to the Authority within a minimum of five days prior to the occupation of the dwelling or five days after the completion, whichever comes first. Dwellings includes Flats.
The person carrying out the building work must display in the dwelling, as soon as practicable after the energy rating has been calculated, a notice of that rating. The notice must be displayed in a conspicuous place in the dwelling. Should occupation of the dwelling take place before physical completion, then the energy rating notice should be given to the occupier.
The requirement to display an energy rating notice in the dwelling, or give a notice to the occupier, is not necessary where the person carrying out the building work intends to occupy, or occupies, the dwelling as his/her residence.

4 Where the proposed work involves the insertion of insulating material into the cavity walls of a building this building notice shall be accompanied by a statement as to:

4.1 the name and type of insulating material to be used;

4.2 the name and type of insulating material is approved by the British Board of Agrément or conforms to a British Standard specification;

4.3 whether or not the installer is a person who is subject of a British Standards Institution Certificate of Registration or has been approved by the British Board of Agrément for the insertion of that material.

5 Where the proposed work involves the provision of an un-vented hot water storage system, this building notice shall be accompanied by a statement as to:

5.1 the name and type of system to be provided;

5.2 whether or not the system is approved by the British Board of Agrément

5.3 whether or not the installer has been approved by the British Board of Agrément for the provision of that system.

6 Persons carrying out building work must give written notice of the commencement of the work at least 48 hours beforehand.

7 A fee is usually payable to contribute towards the cost of site inspections, being a single payment which covers all necessary site visits until satisfactory completion of the work in accordance with the Building Regulations.

8 The building notice fee is calculated in accordance with current fees regulations and paid at the time of deposit of application. A Guidance Note of Fees is available on request.

Table A prescribes the fees payable for small domestic buildings. Table B prescribes the fees payable for small alterations and extensions to a dwelling house, and the addition of a small garage or carport.
Table C prescribes the fees payable for all other cases.

9 Subject to certain provisions of the Public Health Act 1936 owners and occupiers of premises are entitled to have their private foul and surface water drains and sewers connected to the public sewers, where available. Special arrangements apply to trade effluent discharge. Persons wishing to make such connections must give not less than 21 days' notice to the appropriate authority.

10 These notes are for general guidance only, particulars regarding the submission of Building Notices are contained in Regulation 12 of the Building Regulations 2000 and, in respect of fees, in the Building (Prescribed Fees etc.) Regulations 1991 - as amended.

11 Persons proposing to carry out building work or make a material change of use are reminded that permission may be required under the Town and Country Planning Acts.

12 Further information and advice concerning the Building Regulations and planning matters may be obtained from your local authority.

13 A building notice shall cease to have effect on the expiry of 3 years from the date on which that notice was given to the local authority, unless the work has commenced.

14 A Building Notice cannot be used where the building is proposed to be put, or is currently put, to a use as a Workplace to which the Fire Precautions (Workplace) Regulations 1997 applies (and/or is designated as defined in the Fire Precautions Act 1971).

15 All Electrical work required to meet the requirements of Part P (Electrical Safety) must be designed, installed, inspected and tested by a person competent to do so.

15.1 Prior to completion the Council should be satisfied that Part P has been complied with. This may require an appropriate BS 7671:2001 electrical installation certificate to be issued for the work by a person competent to do so.

15.2 Compliance with Part P2
(a) The electrical installation certificate will normally include all the details necessary to identify and give details of each circuit.
(b) The person signing the certificate will also ensure that all appropriate notices required by BS 7671:2001 (for example those warning of differing wiring colours) are correctly provided and fixed.
(c) Provided the information in (a) and (b) is provided electrical installation diagrams should not normally be necessary. In cases where the works are of unusual complexity (perhaps in mixed use developments for example), the right to require such diagrams will remain at the discretion of the Building Control Surveyor.

Appendix 2B Example planning permission form

RIDDIFORD DISTRICT COUNCIL

Building Control Service

Building Act 1984 – Building Regulations 2000

DEPOSIT OF FULL PLANS

Local Authority **BUILDING CONTROL**

Submit **Two** Copies of **Forms** To:	Complete in BLOCK Capitals and BLACK Ink.
Director of Planning and Technical Services **Riddiford District Council** **Riddiford House** **Riddiford, Devon** **EX19 8DW**	APPLICATION NUMBER:.................... EXPIRY DATE: .... **PLAN FEE**　　　　　　　　　　**INSPECTION FEE** Cheque/PO/Cash: .... VAT:　　　　　　　.......................... Net Amount .... Total Fee:　　　　　　.......................... VAT Accepted:　　　　　　.......................... Total Fee　..................... *For Office Use Only*

Please Read The Notes Before Completing This Form

1. **Applicants Details (See Note 1):**　　　　　　　Owners Details (if different)
   Name:
   Address:

   Post Code:　　　　　　　　Tel. No:
   Fax No:　　　　E-Mail:

2. **Agents Details** (if any) to whom correspondence should be sent
   Name:
   Address:

   Post Code:　　　　　　　　Tel. No:
   Fax No:　　　　E-Mail:

3. **Location** of building to which work relates
   Address:

   Post Code:　　　　Tel. No:

4. **Proposed Work**
   Description:
   Date of commencement:

5. **Use of building**
   a. If new building or extension state proposed use:
   b. If existing building state present use:

6. **Method of Drainage**
   a. Foul water:
   b. Surface water:

7. **Means of water supply:**

8. Have you received Planning Consent for this work?　　　　　　　YES/NO
   If yes, please give Consent No. and Date of Decision:

9. **Conservation of Fuel & Power** (See Note 5.3) Please indicate the method used to show compliance with Part L

   For Dwellings　　　　　　　　　For Other Buildings
   ☐ Elemental Method　　　　　　　☐ Elemental Method
   ☐ Target U-Value Method　　　　　☐ Whole Building Method
   ☐ Carbon Index Method (Calculations to be enclosed)　☐ Carbon Emission Calculation Method
   ☐ Material Alteration / Change of Use　　☐ Material Alteration / Change of Use

10. **Fire Safety Requirements**
    Is the building proposed to be put, or is currently put, to a use as a Workplace to which the Fire Precautions
    (Workplace) Regulations 1997 applies and/or is designated as defined in the Fire Precautions Act 1971) ?　　Yes ☐ /
    No ☐

    If Yes, please indicate the design method you have used to show compliance with the Requirements of Part B of
    Schedule 1 to the Building Regulations.
    An additional copy of all plans/specifications are to be submitted if you have answered Yes to the above.

PLEASE TURN OVER	**JANUARY 2005**

11.   Conditions

Do you consent to the plans being passed subject to conditions where appropriate Yes ☐ / No ☐

12.   Extension of time

Do you consent to the prescribed period being extended to two months Yes ☐ / No ☐

13.   Fees (See Guidance Note on Fees for information)

    a.   If Table A work - please state number of small domestic buildings:

                  - please state number of different types of buildings:

    b.   If Table B work - please state floor area: m^2

    If Table C work - please state estimated cost of work excluding VAT £:

14.   Statement

This notice is given in relation to the building work as described, and is submitted in accordance with Regulation 11(1)(b) and is accompanied by the appropriate fee. I understand that further fees will be payable following the first inspection by the Local Authority.

Name: Signature: Date:

Should you have difficulty in filling out these forms please contact this office.

NOTES

1 The applicant is the person proposing to carry out the building works, e.g. the building's owner.

2 Two copies of this notice should be completed and submitted with two copies of plans and particulars in accordance with the provisions of Building Regulation 13. If a sewer connection is proposed an additional set of plans will be required.

3 Subject to certain exceptions a Full Plans Submission attracts fees payable by the person by whom or on whose behalf the work is to be carried out. Fees are payable in two stages. The first must accompany the deposit of plans and the second fee is payable after the first site inspection of work in progress. This second fee is a single payment in respect of each individual building, to cover all site visits and consultations, which may be necessary, until the work is satisfactorily completed.

3.1 Table A prescribes the plan and inspection fees payable for small domestic buildings. Table B prescribes the fees payable for small alterations and extensions to a dwelling home, and the addition of a small garage or carport. Table C prescribes the fees payable for all other cases.

3.2 The appropriate fee is dependent upon the type of work proposed. Fee scales and methods of calculation are set out in the Guidance Notes on Fees, which is available on request.

4 Subject to certain provisions of the Public Health Act 1936 owners and occupiers of premises are entitled to have their private foul and surface water drains and sewers connected to the public sewers, where available. Special arrangements apply to trade effluent discharge. Persons wishing to make such connections must give not less than 21 days' notice to the appropriate authority.

5 The requirements of Part L (Conservation of Fuel and Power) must be considered for all dwellings and other buildings. The requirements also apply to material alterations and change of use.

5.1 Extensions to dwellings less than 6 square metres would only need to be as energy efficient as the existing dwelling. Therefore question 9 is not applicable.

5.2 For other buildings with a floor area exceeding 100 m^2 floor area, details of artificial lighting systems will be required with the application.

5.3 Regulation 16 requires the provision of Energy Ratings calculated by the Government's Standard Assessment Procedure (SAP) for new dwellings and dwellings created as a result of material changes of use.

The SAP calculation must be provided to the Authority within a minimum of five days prior to the occupation of the dwelling or five days after the completion, whichever comes first. Dwellings include Flats. The person carrying out the building work must display in the dwelling, as soon as practicable after the energy rating has been calculated, a notice of that rating. The notice must be displayed in a conspicuous place in the dwelling. Should occupation of the dwelling take place before physical completion, then the energy rating notice should be given to the occupier.
The requirement to display an energy rating notice in the dwelling, or give a notice to the occupier, is not necessary where the person carrying out the building work intends to occupy, or occupies, the dwelling as his/her residence.

6 Section 16 of the Building Act 1984 provides for the passing of plans subject to conditions. The conditions may specify modifications to the deposited plans and/or that further plans shall be deposited.

7 These notes are for general guidance only, particulars regarding the deposit of plans are contained in Regulation 13 of the Building Regulations 2000 and, in respect of fees, in the Building (Prescribed Fees, etc.) Regulations 1991 as amended.

8 Persons proposing to carry out building work or make a material change of use of a building are reminded that permission may be required under the Town and Country Planning Acts.

9 Further information concerning the Building Regulations and Planning matters may be obtained from your Local Authority.

10 All Electrical work required to meet the requirements of Part P (Electrical Safety) must be designed, installed, inspected and tested by a person competent to do so.

10.1 Prior to completion the Council should be satisfied that Part P has been complied with. This may require an appropriate BS 7671:2001 electrical installation certificate to be issued for the work by a person competent to do so.

10.2 Compliance with Part P2

    (a)   The electrical installation certificate will normally include all the details necessary to identify and give details of each circuit.

    (b)   The person signing the certificate will also ensure that all appropriate notices required by BS 7671:2001 (for example those warning of differing wiring colours) are correctly provided and fixed.

    (c)   Provided the information in (a) and (b) is provided electrical installation diagrams should not normally be necessary. In cases where the works are of unusual complexity (perhaps in mixed use developments for example), the right to require such diagrams will remain at the discretion of the Building Control Surveyor.

Appendix 2C   Example of an application
for listed building consent

Planning and Technical Services Department

RIDDIFORD DISTRICT COUNCIL

Application For
Listed Building Consent

Local Authority
BUILDING CONTROL

Planning (Listed Buildings and Conservation Areas) Act 1990

Submit Three Copies of Plans and Forms To: (See Note 1.)	Complete in BLOCK Capitals and BLACK Ink.
Director of Planning and Technical Services Riddiford District Council Riddiford House Riddiford, Devon EX19 8DW	APPLICATION No: .... DATE RECEIVED: .... *For Office Use Only*

Please Read The Notes Before Completing This Form

1.   **Applicants Details**
Name:
Address:

Post Code: Tel. No.:

Are you, or your partner related to or connected with an elected member or officer of the Riddiford District Council? If yes give details:

2.   **Agents Details** (if any) to whom correspondence should be sent
Name:
Address:

Post Code: Tel. No.:

3.   Full address or location of the building to which this application relates:
Address:

Post Code: Tel. No.:

4.   Applicants interest in building (e.g. owner, lease, prospective purchaser, etc.)

5.   Proposed work e.g. demolition, alteration, extension. Description:

6.   Justification why the works are considered desirable or necessary.

7.   Drawings and plans submitted with application.
Details:

Note: The plans should be sufficient to identify the buildings and all the alterations and extensions should be shown in detail; the works should also be shown in relation to any adjacent buildings.

PLEASE TURN OVER

8. * I/We hereby apply for listed building consent to execute the works described in the application and all the accompanying plans and drawings and in accordance therewith.

DATE: SIGNED: ....

On behalf of:

* Delete where appropriate   (insert applicants name if signed by agent)

Planning (Listed Buildings and Conservation Areas) Act 1990 provide that an application for listed building consent shall not be entertained unless it is accompanied by one of four certificates. If you are the sole owner (see(a) below) of all the land, Certificate A, which is printed below, is appropriate: only one copy need be completed. If you can not complete certificate A you will have to give notice to the other owners and complete certificate B. Certificate C and D are appropriate only if you have made efforts to trace the other owners and have failed.
The forms of these notices and certificates are prescribed in the regulations.

9. Planning (Listed Buildings and Conservation Areas) Act 1990.

CERTIFICATE A

I hereby certify that no person other than *myself / the applicant was an owner (a) of the building to which the application relates at the beginning of the period of 20 days before the date of the accompanying application.

SIGNED: ....
*On behalf of: Date:
• Delete where appropriate   (insert applicants name if signed by agent)

NOTE: (a) "owner" means a person having freehold interest or a leasehold interest the unexpired term of which was not less than 7 years.

NOTES:

1. Planning Policy Guidance Note 15 (para. 3.4) requires that the application must make clear the justification for the works (see question 6).

2. In support of the application, please provide 3no. plans / drawings showing , in red, the location of the property and the work to be carried out. The drawings should preferably be of a scale of no less than 1:100 and should clearly indicate existing and new work

3. Any object or structure fixed to the listed building or forming part of the land and comprised within the curtilage of the building is treated as part of the listed building.

4. If an appeal is made to the Secretary of State concerning this application, the regulations require that a copy of the following documents shall be furnished to the Secretary of State by the applicant:

(a) the application made to the Local Authority Planning Department together with all the relevant plans, drawings, particulars and documents (including a copy of the certificate) submitted with it.
(b) the notice of decision (if any) and all other relevant correspondence with the Local Planning Authority.

5. If consent is granted for the demolition of a listed building, the effect of section 8(2)(b) of the 1990 Act is that demolition may not be undertaken until notice of the proposal has been given to the Royal Commission and the Commission subsequently have either been given reasonable access to the building for at least one month following the grant of consent and before works commenced or have stated that they have completed their record of the building or that they do not wish to record it.

Appendix 2D   Typical application for agricultural/forestry determination

RIDDIFORD DISTRICT COUNCIL

Planning and Technical Services Department

Application For
Agricultural/Forestry Determination

Local Authority
BUILDING CONTROL

Town and Country Planning (General Permitted Development) Order 1995 Part 6 & 7, Class A, Schedule 2

Submit Two Copies of Plans and Forms To:	Complete in BLOCK Capitals and BLACK Ink.
Director of Planning and Technical Services Riddiford District Council Riddiford House Riddiford, Devon EX19 8DW	APPLICATION No: .... DATE RECEIVED: .... *For Office Use Only*

Please Read The Notes Before Completing This Form

1. **Applicants Details**
   Name:
   Address:
   　　　　　　　　　　　Post Code:　　　　　　　Tel. No.:

2. **Agents Details** (if any) to whom correspondence should be sent
   Name:
   Address:
   　　　　　　　　　　　Post Code:　　　　　　　Tel. No.:

3. **Full Postal address & Location of Works / Building(s)**

4. **Description of proposed works / building(s) [see note 4]**
   Details:

5. Please state:
   (a) Does this holding consist of more that one separate block of land?　　Yes ☐ / No ☐
   If yes, please give the size of each block of land and indicate which block the proposal relates to :-

   (b) Size of holding　　　　　　　　　　　　　　　　　　　　　　[see note 4]

   (c) Size of building(s) in metres
   (d) Height (e.g. to ridge) in metres
   (e) Materials to be used
   Purpose to which building(s) works will be put

PLEASE TURN OVER　　　　　　　　　　JANUARY 2005

6. Please give details on any Agricultural Building(s)/extensions erected within the previous two years

*   I / We enclose the appropriate fee of £35.00

DATE: **SIGNED:**

*   Delete where appropriate   **On behalf of:**

                                             (insert applicants name if signed by agent)

In accordance with the scale of charges, I enclose a remittance of £

NOTES:

1.
   An application for Agricultural Determination is required for development consisting of the erection of a building or the significant alteration of a private way. (Note: "Significant extension and significant alteration" mean any extension or alteration of the building where the cubic content of the original building will be exceeded by more than 10%, or the height of the building as extended or altered would exceed the height of the original building.)

2. The accompanying plans must show the siting, design and external appearance of the building, or as the case may be, siting and means of construction of the private way.

3. Please ensure that any dimensions shown on any drawings/plans accompanying this application are given in metric measurement.

4. If you intend to erect a new building as opposed to altering or extending one, it will be necessary to provide a plan demonstrating that the size of the holding exceeds 5 hectares.

5. If the holding is less than 5 hectares, planning permission will always be required. In these cases this application form should NOT be used.

6. The application must be accompanied by a fee of £35.00

Appendix 2E   Example of an application
for consent to display advertisements

**RIDDIFORD
DISTRICT
COUNCIL**

Planning and Technical Services
Department

Application For Consent
To Display
Advertisements

Local Authority
**BUILDING
CONTROL**

Town and Country Planning (Control of Advertisements) Regulations
Town and Country Planning Act 1990

Submit Three Copies of Plans and Forms To:	Complete in BLOCK Capitals and BLACK Ink.
Director of Planning and Technical Services Riddiford District Council Riddiford House Riddiford, Devon EX19 8DW	APPLICATION No: ………………………… DATE RECEIVED: ………………………… *For Office Use Only*

Please Read The Notes Before Completing This Form

1. Applicants Details
   Name:
   Address:

   Post Code: Tel. No. :

2. Agents Details (if any) to whom correspondence should be sent
   Name:
   Address:

   Post Code: Tel. No. :

3. Full Postal Address & Location of Land or Building on which the advertisement is to be
   displayed

4. State the purpose for which the land or building is now being used
   Details:

5. (A) Has the applicant an interest in the land or building?

   (B) If not, has the permission of the owner or of any other persons entitled to give permission
       for the display of advertisement been obtained? (see note 4).

6. (A) State the nature of the advertisement (e.g. hoarding, shop sign, projecting sign etc.)

   (B) Is the advertisement already being displayed?   Yes ☐ / No ☐

PLEASE TURN OVER   **JANUARY 2005**

7.	(A) Will the advertisement be illuminated?	Yes ☐ / No ☐
	(B) If so state the type of illumination (e.g. internally, externally, floodlighting etc.)	
	(C) Will the illumination be static or intermittent?	
8.	Period for which consent is sought (see note 2).	
9.	Any additional information that the applicant may wish to supply.	

* I / We hereby apply for consent to display the advertisement as described in the application and all the accompanying plans and drawings and in accordance therewith.

DATE: **SIGNED:**

* Delete where appropriate   **On behalf of:**

(insert applicants name if signed by agent)

In accordance with the scale of charges, I enclose a remittance of £

Notes:

1. General. Under the Town and Country Planning (Control of Advertisements) Regulations many outdoor advertisements require express consent before they can be lawfully displayed. Applicants should refer to the regulations for details or call the Local Planning Authority.

2. Period of Consent. Normally the maximum period for which consent may be granted is 5 years. The Council may not grant for a longer period without special approval from the Secretary of State. If consent is required for a specific period of less than 5 years this should be stated in reply to question 8.

3. Drawings Required. The drawing can be in black and white on paper. It should show the size of the advertisement and its position on the land or the building in question. In the case of a sign it should also give the materials to be used, fixings, colours, height above the ground and where it would project from a building, the amount of the projection. The drawing should include a site location plan which should be of sufficient detail to enable the site to be identified.

4. Owners Consent. It is a condition of every consent by or under the regulations that before the advertisement to which the consent relates is displayed, the permission of the owner of the land or other persons entitled to grant permission shall be obtained.

5. Other Consents. Consent under the Town and Country Planning (Control of Advertisements) Regulations does not relieve the applicant from obtaining any other consents which may be necessary, under Listed Building & Conservation Areas Act etc.

6. Fees for advertisement applications.

Adverts Relating to the Business on the Premises	£60.00
Advance Signs Directing the Public to a Business	£60.00
(Unless Business can be seen from the signs position)	£220.00
Other Advertisements (e.g. Hoardings)	£220.00

Appendix 2F   Regularization (Example)

Building Control Service

RIDDIFORD DISTRICT COUNCIL

Building Act 1984 – Building Regulations 2000

REGULARIZATION

Local Authority
BUILDING CONTROL

Submit **Two** Copies of **Forms** To:	Complete in BLOCK Capitals and BLACK Ink.
Director of Planning and Technical Services Riddiford District Council Riddiford House Riddiford, Devon EX19 8DW	Application No: .... FEE: .... Cheque/PO/Cash: .... VAT: .... Total Fee: .... Accepted: .... *For Office Use Only*

Please Read The Notes Before Completing This Form

1. **Applicants Details (See Note 1):**   **Owners Details (if different)**
   Name:
   Address:

   Post Code: Tel. No.:
   Fax No: E-Mail:

2. **Agents Details** (if any) to whom correspondence should be sent
   Name:
   Address:

   Post Code: Tel. No.:
   Fax No: E-Mail:

3. **Location** of building to which work relates
   Address:

   Post Code: Tel. No.:

4. **Work requiring Regularization certificate (See Note 3):**
   Description:

5. **Use of building**
   a.   Previous use:
   b.   Present use:

6. **Services**
   a.   Foul water:
        Surface water:
   c.   Means of water supply:

7. Have you received Planning Consent for this work?   Yes / No
   If yes, please give Consent No. & Date of Decision:

8. **Conservation of Fuel & Power** (See Note 3.3) Please indicate the method used to show compliance with Part L
   For Dwellings   For Other Buildings
   ☐ Elemental Method   ☐ Elemental Method
   ☐ Target U-Value Method   ☐ Whole Building Method
   ☐ Carbon Index Method (Calculations to be enclosed)   ☐ Carbon Emission Calculation Method
   ☐ Material Alteration / Change of Use   ☐ Material Alteration / Change of Use

9. **Particulars of building work**
   Date work commenced: Date work completed:

   Name of Builder: Address:

   Postcode: Tel. No.:

PLEASE TURN OVER   **JANUARY 2005**

10.   Fire Safety Requirements

Is the Building proposed to be put, or is currently put, to a use as a Workplace to which the Fire Precautions (Workplace) Regulations 1997 applies and/or is designated as defined in the Fire Precautions Act 1971? Yes ☐ / No ☐

If Yes, please indicate the method you have used to show compliance with the Requirements of Part B of Schedule 1 to the Building Regulations.

An additional copy of all plans/specifications are to be submitted if you have answered Yes to the above.

11.   Fees (See Guidance Note on Fees for information)

a.   If Table A work - please state number of small domestic buildings:

- please state number of different types of buildings:

b.   If Table B work - please state floor area m^2:

c.   If Table C work - please state estimated cost of work excluding VAT £:

12.   Statement

This notice is given in relation to the building work as described, and is submitted in accordance with Regulation 13(a) and is accompanied by the appropriate fee.

Name: Signature: Date:

Should you have difficulty in filling out these forms please contact this office.

NOTES

1 The applicant is the building's owner.

2 Two copies of this notice should be completed and submitted.

3 Where the work includes the erection of a new building, material alteration, material change of use or extension, this notice shall be accompanied, so far as is reasonably practicable, by the following:

3.1 a block plan to a scale of not less than 1:1250 showing:

3.1.1 a plan of the unauthorized work;

3.1.2 the provision made for the drainage of the building or extension;

3.1.3 a plan showing any additional work required to be carried out to secure the unauthorized work complies with the requirements relating to building work in the building regulations which were applicable to that work when it was carried out.

3.2 where the building or extension has been erected over a sewer or drain shown on the relative map of public sewers, the precautions taken in building over a sewer or drain.

3.3 the requirements of Part L (Conservation of Fuel and Power) must be considered for all dwellings and all other buildings. The requirements also apply to material alterations and changes of use.

3.3.1 for some small extensions to dwellings, these would only need to be as energy efficient as the existing dwelling. Therefore question 8 is not applicable.

3.3.2 for other buildings with a floor area exceeding 100 m^2 floor area, details of artificial lighting systems will be required with the application.

3.3.3 Regulation 16 requires the provision of Energy Ratings calculated by the Government's Standard Assessment Procedure (SAP) for new dwellings and dwellings created as the result of material changes of use. The SAP calculations must be provided to the Authority as soon as possible. Dwellings include Flats. The Person who carried out the building work must display in the dwelling, as soon as practicable after the energy rating has been calculated, a notice of that rating. The notice must be displayed in a conspicuous place in the dwelling. Should occupation take place before physical completion, then the energy rating notice should be given to the occupier.

The requirement to display an energy rating notice in the dwelling, or give a notice to the occupier, is not necessary where the person who carried out the building work intends to occupy, or occupies, the dwelling as his/her residence.

4 Where the work involved the insertion of insulating material into the cavity walls of a building this application shall be accompanied by a statement as to:

4.1 the name and type of insulating material used;

4.2 whether or not the insulating material is approved by the British Board of Agrément or conforms to a British Standard specification;

4.3 whether or not the installer was a person who is subject of a British Standards Institution Certification of Registration or has been approved by the British Board of Agrément for the insertion of that material.

5 Where the work involved the provision of an unvented hot water storage system, this application shall be accompanied by a statement as to:

5.1 the name and type of system to be provided;

5.2 whether or not the system is approved by the British Board of Agrément;

5.3 whether or not the British Board of Agrément has approved the installer for the provision of that system.

6 In accordance with Regulation 13(a), the Council may require an applicant to take such reasonable steps, including laying open the unauthorized work for inspection, making tests and taking samples as the Authority think appropriate to ascertain what work, if any, is required to secure compliance with the relevant Regulations.

7 The **Regularization Fee** is payable at the time the submission is made. A Guidance Note on Fees is available on request.

8 These notes are for general guidance only, particulars regarding the submission of Regularization applications are contained in Regulation 13(a) of the Building Regulations 2000 and, in respect of fees, in the Building (Prescribed Fees etc.) Regulations 1991 – as amended.

9 Further information and advice concerning the Building Regulations and planning matters may be obtained from your Local Authority.

10 The responsibility for demonstrating that the electrical work complies with Part P lies squarely with the person requesting a Regularization certificate, and the Authority is under no obligation to issue a certificate until satisfied on that point.

10.1 The applicant should provide any electrical certification issued at the time the work was carried out. If not, a periodic Inspection Report will be necessary to show compliance together with approximate exposure/tracing of the cables as is considered necessary.

3

The requirements of the Building Regulations

Statutory Instrument	Made	Laid before Parliament	Coming into force
SI 2001 No 3335	4 Oct 2001	11 Oct 2001	1 Apr 2002
SI 2002 No 440	28 Feb 2002	5 Mar 2002	1 Apr 2002
SI 2002 No 2871	16 Nov 2002	25 Nov 2002	1 Jul 2003 (less sound insulation)
			1 Jan 2004 (sound insulation)
SI 2003 No 2692	17 Oct 2003	27 Oct 2003	1 Dec 2003 (Regulations 1, 2(1) and (8) plus 3(5))
			1 May 2003 (remainder)
SI 2004 No 1465	28 May 2004	8 Jun 2004	1 Jul 2004 Regulations 1(1), (2), (4) and (5)
			1 Dec 2004 (remainder)
SI 2004 No 3210	6 Dec 2004	10 Dec 2004	31 Dec 2004

 Note: Copies of the above documents are available from TSO (☎ 0870 600 5522) and through booksellers. They can also be viewed on the OPDM website at www.safety.odpm. gov.uk/bregs/brpub/01.htm.

3.1 Part A – Structure

Number	Title	Regulation	Requirement (in a nutshell)
A1	Loading	(1) The building shall be constructed so that the combined dead, imposed and wind loads are sustained and transmitted by it to the ground – (a) safely; and (b) without causing such deflection or deformation of any part of the building, or such movement of the ground, as will impair the stability of any part of another building. (2) In assessing whether a building complies with sub-paragraph (1) regard shall be had to the imposed and wind loads to which it is likely to be subjected in the ordinary course of its use for the purpose for which it is intended.	The safety of a structure depends on: • the loading (see BS 6399, Parts 1 and 3); • properties of materials; • design analysis; • details of construction; • safety factors; • workmanship.
A2	Ground movement	The building shall be constructed so that ground movement caused by – (a) swelling, shrinkage or freezing of the subsoil; or (b) land-slip or subsidence (other than subsidence arising from shrinkage), in so far as the risk can be reasonably foreseen, will not impair the stability of any part of the building.	• Horizontal and vertical ties should be provided.

3.2 Part B – Fire safety

Number	Title	Regulation	Requirement (in a nutshell)
B1	**Means of warning and escape**	*The building shall be designed and constructed so that there are appropriate provisions for the early warning of fire, and appropriate means of escape in case of fire from the building to a place of safety outside the building capable of being safely and effectively used at all material times.* Requirement B1 does not apply to any prison provided under Section 33 of the Prisons Act 1952 (power to provide prisons etc.).	For a typical one or two storey dwelling, the requirement is limited to the provision of smoke alarms and to the provision of openable windows for emergency exit. For all other types of buildings, in case of fire, escape routes should be provided that: • are sufficient in number and capacity according to the size and use of the building; • are suitably located to enable persons to escape to a place of safety in the event of fire; • are sufficiently protected from the effects of fire (by enclosure where necessary); • are adequately lit; • are suitably signed; • either limit the ingress of smoke to the escape route(s) or restrict the fire and remove smoke.
B2	**Internal fire spread (linings)**	*(1) To inhibit the spread of fire within the building the internal linings shall –* *(a) adequately resist the spread of flame over their surfaces; and* *(b) have, if ignited, a rate of heat release which is reasonable in the circumstances.* *(2) In this paragraph 'internal linings' means the materials or products used in lining any partition, wall, ceiling or other internal structure.*	As a fire precaution, all materials used for internal linings of a building should have a low rate of surface flame spread and (in some cases) a low rate of heat release.

B3	Internal fire spread (structure)		

(1) *The building shall be designed and constructed so that, in the event of fire, its stability will be maintained for a reasonable period.*

(2) *A wall common to two or more buildings shall be designed and constructed so that it adequately resists the spread of fire between those buildings. For the purposes of this sub-paragraph a house in a terrace and a semi-detached house are each to be treated as a separate building.*

(3) *To inhibit the spread of fire within the building, it shall be sub-divided with fire-resisting construction to an extent appropriate to the size and intended use of the building.*

(4) *The building shall be designed and constructed so that the unseen spread of fire and smoke within concealed spaces in its structure and fabric is inhibited.*

Requirement B3(3) does not apply to material alterations to any prison provided under Section 33 of the Prisons Act 1952.

- All structural, loadbearing elements of a building shall be capable of withstanding the effects of fire for an appropriate period without loss of stability.
- Ideally the building should be sub-divided by elements of fire-resisting construction into compartments.
- All openings in fire-separating elements shall be suitably protected in order to maintain the integrity of the continuity of the fire separation.
- Any hidden voids in the construction shall be sealed and sub-divided to inhibit the unseen spread of fire and products of combustion, in order to reduce the risk of structural failure, and the spread of fire.

B4	External fire spread		

(1) *The external walls of the building shall adequately resist the spread of fire over the walls and from one building to another, having regard to the height, use and position of the building.*

(2) *The roof of the building shall adequately resist the spread of fire over the roof and from one building to another, having regard to the use and position of the building.*

- External walls shall be constructed so that the risk of ignition from an external source, and the spread of fire over their surfaces, is restricted.
- The amount of unprotected area in the side of the building shall be restricted so as to limit the amount of thermal radiation that can pass through the wall.

(Continued)

Part B (*Continued*)

Number	Title	Regulation	Requirement (in a nutshell)
			• The roof shall be constructed so that the risk of spread of flame and/or fire penetration from an external fire source is restricted. • The risk of a fire spreading from the building to a building beyond the boundary, or vice versa shall be limited.
B5	**Access and facilities for the fire service**	*(1) The building shall be designed and constructed so as to provide reasonable facilities to assist fire fighters in the protection of life.* *(2) Reasonable provision shall be made within the site of the building to enable fire appliances to gain access to the building.*	For dwellings and other small buildings, it is usually only necessary to ensure that the building is sufficiently close to a point accessible to fire brigade vehicles. In more detail this includes: • vehicle access for fire appliances; • access for fire-fighting personnel; • the provision of fire mains within the building (for non-domestic buildings); • venting for heat and smoke from basement areas.

3.3 Part C – Site preparation and resistance to contaminants and moisture

Number	Title	Regulation	Requirement (in a nutshell)
C1	**Preparation of site and resistance to moisture**	*(1) The ground to be covered by the building shall be reasonably free from any material that might damage the building or affect its stability, including vegetable matter, topsoil and pre-existing foundations.*	Buildings should be safeguarded from the adverse effects of: • vegetable matter; • contaminants on or in the ground to be covered by the building; • ground water.
		(2) Reasonable precautions shall be taken to avoid danger to health and safety caused by contaminants on or in the ground covered, or to be covered by the building and any land associated with the building.	
		(3) Adequate subsoil drainage shall be provided if it is needed to avoid *(a) the passage of the ground moisture to the interior of the building;* *(b) damage to the building, including damage through the transport of water-borne contaminants to the foundations of the building.*	
		Note: For the purpose of this requirement, 'contaminant' means any substance which is or may become harmful to persons or buildings including substances, which are corrosive, explosive, flammable, radioactive or toxic.	

(Continued)

Part C (*Continued*)

Number	Title	Regulation	Requirement (in a nutshell)
C2	**Resistance to moisture**	*The floors, walls and roof of the building shall adequately protect the building and people who use the building from harmful effects caused by:* *(a) ground moisture;* *(b) precipitation and wind-driven spray;* *(c) interstitial and surface condensation; and* *(d) spillage of water from or associated with sanitary fittings or fixed appliances.*	• A solid or suspended floor shall be built next to the ground to prevent undue moisture from reaching the upper surface of the floor. • A wall shall be erected to prevent undue moisture from the ground reaching the inside of the building, and (if it is an outside wall) adequately resisting the penetration of rain and snow to the inside of the building. • The roof of the building shall be resistant to the penetration of moisture from rain or snow to the inside of the building. • All floors next to the ground, walls and roof shall not be damaged by moisture from the ground, rain or snow and shall not carry that moisture to any part of the building which it would damage.

3.4 Part D – Toxic substances

Number	Title	Regulation	Requirement (in a nutshell)
D1	**Cavity insulation**	*If insulating material is inserted into a cavity in a cavity wall reasonable precautions shall be taken to prevent the subsequent permeation of any toxic fumes from that material into any part of the building occupied by people.*	Fumes given off by insulating materials such as by urea formaldehyde (UF) foams should not be allowed to penetrate occupied parts of buildings to an extent where they could become a health risk to persons in the building by reaching an irritant concentration.

3.5　Part E – Resistance to the passage of sound

Number	Title	Regulation	Requirement (in a nutshell)
E1	**Protection against sound from other parts of the building and adjoining buildings**	*Dwelling-houses, flats and rooms for residential purposes shall be designed and constructed in such a way that they provide reasonable resistance to sound from other parts of the same building and from adjoining buildings.*	Dwellings shall be designed so that the noise from domestic activity in an adjoining dwelling (or other parts of the building) is kept to a level that: • does not affect the health of the occupants of the dwelling; • will allow them to sleep, rest and engage in their normal activities in satisfactory conditions. Dwellings shall be designed so that any domestic noise that is generated internally does not interfere with the occupants' ability to sleep, rest and engage in their normal activities in satisfactory conditions.
E2	**Protection against sound within a dwelling-house etc.**	*Dwelling-houses, flats and rooms for residential purposes shall be designed and constructed in such a way that:* *(a) internal walls between a bedroom or a room containing a water closet, and other rooms* *and* *(b) internal floors* *provide reasonable resistance to sound.*	**Note:** Requirement E2 does not apply to: (a) an internal wall which contains a door; (b) an internal wall which separates an en suite toilet from the associated bedroom; (c) existing walls and floors in a building which is subject to a material change of use.

(Continued)

Part E (*Continued*)

Number	Title	Regulation	Requirement (in a nutshell)
E3	**Reverberation in the common internal parts of buildings containing flats or rooms for residential purposes**	*The common internal parts of buildings which contain flats or rooms for residential purposes shall be designed and constructed in such a way as to prevent more reverberation around the common parts than is reasonable.*	Suitable sound absorbing material shall be used in domestic buildings so as to restrict the transmission of echoes. Requirement E3 only applies to corridors, stairwells, hallways and entrance halls which give access to the flat or rooms for residential purposes.
E4	**Acoustic conditions in schools**	*(1) Each room or other space in a school building shall be designed and constructed in such a way that it has the acoustic conditions and the insulation against disturbance by noise appropriate to its intended use.* *(2) For the purposes of this part – 'school' has the same meaning as in section 4 of the Education Act 1996; and 'school building' means any building forming a school or part of a school.*	Suitable sound insulation materials shall be used within a school building so as to reduce the level of ambient noise (particularly echoing in corridors etc.).

3.6 Part F – Ventilation

Number	Title	Regulation	Requirement (in a nutshell)
F1	**Means of ventilation**	*There shall be adequate means of ventilation provided for people in the building.* Requirement F1 does not apply to a building or space within a building – (a) into which people do not normally go; or (b) which is used solely for storage; or (c) which is a garage used solely in connection with a single dwelling.	• Ventilation (mechanical and/or air-conditioning systems designed for domestic buildings) shall be capable of restricting the accumulation of moisture and pollutants originating within a building.
F2	**Condensation in roofs**	*Adequate provision shall be made to prevent excessive condensation –* (a) *in a roof; or* (b) *in a roof void above an insulated ceiling.*	Condensation in a roof and in spaces above insulated ceilings shall be limited (by the ventilation of cold deck roofs) so that, under normal conditions: • the thermal performance of the insulating materials; and • the structural performance of the roof construction will not be substantially and permanently reduced.

3.7 Part G – Hygiene

Number	Title	Regulation	Requirement (in a nutshell)
G1	Sanitary conveniences and washing facilities	(1) Adequate sanitary conveniences shall be provided in rooms provided for that purpose, or in bathrooms. Any such room or bathroom shall be separated from places where food is prepared.	**All** dwellings (house, flat or maisonette should have at least one closet and one washbasin:
		(2) Adequate washbasins shall be provided in –	• closets (and/or urinals) should be separated by a door from any space used for food preparation or where washing-up is done;
		(a) rooms containing water closets; or	• washbasins should, ideally, be located in the room containing the closet;
		(b) rooms or spaces adjacent to rooms containing water closets.	• the surfaces of a closet, urinal or washbasin should be smooth, non-absorbent and capable of being easily cleaned;
		Any such room or space shall be separated from places where food is prepared.	• closets (and/or urinals) should be capable of being flushed effectively;
		(3) There shall be a suitable installation for the provision of hot and cold water to washbasins provided in accordance with paragraph (2).	• closets (and/or urinals) should only be connected to a flush pipe or discharge pipe;
		(4) Sanitary conveniences and washbasins to which this paragraph applies shall be designed and installed so as to allow effective cleaning.	• washbasins should have a supply of hot and cold water;
			• closets fitted with flushing apparatus should discharge through a trap and discharge pipe into a discharge stack or a drain.
G2	Bathrooms	A bathroom shall be provided containing either a fixed bath or shower bath, and there shall be a suitable installation for the provision of hot and cold water to the bath or shower bath.	All dwellings (house, flat or maisonette) should have at least one bathroom with a fixed bath or shower and the bath or shower should:
			• have a supply of hot and cold water;

• discharge through a grating, a trap and branch discharge pipe to a discharge stack or (if on a ground floor);
• discharge into a gully or directly to a foul drain;
• be connected to a macerator and pump (of an approved type) if there is no suitable water supply or means of disposing of foul water.

A hot water storage system shall:
• be installed by a competent person;
• not exceed 100°C;
• discharge safely;
• not cause danger to persons in or about the building.

 Requirement G2 applies only to dwellings.

G3   Hot water storage

A hot water storage system that has a hot water storage vessel which does not incorporate a vent pipe to the atmosphere shall be installed by a person competent to do so, and there shall be precautions:

(a) to prevent the temperature of stored water at any time exceeding 100°C; and

*(b) to ensure that the hot water discharged from safety devices is safely conveyed to where it is visible but will not cause danger **to** persons in or about the building.*

Requirement G3 does not apply to:
(a) a hot water storage system that has a storage vessel with a capacity of 15 litres or less;
(b) a system providing space heating only;
(c) a system that heats or stores water for the purposes only of an industrial process.

3.8 Part H – Drainage and waste disposal

Number	Title	Regulation	Requirement (in a nutshell)

H1	**Foul water drainage**	(1) An adequate system of drainage shall be provided to carry foul water from appliances within the building to one of the following, listed in order of priority –
		
		   (a) a public sewer; or, where that is not reasonably practicable,
		
		   (b) a private sewer communicating with a public sewer; or, where that is not reasonably practicable,
		
		   (c) either a septic tank which has an appropriate form of secondary treatment or another wastewater treatment system; or, where that is not reasonably practicable,
		
		   (d) a cesspool.
		
		(2) In this Part 'foul water' means wastewater which comprises or includes
		
		   (a) waste from a sanitary convenience, bidet or appliance used for washing receptacles for foul waste; or
		
		   (b) water which has been used for food preparation, cooking or washing.
		
		Requirement H1 does not apply to the diversion of water which has been used for personal washing or for the washing of clothes, linen or other articles to collection systems for reuse.
			The foul water drainage system shall:
			
			• convey the flow of foul water to a foul water outfall (i.e. sewer, cesspool, septic tank or settlement (i.e. holding) tank);
			• minimize the risk of blockage or leakage;
			• prevent foul air from the drainage system from entering the building under working conditions;
			• be ventilated;
			• be accessible for clearing blockages;
			• not increase the vulnerability of the building to flooding.
			
			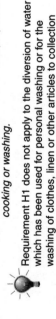H1 is applicable to domestic buildings and small non-domestic buildings. Further guidance on larger buildings is provided in Appendix A to Approved Document H. Complex systems in larger buildings should be designed in accordance with BS EN 12056.

H2	Wastewater treatment systems and cesspools	(1)	Any septic tank and its form of secondary treatment, other wastewater treatment system or cesspool, shall be so sited and constructed that – (a) it is not prejudicial to the health of any person; (b) it will not contaminate any watercourse, underground water or water supply; (c) there are adequate means of access for emptying and maintenance; and (d) where relevant, it will function to a sufficient standard for the protection of health in the event of a power failure.	Wastewater treatment systems shall: • have sufficient capacity to enable breakdown and settlement of solid matter in the wastewater from the buildings; • be sited and constructed so as to prevent overloading of the receiving water. Cesspools shall have sufficient capacity to store the foul water from the building until they are emptied. Wastewater treatment systems and cesspools shall be sited and constructed so as not to: • be prejudicial to health or a nuisance; • adversely affect water sources or resources; • pollute controlled waters; • be in an area where there is a risk of flooding.
		(2)	Any septic tank, holding tank which is part of a wastewater treatment system or cesspool shall be – (a) of adequate capacity; (b) so constructed that it is impermeable to liquids; and (c) adequately ventilated.	Septic tanks and wastewater treatment systems and cesspools shall be constructed and sited so as to: • have adequate ventilation; • prevent leakage of the contents and ingress of subsoil water; • having regard to water table levels at any time of the year and rising groundwater levels.
		(3)	Where a foul water drainage system from a building discharges to a septic tank, wastewater treatment system or cesspool, a durable notice shall be affixed in a suitable place in the building containing information on any continuing maintenance required to avoid risks to health.	Drainage fields shall be sited and constructed so as to: • avoid overloading of the soakage capacity; and • provide adequately for the availability of an aerated layer in the soil at all times.

(Continued)

Part H (*Continued*)

Number	Title	Regulation	Requirement (in a nutshell)
H3	**Rainwater drainage**	(1) *Adequate provision shall be made for rainwater to be carried from the roof of the building.* (2) *Paved areas around the building shall be so constructed as to be adequately drained.* (3) *Rainwater from a system provided pursuant to sub-paragraphs (1) or (2) shall discharge to one of the following, listed in order of priority –* (a) *an adequate soakaway or some other adequate infiltration system; or, where that is not reasonably practicable,* (b) *a watercourse; or, where that is not reasonably practicable,* (c) *a sewer.* Requirement H3(2) applies only to paved areas – (a) which provide access to the building pursuant to paragraph M2 of Schedule 1 (access for disabled people); (b) which provide access to or from a place of storage pursuant to paragraph H6(2) of Schedule 1 (solid waste storage); or (c) in any passage giving access to the building, where this is intended to be used in common by the occupiers of one or more other buildings. Requirement H3(3) does not apply to the gathering of rainwater for reuse.	Rainwater drainage systems shall: • minimize the risk of blockage or leakage; • be accessible for clearing blockages; • ensure that rainwater soaking into the ground is distributed sufficiently so that it does not damage foundations of the proposed building or any adjacent structure; • ensure that rainwater from roofs and paved areas is carried away from the surface either by a drainage system or by other means; • ensure that the rainwater drainage system carries the flow of rainwater from the roof to an outfall (e.g. a soakaway, a watercourse, a surface water or a combined sewer).

H4	**Building over sewers**	*(1)*	*The erection or extension of a building or work involving the underpinning of a building shall be carried out in a way that is not detrimental to the building or building extension or to the continued maintenance of the drain, sewer or disposal main.*	Building or extension or work involving underpinning shall:
		(2)	*In this paragraph 'disposal main' means any pipe, tunnel or conduit used for the conveyance of effluent to or from a sewage disposal works, which is not a public sewer.*	• be constructed or carried out in a manner which will not overload or otherwise cause damage to the drain, sewer or disposal main either during or after the construction;
		(3)	*In this paragraph and paragraph H5 'map of sewers' means any records kept by a sewerage undertaker under Section 199 of the Water Industry Act 1991.*	• not obstruct reasonable access to any manhole or inspection chamber on the drain, sewer or disposal main;
			Requirement H4 applies only to work carried out –	• in the event of the drain, sewer or disposal main requiring replacement, not unduly obstruct work to replace the drain, sewer or disposal main, on its present alignment;
			(a) over a drain, sewer or disposal main which is shown on any map of sewers; or	• reduce the risk of damage to the building as a result of failure of the drain, sewer or disposal main.
			(b) on any site or in such a manner as may result in interference with the use of, or obstruction of the access of any person to, any drain, sewer or disposal main which is shown on any map of sewers.	
H5	**Separate systems of drainage**		*Any system for discharging water to a sewer which is provided pursuant to paragraph H3 shall be separate from that provided for the conveyance of foul water from the building.*	Separate systems of drains and sewers shall be provided for foul water and rainwater where:
			Requirement H5 applies only to a system provided in connection with the erection or extension of a building	(a) the rainwater is not contaminated; and
				(b) the drainage is to be connected either directly or indirectly to the public sewer system and either –

(Continued)

Part H (*Continued*)

Number	Title	Regulation	Requirement (in a nutshell)
		where it is reasonably practicable for the system to discharge directly or indirectly to a sewer for the separate conveyance of surface water which is – (a) shown on a map of sewers; or (b) under construction either by the sewerage undertaker or by some other person (where the sewer is the subject of an agreement to make a declaration of vesting pursuant to Section 104 of the Water Industry Act 1991).	(i) the public sewer system in the area comprises separate systems for foul water and surface water; or (ii) a system of sewers which provides for the separate conveyance of surface water is under construction either by the sewerage undertaker or by some other person (where the sewer is the subject of an agreement to make a declaration of vesting pursuant to Section 104 of the Water Industry Act 1991).
H6	**Solid waste storage**	(1) *Adequate provision shall be made for storage of solid waste.* (2) *Adequate means of access shall be provided –* (a) *for people in the building to the place of storage; and* (b) *from the place of storage to a collection point (where one has been specified by the waste collection authority under Section 46 (household waste) or Section 47 (commercial waste) of the Environmental Protection Act 1990 or to a street (where no collection point has been specified)).*	Solid waste storage shall be: • designed and sited so as not to be prejudicial to health; • of sufficient capacity having regard to the quantity of solid waste to be removed and the frequency of removal; • sited so as to be accessible for use by people in the building and of ready access from a street for emptying and removal.

3.9 Part J – Combustion appliances and fuel storage systems

Number	Title	Regulation	Requirement (in a nutshell)
J1	**Air supply**	*Combustion appliances shall be so installed that there is an adequate supply of air to them for combustion, to prevent over-heating and for the efficient working of any flue.*	The building shall: • enable the admission of sufficient air for: – the proper combustion of fuel and the operation of flues; and – the cooling of appliances where necessary; • enable normal operation of appliances without the products of combustion becoming a hazard to health; • enable normal operation of appliances without their causing danger through damage by heat or fire to the fabric of the building; • have been inspected and tested to establish suitability for the purpose intended; • have been labelled to indicate performance capabilities.
		Requirement J1 only applies to fixed combustion appliances (including incinerators).	
J2	**Discharge of products of combustion**	*Combustion appliances shall have adequate provision for the discharge of products of combustion to the outside air.*	Oil and LPG fuel storage installations shall be located and constructed so that they are reasonably protected from fires that may occur in buildings or beyond boundaries.
		Requirement J2 only applies to fixed combustion appliances (including incinerators).	Oil storage tanks used wholly or mainly for private dwellings shall be: • reasonably resistant to physical damage and corrosion;
J3	**Protection of building**	*Combustion appliances and flue-pipes shall be so installed, and fireplaces and chimneys shall be so constructed and installed, as to reduce to a reasonable level the risk of people suffering burns or the building catching fire in consequence of their use.*	• designed and installed so as to minimize the risk of oil escaping during the filling or maintenance of the tank;
J4	**Provision of information**	Requirement J3 only applies to fixed combustion appliances (including incinerators).	• incorporate secondary containment when there is a significant risk of pollution; • be labelled with information on how to respond to a leak.
J5	**Protection of liquid fuel storage systems**	*Where a hearth, fireplace, flue or chimney is provided or extended, a durable notice containing information on the performance capabilities of the hearth, fireplace, flue or chimney shall be affixed in a suitable place in the building for the purpose of enabling combustion appliances to be safely installed.*	

(Continued)

Part J (*Continued*)

Number	Title	Regulation	Requirement (in a nutshell)
J6	**Protection against pollution**	*Liquid fuel storage systems and the pipes connecting them to combustion appliances shall be so constructed and separated from buildings and the boundary of the premises as to reduce to a reasonable level the risk of the fuel igniting in the event of fire in adjacent buildings or premises.* Requirement J5 applies only to – (a) fixed oil storage tanks with capacities greater than 90 litres and connecting pipes; and (b) fixed liquefied petroleum gas storage installations with capacities which are located outside the building and which serve fixed combustion appliances (including incinerators) in the building. *Oil storage tanks and the pipes connecting them to combustion appliances shall –* *(a) be so constructed and protected as to reduce to a reasonable level the risk of the oil escaping and causing pollution; and* *(b) have affixed in a prominent position a durable notice containing information on how to respond to an oil escape so as to reduce to a reasonable level the risk of pollution.* Requirement J6 applies only to fixed oil storage tanks with capacities of 3500 litres or less, and connecting pipes, which are – (a) located outside the building; and (b) serve fixed combustion appliances (including incinerators) in a building used wholly or mainly as a private dwelling but does not apply to buried systems.	

3.10 Part K – Protection from falling, collision and impact

Number	Title	Regulation	Requirement (in a nutshell)
K1	**Stairs, ladders and ramps**	*Stairs, ladders and ramps shall be so designed, constructed and installed as to be safe for people moving between different levels in or about the building.* Requirement K1 applies only to stairs, ladders and ramps which form part of the building.	All stairs, steps and ladders shall provide reasonable safety between levels in a building. In a public building the standard of stair, ladder or ramp may be higher than in a dwelling, to reflect the lesser familiarity and greater number of users.
K2	**Protection from falling**	*(a) Any stairs, ramps, floors and balconies and any roof to which people have access, and* *(b) any light well, basement area or similar sunken area connected to a building,* *shall be provided with barriers where it is necessary to protect people in or about the building from falling.* Requirement K2 (a) applies only to stairs and ramps which form part of the building.	Pedestrian guarding should be provided for any part of a floor, gallery, balcony, roof, or any other place to which people have access and any light well, basement area or similar sunken area next to a building.

(Continued)

Part K (*Continued*)

Number	Title	Regulation	Requirement (in a nutshell)
K3	**Vehicle barriers and loading bays**	*(1) Vehicle ramps and any levels in a building to which vehicles have access, shall be provided with barriers where it is necessary to protect people in or about the building.* *(2) Vehicle loading bays shall be constructed in such a way, or be provided with such features, as may be necessary to protect people in them from collision with vehicles.*	Vehicle barriers should be provided that are capable of resisting or deflecting the impact of vehicles. Loading bays shall be provided with an adequate number of exits (or refuges) to enable people to avoid being crushed by vehicles.
K4	**Protection from collision with open windows etc.**	*Provision shall be made to prevent people moving in or about the building from colliding with open windows, skylights or ventilators.* Requirement K4 does not apply to dwellings.	All windows, skylights, and ventilators shall be capable of being left open without danger of people colliding with them.
K5	**Protection against impact from and trapping by doors**	*(1) Provision shall be made to prevent any door or gate –* *(a) which slides or opens upwards, from falling onto any person; and* *(b) which is powered, from trapping any person.* *(2) Provision shall be made for powered doors and gates to be opened in the event of a power failure.* *(3) Provision shall be made to ensure a clear view of the space on either side of a swing door or gate.*	Requirement K5 does not apply to – (a) dwellings, or (b) any door or gate that is part of a lift.

3.11 Part L1 – Conservation of fuel and power in dwellings

Number	Title	Regulation	Requirement (in a nutshell)
L1	**Dwellings**	Reasonable provision shall be made for the conservation of fuel and power in dwellings by – (a) limiting the heat loss: 　(i) through the fabric of the building; 　(ii) from hot water pipes and hot air ducts used for space heating; 　(iii) from hot water vessels; (b) providing space heating and hot water systems which are energy-efficient; (c) providing lighting systems with appropriate lamps and sufficient controls so that energy can be used efficiently; (d) providing sufficient information with the heating and hot water services so that building occupiers can operate and maintain the services in such a manner as to use no more energy than is reasonable in the circumstances. The requirement for sufficient controls in paragraph L1(c) applies only to external lighting systems fixed to the building.	Energy efficiency measures shall be provided.

3.12 Part L2 – Conservation of fuel and power in buildings other than dwellings

Number	Title	Regulation	Requirement (in a nutshell)

L2	**Buildings other than dwellings**	Reasonable provision shall be made for the conservation of fuel and power in buildings other than dwellings by –
		
		(a) limiting the heat losses and gains through the fabric of the building;
		(b) limiting the heat loss:
		   (i) from hot water pipes and hot air ducts used for space heating;
		   (ii) from hot water vessels and hot water service pipes;
		(c) providing space heating and hot water systems that are energy-efficient;
		(d) limiting exposure to solar overheating;
		(e) making provision where air conditioning and mechanical ventilation systems are installed, so that no more energy needs to be used than is reasonable in the circumstances;
		(f) limiting the heat gains by chilled water and refrigerant vessels and pipes and air ducts that serve air conditioning systems;
		(g) providing lighting systems that are energy-efficient;
		(h) providing sufficient information with the relevant services so that the building can be operated and maintained in such a manner as to use no more energy than is reasonable in the circumstances.
			Requirements L2(e) and (f) apply only within buildings and parts of buildings where more than 200 m^2 of floor area is to be served by air conditioning or mechanical ventilation systems.
			
			Requirement L2(g) applies only within buildings and parts of buildings where more than 100 m^2 of floor area is to be served by artificial lighting.

3.13 Part M – Access to and use of buildings

Number	Title	Regulation	Requirement (in a nutshell)
M1	**Access and use**	*Reasonable provision shall be made for people to:* *(a) gain access to and* *(b) use* *the buildings and its facilities.* The requirements of this part do not apply to: (a) an extension of, or material alteration of, a dwelling; or (b) any part of a building which is used solely to enable the building or service or fitting in the building to be inspected, repaired or maintained.	(1) In addition to the requirements of the Disability Discrimination Act 1995, Approved Document M also requires that: precautions need to be taken to ensure that: • new non-domestic buildings and/or dwellings (e.g. houses and flats used for student living accommodation etc.); • extensions to existing non-domestic buildings; • non-domestic buildings that have been subject to a material change of use (e.g. so that they become a hotel, boarding house, institution, public building or shop); (2) Are capable of allowing people, regardless of their disability, age or gender to: (a) gain access to buildings; (b) gain access within buildings; (c) be able to use the facilities of the buildings (both as visitors and as people who live or work in them).
M2	**Access to extensions of buildings other than dwellings**	*Suitable independent access shall be provided in any building that is to be extended. Reasonable provision shall be made within the extension for sanitary convenience.* Requirement M2 does not apply where suitable access to the extension is provided through the building that is extended.	

(Continued)

Part M (*Continued*)

Number	Title	Regulation	Requirement (in a nutshell)
M3	**Sanitary conveniences in extensions to buildings other than dwellings**	*If sanitary conveniences are provided in any building that is to be extended, reasonable provision shall be made within the extension for sanitary conveniences.*	(3) Use sanitary conveniences in the principal storey of any new dwelling.
		Requirement M3 does not apply where there is reasonable provision for sanitary conveniences elsewhere in the building, such that people occupied in, or otherwise having occasion to enter the extension, can gain access to and use those sanitary conveniences.	
M4	**Sanitary conveniences in dwellings**	*Reasonable provision shall be made in the entrance storey for sanitary conveniences, or where the entrance contains no habitable rooms, reasonable provision for sanitary convenience shall be made in either the entrance storey or principal storey.*	
		In requirement M4: 'entrance storey' means the storey which contains the principal entrance 'principal storey' means the storey nearest to the entrance storey which contains a habitable room, or if there are two stories equally near, either such storey.	

3.14   Part N – Glazing – safety in relation to impact, opening and cleaning

Number	Title	Regulation	Requirement (in a nutshell)
N1	**Protection against impact**	*Glazing with which people are likely to come into contact whilst moving in or about the building shall –* *(a) if broken on impact, break in a way which is unlikely to cause injury; or* *(b) resist impact without breaking; or* *(c) be shielded or protected from impact.*	All glazing installed in buildings shall be: • sufficiently robust to withstand impact from a falling or passing person; or • protected from a falling or passing person.
N2	**Manifestation of glazing**	*Transparent glazing with which people are likely to come into contact while moving in or about the building, shall incorporate features which make it apparent.*	Requirement N2 does not apply to dwellings.
N3	**Safe opening and closing of windows etc.**	*Windows, skylights and ventilators which can be opened by people in or about the building shall be so constructed or equipped that they may be opened, closed or adjusted safely.*	Requirement N3 does not apply to dwellings.
N4	**Safe access for cleaning windows etc.**	*Provision shall be made for any windows, skylights, or any transparent or translucent walls, ceilings or roofs to be safely accessible for cleaning.*	Requirement N4 does not apply to – (a) dwellings; or (b) any transparent or translucent elements whose surfaces are not intended to be cleaned.

3.15 Part P – Electrical safety

Number	Title	Regulation	Requirement (in a nutshell)
P1	**Design, installation, inspection and testing**	Reasonable provision shall be made in the design, installation, inspection and testing of electrical installations in order to protect persons from fire or injury.	These requirements only apply to electrical installations that are intended to operate at low or extra-low voltage in: (a) a dwelling; (b) common parts of a building serving one or more dwellings (but excluding power supplies to lifts);
P2	**Provision of information**	Sufficient information shall be provided so that persons wishing to operate, maintain or alter an electrical installation can do so with reasonable safety.	(c) a building that receives its electricity from a source located within (or shared with) a dwelling; (d) a garden associated with a building where the electricity is from a source located within or shared with a dwelling; (e) or on land associated with a building where the electricity is from a source located within or shared with a dwelling.

4

Planning permission

Before undertaking any building project, you must first obtain the approval of local government authorities. Many people (particularly householders) are initially reluctant to approach local authorities because, according to local gossip, they are 'likely to be obstructive'. In fact the reality of it is quite the reverse as their purpose is to protect all of us from irresponsible builders and developers and they are normally most sympathetic and helpful to any builder and/or DIY person who wants to comply with the statutory requirements and has asked for their advice.

There are two main controls that districts rely on to ensure that adherence to the local plan is ensured, namely planning permission and Building Regulation approval. Quite a lot of people are confused as to their exact use and whilst both of these controls are associated with gaining planning permission, actually receiving planning permission does not automatically confer Building Regulation approval and vice versa. You **may** require **both** before you can proceed. Indeed, there may be a variation in the planning requirements (and to some extent the Building Regulations) from one area of the country to another. Consequently, the information given on the following pages should be considered as a guide only and not as an authoritative statement of the law.

You are allowed to make certain changes to your home without having to apply to the local council for permission **provided** that it does not affect the external appearance of the building. These are called permitted development rights. The majority of building work that you are likely to complete will, however, probably require you to have planning permission and it is the nation's planning system that plays an important role in today's society by helping to protect the environment in our towns, cities and the countryside.

For example, if you are thinking about carrying out work on a listed building or work that requires the pruning or felling of a tree protected by a Tree Preservation Order, then you will need to contact your local authority planning department before carrying out any work. You never know, you might even require listed building consent or be required to follow certain procedures if carrying out work to trees.

 You do **not** require planning permission to carry out any internal alterations to your home, house, flat or maisonette, provided that it does not affect the external appearance of the building.

4.1 Planning controls

Planning controls exist primarily to regulate the use and siting of buildings and other constructions – as well as their appearance. What might seem to be a minor development in itself, could have far-reaching implications that you had not previously considered (for example, erecting a structure that would ultimately obscure vision at a busy junction and thereby constitute a danger to traffic). Equally, the local authority might refuse permission on the grounds that the planned scheme would not blend sympathetically with its surroundings. Your property could also be affected by legal restrictions such as a right of way, which could prejudice planning permission.

The actual details of planning requirements are complex but in respect of domestic developments, the planning authority is concerned primarily with the construction work such as an extension to the house or the provision of a new garage or new outbuildings that is being carried out. Structures like walls and fences also need to be considered because their height or siting might well infringe the rights of neighbours and other members of the community. The planning authority will also want to approve any change of use, such as converting a house into flats or running a business from premises previously occupied as a dwelling only.

4.1.1 Why are planning controls needed?

The purpose of the planning system is to protect the environment as well as public amenities and facilities. It is not designed to protect the interests of one person over another. Within the framework of legislation approved by Parliament, councils are tasked to ensure that development is allowed where it is needed, while ensuring that the character and amenity of the area are not adversely affected by new buildings or changes in the use of existing buildings or land.

Some people think the planning system should be used to prevent any change in their local environment, while others may think that planning controls are an unnecessary interference on their individual rights. The present position is that **all** major works need planning permission from the council but many minor works do not. Parliament thinks this is the right balance as it enables councils to protect the character and amenity of their area, while individuals have a reasonable degree of freedom to alter their property.

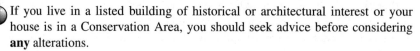 If you live in a listed building of historical or architectural interest or your house is in a Conservation Area, you should seek advice before considering **any** alterations.

4.2 Who requires planning permission?

Although the rules and requirements vary according to whether you actually own a house or a flat/maisonette, generally speaking, the principles and procedures for making planning applications are exactly the same for owners of houses and for freeholders (or leaseholders) of flats and maisonettes. Planning regulations, however, have to cover many different situations and so even the provisions that affect the average householder are quite detailed.

 You will not need to apply for planning permission to paint your flat or maisonette but, if you are a leaseholder, you may first need to get permission from your landlord or management company.

4.3 Who controls planning permission?

The planning system is made up of a cascade of documents. Currently, under the provisions of the Building Act 1984, national policy is mainly set out in Planning Policy Guidance notes (PPGs). Regions set out regional policy through Regional Planning Guidance notes (RPGs). Structure Plans establish broad planning policies at County Council level, and finally Local Plans set out detailed planning policy at District Council level (where Unitary Councils exist these two documents are generally combined into a Unitary Development Plan). Each layer has to be in conformity with the policies above it in the hierarchy.

4.3.1 Planning and Compulsory Purchase Act 2004

This system will be changing significantly thanks to the passing of the Planning and Compulsory Purchase Act 2004. This creates a new hierarchy of policies, and includes complex guidance on how local authorities are supposed to move from the old system to the new.

Under the 2004 Act County Structure Plans are abolished entirely, and the regional tier of policies is reclassified as a Regional Spatial Strategy (RSS). Local Plans are changed radically, and renamed Local Development Frameworks (LDFs).

These Frameworks will be made up of Local Development Documents (LDDs) which set out specific policies for the whole area or which give detailed guidance for a particular site. LDDs can include Development Plan Documents (DPDs), Supplementary Planning Documents (SPDs), a Statement of Community Involvement (SCI), an Annual Monitoring Report and a Local Development Scheme.

The transitional arrangements will allow Councils to redesignate their current Local Plans as part of their Local Development Framework, but there may well be many cases where Council's will have to shelve Local Plans that are currently being developed and revert to older documents. The potential for confusion is substantial and readers are advised to talk to their local council planning officials.

4.3.2 District Local Plan

Local Plans are prepared by district councils for their areas (except local plans concerning waste and minerals, which are prepared by the county council), and they set out the planning policies for the whole of the district and are used as the basis for assessing **all** planning applications. The district council is responsible

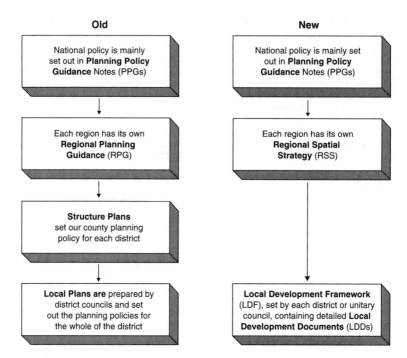

Figure 4.1 Planning responsibilities (old and new)

for keeping their plan under constant review and for making it available to everybody (usually via their council website).

The published Local Plan is used as a guide to the location of development over a ten-year period. For example, they:

- will identify where new homes, jobs and other types of development may be built;
- may require related development to be provided, such as children's play areas, parking facilities and road improvements;
- will outline restrictions where certain types of development are unacceptable.

In preparing Local Plans, districts are responsible for consulting local people and for ensuring that their views are taken into account, thereby giving them a chance to influence the way in which their area is affected.

At all times, Local Plans must also take account of national, regional and county planning policy.

A Local Plan consists of a written statement (which sets out and explains the policies and proposals) and the proposals map, which shows where they apply. Together, these elements of the plan:

- allow local people to clearly see if their homes, businesses or other property would be affected by what is proposed;

- give guidance to anyone who wants to build on a piece of land or change the use of a building in the area; and
- provide a basis for decisions on planning applications.

Planning permission is required for most building works, engineering works and use of land, and the following are some common examples of when you would need to apply for planning permission:

- you want to make additions or extensions to a flat or maisonette (including those converted from houses);

But you do **not** need planning permission to carry out internal alterations or work, provided that it does not affect the external appearance of the building.

- you want to divide off part of your house for use as a separate home (for example, a self-contained flat or bed-sit) or use a caravan in your garden as a home for someone else;

But you do **not** need planning permission to let one or two of your rooms to lodgers.

- you want to divide off part of your home for business or commercial use (for example, a workshop);
- you want to build a parking place for a commercial vehicle;
- you want to build something which goes against the terms of the original planning permission for your house – for example, your house may have been built with a restriction to stop people putting up fences in front gardens because it is on an 'open plan' estate;
- the work you want to complete might obstruct the view of road users;
- the work would involve a new or wider access to a trunk or classified road.

If you have any queries about a particular case, the first thing to do is to ask the planning department of your local council who will have records of all planning permissions in its area. You may also be able to find out more about planning law in your local library. If you are concerned about a legal problem involving planning, you may need to get professional advice or ask your local Citizens Advice Bureau.

A DETR booklet (*Planning Permission, A Guide for Business*) giving advice about working from home and whether planning permission is likely to be required is available from councils.

4.4 What is planning permission?

The planning control process is administered by your local authority and the system '*exists to control the development and use of land and buildings for the best interests of the community*'.

The process is intended to make the environment better for everyone and acts as a service to manage the types of constructions, modifications of premises, uses of land, and ensures the right mix of premises in any one vicinity (that individuals may plan to make) is maintained. The key feature of the process is to allow a party to propose a plan and for other parties to object if they wish to, or are qualified to.

4.5　What types of planning permission are available?

There are three types of planning permission available: outline, reserved and full.

4.5.1　Outline

This is an application for a development 'in principle' without giving too much detail on the actual building or construction. It basically lets you know, in advance, whether the development is likely to be approved. Assuming permission is granted under these circumstances you will then have to submit a further application in greater detail. In the main, this applies to large-scale developments only and you will probably be better off making a full application in the first place.

4.5.2　Reserved matters

This is the follow-up stage to an outline application to give more substance and more detail.

4.5.3　Full planning permission

Is the most widely used and is for erection or alteration of buildings or changes of use. There are no preliminary or outline stages and when consent is granted it is for a specific period of time. If this is due to lapse, a renewal of limited permission can be applied for.

4.6　How do I apply for planning permission?

There are fees to pay for each application for planning permission and your local planning office can provide you with the relevant details. You must make sure you have paid the correct fee – as permission can be refused if there is a discrepancy on fees paid.

When your forms and plans are ready, they need to be submitted to the planning office. The planning office will arrange for them to be listed in the local newspaper under 'latest planning applications' and will write to each neighbouring property and (normally) give them 21 days in which to raise any objections.

At the planning office, officials will produce a file after the 21 days have expired, with any objections or supporting information, and will make a recommendation on the application, ready for presenting it at the next planning committee or subcommittee meeting. At this meeting, they will discuss the case, reject it, ask for modifications or accept it. Whichever the decision the planning officer will feed back the decision to the applicant.

 There is an appeal procedure, which your local authority planning officer can advise you about.

4.7 Do I really need planning permission?

Most alterations and extensions to property and changes of use of land need to have some form of planning permission, which is achieved by submitting a planning application to the local authority. The purpose of this control is to protect and enhance our surroundings, to preserve important buildings and natural areas and strengthen the local economy.

However, not all extensions and alterations to dwelling houses require planning permission. Certain types of development are permitted without the need to make an official request, and it is always wise to contact the local authority before commencing any work.

Whether or not planning permission is required, good design is always important. Extensions and alterations should be in scale and in harmony with the remainder of the house. The builder should ensure that details such as window openings and matching materials are taken into account.

 Householders are encouraged (by councils) to employ a skilled designer when preparing plans for extensions and alterations. Alternatively, the authority's planning officers are able to offer general design guidance prior to the submission of your scheme.

Table 4.1 provides an indication of the basic requirements for planning permission and building regulation approval.

 Table 4.1 is only meant as guidance. A more complete description of the above synopsis is contained in Chapter 5. In all circumstances it is recommended that you talk to your local planning officer before contemplating any work. The cost of a local phone call could save you a lot of money (and stress) in the long term!

4.8 How should I set about gaining planning permission?

If you are in the planning stages for your work and you know planning permission will be required, it is wise to get the plans passed before you go to any expense or make any decisions that you may find hard to reverse – such as signing a contract for work. If your plans are rejected, you will still have to pay

Table 4.1 Basic requirements for planning permission and building regulation approval

Type of work	Planning permission		Building Regulation approval	
Decoration and repair inside and outside a building	No	Unless it is of a listed building or within a Conservation Area. Consult your local authority.	No	Unless it is a listed building or within a Conservation Area. Consult your local authority.
Structural alterations inside	No	As long as the use of the house is not altered.	Possibly	Consult your local authority.
	Yes	If the alterations are major such as removing or part removing of a load bearing wall or altering the drainage system.	Yes	
	Yes	If they are to an office or shop.	Yes	
Replacing windows and doors	No	Unless: • they project beyond the foremost wall of the house facing the highway • the building is a listed building • the building is in a Conservation Area	Possibly	Consult your local authority.
	Yes	To replace shop windows.	Yes	
Electrical work	No		Probably	All proposals to carry out electrical installation work **must** be notified to the local authority's building control body before work begins, **unless** the proposed installation work is undertaken by a competent person who is a person registered with an electrical self-certification scheme and does not include the provision of a new circuit.

Item			Notes
Plumbing	No	No	For replacements (but you will need to consult the Technical Services Department for any installation which alters present internal or external drainage).
Central heating	No	Yes	For an unvented hot water system.
		No	If electric.
		Yes	If gas, solid fuel or oil.
Oil-storage tank	No	No	Provided that it is in the garden and has a capacity of not more than 3500 litres (778 gallons) and **no** point is more than 3 m (9 ft 9'') high and **no** part projects beyond the foremost wall of the house facing the highway.
Planting a hedge	No	No	Unless it obscures view of traffic at a junction or access to a main road.
Building a garden wall or fence	Yes	Yes	If it is more than 1 m (3 ft 3'') high and is a boundary enclosure adjoining a highway.
		Yes	If it is more than 2 m (6 ft 6'') high elsewhere.
Felling or lopping trees	No	No	Unless the trees are protected by a Tree Preservation Order or you live in a Conservation Area.
Laying a path or a driveway	No	No	Unless it provides access to a main road.
Building a hard standing for a car	No	No	Provided that it is within your boundary and is not used for a commercial vehicle.
Installing a swimming pool	Possibly	Yes	Consult your local planning officer. For an indoor pool.
Erecting aerials, satellite dishes and flagpoles	No	No	Unless it is a stand-alone antenna or mast greater than 3 m in height.

(Continued)

Table 4.1 (*Continued*)

Type of work	Planning permission	Building Regulation approval
	Possibly — If erecting a satellite dish, especially in a Conservation Area or if it is a listed building (consult your local planning officer).	No
Advertising	No — If the advertisement is less than 0.3 m² and not illuminated.	Possibly — Consult your local planning officer.
Building a porch	No — Unless: • the floor area exceeds 3 m² (3.6 y²) • any part is more than 3 m (9 ft 9'') high • any part is less than 2 m (6 ft 6'') from a boundary adjoining a highway or public footpath	Yes — If area exceeds 30 m² (35.9 y²).
Constructing a small outbuilding	Possibly — Provided the building is less than 10 m³ (13.08 y³) in volume, not within 5 m (16 ft 3'') of the house or an existing extension. Erecting 'outbuildings' can be a potential minefield and it is best to consult the local planning officer before commencing work.	Yes — If area exceeds 30 m² (35.9 y²). If it is within 1 m (3 ft 3'') of a boundary, it must be built from incombustible materials.
Building a garage	Possibly — You can build a garage up to 10 m³ (13.08 y³) in volume without planning permission, if it is within 5 m (16 ft 3'') of the house or an existing extension. Further away than this, it can be up to half the area of the garden, but the height must not exceed 4 m (13 ft).	Yes
Building a conservatory	Possibly — You can extend your house by building a conservatory, provided that the total of both previous and new extensions does not exceed the permitted volume.	Yes — If area exceeds 30 m² (35.9 y²).

Loft conversions and roof extensions	No	Provided the volume of the house is unchanged and the highest part of the roof is not raised.		
	Yes	For front elevation dormer windows or rear ones over a certain size.		
Building an extension	Possibly	You can extend your house by building an extension, provided that the total of both previous and new extensions does not exceed the permitted volume.	Yes	If area exceeds 30 m² (35.9 y²).
		'Building extensions' can be a potential minefield and it is best to consult the local planning officer before contemplating any work.		
Converting a house to business premises (including bedsitters)	Yes	Even where construction work may not be intended.	Yes	Unless you are not proposing any building work to make the change.
Converting an old building	Yes		Yes	
Material change of use	Possibly	Even if no building or engineering work is proposed.	Yes	
Building a new house	Yes		Yes	
Infilling	Possibly	Consult your local planning officer.	Yes	If a new development.
Demolition	Yes	If it is a listed building or in a Conservation Area.	No	For a complete detached house.
		If the whole house is to be demolished.		
	Possibly	For partial demolition (seek advice from your local planning officer before proceeding).	Yes	For a partial demolition to ensure that the remaining part of the house (or adjoining buildings/extensions) is structurally sound.

The above table is only meant as guidance. A more complete description of the above synopsis is contained in Chapter 5. In all circumstances it is recommended that you talk to your local planning officer before contemplating any work. The cost of a local phone call could save you a lot of money (and stress) in the long term!

your architect or whoever prepared your plans for submission but you won't have to pay any penalty clauses to the building contractor.

 It is always best to submit an application in the early stages – if you try to be clever by submitting plans at the last minute (in the hope that neighbours will not have time to react) then you could be in for an expensive mistake! It's much better to do things properly and up-front.

An architect (surveyor or general contractor) can be asked to prepare and submit your plans on your behalf if you like, but as the owner and person requiring the development, it will be your name that goes on the application, even if all the correspondence goes between your architect and the planning department.

 You don't have to own the land to make a planning application, but you will need to disclose your interest in the property. This might happen if you plan to buy land, with the intention of developing it, subject to planning approval. It would, therefore, be in your best interest to obtain the consent before the purchase proceeds.

To submit your application you will need to use the official forms, available from the local authority planning department. It's a good idea to collect these personally, as you may get the opportunity to talk through your ideas with a planning officer and in doing so probably get some useful feedback. You will also need to include detailed plans of the present and proposed layout as well as the property's position in relation to other properties and roads or other features.

 New work requires details of materials used, dimensions and all related installations, similar to that required for Building Regulations.

4.9 What sort of plans will I have to submit?

There are three types of plans (namely site, block and building) that can accompany your application and, as indicated above, the choice will depend on the work proposed.

4.9.1 Site plan

A site plan indicates the development location and relationship to neighbouring property and roads etc. Minimum scale is 1:2500 (or 1:1250 in a built-up area). The land to which the application refers is outlined in red ink. Adjacent land, if owned by the applicant, is outlined in blue ink.

Block plan

A block plan is a detailed plan of a construction or structural alteration that shows the existing and proposed building, all trees, waterways, ways of access, pipes and drainage and any other important features. Minimum scales are 1:1500.

Building plans

Building plans are the detailed drawings of the proposed building works and would show plans, elevations and cross-sections that accurately describe every feature of the proposal. These plans are normally very thorough and include types of material, colour and texture, the layers of foundations, floor constructions, and roof constructions etc.

4.10 What is meant by 'building works'?

In the context of the Building Regulations, 'building works' means:

(a) the erection or extension of a building;
(b) the provision or extension of a controlled service or fitting;
(c) the material alteration of a building, or a controlled service or fitting;
(d) work required by Regulation 6 (requirements relating to material change of use);
(e) the insertion of insulating material into the cavity wall of a building;
(f) work involving the underpinning of a building.

4.11 What important areas should I take into consideration?

The following are some of the most important areas that should be considered before you submit a planning application.

4.11.1 Advertisement applications

If your proposal is to display an advertisement, you will need to make a separate application on a special set of forms. Three copies of the forms and the relevant drawings must be supplied. These must include a location plan and sufficient detail to show the size, materials and colour of the sign and its position. No certificate of ownership is needed, but it is illegal to display signs on the property without the consent of the owner.

4.11.2 Conservation Area consent

If you live in a Conservation Area, you will need Conservation Area consent to do the following:

- demolish a building with a volume of more than $115\,m^3$ (there are a few exceptions and further information will be available from your council);

- demolish a gate, fence, wall or railing over 1 m high if it is next to a highway (including a public footpath or bridleway) or public open space; or over 2 m high elsewhere.

4.11.3 Listed building consent

You will need to apply for listed building consent if either of the following cases apply:

- you want to demolish a listed building;
- you want to alter or extend a listed building in a manner which would affect its character as a building of special architectural or historic interest.

 You may also need listed building consent for any works to separate buildings within the grounds of a listed building. Check the position carefully with the council – it is a criminal offence to carry out work which needs listed building consent without obtaining it beforehand.

4.11.4 Trees

Many trees are protected by Tree Preservation Orders (TPOs), which mean that, in general, you need the council's consent to prune or fell them. In addition, there are controls over many other trees in Conservation Areas.

 Ask the council for a copy of the department's free leaflet *Protected Trees: a guide to tree preservation procedures*.

4.12 What are the government's restrictions on planning applications?

All applications for planning permission will have to take into account the following Acts and regulations.

Planning (Listed Buildings and Conservation Areas) Act 1990

Under the terms of the Planning (Listed Buildings and Conservation Areas) Act 1990, local councils must maintain a list of buildings within their boroughs, which have been classified as being of special architectural or historic interest. Councils are also required to keep maps showing which properties are within Conservation Areas.

Town and Country Planning (Control of Advertisements) Regulations 1992

In accordance with the Town and Country Planning (Control of Advertisements) Regulations 1992, councils need to maintain a publicly available register of applications and decisions for consent to display advertisements.

The Local Government (Access to Information) (Variation) Order 1992

The Local Government (Access to Information) (Variation) Order 1992 ensures that information relating to proposed development by councils cannot be treated as exempt when the planning decision is made.

Town and Country Planning (General Development Procedure) Order 1995

Every council must keep the following registers available for public inspection in accordance with the Town and Country Planning (General Development Procedure) Order 1995:

- planning applications, including accompanying plans and drawings;
- applications for a certificate of lawfulness of existing or proposed use or development;
- Enforcement Notices and any related stop notices.

All applications for planning permission must receive publicity.

Other areas

As well as the legal requirement to make the planning register available for public inspection, councils will also allow the public to have access to all other relevant information such as letters of objection/support for an application or correspondence about considerations. Three clear days before any committee meeting, the file will normally be made available for public inspection and this file will remain available (i.e. for further public inspection) after the committee meeting. Although commercial confidentiality could well be a valid consideration, the council will not use it so as to prevent important information about materials and facilities also being available.

4.13 How do I apply for planning permission?

Once you have established that planning permission is required, you will need to submit a planning application. Remember, it may take up to eight weeks, or even longer, to get planning permission, so apply early.

You will have to prepare a plan showing the position of the site in question (i.e. the site plan) so that the authority can determine exactly where the building is located. You must also submit another, larger-scale, plan to show the relationship of the building to other premises and highways (i.e. the block plan). In addition, it would help the council if you also supplied drawings to give a clear idea of what the new proposal will look like, together with details of both the colour and the kind of materials you intend using. You may prepare the drawings yourself, provided you are able to make them accurate.

Under normal circumstances you will have to pay a fee in order to seek planning permission, but there are exceptions. The planning department will advise you.

4.13.1 Application forms and plans

It is important to make sure that you make your planning application correctly. The following checklist may help:

- Obtain the application forms from the planning department or from the local council's website.
- Read the 'Notes for Applicants' carefully – again available from the planning department, or the local council's website.
- Fill in the relevant parts of the forms and remember to sign and date them.
- Submit the correct number and type of supporting plans. Each application should be accompanied by a site plan of not less than 1:2500 scale and detailed plans, sections and elevations, where relevant.
- Fill in and sign the relevant certificate relating to land ownership.

It is in your own interest to provide plans of good quality and clarity and so it is probably advisable to get help from an architect, surveyor, or similarly qualified person to prepare the plans and carry out the necessary technical work for you. You can obtain the necessary application form from the planning department of your local council and you will find that this is laid out simply, with guidance notes to help you fill it in. Alternatively, you can ask a builder or architect to make the application on your behalf. This is sensible if the development you are planning is in any way complicated, because you will have to include measured drawings with the application form.

4.14 What is the planning permission process?

If you think you might need to apply for planning permission, then this is the process to follow:

Step 1

Contact the planning department of your council. Tell the planning staff what you want to do and ask for their advice.

Step 2

If they think you need to apply for planning permission, ask them for an application form. They will tell you how many copies of the form you will need to send back and how much the application fee will be. Ask if they foresee any difficulties which could be overcome by amending your proposal. It can save time or trouble later if the proposals you want to carry out also reflect what the council would like to see. The planning department will also be able to tell you if Building Regulations approval will also be required.

Step 3

Decide what type of application you need to make. In most cases this will be a full application but there are a few circumstances when you may want to make an outline application – for example, if you want to see what the council thinks of the building work you intend to carry out before you go to the trouble of making detailed drawings (but you will still need to submit details at a later stage).

Step 4

Send the completed application forms and supporting documents to your council, together with the correct fee. Each form must be accompanied by a plan of the site and a copy of the drawings showing the work you propose to carry out. (The council will advise you on what drawings are needed.)

Extracts from Ordnance Survey maps can be supplied for planning applications submitted by private individuals and for school/college use. There is usually a charge for this service.

Step 5

The planning department will acknowledge receipt of your application, and publicly announce it – via letters to the neighbourhood parish council and anyone directly affected by the proposal, by publishing details of the application in the local press, notifying your neighbours and/or putting up a notice on or near the site. The council may also consult other organizations, such as the highway authority or the parish council (or community council in Wales).

A copy of the application will also be placed on the planning register at the council offices so that it can be inspected by any interested member of the public. Anyone can object to the proposal, but there is a limited period of time in which to do this and they must specify the grounds for objection.

 Under the Local Government Act 1972 (as amended), the public have the right to inspect and copy the following documents:

- the agenda for a council committee or sub-committee meeting; reports for the public part of the meeting;
- the minutes of such meetings and any background papers, including planning applications, used in preparing reports.

These documents can be inspected and copied from three clear days before a meeting. There is no charge to inspect a document but councils will charge for making photocopies.

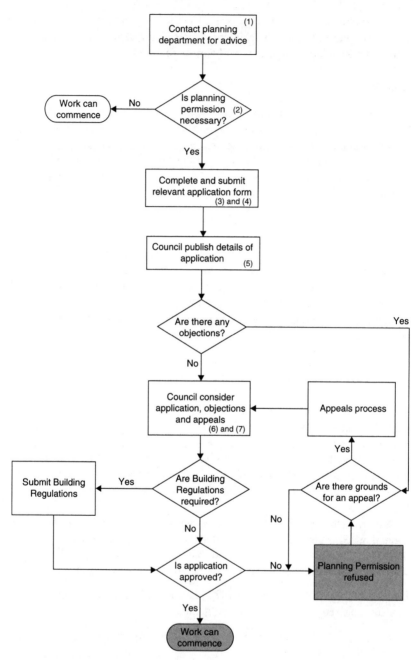

Figure 4.2 Planning permission

Step 6

The planning department may prepare a report for the planning committee, which is made up of elected councillors. Or the council may give a senior officer in the planning department the responsibility for deciding your application on its behalf. If a report has been made, then this will be presented to a meeting of the council committee, with recommendations on the decision to be made, based on the implications and objections received.

 You are entitled to see and have a copy of any report submitted to a local government committee. You are also entitled to see certain background papers used in the preparation of reports. The background papers will generally include the comments of consultees, objectors and supporters which are relevant to the determination of your application. Such material should normally be made available at least three working days before the committee meeting.

Step 7

The councillors or council officers who decide your application must consider whether there are any good planning reasons for refusing planning permission or for granting permission subject to conditions. The council cannot reject a proposal simply because many people oppose it. It will also look at whether your proposal is consistent with the development plan for the area.

The committee will consider the merits of a proposal; ensure the proposed work meets all the conditions of any local plan or requirements for a district and that the process has been followed properly. The kinds of planning issue it can also consider include potential traffic problems; the effect on amenity and the impact the proposal may have on the appearance of the surrounding area. Moral issues, the personal circumstances of the applicant or the effect the development might have on nearby property prices are not relevant to planning and will not normally be taken into account by the council. The committee will arrive at its decision and the result will be communicated back to the applicants via the planning department.

4.14.1 How long will the council take?

You can expect to receive a decision from the planning department within eight weeks and, **once granted, planning permission is valid for five years**. If the work is not begun within that time, you will have to apply for planning permission again.

If the council cannot make a decision within eight weeks then it must obtain your written consent to extend the period. If it has not done so, you can appeal to the Secretary of State for the Environment, Transport and the Regions, or, in Wales, to the National Assembly for Wales (see below). But appeals can take several months to decide and it may be quicker to reach agreement with the council.

Do not be afraid to discuss the proposal with a representative of the planning department **before** you submit your application. They will do their best to help you meet the requirements.

4.14.2 What can I do if my application is refused?

If the council refuses permission or imposes conditions, it must give reasons. If you are unhappy or unclear about the reasons for refusal or the conditions imposed, talk to staff at the planning department. Ask them if changing your plans might make a difference. If your application has been refused, you may be able to submit another application with modified plans free of charge within 12 months of the decision on your first application.

The planning department will always grant planning permission unless there are very sound reasons for refusal, in which case the department must explain the decision to you so that you can amend your plans accordingly and resubmit them for further consideration.

A second application is normally exempt from a fee.

The following are some of the main objection areas that your application may meet.

The property is a listed building

Listed buildings are protected for their special architectural or historical value. A *Listed Building Consent* may be needed for alterations but grants could be available towards repair and restoration!

If it's a listed building, it probably has some historic importance and will have been listed by the Department of the Environment. This could apply to houses, factories, warehouses and even walls or gateways. Most alterations which affect the external appearance or design will require listed building consent in addition to other planning consents.

The property is in a Conservation Area

This is an area defined by the local authority, which is subject to special restrictions in order to maintain the character and appearance of that area. Again, other planning consents may be needed for areas designated as green belt, areas of outstanding natural beauty, national parks or sites of specific scientific interest.

The application does not comply with the local development plan

Local authorities often publish a development plan, which sets out policies and aims for future development in certain areas. These are to maintain specific environmental standards and can include very detailed requirements such as minimum or maximum dimensions of plot sizes, number of dwellings per acre, height and style of dwellings etc. It is important to check if a plan exists for your area, as proposals can meet with some fierce objections from residents protecting their environment.

The property is subject to a covenant

This is an agreement between the original owners of the land and the persons who acquired it for development. They were implemented to safeguard residential standards and can include things like the size of outbuildings, banning use of front gardens for parking cars, or even just the colours of exterior paintwork.

Is there existing planning permission?

A previous resident or owner may have applied for planning permission, which may not have expired yet. This could save time and expense if a new application can be avoided. If you are considering a planning application, you should consider the above questions. Normally your retained expert – architect, surveyor or builder – can advise and help you to get an application passed. Most information can be collected from your local planning department, or if you need to find out about covenants, look for the appropriate land registry entry.

The proposal infringes a right of way

If your proposed development would obstruct a public path that crosses your property, you should discuss the proposals with the council at an early stage. The granting of planning permission will **not** give you the right to interfere with, obstruct or move the path. A path cannot be legally diverted or closed unless the council has made an order to divert or close it to allow the development to go ahead. The order must be advertised and anyone may object. You must not obstruct the path until any objections have been considered and the order has been confirmed. You should bear in mind that confirmation is not automatic; for example, an alternative line for the path may be proposed, but not accepted.

4.14.3 What matters cannot be taken into account?

- Competition
- Disturbance from construction work
- Loss of property value
- Loss of view
- Matters controlled under other legislation such as Building Regulations (e.g. structural stability, drainage, fire precautions etc.)
- Moral issues
- Need for development
- Private issues between neighbours (e.g. land and boundary disputes, damage to property, private rights of way, deeds, covenants etc.)

- Sunday trading
- The identity or personal characteristics of the applicant.

4.14.4 What are the most common stumbling blocks?

In no particular order of priority, these are:

- Adequacy of parking
- Archaeology
- Design, appearance and materials
- Effect on listed building or Conservation Area
- Government advice
- Ground contamination
- Hazardous materials
- Landscaping
- Light pollution
- Local planning policies
- Nature conservation
- Noise and disturbance from the use (but not from construction work)
- Overlooking and loss of privacy
- Previous planning decisions
- Previous appeal decisions
- Road access
- Size, layout and density of buildings
- The effect on the street or area (but not loss of private view)
- Traffic generation and overall highway safety.

4.15 Can I appeal if my application is refused?

If you think the council's decision is unreasonable, you can appeal to the Secretary of State or (in Wales) to the National Assembly for Wales. Appeals must be made within six months of the date of the council's notice of decision. You can also appeal if the council does not issue a decision within eight weeks.

 A free booklet *Planning Appeals – A Guide* is available from the Planning Inspectorate, Tollgate House, Houlton Street, Bristol BS2 9DJ or Crown Buildings, Cathays Park, Cardiff CF10 3NQ.

Appeals are intended as a last resort and they can take several months to decide. It is often quicker to discuss with the council whether changes to your proposal would make it more acceptable. The planning authority will supply you with the necessary appeal forms.

 Be careful not to proceed without approval, as you might find yourself obliged to restore the property to its original condition.

4.16 Before you start work

There are many kinds of alterations and additions to houses and other buildings which do not require planning permission. Whether or not you need to apply, you should think about the following before you start work.

4.16.1 What about neighbours?

Have the neighbours any rights to complain?
Many of us live in close proximity to others and your neighbours should be the first individuals you talk to.

What if your alteration infringes their access to light, or a view? Such disputes are notorious for causing bad feeling but with a little consideration, at an early stage, you can avoid a good deal of unpleasantness later.

Plans for the local area can normally be viewed at the local town hall, but most planning applications will involve consultation with neighbours and statutory consultees such as the Highways Authority and the drainage authorities. The extent of consultation will, quite naturally, reflect on the nature and scale of the proposed development – together with its location. Applications to make an alteration to your property can also be refused because you live in an area of outstanding natural beauty, a national park, or a Conservation Area or your property is listed. Any alterations to public utilities such as drains or sewers, or changes to public access such as footpaths will require consultation with the local council. They will have to approve your plans. Even a sign on or above your property may need to be of a certain size or shape.

Some properties may also be the home of a range of protected species such as bats or owls. These animals are protected by the Wildlife and Countryside Act 1981 and the Nature Conservancy Council must give approval to any work that may potentially disturb them. Likewise many members of the public are extremely defensive of trees that grow where they live. Tree Preservation Orders may control the extent to which you can fell or even prune a tree, even if it is on your property. Trees in Conservation Areas are particularity protected, and you will need to supply at least six weeks' notice before working upon them.

New street names and house numbers need approval from the council.

Let your neighbours know about the work you intend to carry out to your property. They are likely to be as concerned about work which might affect them as you would be about changes which might affect your enjoyment of your own property. For example, your building work could take away some of their light or spoil a view from their windows. If the work you carry out seriously overshadows a neighbour's window and that window has been there for 20 years or more, you may be affecting his or her 'right to light' and you could be open to legal action. It is best to consult a lawyer if you think you need advice about this.

You may be able to meet some of your neighbour's worries by modifying your proposals. Even if you decide not to change what you want to do, it is usually better to have told your neighbours what you are proposing before you apply for planning permission and before any building work starts.

If you do need to make a planning application for the work you want to carry out, the council will ask your neighbours for their views. If you or any of the people you are employing to do the work need to go on to a neighbour's property, you will, of course, need to obtain their consent before doing so.

4.16.2 What about design?

Everybody's taste varies and different styles will suit different types of property. Nevertheless, a well-designed building or extension is likely to be much more attractive to you and to your neighbours. It is also likely to add more value to your house when you sell it. It is therefore worth thinking carefully about how your property will look after the work is finished.

Extensions often look better if they use the same materials and are in a similar style to the buildings that are there already – but good design is impossible to define and there may be many ways of producing a good result. In some areas, the council's planning department issues design guides or other advisory leaflets that may help you.

4.16.3 What about crime prevention?

You may feel that your home is secure against burglary and you may already have taken some precautions such as installing security locks to windows. However, alterations and additions to your house may make you more vulnerable to crime than you realize. For example, an extension with a flat roof, or a new porch, could give access to upstairs windows which previously did not require a lock. Similarly, a new window next to a drainpipe could give access. Ensure that all windows are secure. Also, your alarm may need to be extended to cover any extra rooms or a new garage. The crime prevention officer at your local police station can provide helpful advice on ways of reducing the risk.

4.16.4 What about lighting?

If you are planning to install external lighting for security or other purposes, you should ensure that the intensity and direction of light does not disturb others. Many people suffer extreme disturbance due to excessive or poorly designed lighting. Ensure that beams are **not** pointed directly at windows of other houses. Security lights fitted with passive infra-red detectors (PIRs) and/or timing devices should be adjusted so that they minimize nuisance to neighbours and are set so that they are not triggered by traffic or pedestrians passing outside your property.

4.16.5 What about covenants?

Covenants or other restrictions in the title to your property or conditions in the lease may require you to get someone else's agreement before carrying out some kinds of work to your property. This may be the case even if you do not need to apply for planning permission. You can check this yourself or consult a lawyer.

You will probably need to use the professional services of an architect or surveyor when planning a loft conversion. Their service should include considerations of planning control rules.

4.16.6 What about listed buildings?

Buildings are listed because they are considered to be of special architectural or historic interest and as a result require special protection. Listing protects the **whole** building, both inside and out and possibly also adjacent buildings if they were erected before 1 July 1948.

The prime purpose of having a building listed is to protect the building and its surroundings from changes that will materially alter the special historic or architectural importance of the building or its setting.

The list of buildings is prepared by the Department of Culture, Media and Sport and properties are scheduled into one of three grades, Grade I, Grade II* and Grade II, with Grade I being the highest grade. Over 90% of all listed properties fall within Grade II. (In Scotland the grades are A, B and C.)

All buildings erected prior to 1700 and substantially intact are listed, as are most buildings constructed between 1700 and 1840, although some selection does take place. The selection process is more discriminating for buildings erected since 1840 because so many more properties remain today. Buildings less than 30 years old are generally only listed if they are of particular architectural or historic value and are potentially under threat. Your district council holds a copy of the statutory list for public inspection and this provides details on each of the listed properties.

 See *Planning Policy Guidance Note 15 (PPG.15) – Planning and the Historic Environment*, which provides a practical understanding of the Planning (Listed Buildings and Conservation Areas) Act 1990 which can be viewed at your planning office or in main libraries, or purchased from The Stationery Office, 29 Duke Street, Norwich, NR3 1GN (Tel: 0870 600 5522; Fax: 0870 600 5533, www.tso.co.uk).

Owner's responsibilities?

If you are the owner of a listed building or come into possession of one, you are tasked with ensuing that the property is maintained in a reasonable state of repair. The council may take legal action against you if they have cause to believe that you are deliberately neglecting the property, or have carried out works without consent. Enforcement action may be instigated.

There is no statutory duty to effect improvements, but you must not cause the building to fall into any worse state than it was in when you became its owner. This may necessitate some works, even if they are just to keep the building wind and watertight. However, you may need listed building consent in order to carry these works out!

A photographic record of the property when it came into your possession may be a useful asset, although you may also have inherited incomplete or unimplemented works from your predecessor, which you will become liable for.

If you are selling a listed building you may wish to indemnify yourself against future claims: speak to your solicitor.

4.16.7 What about Conservation Areas?

Tighter regulations apply to developments in Conservation Areas and to developments affecting listed buildings. Separate Conservation Area consent and/or listed building consent may be needed in addition to planning consent and Building Regulation consent.

 Conservation Areas are *'areas of special architectural or historic interest the character and appearance of which it is desirable to preserve or enhance'*. (Civic Amenities Act 1967)

As the title indicates these designations cover more than just a building or property curtilage and most local authorities have designated Conservation Areas within their boundary. Although councils are not required to keep any statutory lists, you can usually identify Conservation Areas from a local plan's 'proposals maps' and appendices. Some councils may keep separate records or even produce leaflets for individual areas.

   The purpose of designating a Conservation Area is to provide the council with an additional measure of control over an area that they consider being of special historic or architectural value. This does not mean that development proposals cannot take place, or that works to your property will be automatically refused. It means however that the council will have regard to the effect of your proposals on the designation in addition to their normal assessment. The council may also apply this additional tier of assessment to proposals that are outside the designated Conservation Area boundary, but which may potentially affect the character and appearance of the area.

   As a result, local planning authorities may ask for more information to accompany your normal planning application concerning proposals within (or adjoining) a Conservation Area. This may include:

- a site plan to 1:1250 or 1:2500 scale showing the property in relation to the Conservation Area;
- a description of the works and the effect (if any) you think they may have on the character and appearance of the Conservation Area;
- a set of scale drawings showing the present and proposed situation, including building elevations, internal floor plans and other details as necessary.

 If you live or work in a Conservation Area, grants may be available towards repairing and restoring your home or business premises.

For major works you may need to involve an architect with experience of works affecting Conservation Areas.

4.16.8 What is Conservation Area consent?

Development within Conservation Areas is dealt with under the normal planning application process, except where the proposal involves demolition.

In this case you will need to apply for Conservation Area consent on the appropriate form obtainable from the planning department.

Here again the council will assess the proposal against its effect upon the special character and appearance of the designated area.

 More details can be obtained by reference to *Planning Policy Guidance Note 15 (PPG.15) – Planning and the Historic Environment*, which provides a practical understanding of the Planning (Listed Buildings and Conservation Areas) Act 1990. These can be viewed at your planning office or in main libraries, or purchased from The Stationery Office, 29 Duke Street, Norwich, NR3 1GN (Tel: 0870 600 5522 Fax: 0870 600 5533, www.tso.co.uk).

4.16.9 What about trees in Conservation Areas?

Nearly all trees in Conservation Areas are automatically protected.

Trees in Conservation Areas are generally treated in the same way as if they were protected by a Tree Preservation Order, i.e. it is necessary to obtain the council's approval for works to trees in Conservation Areas before they are carried out. There are certain exceptions (where a tree is dead or in a dangerous condition) but it is always advisable to seek the opinion of your council's tree officer to ensure your proposed works are acceptable. Even if you are certain that you do not need permission, notifying the council may save the embarrassment of an official visit if a neighbour contacts them to tell them what you are doing.

If you wish to lop, top or fell a tree within a Conservation Area you must give six weeks' notice, in writing, to the local authority. This is required in order that they can check to see if the tree is already covered by a Tree Preservation Order (TPO), or consider whether it is necessary to issue a TPO to control future works on that tree.

 Contact your council's landscape or tree officer for further information.

4.16.10 What are Tree Preservation Orders?

Trees are possibly the biggest cause of upset in town and country planning and many neighbours fall out over tree related issues. They may be too tall, may block out natural light, have overhanging branches, shed leaves on other property or the roots may cause damage to property. When purchasing a property the official searches carried out by your solicitor should reveal the presence of a TPO on the property or whether your property is within a Conservation Area within which trees are automatically protected.

However not all trees are protected by the planning regulations system – but trees that have protection orders on them must not be touched unless specific approval is granted. Don't overlook the fact that a preservation order could have been put on a tree on your land before you bought it and is still enforceable.

Planning authorities have powers to protect trees by issuing a TPO and this makes it an offence to cut down, top, lop, uproot, wilfully damage or destroy any protected tree(s) without first having obtained permission from the local authority. All types of tree can be protected in this way, whether as single trees or as part of a woodland, copse or other grouping of trees. Protection does not however extend to hedges, bushes or shrubs.

TPOs are recorded in the local land charges register which can be inspected at your council offices. The local authority regularly checks to see if trees on their list still exist and are in good condition. Civic societies and conservation groups also keep a close eye on trees. Before carrying out work affecting trees, you should check if the tree is subject to a TPO. If it is, you will need permission to carry out the work.

All trees in a Conservation Area are protected, even if they are not individually registered. If you intend to prune or alter a tree in any way you must give the local authority plenty of notice so they can make any necessary checks.

Even with a preservation order it is possible to have a tree removed, if it is too decayed or dangerous, or if it stands in the way of a development, the local authority may consider its removal, but will normally want a similar tree put in or near its place.

A TPO will not prevent planning permission being granted for development. However, the council will take the presence of TPO trees into account when reaching their decision.

If you have a tree on your property that is particularly desirable – either an uncommon species or a mature specimen – then you can request a preservation order for it. However, this will mean that in years to come you, and others, will be unable to lop it, remove branches or fell it unless you apply for permission.

What are my responsibilities?

Trees covered by TPOs remain the responsibility of the landowner, both in terms of any maintenance that may be required from time to time and for any damage they may cause. The council must formally approve any works to a TPO tree. If you cut down, uproot or wilfully damage a protected tree or carry out works such as lopping or topping which could be likely to seriously damage or destroy the tree then there are fines on summary conviction of up to £20 000, or, on indictment, the fines are unlimited. Other offences concerning protected trees could incur fines of up to £2500.

What should I do if a protected tree needs lopping or topping?

Although there are certain circumstances in which permission to carry out works to a protected tree are not required, it is generally safe to say that you should always write to your council seeking their permission before undertaking any works. You should provide details of the trees on which you intend to do work, the nature of that work – such as lopping or topping – and the reasons why you think this is necessary. The advice of a qualified tree surgeon may also be helpful, see *Yellow Pages*.

 You may be required to plant a replacement tree if the protected tree is to be removed.

4.16.11 What about nature conservation issues?

Many traditional buildings, particularly farm buildings, provide valuable wildlife habitats for protected species such as barn owls and bats.

Planning permission will not normally be granted for conversion and reuse of buildings if protected species would be harmed. However, in many cases, careful attention to the timing and detail of building work can safeguard or re-create the habitat value of a particular building. Guidance notes prepared by English Nature are available from councils or from their website www.english-nature.org.uk.

4.16.12 What about bats and their roosts?

Bats make up nearly one-quarter of the mammal species throughout the world. Some houses may hold roosts of bats or provide a refuge for other protected species. The Wildlife and Countryside Act 1981 gives special protection to **all** British bats because of their roosting requirements. English Nature (EN) or the Countryside Council for Wales (CCW) must be notified of any proposed action (e.g. remedial timber treatment, renovation, demolition and extensions) which is likely to disturb bats or their roosts. EN or CCW must then be allowed time to advise on how best to prevent inconvenience to both bats and householders.

 Information on bats and the law is included in the booklet *Focus on Bats* which can be obtained free of charge from your local EN office. Similar booklets can be obtained from CCW local offices.

The type of stone barns and traditional buildings found in the UK have lots of potential bat roosting sites; the most likely places being gaps in stone rubble walls, under slates or within beam joints. These sites can be used throughout the year by varying numbers of bats, but could be particularly important for winter hibernation. As a result, the following points should be followed when considering or undertaking any work on a stone barn or similar building, particularly where bats are known to be in the area.

A survey for the presence of bats should be carried out by a member of the local bat group (contact via English Nature) before any work is done to a suitable barn during the summer bat breeding period.

 The pipistrelle, the smallest of the European bats, has been found lurking in many strange places including vases, under floorboards, and between the panes of double glass.

Any pointing of walls should not be carried out between mid-November and mid-April to avoid potentially entombing any bats. When walls are to be pointed, areas of the walls high up on all sides of the building should be left unpointed to preserve some potential roosting sites. If any bats are found whilst work is in progress, work should be stopped and English Nature contacted for advice on how to proceed.

If any timber treatment is carried out, only chemicals safe for use in bat roosts should be used. A list of suitable chemicals is available from English Nature on request. Any pre-treated timber used should have been treated using the CCA method (copper chrome arsenic) which is safe for bats.

Work should not be commenced during the winter hibernation period (mid-November to mid-April). Any bats present during the winter are likely to be torpid, i.e. unable to wake up and fly away and are therefore particularly vulnerable.

 If these guidelines are followed, then the accidental loss of bat roosts and death or injury to bats will be reduced.

 Although vampire bats feed primarily on domestic animals, they have been known to feed on sleeping humans on rare occasions! Vampire bats have chemicals in their saliva that prevent the blood they are drinking from clotting. They consume five teaspoons of blood each day. The vampire bat has been known to transmit rabies to livestock and to man.

4.16.13 What about barn owls?

Barn owls also use barns and similar buildings as roosting sites in some areas. These are more obvious than bats and, therefore perhaps easier to take into account. Barn owls are also fully protected by law and should not be disturbed during their breeding season. Special owl boxes can be incorporated into walls during building work, details of which can be obtained from English Nature.

4.17 What could happen if you don't bother to obtain planning permission?

If you build something which needs planning permission without obtaining permission first, you may be forced to put things right later, which could prove troublesome and extremely costly. You might even have to remove an unauthorized building.

4.17.1 Enforcement

If you think that works are being carried out without planning permission, or not in accord with approved plans and/or conditions of consent, then seek the advice of the local planning officer who will then investigate, and if necessary take appropriate steps to deal with the problem. Conversely, if you are carrying out development works, it is important that you stick to the approved plans and condition. If changes become necessary please contact the development control staff before they are made.

4.18 How much does it cost?

A fee is required for the majority of planning applications and the council cannot deal with your application until the correct fee is paid. The fee is not refundable if your application is withdrawn or refused.

In most cases you will also be required to pay a fee when the work is commenced. These fees are dependent on the type of work that you intend to carry out. The fees outlined in Sections 14.18.1–10 are typical of the charges made by Local Authorities during 2004, when submitting an application.

 Work to provide access and/or facilities for disabled people to existing dwellings are exempt from these fees.

4.18.1 Householder applications

Outline applications (most types)	£220 per 0.1 ha (or part thereof) of site area, maximum £5550 (2.5 ha)
Full applications and reserved matters	
Dwellings – erection of new	£220 per dwelling house, maximum £11 000 (=50)
Dwellings – alteration (including outline)	£110 per dwelling house, maximum
Approval of reserved matters where flat rate does **not** apply	A fee based upon the amount of floorspace and/or number of dwelling houses involved
Flat rate (only when maximum fee has been paid)	£220

4.18.2 Industrial/retail and other buildings applications

Industrial retail buildings	<ul><li>Where no additional floorspace is created £110</li><li>Works not creating more than 40 m² of additional floorspace £110</li><li>More than 40 m² but not more than 75 m² of additional floorspace £220</li><li>Each additional 75 m² (or part thereof) £220, maximum £11 000 (=3750 m²)</li></ul>
Outline applications (see above)	

| Plant and machinery (erection, alteration, replacement) | £190 per 0.1 ha (0.24 acre), or part thereof, of the site area, maximum £9500 (5 ha) |

4.18.3 Prior notice applications

| Approvals for agricultural/forestry buildings/operations and demolition of buildings and telecommunications works | £40 |

4.18.4 Agricultural applications

| Agricultural buildings | • Buildings not exceeding 465 m^2 £40
• Buildings exceeding 465 m^2 but not more than 540 m^2 £220
• More than 540 m^2 £220 for first 540 m^2 and £220 for each additional 75 m^2 (or part thereof), maximum £11 000 |
| Erection of glasshouses/ polytunnels (on land used for agriculture) | • Works not creating more than 465 m^2 £40
• For each additional 75 sq. ft (if within 90 m of an existing building) £220
• Works creating more than 465 m^2 £1235 |

 With a permitted development, there is a possibility of having two glasshouses separated by greater than 90 m!

4.18.5 Legal applications

Application for a certificate of lawfulness for an **existing** use or operation	Same fee payable as if making a planning application
Application for a certificate of lawfulness for an **existing** activity in breach of planning condition(s)	£110
Application for a certificate of lawfulness for a **proposed** use or operation	Half the fee payable as if making a planning application

4.18.6 Advertisement applications

Adverts relating to the business on the premises	£60
Advance signs directing the public to a business	£60
Advance signs directing the public to a business (unless business can be seen from the sign's position)	£220
Other advertisements (e.g. hoardings)	£220

4.18.7 Other applications

Exploratory drilling for oil or natural gas	£440 per 0.1 ha (or part thereof) of site area, maximum £33 000 (=7.5 ha)
Storage of minerals etc. and waste disposal	£220 per 0.1 ha (or part thereof) of site area, maximum £33 000 (=15 ha)
Car parks, service roads or other accesses (existing uses only)	£110
Other operations on land	£110 per 0.1 ha (or part thereof) of site area, maximum £1100 (=1 ha)
Non-compliance with conditions	£110
Renewal of temporary permissions	£110
Removal or variation of conditions including renewal of unimplemented consents that have not lapsed	£110 (full fee if consent has lapsed)
Change of use to subdivision of dwellings	£220 per additional dwelling created (maximum £11 000)
Other changes of use except waste or minerals	£220

4.18.8 Concessionary fees and exemptions

Works to improve the disabled persons' access to a public building, or to improve their access, safety, health or comfort at their dwelling house.

Applications by parish councils (all types)	No fee
Applications required by an Article 4 direction or removal of permitted development rights	Half the normal fee

Playing fields (for sports clubs etc.)	£220
Revised or fresh applications of the same character or description within 12 months of refusal, or the expiry of the statutory 8 week period where the applicant has appealed to the Secretary of State on grounds of non-determination. Withdrawn applications of the same character or description must be made within 12 months of making the earlier one	No fee
Revised or fresh application of the same character or description within 12 months of receiving permission	No fee
Duplicate applications made by the same applicant submitted within 28 days of each other	Full fee for each application
Alternative applications for one site submitted at the same time	Highest of the fees applicable for each alternative and a sum equal to half the rest
Development crossing local authority boundaries	Only one fee paid to the authority having the larger site but calculated for the whole scheme and subject to a special ceiling

4.18.9 Hazardous substances applications

Application for new consent	£200.00
New consent where maximum quantity specified exceeds twice the controlled quantity	£400.00
All other types of application	£250.00
Continuation of hazardous consent under Section 17(1) of the 1992 Regulations	£200.00

4.18.10 Examples of mixed development

(1) **Outline application for 1000 m² of office floorspace and five flats on a site of 0.5 ha:**
    Since the application is in outline and the sites are one and the same, the fee is the same under either category of development

0.5 ha at £220 per 0.1 ha (or part thereof) = 5 × £220 = £1100

(2)   **Application for approval of reserved matters, if not subject to a flat-rate fee, for a development of 4000 m² of mixed shops and offices and 60 dwellings:**
Since the floorspace exceeds 3750 m² the maximum fee of £11 000 applies, and since the number of dwelling houses exceeds 50, the maximum fee of £11 000 also applies. Fees for this kind of mixed development are additions

The total fee is therefore £22 000

(3)   **Application for full permission for a building incorporating 200 m² of shops, four flats with a total area of 30 m² and of common service floor-space:**
The total floorspace of the development occupies 200 + 370 + 30 m² = 600 m². The proportion occupied by shops is 200 out of 600 = one third. One third of the common service area (1/3 × 30 m² = 10 m²) is added to the non-residential floorspace

Fee calculation is 200 + 10 = 210 m² of shops at £220 per 75 m² (or part thereof)
4 Dwellings at £220

Total £1540

(4)   **Application for use of land as a caravan site of 2 ha incorporating roads, hardstandings, a shop and service building of 150 m² of floorspace**
(Another example in this category would be development of a golf course)

Change of use = £220
2 ha of 'other operations' (because the site exceeds 1 ha), the maximum fee of £1100 applies 150 m² of non-residential floorspace at £220 per 75 m² (or part thereof) = £440
Of these the highest fee, £1100, applies

Extracted from *The Town & Country Planning (Fees for Applications & Deemed Applications) (Amendment) Regulations 2003. (Amended 2005.)*

5

Requirements for planning permission and Building Regulations approval

Before undertaking any building project, you must first obtain the approval of local-government authorities. There are two main controls that districts rely on to ensure that adherence to the local plan is ensured, namely planning permission and Building Regulation approval.

Whilst both of these controls are associated with gaining planning permission, actually receiving planning permission does not automatically confer Building Regulation approval and vice versa. You **may** require **both** before you can proceed. Indeed, there may be variations in the planning requirements, and to some extent the Building Regulations, from one area of the country to another.

Provided, however, that the work you are completing does not affect the external appearance of the building, you are allowed to make certain changes to your home without having to apply to the local council for permission. These are called permitted development rights, but the majority of building work that you are likely to complete will still require you to have planning permission – so be warned!

The actual details of planning requirements are complex but for most domestic developments, the planning authority is only really concerned with construction work such as an extension to the house or the provision of a new garage or new outbuildings that is being carried out. Structures like walls and fences also need to be considered because their height or siting might well infringe the rights of your neighbours and other members of the community. The planning authority will also want to approve any change of use, such as converting a house into flats or running a business from premises previously occupied as a dwelling only.

 Planning consent **may** be needed for minor works such as television satellite dishes, dormer windows, construction of a new access, fences, walls, and garden extensions. You are advised to consult with Development Control staff before going ahead with such minor works.

5.1 Decoration and repairs inside and outside a building

Planning permission		Building Regulation approval	
No	Unless it is of a listed building or within a Conservation Area Consult your local authority	No	Unless it is a listed building or within a Conservation Area Consult your local authority
No	As long as the use of the house is not altered	Possibly	Consult your local authority
Yes	If the alterations are major such as removing or part removing of a load bearing wall or altering the drainage system	Yes	

Generally speaking, you do not need to apply for planning permission:

- for repairs or maintenance;
- for minor improvements, such as painting your house or replacing windows;
- for internal alterations;
- for the insertion of windows, skylights or roof lights (but, if you want to create a new bay window, this will be treated as an extension of the house);
- for the installation of solar panels which do not project significantly beyond the roof slope (rules for listed buildings and houses in Conservation Areas are different however);
- to re-roof your house (but additions to the roof are treated as extensions to the house).

Occasionally, you may need to apply for planning permission for some of these works because your council has made an Article 4 direction withdrawing permitted development rights.

Do I need approval to carry out repairs to my house, shop or office?

No – if the repairs are of a minor nature – e.g. replacing the felt to a flat roof, repointing brickwork, or replacing floorboards.
Yes – if the repair work is major in nature – e.g. removing a substantial part of a wall and rebuilding it, or underpinning a building.

Do I need to apply for planning permission for internal decoration, repair and maintenance?

No.

Do I need to apply for planning permission for external decoration, repair and maintenance?

No – external work in most cases doesn't need permission, provided it does not make the building any larger.

Do I need approval to or alter the position of a WC, bath, etc. within my house, shop or flat?

No – unless the work involves new or an extension of drainage or plumbing.

Do I need approval to alter in any way the construction of fireplaces, hearths or flues within my house, shop or flat?

Yes.

Do I need to apply for planning permission if my property is a listed building?

Yes – if your property is a listed building consent will probably be needed for **any** external work, especially if it will alter the visual appearance, or use alternative materials. You also may need planning permission to alter, repair or maintain a gate, fence, wall or other means of enclosure.

Do I need to apply for planning permission if my property is in a Conservation Area?

Yes – if the building undergoing repair or decoration is in a Conservation Area, or comes under any type of covenant restricting changes you will probably need planning permission. You may also be restricted to replacing items such as roof tiles with the approved material, colour and texture, and have to use cast iron guttering rather than plastic etc.

5.2 Structural alterations inside

Planning permission		Building Regulation approval	
No	As long as the use of the house is not altered	Possibly	Consult your local authority
Yes	If the alterations are major such as removing or part removing of a load bearing wall or altering the drainage system	Yes	
Yes	If they are to an office or shop	Yes	

Do I need approval to make internal alterations within my house?

Yes – if the alterations are to the structure such as the removal or part removal of a load bearing wall, joist, beam or chimney breast, or would affect fire precautions of a structural nature either inside or outside your house. You also need approval if, in altering a house, work is necessary to the drainage system or to maintain the means of escape in case of fire.

Do I need approval to make internal alterations within my shop or office?

Yes.

Do I need approval to insert cavity wall insulation?

Yes.

Do I need approval to apply cladding?

Yes – if you live in a Conservation Area, a national park, an area of outstanding natural beauty or the Norfolk Broads. You will need to apply for planning permission before cladding the outside of your house with stone, tiles, artificial stone, plastic or timber.

 If you are in any doubt about whether you need to apply for permission, you should contact your local authority planning department before commencing any work to your property. They will usually give you advice but if you want to obtain a formal ruling you can apply, on payment of a fee, for a lawful development certificate. You may also require Building Regulation approval.

5.3 Replacing windows and doors

Planning permission		Building Regulation approval	
Yes	If they are to an office or shop	Yes	
No	Unless: • they project beyond the foremost wall of the house facing the highway • the building is a listed building • the building is in a Conservation Area	Possibly	Consult your local authority
Yes	To replace shop windows	Yes	

Do I need approval to install replacement windows in my house, shop or office?

No – provided:

- the window opening is not enlarged. If a larger opening is required, or if the existing frames are load-bearing, then a structural alteration will take place and approval will be required.
- the installation (as a replacement) is carried out by a person who is registered under the Fenestration Self-Assessment Scheme by Fensa Ltd.
- you do not remove those opening windows which are necessary as a means of escape in case of fire.

Do I need approval to replace my shop front?

Yes.

Anyone who installs replacement windows or doors has to comply with strict thermal performance standards and when a property is sold, the purchaser's surveyors will normally ask for evidence that *'any replacement glazing installed after April 2002 complies with the new Building Regulations'*.

There will be two ways to prove compliance:

1. A certificate showing that the new work has been done by an installer who is registered under the FENSA Scheme (a scheme which allows installation companies to self-certify that their work complies with the Building Regulations), or
2. A certificate from the local authority saying that the installation has approval under the Building Regulations.

Note: Further information is available from your local building control or from the Glass and Glazing Federation (GGF) website www.ggf.org.uk.

5.4 Electrical work

Planning permission	Building Regulation approval	
No	Probably	But it must comply with IEE Regulations

Do I need approval to replace electric wiring?

No – but:

- you must comply with current IEE Regulations;
- your contract with the electricity supply company has conditions about safety which must not be broken. In particular, you should not interfere with the company's equipment which includes the cables to your consumer unit or up to and including the separate isolator switch if provided.

Do I need approval to replace an existing electrical fitting?

No – non-notifiable work (such as replacing an electrical fitting) can be completed by a DIY enthusiast (family member or friend) but needs to be installed in accordance with manufacturers' instructions and done in such a way that they do not present a safety hazard.

This work does **not** need to be notified to a local authority building control body (unless it is installed in an area of high risk such as a kitchen or a bathroom, etc.) **but** all DIY electrical work (unless completed by a qualified professional) will still need to be checked, certified and tested by a competent electrician.

Do I need approval to install a new electrical circuit?

Probably – any work that involves adding a new circuit to a dwelling will need to be either notified to the building control body (who will then inspect the work) or needs to be carried out by a competent person who is registered under a Government Approved Part P Self-Certification Scheme.

Work involving any of the following will also have to be notified:

- locations containing a bath tub or shower basin;
- swimming pools or paddling pools;
- hot air saunas;
- electric floor or ceiling heating systems;
- garden lighting or power installations;
- solar photovoltaic (PV) power supply systems;
- small-scale generators such as microCHP units;
- extra-low voltage lighting installations, other than pre-assembled, CE-marked lighting sets.

 Note: Where a person who is **not** registered to self-certify intends to carry out the electrical installation, then a Building Regulation (i.e. a building notice or full plans) application will need to be submitted together with the appropriate fee, based on the estimated cost of the electrical installation. The building control body will then arrange to have the electrical installation inspected at first fix stage and tested upon completion.

5.5 Plumbing

Planning permission	Building Regulation approval	
No	No	For replacements (but you will need to consult the technical services department for any installation which alters present internal or external drainage)
	Yes	For an unvented hot water system

Do I need approval to install hot water storage within my house, shop or flat?

Yes – if the water heater is unvented (i.e. supplied directly from the mains without an open expansion tank and with no vent pipe to atmosphere) and has storage capacity greater than 15 litres.

5.6 Central heating

Planning permission	Building Regulation approval
No	No If electric Yes If gas, solid fuel or oil

Do I need approval to alter the position of a heating appliance within my house, shop or flat?

- **Gas**: Yes, unless the work is supervised by an approved installer under the Gas Safety (Installation and Use) Regulations 1984.
- **Solid fuel**: Yes.
- **Oil**: Yes.
- **Electric**: Yes, unless the work is carried out by a competent person who is registered under a Government Approved Part P Self-Certification Scheme.

5.7 Oil-storage tank

Planning permission	Building Regulation approval
No Provided that it is in the garden and has a capacity of not more than 3500 litres (778 gallons) and **no** point is more than 3 m (9 ft 9″) high and **no** part projects beyond the foremost wall of the house facing the highway	No

Oil storage tanks, and the pipes connecting them to combustion appliances, should be constructed and protected so as to reduce the risk of the oil escaping and causing pollution.

5.8 Planting a hedge

Planning permission	Building Regulation approval
No Unless it obscures view of traffic at a junction or access to a main road	No

You do not need planning permission for hedges or trees. However, if there is a condition attached to the planning permission for your property which restricts the planting of hedges or trees (for example, on an 'open plan' estate or where a sight line might be blocked), you will need to obtain the council's consent to relax or remove the condition before planting a hedge or tree screen. If you are unsure about this, you can check with the planning department of your council.

Hedges should not be allowed to block out natural light, and the positioning of fast growing hedges should be checked with your local authority. Recent incidents regarding hedging of the fast growing Leylandii trees have led to changes in the planning rules, where hedges previously had no restrictive laws.

5.9 Building a garden wall or fence

Planning permission	Building Regulation approval
Yes If it is more than 1 m (3 ft 3″) high and is a boundary enclosure adjoining a highway	No
Yes If it is more than 2 m (6 ft 6″) high elsewhere	No

Do I need approval to build or alter a garden wall or boundary wall?

No – subject to size.

You will need to apply for planning permission if:

- your house is a listed building or in the curtilage of a listed building; or
- the fence, wall or gate would be over 1 metre high and next to a highway used for vehicles; or over 2 metres high elsewhere.

In normal circumstances, the only restriction on walls and fences is the height allowed. This is 2 metres or no more than 1 metre if the walls or fence is near a highway or road junction, where its height might obscure a driver's view of other traffic, pedestrians or road users.

If there is a valid reason for a wall or fence higher than the prescribed dimensions, then it is possible to get planning consent. There may be security issues that would support an application for a high fence. If it has no affect on other people's valid interests and does not impair any amenity qualities in an area, there is no reason why a request should be refused.

Some walls have historic value and they, as well as arches and gateways, can be listed. Modifications, extensions and removal of these must have planning consent.

5.10 Felling or lopping trees

Planning permission	Building Regulation approval
No Unless the trees are protected by a Tree Preservation Order or you live in a Conservation Area	No

Many trees are protected by Tree Preservation Orders (TPOs), which mean that, in general, you need the council's consent to prune or fell them. Nearly all trees in Conservation Areas are automatically protected.

Ask the council for a copy of the free leaflet *Protected Trees: a guide to tree preservation procedures.*

5.11 Laying a path or a driveway

Planning permission	Building Regulation approval
No Unless it provides access to a main road	No

Do I need to apply for planning permission to install a pathway?

Generally no – but you may need approval from the highways department if the pathway crosses a pavement.

Do I need to apply for planning permission to lay a driveway?

No – unless it adjoins the main road.

Driveways

Provided a pathway or drive does not meet a public thoroughfare you will not need planning consent. There are no restrictions on the area of land around your house that you can cover with hard surfaces.

You will need to apply for planning permission only if the hard surface is not to be used for domestic purposes and is to be used instead, for example, for parking a commercial vehicle or for storing goods in connection with a business.

In the case of hardstanding you do not need permission to gain access to it within the confines of your land, but you would need permission for a hardstanding leading on to a public highway.

You must obtain the separate approval of the highways department of your council if you want to make access to a roadway or if a new driveway would cross a pavement or verge. The exception is if the roadway is unclassified and the drive or footway is related to a development that does not require planning permission. Your local authority highways department will be able to tell you if a road is classified or unclassified. If the road is classified then, depending on the volumes of traffic, it is harder to get permission. The busier the road the less likely a new driveway or footway will be allowed to meet it.

If a driveway crosses a pedestrian access, pavement or roadside verge, then the planning department will gain approval from the highways department. If this is the case, highways approval is required in addition to planning consent. The basic principle is to maintain safety and eliminate hazards.

You will also need to apply for planning permission if you want to make a new or wider access for your driveway onto a trunk or other classified road. The highways department of your council can tell you if the road falls into this category.

Pathways

Pathways do not normally need planning permission and you can lay paths however you like in the confines of your own property. The exception is for any path making access to a highway or public thoroughfare, in which case certain safety aspects arise. You may also need permission if your building is listed or is in a Conservation Area, so the style and size is suitable for the area.

If a pathway crosses a pedestrian access, pavement or roadside verge, then the planning department will gain approval from the highways department. If this is the case, highways approval is required in addition to planning consent. The basic principle is to maintain safety and eliminate hazards.

5.12 Building a hardstanding for a car, caravan or boat

Planning permission	Building Regulation approval
No Provided that it is within your boundary and is not used for a commercial vehicle	No

Do I need to apply for planning permission to build a hardstanding for a car?

No – provided that it is within your boundary and is not used for a commercial vehicle.

Check local council rules.

Access from a new hardstanding to a highway requires planning consent. The exception is if the roadway is unclassified and the access to the hardstanding is related to a development that does not require planning permission. Your local authority highways department will be able to tell you if a road is classified or unclassified. If the road is classified then, depending on the volumes of traffic, it is harder to get permission. The busier the road the less likely a new driveway or footway will be allowed to meet it.

If the access crosses a pedestrian thoroughfare, pavement or roadside verge, then the planning department will gain approval from the highways department. If this is the case, highways approval is required in addition to planning consent. The basic principle is to maintain safety and eliminate hazards.

For a hardstanding on your own land, you do not need permission to gain access to it within the confines of your land, but you would need permission for a hardstanding leading on to a public highway.

There are different rules depending on what you use a hardstanding for. Planning permission is generally not needed provided there are no covenants limiting the installation of hardstanding for parking of cars, caravans or boats. There are still rules for commercial parking, however (e.g. taxis or commercial delivery vans) and a 'change of use' as a trade premises would probably need to be granted for this to be allowed.

You should check if there are any local covenants limiting changes in access to your premises or for hardstanding and parking of vehicles on it. If in doubt, contact the relevant local authority planning department for specific advice.

*Do I need to apply for planning permission to build
a hardstanding for a caravan and/or boat?*

Some local authorities do not allow the parking of caravans or boats on drive-
ways or hardstandings in front of houses. Check what the local rules are with
your planning department, and if there's no restriction then you don't need to
apply for permission.

There are no laws to prevent you, or your family from making use of a parked
caravan while it's on your land or drive, but you cannot actually live in it as this
would be classed as an additional dwelling. In addition, you cannot use a parked
caravan for business use as this would constitute a change of use of the property.

If you want to put a caravan on your land to lease out as holiday accommodation or for friends or family to stay in while they visit you, then this would
require planning permission. Rules on siting of static caravans or mobile homes
are quite stringent.

5.13 Installing a swimming pool

Planning permission		Building Regulation approval	
Possibly	Consult your local planning officer	Yes	For an indoor pool

 Swimming pools and saunas are subject to special requirements specified in
Part 6 of BS 7671:2001.

5.14 Erecting aerials, satellite dishes and flagpoles

Planning permission		Building Regulation approval
No	Unless it is a standalone antenna or mast greater than 3 m in height	No
Possibly	If erecting a satellite dish, especially in a Conservation Area or if it is a listed building (consult your local planning officer)	No

*Do I need to apply for planning permission to erect satellite
dishes, television and radio aerials and flagpoles?*

No – unless it is a stand-alone antenna or flagpole greater than 3 m in height.

Flagpoles etc. erected in your garden are treated under the same rules as outbuildings, and cannot exceed 3 metres in height.

Normally there is no need for planning permission for attaching an aerial or
satellite dish to your house or its chimneys. However, if it rises significantly

higher than the roof's highest point then it may contravene local regulations or covenants.

You should get specific advice if you plan to install a large satellite dish or aerial, such as a short wave mast, as the rules differ between authorities.

In certain circumstances, you will need to apply for planning permission to install a satellite dish on your house (see DTE's free booklet *A Householder's Planning Guide for the Installation of Satellite Television Dishes*, which can be obtained from your local council).

Conservation Areas have specific local rules on aerials and satellite dishes, so you need to approach your local planning department to find out the particular rules for your area. Certainly, if your house is a listed building, you may need listed building consent to install a satellite dish on your house.

Remember, if you are a leaseholder, you may need to obtain permission from the landlord.

5.15 Advertising

Planning permission		Building Regulation approval	
No	If the advertisement is less than 0.3 m² and not illuminated	Possibly	Consult your local planning officer

Do I need to apply for planning permission to erect an advertising sign?

Advertisement signs on buildings and on land often need planning consent. Some smaller signs and non-illuminated signs may not need consent, but it is always advisable to check with development control staff.

You are allowed to display certain small signs at the front of residential premises such as election posters, notices of meetings, jumble sales, car for sale etc. but business types of display and permanent signs may need to have planning permission granted. They may come under the category of 'advertising control' for which planning consent is required.

You may need to apply for advertisement consent to display an advertisement bigger than 0.3 m² on the front of, or outside, your property. This includes your house name or number or even a sign saying 'Beware of the dog'. Temporary notices up to 0.6 m² relating to local events, such as fêtes and concerts, may be displayed for a short period. There are different rules for estate agents' boards, but, in general, these should not be bigger than 0.5 m² on each side.

It is illegal to post notices on empty shops' windows, doors, and buildings, and also on trees. This is commonly known as 'fly posting' and can carry heavy fines under the Town and Country Planning Act.

Illuminated signs and all advertising signs outside commercial premises need to be approved. Most local authorities can give advice, by way of booklets or leaflets on what kinds of sign are allowed, not allowed or need approval.

 You can get advice from the planning department of your local council; ask for a copy of the free booklet *Outdoor advertisements and signs*.

5.16 Building a porch

Planning permission	Building Regulation approval
No Unless: • the floor area exceeds 3 m² (3.6 y²) • any part is more than 3 m (9 ft 9″) high • any part is less than 2 m (6 ft 6″) from a boundary adjoining a highway or public footpath	Yes If area exceeds 30 m² (35.9 y²)

Do I need planning permission for a porch?

Yes – depending on its size and position.

You will need to apply for planning permission:

- if your house is listed or is in a Conservation Area, national park, area of outstanding natural beauty;
- if the porch would have a ground area (measured externally) of more than 3 m²;
- if the porch would be higher than 3 m above ground level;
- if the porch would be less than 2 m away from the boundary of a dwelling house with a highway (which includes all public roads, footpaths, bridleways and byways).

 All measurements are taken externally.

However, a porch or conservatory built at ground level and under 30 m² in floor area is exempt provided that the glazing complies with the safety glazing requirements of the Building Regulations (Part N). Your local authority building control department or an approved inspector can supply further information on safety glazing. It is advisable to ensure that a conservatory is not constructed so that it restricts ladder access to windows serving a room in the roof or a loft conversion, particularly if that window is needed as an emergency means of escape in the case of fire.

 The regulations are quite complicated and depend on previous works on the site, if any, so you should always check with development control staff.

5.17 Outbuildings

Planning permission		Building Regulation approval	
Possibly	Provided the building is less than 10 m³ (13.08 y³) in volume, not within 5 m (16 ft 3″) of the house or an existing extension	Yes	If area exceeds 30 m² (35.9 y²)
	Erecting outbuildings can be a potential minefield and it is best to consult the local planning officer before commencing work		If it is within 1 m (3 ft 3″) of a boundary, it must be built from incombustible materials

Many kinds of buildings and structures can be built in your garden or on the land around your house without the need to apply for planning permission. These can include sheds, garages, greenhouses, accommodation for pets and domestic animals (e.g. chicken houses), summer houses, swimming pools, ponds, sauna cabins, enclosures (including tennis courts) and many other kinds of structure.

Outbuildings intended to go in the garden of a house do not normally require any planning permission, so long as they are associated with the residential amenities of the house and a few requirements are adhered to such as position and size.

You can build an outbuilding up to 10 m³ (13.08 y³) in volume without planning permission if it is within 5 m (16 ft 3″) of the house or an extension. Further away than this, it can be up to half the area of the garden, but the height must not exceed 4 m (13 ft).

If your new building exceeds 10 m³ (and/or comes within 5 m of the house) it would be treated as an extension and would count against your overall volume entitlement.

There are a few conditions to follow in order to avoid the need for planning consent:

- The structure should not result in more than half the original garden space being covered by the building.
- No part of the structure should extend beyond the original house limits on any side facing a public highway or footpath or service road.
- The height should not exceed 3 m (or 4 m if it has a ridged roof).

If your house is listed or is in a Conservation Area, national park, or area of outstanding natural beauty, then you will more than likely need to obtain planning consent. If in doubt, contact the relevant local authority planning department for specific advice.

Permission is required, however, for:

- any building/structure nearer to a highway than the nearest part of the original house, unless more than 20 m away from a highway;
- structures not required for domestic use;

- structures over 3 m high (or 4 m if it has a ridged roof);
- propane gas (LPG) tank;
- storage tank holding more than 3500 litres;
- a building or structure which would result in more than half of the grounds of your house being covered by buildings/structures.

You will also need to apply for planning permission if any of the following cases apply:

- You want to put up a building or structure which would be nearer to any highway than the nearest part of the original house, unless there would be at least 20 m between the new building and any highway. The term 'highway' includes public roads, footpaths, bridleways and byways.
- More than half the area of land around the original house would be covered by additions or other buildings.
- The building or structure is not to be used for domestic purposes and is to be used instead, for example, for parking a commercial vehicle, running a business or for storing goods in connection with a business.
- You want to put up a building or structure which is more than 3 m high, or more than 4 m high if it has a ridged roof (measured from the highest ground next to it).
- If your house is a listed building and you want to put up a building or structure with a volume of more than $10\,m^3$.

External water storage tanks

Many years ago the demand for external tanks for capturing rainwater made their installation quite commonplace. But it is rare today to need extra storage tanks, unless you are in a rural position.

If you are considering installing an external water tank you should seek guidance from your local authority, especially if the tank is to be mounted on a roof.

Fuel storage tanks

Storage of oil, or any other liquids, especially petrol, diesel and chemicals is strictly controlled and would not be allowed on residential premises. If you are considering installing an external oil storage tank for central heating use, then no planning permission is required, provided its capacity is no more than 3500 litres, it is no more than 3 m from the ground and it does not project beyond any part of a building facing a public thoroughfare.

You will need to apply for planning permission in the following circumstances:

- You want to install a storage tank for domestic heating oil with a capacity of more than 3500 litres or a height of more than 3 m above ground level.
- You want to install a storage tank, which would be nearer to any highway than the nearest part of the 'original house', unless there would be at least

20 m between the new storage tank and any highway. The term 'highway' includes public roads, footpaths, bridleways and byways.

- You want to install a tank to store Liquefied Petroleum Gas (LPG) or any liquid fuel other than oil.

 Erecting any type of outbuilding can be a potential minefield and it is best to consult with the local planning officer before commencing work.

5.18 Garages

Planning permission		Building Regulation approval
Possibly	You can build a garage up to 10 m³ (13.08 y³) in volume without planning permission, if it is within 5 m (16 ft 3″) of the house or an existing extension. Further away from this, it can be up to half the area of the garden, but the height must not exceed 4 m (13 ft)	Yes

Do I need approval to build a garage extension to my house, shop or office?

Yes – but a carport extension built at ground level, open on at least two sides and under 30 m² in floor area, is exempt.

Do I need approval for a detached garage?

Yes – but a single storey garage at ground level, under 30 m² in floor area and with no sleeping accommodation, is exempt provided it is either built mainly using non-combustible material or, when built, it has a clear space of at least 1 m from the boundary of the property.

Garages planned to go in the garden of a house do not normally require any planning permission, so long as they are associated with the residential amenities of the house and a few requirements are adhered to such as position and size.

There are a few conditions to follow in order to avoid the need for planning consent, such as:

- the structure should not result in more than half the original garden space being covered by the building;
- no part of the structure should extend beyond the original house limits, on any side facing a public highway or footpath or service road;
- the height should not exceed 3 m (or 4 m if it has a ridged roof).

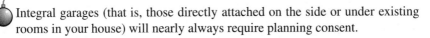 Integral garages (that is, those directly attached on the side or under existing rooms in your house) will nearly always require planning consent.

5.19 Building a conservatory

Planning permission		Building Regulation approval	
Possibly	You can extend your house by building a conservatory, provided that the total of both previous and new extensions does not exceed the permitted volume	Yes	If area exceeds $30\,m^2$ $(35.9\,y^2)$

Do I need permission to erect a conservatory?

Possibly – see below.

Conservatories and sun lounges attached to a house are classed as extensions. If you want a conservatory or sun lounge separated from the house, this needs planning consent under similar rules for outbuildings.

If the answer to all the following questions is **no** then it is quite likely that planning permission will not be required:

- Is the conservatory going to be used as a separate dwelling, i.e. self-contained accommodation?
- Is your property listed, in a Conservation Area, national park or area of outstanding natural beauty?
- Will the conservatory cover more than half of the original garden space?
- Will any part of the conservatory within 2 m of the plot boundary be more than 4 m above ground level?
- Will any part of the conservatory be higher than the original roof of the main building?
- Will any part of the conservatory be nearer to a service road, public road or footpath than any part of the original building?
- Will the volume of the original house be increased so that any part of the conservatory within 2 metres of the plot boundary is more than 4 metres above ground level?
- Will the conservatory be behind the building line?
- Will your conservatory be 1 m away from the boundary (although most buildings tend to be nearer than this)?
- Will the conservatory be more than 50 ft away from the nearest road?

A conservatory has to be separated from the rest of the house to be exempt (i.e. patio doors).

Another thing to keep in mind is your neighbours' reaction – always keep them informed of what's happening and be prepared to alter the plans you had for locating the building if they object – it's better in the long run, believe me.

Will you need planning permission therefore? Generally no, as the building is classed as a 'portable building', nevertheless it is *your* responsibility to check with your local planning office.

However, a porch or conservatory built at ground level and under $30\,\mathrm{m}^2$ in floor area is exempt provided that the glazing complies with the safety glazing requirements of the Building Regulations (Part N). Your local authority building control department or an approved inspector can supply further information on safety glazing. It is advisable to ensure that a conservatory is not constructed so that it restricts ladder access to windows serving a room in the roof or a loft conversion, particularly if that window is needed as an emergency means of escape in the case of fire.

The regulations are quite complicated and depend on previous works on the site, if any, so you should always check with development control staff.

5.20 Loft conversions, roof extensions and dormer windows

Planning permission		Building Regulation approval
No	Provided the volume of the house is unchanged and the highest part of the roof is not raised	Yes
Yes	For front elevation dormer windows or rear ones over a certain size	Yes

Do I need approval for a loft conversion?

Yes – see below.

Do I need to apply for planning permission to re-roof my house?

No – unless you live in a Conservation Area, a national park, an area of outstanding natural beauty or the Norfolk Broads.

Do I need to apply for permission to insert roof lights or skylights?

No.

Do I need to apply for planning permission to extend or add to my house?

Yes – in the following circumstances:

- If you want to build an addition or extension to any roof slope which faces a highway.
- If the roof extension would add more than $40\,\mathrm{m}^3$ to the volume of a terraced house or more than $50\,\mathrm{m}^3$ to any other kind of house.

These volume limits count as part of the allowance for extending the property (see extensions above).

- If the work would increase the height of the roof.
- If the intention is to create a separate dwelling, such as self-contained living accommodation or a granny flat.

If the answer to all the following questions is **no**, therefore, then it is quite likely that planning permission will not be required:

- Is the loft conversion going to be used as a separate dwelling, i.e. a self-contained flat?
- Is your property listed, in a Conservation Area, national park or area of outstanding natural beauty?
- Will any part of the loft conversion within 2 m of the plot boundary be more than 4 m above ground level?
- Will any part of the loft conversion be higher than the original roof of the main building?
- Will any part of the loft conversion be nearer to a service road, public road or footpath than any part of the original building?
- Will the roof be extended where it faces a public highway?
- Will the volume of the original house be increased beyond the following limits?
  - if the house is in a terrace, a Conservation Area, a national park, or an area of outstanding natural beauty 40 m³ or 10% whichever is the greater, up to a maximum of 115 m³.
  - for any other kind of house 50 m³ or 15% whichever is the greater, up to a maximum of 115 m³.

 The volume is calculated from external measurement.

Do I need to apply for planning permission to alter a roof?

You will need to apply for planning permission if you live in a Conservation Area, a national park, an area of outstanding natural beauty or the Norfolk Broads and you want to build an extension to the roof of your house or any kind of addition that would materially alter the shape of the roof.

Roofs are expected to match those of the surrounding area, so consider this if you live in a protected area. Some areas require that the colour and style of the roof covering matches the original, and the pitch and construction should be the same. If you plan to save the expense of matching the roof, by opting for a flat roof, be sure that your local authority will accept this. Often high flat roofs are not desirable, due to the appearance of the house elevation. Provided the alterations to your roof do not make a noticeable change or don't increase its height you would normally not need to obtain planning permission.

 You do not normally need to apply for planning permission to re-roof your house or for the insertion of roof lights or skylights.

In the case of re-roofing, if the tiles are the same type then no approval is needed. If the new tiling or roofing material is substantially heavier or lighter than the existing material, or if the roof is thatched or is to be thatched where previously it was not, then an approval under Building Regulations is probably required.

5.21 Building an extension

Planning permission		Building Regulation approval	
Possibly	You can extend your house by building an extension, provided that the total of both previous and new extensions does not exceed the permitted volume	Yes	If area exceeds 30 m² (35.9 y²)

Do I need approval to build an extension to my house?

Yes – if it would '*materially alter the appearance of the building*'.

Major alteration and extension nearly always need approval. However some small extensions such as porches, garages and conservatories may be 'Permitted Development' and, therefore, do not need planning consent.

Building extensions can be a potential minefield and it is best to consult the local planning officer before contemplating any work.

If the answer to all the following questions is **no**, then it is quite likely that planning permission will not be required to build an extension to your house:

- Is the extension going to be used as a separate dwelling, i.e. a self-contained flat?
- Is the property listed?
- Will the extension cover more than half of the original garden space?
- Will any part of the extension within 2 m of the plot boundary be more than 4 m above ground level?
- Will any part of the extension be higher than the original roof of the main building?
- Will any part of the extension be nearer to a service road, public road or footpath than any part of the original building?
- Will the volume of the original house be increased beyond the following limits?
  - If the house is in a terrace, a Conservation Area, a national park, or an area of outstanding natural beauty – 50 m³ (or 10% whichever is the greater) up to maximum of 115 m³.
  - For any other kind of house, 70 m³ (or 15% whichever is the greater) up to a maximum of 115 m³.

The volume is calculated from external measurements.

Note: If a building is extended, or undergoes a material alteration, the completed building must comply with the relevant requirements of the Approved Documents or, where this is not feasible, be '*no more unsatisfactory than before*'.

You **will** need to apply for planning permission to extend or add to your house. You may also require planning permission if your house has previously been added to or extended. You may also require planning permission if the original planning permission for your house imposed restrictions on future development. (i.e. permitted development rights may have been removed by an 'Article 4 direction'. This is often the case with more recently constructed houses.)

You **will** also require planning permission if you want to make additions or extensions to a flat or maisonette.

 Check with your local authority planning department if you are not sure.

You will, therefore, need to apply for planning permission:

- if an extension to your house comes within 5 m of another building belonging to your house (i.e. a garage or shed). The volume of that building counts against the allowance given above.
- for all additional buildings which are more than $10\,m^3$ in volume, if you live in a Conservation Area, a national park, an area of outstanding natural beauty or the Norfolk Broads. Wherever they are in relation to the house, these buildings will be treated as extensions of the house and reduce the allowance for further extensions.
- for a terraced house, end-of-terrace house, or any house in a Conservation Area, national park, an area of outstanding natural beauty or the Broads – where the volume of the original house would be increased by more than 10% or $50\,m^3$ (whichever is the greater).
- for any other type of house (i.e. detached or semi-detached) the volume of the original house would be increased by more than 15% or $70\,m^3$ (whichever is the greater). In any case the volume of the original house would be increased by more than $115\,m^3$.
- for alterations to the roof, including dormer windows (but permission is not normally required for skylights).
- for extensions nearer to a highway than the nearest part of the original house (unless the house, as extended, would be at least 20 m away from the highway).
- to extend or add to your house so as to create a separate dwelling, such as self-contained living accommodation or a granny flat.
- if an extension to your house comes within 5 m of another building belonging to your house.
- to build an addition, which would be nearer to any highway than the nearest part of the original house, unless there would be at least 20 m between your house (as extended) and the highway. The term 'highway' includes all public roads, footpaths, bridleways and byways.

- if more than half the area of land around the original house would be covered by additions or other buildings – although you may not have built an extension to the house, a previous owner may have done so.
- if the extension or addition exceeds the certain limits on height and volume.
- if the extension is higher than the highest part of the roof of the original house or any part of the extension is more than 4 m high and is within 2 m of the boundary of your property.

Any building which has been added to your property and which is more than 10 m³ in volume and which is within 5 m of your house is treated as an extension of the house and so reduces the allowance for further extensions without planning permission.

Where the word 'original' is used above, in planning regulations terms this means the house as it was first built or as it was on 1 July 1948. Any extensions added since that date are counted towards the allowances.

Limitations
Planning permission is required if:

- **Height** – Any part is higher than the highest part of the house roof.
- **Projections** – Any part projects beyond the foremost wall of the house facing a highway.
- **Boundary** – Any part within 2 m (6 ft 6″) of a boundary is more than 4 m (13 ft) high.
- **Area** – It will cover more than half the original area of the garden.
- **Dwelling** – It is to be an independent dwelling.

You should measure the height of buildings from the ground level immediately next to it. If the ground is uneven, you should measure from the highest part of the surface, unless you are calculating volume.

- **Volume** – Planning permission is required if the extension results in an increase in volume of the original house by whichever is the greater of the following amounts:
  - for terraced houses 50 m³ (65.5 y³) or 10% up to a maximum of 115 m³ (150.4 y³);
  - other houses 70 m³ (91.5 y³) or 15% up to a maximum of 115 m³ (150.4 y³);
  - in Scotland, general category 24 m² (28.7 y²) or 20%.

The volume of other buildings which belong to your house (such as a garage or shed) will count against the volume allowances. In some cases, this can include buildings that were built at the same time as the house or that existed on 1 July 1948.

5.21.1 Extensions to non-domestic buildings

With the new revision of Part M, an extension to a non-domestic building should now be treated in the same manner as a new building for compliance, which means that:

- there must be *'suitable independent access to the extension where reasonably practicable'*;
- if a building is to be extended, *'reasonable provision must be made within the extension for sanitary conveniences'*.

 Note: This requirement does not apply if it is possible for people using the extension to gain access to and be able to use sanitary conveniences in the existing building.

5.22 Conversions

Planning permission	Building Regulation approval
Yes For flats – even where construction works may not be intended	Yes Unless you are not proposing any building work to make the change
Yes For shops and offices unless no building work is envisaged	

Do I need approval to convert my house into flats?

Yes – even where construction works may not be intended.

Do I need approval to convert my house to a shop or office?

No – if you are not proposing any building work to make the change.

Do I need approval to convert part or all of my shop or office to a flat or house?

Yes.

Where building work is proposed you will probably need approval if it affects the structure or means of escape in case of fire. But you should check with the local fire authority and the county council, to see whether a fire certificate is actually required.

 You will probably also need planning permission whether or not building work is proposed.

5.22.1 Converting an old building

Planning permission	Building Regulation approval
Yes	Yes

Do I need planning permission to convert an old building?

Yes.

Throughout the UK there are many under-used or redundant buildings, particularly farm buildings which may no longer be required, or suitable, for agricultural use. Such buildings of weathered stone and slate contribute substantially to the character and appearance of the landscape and the built environment. Their interest and charm stems from an appreciation of the functional requirements of the buildings, their layout and proportions, the type of building materials used and their display of local building methods and skills.

In most cases traditional buildings are best safeguarded if their original use can be maintained. However with changing patterns of land use and farming methods, changes of use or conversion may have to be considered.

The conversion or re-use of traditional buildings may, in the right locations, assist in providing employment opportunities, housing for local people, or holiday accommodation. Applicants and developers are encouraged to refer to the local plan for comprehensive guidance and to seek advice from a Planning Officer if further assistance is necessary.

All councils place the highest priority to good design and proposals. Those that fail to respect the character and appearance of traditional buildings, will not be permitted. Sensitive conversion proposals should ensure that existing ridge and eaves lines are preserved; new openings avoided as far as possible; traditional matching materials are used; and the impact of parking and garden areas is minimized. Buildings that are listed as being of 'special architectural or historic interest' require skilled treatment to conserve internal and external features.

In many instances, traditional buildings that are of simple, robust form with few openings may only be suitable for use as storage or workshops. Other uses, such as residential, may be inappropriate.

5.23 Change of use

Planning permission		Building Regulation approval
Possibly	Even if no building or engineering work is proposed	Yes

The use of buildings or land for a different purpose may need consent even if no building or engineering works are proposed. Again, it is always advisable to check with development control staff.

What is meant by material change of use?

A material change of use is where there is a change in the purposes for which or the circumstances in which a building is used, so that after that change:

(*a*)   *the building is used as a dwelling, where previously it was not*;

(*b*)   *the building contains a flat, where previously it did not*;

(c)   *the building is used as an hotel or a boarding house, where previously it was not;*

(d)   *the building is used as an institution, where previously it was not;*

(e)   *the building is used as a public building, where previously it was not;*

(f)   *the building is not a building described in Classes I to VI in Schedule 2, where previously it was;*

(g)   *the building, which contains at least one dwelling, contains a greater or lesser number of dwellings than it did previously;*

(h)   *the building contains a room for residential purposes, where previously it did not;*

(i)   *the building, which contains at least one room for residential purposes, contains a greater or lesser number of such rooms than it did previously; or*

(j)   *the building is used as a shop, where previously it was not.*

 'Public building' means a building consisting of or containing:

- a theatre, public library, hall or other place of public resort;
- a school or other educational establishment;
- a place of public worship.

Material changes of use

Where there is a material change of use of a **whole** building to a hotel, boarding house, institution, public building or a shop (restaurant, bar or public house) the building must be upgraded, if necessary, so as to comply with Approved Document M1 (Access and use).

If an existing building undergoes a change of use so that **part** of it can be used as a hotel, boarding house, institution, public building or a shop, the work being carried out must ensure that:

- people can gain access from the site boundary and any on-site car parking space;
- sanitary conveniences are provided in that part of the building or it is possible for people (no matter their disability) to use sanitary conveniences elsewhere in the building.

Material alterations of non-domestic buildings

Under regulation 4, where an alteration of a non-domestic building is a material alteration:

- the work itself must comply, where relevant, with Requirement M1;
- reasonable provision must be made for people to gain access to and to use new or altered sanitary conveniences.

Extensions, material alterations or a material change of use

Where any electrical installation work is classified as an extension, a material alteration or a material change of use, the work must consider and include:

- confirmation that the mains supply equipment is suitable and can carry the additional loads envisaged;
- the rating and the condition of existing equipment (belonging to both the consumer and the electricity distributor) are sufficient;
- the amount of additions and alterations that will be required to the existing fixed electrical installation in the building;
- the necessary additions and alterations to the circuits which feed them;
- the protective measures required to meet the requirements;
- the earthing and bonding systems are satisfactory and meet the requirements.

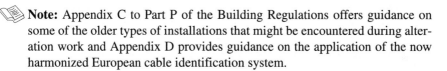 **Note:** Appendix C to Part P of the Building Regulations offers guidance on some of the older types of installations that might be encountered during alteration work and Appendix D provides guidance on the application of the now harmonized European cable identification system.

What are the requirements relating to material change of use?

Where there is a material change of use of the whole of a building, any work carried out shall ensure that the building complies with the applicable requirements of the following paragraphs of Schedule 1:

(a) in all cases:
- means of warning and escape (B1)
- internal fire spread – linings (B2)
- internal fire spread – structure (B3)
- external fire spread – roofs (B4)(2)
- access and facilities for the fire service (B5)
- resistance to moisture (C1)(2)
- dwelling-houses and flats formed by material change of use (E4)
- ventilation (F1)
- sanitary conveniences and washing facilities (G1)
- bathrooms (G2)
- foul water drainage (H1)
- solid waste storage (H6)
- combustion appliances (J1, J2 & J3)
- conservation of fuel and power – dwellings (L1)
- conservation of fuel and power – buildings other than dwellings (L2)
- electrical safety (P1, P2).

In the case of a building exceeding 15 metres in height:
- external fire spread – walls (B4–(1)).

(b) in other cases:

Material change of use	Requirement	Approved Document
The building is used as a dwelling, where previously it was not	Resistance to moisture	C2 E1, E2, E3
The public building consists of a new school	Acoustic conditions in schools	E4
The building contains a flat, where previously it did not	Resistance to the passage of sound	E1, E2 & E3
The building is used as a hotel or a boarding house, where previously it was not	Structure	A1, A2 & A3 E1, E2, E3
The building is used as an institution, where previously it was not	Structure	A1, A2 & A3
The building is used as a public building, where previously it was not		A1, A2 & A3 E1, E2, E3
The building is not a building described in Classes I to VI in Schedule 2, where previously it was	Structure	A1, A2 & A3
The building, which contains at least one room for residential purposes, contains a greater or lesser number of dwellings than it did previously	Structure	A1, A2 & A3 E1, E2, E3
The building, which contains at least one dwelling, contains a greater or lesser number of dwellings than it did previously	Resistance to the passage of sound	E1, E2 & E3

In some circumstances (particularly when a historic building is undergoing a material change of use and where the special characteristics of the building need to be recognized) it may **not** be practical to improve sound insulation to the standards set out in Part E1 or resistance to contaminants and water as set out in Part C. In these cases, the aim should be to improve the insulation and resistance where it is practically possible – always provided that the work does not prejudice the character of the historic building, or increase the risk of long-term deterioration to the building fabric and/or fittings.

 Note: BS 7913:1998 *The principles of the conservation of historic buildings* provides guidance on the principles that should be applied when proposing work on historic buildings.

Mixed use development

In mixed use developments the requirements of the Regulations may differ depending on whether it is part of a building used as a dwelling or part of a building which has a non-domestic use. In these cases the requirements for non-domestic use shall apply in any shared parts of the building.

5.23.1 Buildings suitable for conversion

Most local plans stipulate that conversion proposals '*should relate to buildings of traditional design and construction which enhance the natural beauty of the landscape*' as opposed to '*non-traditional buildings, buildings of inappropriate design, or buildings constructed of materials which are of a temporary nature*'.

Isolated buildings

Planning permission will not normally be granted for the conversion or re-use of isolated buildings. Exceptionally, permission may be given for such buildings to be used for small-scale storage or workshop uses or for camping purposes.

An isolated building is normally:

- *a building, or part of a building, standing alone in the open countryside; or*
- *a building, or part of a building, comprised within a group which otherwise occupies a remote location having regard to the disposition of other buildings within the locality, to the character of the surroundings, and to the nature and availability of access and essential services.*

Assessing whether or not a particular building should be regarded as isolated may not always be straightforward and, in such instances early discussion with a planning officer at the national park authority is advised.

Structural condition

Buildings proposed for conversion should be large enough to accommodate the proposed use without the necessity for major alterations, extension or re-construction. In cases of doubt regarding the structural condition of any particular building, the authority will require the submission of a full structural survey to accompany a planning application. The authority can advise on this requirement and, if necessary, on persons who are suitably qualified to undertake such work and who practise locally.

 Planning permission will not normally be granted for re-construction if substantial collapse occurs during work on the conversion of a building.

 A list of local consulting engineers can be found in *Yellow Pages*.

Workshop conversions

Redundant farm buildings and buildings of historic interest are often well suited to workshop use and such conversions normally require minimal alterations.

Potential problems of traffic generation and unneighbourliness can usually be addressed by the imposition of appropriate conditions.

The local authority will generally favourably consider proposals that make good use of traditional buildings by promoting local employment opportunities. In some instances grants may be available from other agencies to assist the conversion of buildings to workshop use.

Residential conversions

When reviewing proposals for converting a traditional building, the local authority will pay particular attention to the overall objectives of the housing policies of the local plan. If land that can be used for a new housing development is limited, residential conversions can make a valuable contribution to the local housing stock and support the social and economic well being of rural communities.

The local plan will require that residential conversions should, in most instances, contribute to the housing needs of the locality. Permission for such conversions are, in some districts, only granted subject to a condition restricting occupancy to local persons.

 'Local persons' are normally defined as persons working, about to work, or having last worked in the locality or who have resided for a period of three years within the locality.

Renovation

Districts dedicate some areas as Environmentally Sensitive Areas (ESAs) and grants may be available towards the cost of renovating historical and important local buildings that have fallen into disrepair or towards the cost of renovation works to retain agricultural buildings in farming use, so as to retain their importance as landscape features. Further advice on the workings of the scheme may be obtained from the ESA project officers or the authority's building conservation officer.

Applicants are strongly advised to employ qualified architects or designers in preparing conversion proposals. Informal discussions with a planning officer at an early stage in considering design solutions are also encouraged.

5.24 Building a new house

Planning permission	Building Regulation approval
Yes	Yes

Do I need planning permission to erect a new house?

Yes.

All new houses or premises of any kind require planning permission.

Private individuals will normally only encounter this if they intend to buy a plot of land to build on, or buy land with existing buildings that they want to demolish to make way for a new property to be built.

In all cases like this, unless you are an architect or a builder, you **must** seek professional advice. If you are using a solicitor to act on your behalf in purchasing a plot on which to build, he will include the planning questions within all the other legal work, as well as investigating the presence of covenants, existing planning consent together with other constraints or conditions.

The architect, surveyor or contractor you hire will then need to take into account the planning requirements as part of their planning and design procedures. They will normally handle planning applications for any type of new development.

 If you are hiring a professional (or more than one – say a building contractor to do the work and a surveyor or architect to plan and design) be sure to find out exactly who does what and that approval is obtained before going to too much expense, should a refusal arise.

5.25 Infilling

Planning permission	Building Regulation approval
Possibly Consult your local planning officer	Yes If a new development

Can I use an unused, but adjoining, piece of land to build a house (e.g. build a new house on land that used to be a large garden)?

Often there may be no official grounds for denying consent, but residents and individuals can impose quite some delay. It is worth testing the likelihood of a successful application by talking to the neighbours and judging opinions.

Planning consent is often quite difficult to obtain in these cases as this sort of development normally causes a lot of opposition as it is in a settled residential area and people do not like change.

New developments will undoubtedly also need to follow building regulations. This, and all site visits from inspectors, is normally arranged by your building contractor.

 There are plenty of substantial building projects that don't require any planning permission. However, it is undoubtedly a good idea to consult a range of people before you consider any work.

5.26 Demolition

Planning permission		Building Regulation approval	
Yes	If it is a listed building or in a Conservation Area If the whole house is to be demolished	No	For a complete detached house
		Yes	For a partial demolition to ensure that the remaining part of the house (or adjoining buildings/extensions) are structurally sound

You must have good reasons for knocking a building down, such as making way for rebuilding or improvement (which in most cases would be incorporated in the same planning application). Penalties are severe for demolishing something illegally.

You do not need to make a planning application to demolish a listed building or to demolish a building in a Conservation Area. However, you may need listed building or Conservation Area consent.

Elsewhere, you will **not** need to apply for planning permission:

- to demolish a building such as a garage or shed of less than 50 m³; or
- if the demolition is urgently necessary for health and safety reasons; or
- if the demolition is required under other legislation; or
- where the demolition is on land that has been given planning permission for redevelopment; or
- to demolish a gate, fence, wall or other means of enclosure.

In all other cases, such as demolishing a house or block of flats, the council may wish to agree the details of how you intend to carry out the demolition and how you propose to restore the site afterwards. You will need to apply for a formal decision on whether the council wishes to approve these details. This is called a 'prior approval application' and your council will be able to explain what it involves.

You are not allowed to begin any demolition work (even on a dangerous building) unless you have given the local authority notice of your intention and this has either been acknowledged by the local authority or the relevant notification period has expired. In this notice you will have to:

- specify the building to be demolished;
- state the reason(s) for wanting to demolish it;
- show how you intend to demolish it.

Copies of this notice will have to be sent to:

- the local authority;
- the occupier of any building adjacent to the building;
- British Gas;
- the area electricity board in whose area the building is situated.

This regulation does not apply to the demolition of an internal part of an occupied building, or a greenhouse, conservatory, shed or prefabricated garage (that forms part of that building) or an agricultural building defined in Section 26 of the General Rate Act 1967.

5.26.1 What about dangerous buildings? (Building Act 1984 Sections 77 and 78)

If a building, or part of a building or structure, is in such a dangerous condition (or is used to carry loads that would make it dangerous) then the local authority may apply to a magistrates' court to make an order requiring the owner:

- to carry out work to avert the danger;
- to demolish the building or structure, or any dangerous part of it, and remove any rubbish resulting from the demolition.

 The local authority can also make an order restricting its use until such time as a magistrates' court is satisfied that all necessary works have been completed.

These works are controllable by the local authority under Sections 77 and 78 of the Building Act 1984. In inner London the legislation is under the London Building (Amendment) Act 1939.

This involves responding to all reported instances of dangerous walls, structures and buildings within each local authority's area on a 24 hour 365 days a year basis.

 Refer to the relevant local authority building control office during office hours or their local authority emergency switchboard, out of hours.

If the building or structure poses a potential danger to the safety of people, the local authority will take the appropriate action to remove the danger. The local authority has powers to require the owners of buildings or structures to remedy the defects or they can direct their own contractors to carry out works to make the building or structure safe. In addition, the local authority may provide advice on the structural condition of buildings during fire fighting to the fire brigade.

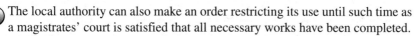 If you are concerned that a building or other structure may be in a dangerous condition, then you should report it to the local council.

Emergency measures

In emergencies the local authority can make the owner take immediate action to remove the danger, or they can complete the necessary action themselves. In these cases, the local authority is entitled to recover from the owner such expenses reasonably incurred by them. For example:

- fencing off the building or structure;
- arranging for the building/structure to be watched.

5.26.2 Can I be made to demolish a dangerous building?
(Building Act 1984 Sections 81, 82 and 83)

If the local authority considers that a building is so dangerous that it should be demolished, they are entitled to issue a notice to the owner requiring the owner/occupier:

- to shore up any building adjacent to the building to which the notice relates;
- to weatherproof any surfaces of an adjacent building that are exposed by the demolition;
- to repair and make good any damage to an adjacent building caused by the demolition or by the negligent act or omission of any person engaged in it;
- to remove material or rubbish resulting from the demolition and clear the site;
- to disconnect, seal and remove any sewer or drain in or under the building;
- to make good the surface of the ground that has been disturbed in connection with this removal of drains etc.;
- in accordance with the Water Act 1945 (interference with valves and other apparatus) and the Gas Act 1972 (public safety), arranging with the relevant statutory undertakers (e.g. water board, British Gas or electricity supplier) for the disconnection of gas, electricity and water supplies to the building;
- to leave the site in a satisfactory condition following completion of all demolition work.

Before complying with this notice, the owner must give the local authority 48 hours' notice of commencement.

In certain circumstances, the owner of an adjacent building may be liable to assist in the cost of shoring up their part of the building and waterproofing the surfaces. It could be worthwhile checking this point with the local authority!

Under Section 80 of the Building Act 1984 anyone carrying out demolition work is required to notify the local authority. The local authority then has 6 weeks to respond with appropriate notices and consultation under Sections 81 and 82 of the Act (this does not apply to inner London).

Replacing a demolished building

If you decide to demolish a building, even one that has suffered fire or storm damage, it does not automatically follow that you will get planning permission to build a replacement.

6

Meeting the requirements of the Building Regulations

Background

The Building Regulations 2000 as amended by the Building Amendment Regulations 2001 (SI 2001/3335) replaced the Building Regulations 1991 (SI 1985 No. 1065). Since then, a series of Approved Documents have been endorsed by the Secretary of State that are intended to provide guidance to some of the more common building situations. They also provide a practical guide to meeting the requirements of Schedule 1 and Regulation 7 of the Building Regulations.

Approved Documents

The 2003 list of Approved Documents is given in Table 6.1 below.

Table 6.1 Approved Documents 2004

Section	Title	Edition	Latest amendment
A	Structure	2004	
B	Fire safety	2000	2002
C	Site preparation and resistance to moisture	2004	
D	Toxic substances	1992	2000
E	Resistance to the passage of sound	2003	2004
F	Ventilation	1995	2000
G	Hygiene	1992	2000
H	Drainage and waste disposal	2002	
J	Combustion and waste disposal	2002	
K	Protection from falling, collision and impact	1998	2000
L1	Conservation of fuel and power in dwellings	2002	
L2	Conservation of fuel and power in buildings other than dwellings	2002	
M	Access and facilities for disabled people	2004	2000
N	Glazing – safety in relation to impact, opening and cleaning	1998	2000
P	Electrical safety	2004	
	Approved Document to support Regulation 7 – Materials and workmanship	1999	2000

Note: All of these documents are published by the Stationery Office. For availability and further details, see www.thestationeryoffice.com

Compliance

There is no obligation to adopt any particular solution that is contained in any of these guidance documents especially if you prefer to meet the relevant requirement in some other way. However, should a contravention of a requirement be alleged, if you have followed the guidance in the relevant Approved Documents, that will be evidence tending to show that you have complied with the Regulations. If you have **not** followed the guidance, then that will be seen as evidence tending to show that you have not complied with the requirements and it will then be up to you, the builder, architect and/or client to demonstrate that you have satisfied the requirements of the Building Regulations.

This compliance may be shown in a number of ways such as using:

- a product bearing CE marking (in accordance with the Construction Products Directive (89/106/EEC) as amended by the CE Marking Directive (93/68/EEC) as implemented by the Construction Products Directive 1994 (SI 1994/3051);
- an appropriate technical specification (as defined in the Construction Products Directive – 89/1 06/EEC);
- a recognized British Standard;
- a British Board of Agrément Certificate;
- an alternative, equivalent national technical specification from any member state of the European economic area;
- a product covered by a national or European certificate issued by a European Technical Approval issuing body.

Limitation on requirements

Parts A to D, F to K (except for paragraphs H2 and J6), N and P of Schedule 1 do not require anything to be done except for the purpose of securing reasonable standards of health and safety for persons in or about buildings (and any others who may be affected by buildings, or matters connected with buildings).

You may show that you have complied with Regulation 7 in a number of ways, for example, by the appropriate use of a product bearing a CE marking in accordance with the Construction Products Directive (89/106/EEC) as amended by the CE Marking Directive (93/68/EEC), or by following an appropriate technical specification (as defined in that Directive), a British Standard, a British Board of Agrément Certificate, or an alternative national technical specification of any member state of the European Community which, in use, is equivalent. You will find further guidance in the Approved Document supporting Regulation 7 on materials and workmanship.

Materials and workmanship

As stated in the Building Regulations, '*Any building work which is subject to requirements imposed by Schedule 1 of the Building Regulations should, in accordance with Regulation 7, be carried out with proper materials and in a workmanlike manner'.*

What materials can I use?

Other than the two exceptions below, provided that the materials and components you have chosen to use are from an approved source and are of approved quality (CE marking in accordance with the Construction Products Directive (89/106/EEC) as amended by the CE Marking Directive (93/68/EEC)) then the choice is fairly unlimited.

Short lived materials

Even if a plan for building work complies with the Building Regulations, if this work has been completed using short lived materials (i.e. materials that are, in the absence of special care, liable to rapid deterioration) the local authority can:

- reject the plans;
- pass the plans subject to a limited use clause (on expiration of which they will have to be removed);
- restrict the use of the building.

(Building Act 1984 Section 19)

Unsuitable materials

If, once building work has begun, it is discovered that it has been made using materials or components that have been identified by the Secretary of State (*or his nominated deputy*) as being unsuitable materials, the local authority have the power to:

- reject the plans;
- fix a period in which the offending work must be removed;
- restrict the use of the building.

(Building Act 1984 Section 20)

If the person completing the building work fails to remove the unsuitable material or component(s), then that person is liable to be prosecuted and, on summary conviction, faces a heavy fine.

Technical specifications

Building Regulations are made for specific purposes such as:

- health and safety;
- conservation of fuel and power;
- prevention of contamination of water;
- welfare and convenience of disabled people.

Although the main requirements for health and safety are now covered by the Building Regulations, there are still some requirements contained in the Workplace (Health, Safety and Welfare) Regulations 1992 that may need to be considered as they could contain requirements which affect building design. For further information see *Workplace (Health, Safety and Welfare) Regulations 1992. Approved Code of Practice L24*, published by HSE Books 1992 (ISBN 0 7176 0413 6).

Standards and technical approvals, as well as providing guidance, also address other aspects of performance such as serviceability and/or other aspects related to health and safety not covered by the Regulations.

When an Approved Document makes reference to a named standard, the relevant version of the standard is the one listed at the end of that particular Approved Document. However, if this version of the standard has been revised or updated by the issuing standards body, the new version may be used as a source of guidance provided it continues to address the relevant requirements of the Regulations.

The Secretary of State has agreed with the British Board of Agrément on the aspects of performance that it needs to assess in preparing its certificates in order that the board may demonstrate the compliance of a product or system that has an Agrément Certificate with the requirements of the Regulations. An Agrément Certificate issued by the board under these arrangements will give assurance that the product or system to which the certificate relates (if properly used in accordance with the terms of the certificate) will meet the relevant requirements.

Independent certification schemes

Within the UK there are many product certification schemes. Such schemes will certify compliance with the requirements of a recognized standard or document that is suitable for the purpose and material to be used.

Standards and technical approvals

Standards and technical approvals provide guidance related to the Building Regulations and address other aspects of performance such as serviceability or aspects which, although they relate to health and safety, are not covered by the Regulations.

European pre-standards (ENV)

The British Standards Institution (BSI) will be issuing Pre-standard (ENV) Structural Eurocodes as they become available from the European Standards Organisation, Comité Europeen de Normalisation Electrotechnique (CEN).

DD ENV 1992-1-1: 1992 Eurocode 2: Part 1 and DD ENV 1993-1-1: 1992 Eurocode 3: Part 1-1 *General Rules* and *Rules for Buildings in concrete and steel* have been thoroughly examined over a period of several years and are considered to provide appropriate guidance when used in conjunction with their national application documents for the design of concrete and steel buildings respectively.

When other ENV Eurocodes have been subjected to a similar level of examination they may also offer an alternative approach to Building Regulation compliance and, when they are eventually converted into fully approved EN standards, they will be included as referenced standards in the guidance documents.

 Note: If a national standard is going to be replaced by a European harmonized standard, then there will be a coexistence period during which either standard may be referred to. At the end of the coexistence period the national standard will be withdrawn.

House – construction

There are two main types of buildings in common use today: those made of brick and those made of timber. There are many different styles of brick-built houses and, equally there are various methods of construction.

Brickwork, as well as giving a building character, provides the main load bearing element of a brick-built house. Timber-framed houses, on the other hand, are usually built on a concrete foundation with a 'strip' or 'raft' construction to spread the weight and differ from their brick-built counterparts in that the main structural elements are timber frames.

6.1 Foundations

To support the weight of the structure, most brick-built buildings are supported on a solid base called foundations. Timber framed houses are usually built on a concrete foundation with a 'strip' or 'raft' construction to spread the weight.

6.1.1 Requirements

The building shall be constructed so that:

- *the combined dead, imposed and wind loads are sustained and transmitted by it to the ground, safely and without causing any building deflection/ deformation or ground movement that will affect the stability of any part of the building;*
- *ground movement caused by swelling, shrinkage or freezing of the subsoil; land-slip or subsidence will not affect the stability of any part of the building.*

(Approved Document A)

Buildings with five or more storeys (each basement level being counted as one storey) shall be constructed so that:

- *in the event of an accident, the building will not collapse to an extent inconsistent to the cause.*

(Approved Document B)

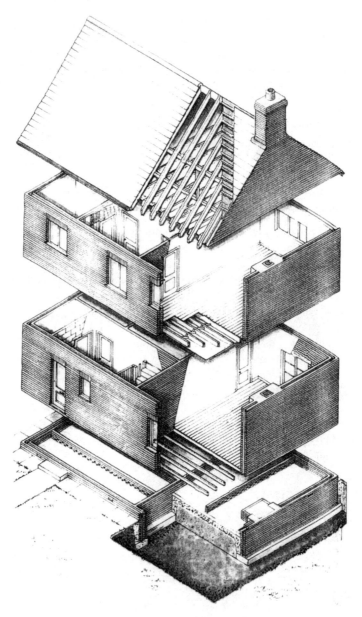

Figure 6.1 Brick built house – typical components

(1) *The ground to be covered by the building shall be reasonably free from any material that might damage the building or affect its stability, including vegetable matter, topsoil and pre-existing foundations.*

(2) *Reasonable precautions shall be taken to avoid danger to health and safety caused by contaminants on or in the ground covered, or to be covered by the building and any land associated with the building.*

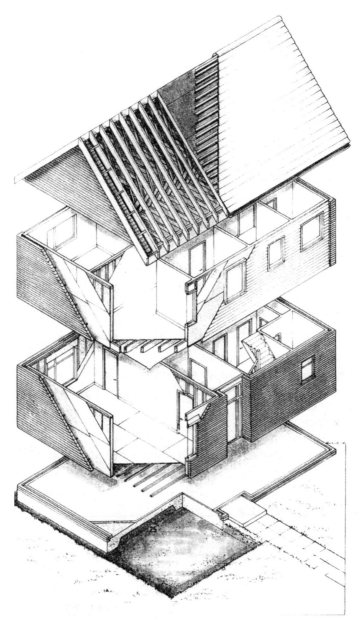

Figure 6.2 Timber framed house – typical components

(3) *Adequate subsoil drainage shall be provided if it is needed to avoid:*
   (a) *the passage of ground moisture to the interior of the building;*
   (b) *damage to the building, including damage through the transport of water-borne contaminants to the foundations of the building.*

(Approved Document C1)

Note: For the purpose of this requirement, 'contaminant' means any substance which is or may become harmful to persons or buildings including substances, which are corrosive, explosive, flammable, radioactive or toxic.

Potential problems

There may be known and/or recorded conditions of ground instability, such as geological faults, landslides or disused mines, or unstable strata of similar nature which affect or may potentially affect a building site or its environs.

There may also be:

- unsuitable material including vegetable matter, topsoil and pre-existing foundations;
- contaminants on or in the ground covered, or to be covered, by the building and any land associated with the building; and
- groundwater.

These conditions should be taken into account before proceeding with the design of a building or its foundations.

What about hazards?

Hazards associated with the ground may include:

- chemical and biological contaminants;
- gas generation from biodegradation of organic matter;
- naturally occurring radioactive radon gas and gases produced by some soils and minerals;
- physical, chemical or biological;
- underground storage tanks or foundations;
- unstable fill or unsuitable hardcore containing sulphate;
- the effects of vegetable matter including tree roots.

In the most hazardous conditions, only the total removal of contaminants from the ground to be covered by the building can provide a complete remedy. In other cases remedial measures can reduce the risks to acceptable levels. These measures should only be undertaken with the benefit of expert advice and where the removal would involve handling large quantities of contaminated materials, then you are advised to seek expert advice.

 Even when these actions have been successfully completed, the ground to be covered by the building will **still** need to have at least 100 mm of concrete laid over it!

What about contaminated ground?

Potential building sites which are likely to contain contaminants can be identified at an early stage from planning records or from local knowledge (e.g. previous uses). In addition to solid and liquid contaminants, problems can also arise from natural contamination such as methane and the radioactive radon gas (and its decay product).

The following list are examples of sites that are most likely to contain contaminants:

- asbestos works;
- ceramics, cement and asphalt manufacturing works;
- chemical works;
- dockyards and dockland;
- engineering works (including aircraft manufacturing, railway engineering works, shipyards, electrical and electronic equipment manufacturing works);
- gas works, coal carbonization plants and ancillary by-product works;
- industries making or using wood preservatives;
- landfill and other waste disposal sites;
- metal mines, smelters, foundries, steelworks and metal finishing works;
- munitions production and testing sites;
- oil storage and distribution sites;
- paper and printing works;
- power stations;
- railway land, especially larger sidings and depots;
- road vehicle fuelling, service and repair: garages and filling stations;
- scrap yards;
- sewage works, sewage farms and sludge disposal sites;
- tanneries;
- textile works and dye works.

If any signs of possible contaminants are present, then the local authority's Environmental Health Officer should be told at once. If he confirms the presence of any of these contaminants (see Table 6.2) then he will require their removal or action to be completed before any planning permission for building work can be sought.

What about gaseous contaminants?

Radon is a naturally occurring radioactive colourless and odourless gas which is formed in small quantities by radioactive decay wherever uranium and radium are found. It can move through the subsoil and then into buildings and exposure to high levels over long periods increases the risk of developing lung cancer. Some parts of the country (in particular the West Country) have higher natural levels than elsewhere and precautions against radon may be necessary.

 Note: Guidance on the construction of dwellings in areas susceptible to radon has been published by the Building Research Establishment as a Report ('*Radon: guidance on protective measures for new dwellings*').

Landfill gas is generated by the action of anaerobic micro-organisms on biodegradable material in landfill sites and generally consists of methane and carbon dioxide together with small quantities of VOCs (Volatile Organic Compounds) which give the gas its characteristic odour. It can migrate under pressure through the subsoil and through cracks and fissures into buildings.

Table 6.2 Examples of possible contaminants

Signs of possible contaminants	Possible contaminant
Vegetation (absence, poor or unnatural growth)	Metals Metal compounds Organic compounds Gases (landfill or natural source)
Surface materials (unusual colours and contours may indicate wastes and residues)	Metals Metal compounds Oily and tarry wastes Asbestos Other mineral fibres Organic compounds including phenols Combustible material including coal and coke dust Refuse and waste
Fumes and odours (may indicate organic chemicals)	Volatile organic and/or sulphurous compounds from landfill or petrol/solvent spillage Corrosive liquids Faecal animal and vegetable matter (biologically active)
Damage to exposed foundations of existing buildings	Sulphates
Drums and containers (empty or full)	Various

Methane and carbon dioxide can also be produced by organically rich soils and sediments such as peat and river silts and a wide range of VOCs can be present as a result of petrol, oil and solvent spillages.

Site preparation

Site investigation is now the recommended method for determining how much unsuitable material should be removed before commencing building work and this will normally consist of a number of well-defined stages, for example:

Planning stage	scope and requirements
Desktop study	historical, geological and environmental information about the site
Site reconnaissance or walkover survey	identification of actual and potential physical hazards and the design of the main investigation
Main investigation and reporting	intrusive and non-intrusive sampling and testing to provide soil parameters

Risk assessment

The site investigation may identify certain risks which will require a risk assessment, of which there are three types:

- Preliminary (once the need for a risk assessment has been identified, and depending on the situation and the outcome);

- Generic Quantitative Risk Assessment (GQRA);
- Detailed Quantitative Risk Assessment (DQRA).

Each risk assessment should include a:

Hazard identification   developing the conceptual model by establishing contaminant sources, pathways and receptors (this is the preliminary site assessment which consists of a desk study and a site walkover in order to gather sufficient information to obtain an initial understanding of the potential risks. An initial conceptual model for the site can then be based on this information.)

Hazard assessment   identifying what pollutant linkages may be present and analysing the potential for unacceptable risks.

Risk estimation   establishing the scale of the possible consequences by considering the degree of harm that may result and to which receptors.

Risk evaluation   deciding whether the risks are acceptable or unacceptable – review all site data to decide whether estimated risks are unacceptable.

6.1.2 Meeting the requirement

General

Where the site is potentially affected by contaminants, a combined geotechnical and geo-environmental investigation should be considered.	C1.3

Hazard identification and assessment

A preliminary site assessment is required to provide information on the past and present uses of the site and surrounding area that may give rise to contamination (see Table 6.2).	C2.10
The site assessment and risk evaluation should pay particular attention to the area of the site subject to building operations.	C2.11
The planning authority should be informed prior to any intrusive investigations or if any substance is found which was not identified in a preliminary statement about the nature of the site.	C2.12

Risks to buildings, building materials and services

The following hazards shall be considered:

- aggressive substances – including inorganic and organic   C2.23a
  acids, alkalis, organic solvents and inorganic chemicals
  such as sulphates and chlorides;
- combustible fill – including domestic waste, colliery   C2.23b
  spoil, coal, plastics, petrol-soaked ground, etc.;
- expansive slags – e.g. blast furnace and steel making slag; C2.23c
- floodwater affected by contaminants – substances in the   C2.23d
  ground, waste matter or sewage.

Contaminated ground

The underlying geology of a potential site has to be   C2.3
considered as natural contaminants may be present, for   and 2.4
example:

- naturally occurring heavy metals (e.g. cadmium and
  arsenic) originating in mining areas;
- gases (e.g. methane and carbon dioxide) originating in
  coal mining areas;
- organic rich soils and sediments such as peat and river
  silts;
- radioactive radon gas – which can also be a problem
  in certain parts of the country.

Possible sulphate attack from some strata on concrete   C2.5
floor slabs and oversite concrete needs to be considered.

Gaseous contaminants

Radon

All new buildings, extensions and conversions (whether   C2.39
residential or non-domestic), which are built in areas where
there may be high radon emissions, may need to incorporate
precautions against radon.

Landfill gas

Methane is an asphyxiant, will burn, and can explode in air.   C2
Carbon dioxide is non-flammable and toxic. Many of the other
components of landfill gas are flammable and some are toxic.
All will require careful analysis.

Risk assessment

A risk assessment should be completed for methane and other
gases particularly:

- on a landfill site or within 250 m of the boundary of a   C2.28a
  landfill site;
- on a site subject to the wide scale deposition of   C2.28b
  biodegradable substances (including made ground or fill);
- on a site that has been subject to a use that could give   C2.28c
  rise to petrol, oil or solvent spillages;
- in an area subject to naturally occurring methane, carbon   C2.28d
  dioxide and other hazardous gases (e.g. hydrogen sulphide).

During a site investigation for methane and other gases:

- measurements should be taken over a sufficiently long   C2.30
  period of time in order to characterize gas emissions fully;
- should include periods when gas emissions are   C2.30
  likely to be higher, e.g. during periods of
  falling atmospheric pressure.

Gas risks (i.e. to human receptors) should be considered for:

- gas entering the dwelling through the substructure (and   C2.32
  building up to hazardous levels);
- subsequent householder exposure in garden areas   C2.32
  including outbuildings (e.g. garden sheds and greenhouses),
  extensions and garden features (e.g. ponds).

When land that is affected by contaminants is being developed, C2.7
'receptors' (i.e. buildings, building materials and building
services, as well as people) are introduced onto the site and it
is necessary to break the pollutant linkages. This can be
achieved by:

- treating the contaminant (e.g. use of physical, chemical or
  biological processes to eliminate or reduce the
  contaminant's toxicity or harmful properties);

- blocking or removing the pathway (e.g. isolating the contaminant beneath protective layers or installing barriers to prevent migration);
- protecting or removing the receptor (e.g. changing the form or layout of the development, using appropriately designed building materials, etc.).

A risk assessment based on the concept of a 'source–pathway– C2.6 receptor' relationship, or pollutant linkage of a potential site (see Figure 6.3) should be carried out to ensure the safe development of land that is affected by contaminants.

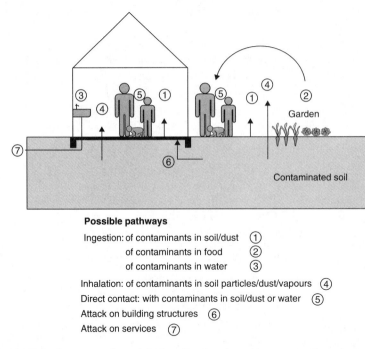

Possible pathways

Ingestion: of contaminants in soil/dust (1)
of contaminants in food (2)
of contaminants in water (3)
Inhalation: of contaminants in soil particles/dust/vapours (4)
Direct contact: with contaminants in soil/dust or water (5)
Attack on building structures (6)
Attack on services (7)

Figure 6.3 Conceptual model of a site showing a source–pathway–receptor

Risk estimation and evaluation

The detailed ground investigation: C2.13

- must provide sufficient information for the confirmation of a conceptual model for the site, the risk assessment and the design and specification of any remedial works;

- is likely to involve collection and analysis of soil, soil gas, surface and groundwater samples by the use of invasive and/or non-invasive techniques.

During the development of land affected by contaminants the health and safety of both the public and workers should be considered.   C2.14

Remedial measures

If the risks posed by the gas are unacceptable then these need to be managed through appropriate building remedial measures.   C2.36

Site-wide gas control measures may be required if the risks on any land associated with the building are deemed unacceptable.   C2.36

Consideration should be given to the design and layout of buildings to maximize the driving forces of natural ventilation.   C2.37

For non-domestic buildings, expert advice concerning gas control measures should be sought as the floor area of such buildings can be large and it is important to ensure that gas is adequately dispersed from beneath the floor.   C2.38

There is a need for continued maintenance and calibration of mechanical (as opposed to passive) gas control systems.   C2.38

Sub-floor ventilation systems should be carefully designed to ensure adequate performance and should not be modified unless subjected to a specialist review of the design.   C2.38

Corrective measures

When building work is undertaken on sites affected by contaminants where control measures are already in place, care must be taken not to compromise these measures.   C2.15

Depending on the contaminant, three generic types of corrective measures can be considered: treatment, containment and removal.

 Note: The containment or treatment of waste may require a waste management licence from the Environmental Agency.

Treatment

The choice of the most appropriate treatment process for a particular site is a highly site-specific decision for which specialist advice should be sought.	C2.16

Containment

In-ground vertical barriers may also be required to control lateral migration of contaminants.	C2.17
Cover systems involve the placement of one or more layers of materials placed over the site and may be used to:	C2.18

- break the pollutant linkage between receptors and contaminants;
- sustain vegetation;
- improve geotechnical properties; and
- reduce exposure to an acceptable level.

Imported fill and soil for cover systems should be assessed at source to ensure that it is not contaminated.	C2.20
The size and design of cover systems (particularly soil-based ones used for gardens) should take account of their long-term performance.	C2.20
Gradual intermixing due to natural effects and activities such as burrowing animals, gardening, etc., needs to be considered.	C2.20

Removal

Imported fill should be assessed at source to ensure that there are no materials that will pose unacceptable risks to potential receptors.	C2.21

Site preparation

Vegetable matter such as turf and roots should be removed from the ground that is going to be covered by the building at least to a depth to prevent later growth.	C1.4
The effects of roots close to the building need to be assessed.	C1.4

Where mature trees are present (particularly on sites with   C1.5
shrinkable clays (see Table 6.3)) the potential damage arising
from ground heave to services and floor slabs and oversite
concrete should be assessed.

Table 6.3 Volume change potential for some common clays

Clay type	Volume change potential
Glacial till	Low
London	High to very high
Oxford and Kimmeridge	High
Lower lias	Medium
Gault	High to very high
Weald	High
Mercian mudstone	Low to medium

Building services such as below-ground drainage should be   C1.6
sufficiently robust or flexible to accommodate the presence of
any tree roots.

Joints should be made so that roots will not penetrate them.   C1.6

Where roots could pose a hazard to building services,   C1.6
consideration should be given to their removal.

On sites previously used for buildings, consideration should be   C1.7
given to the presence of other infrastructure (such as existing
foundations, services and buried tanks, etc.) that could endanger
persons in and about the building and any land associated with
the building.

If the site contains fill or made ground, consideration should   C1.8
be given to its compressibility and its potential to collapse
when wet.

Foundations

Table 6.4 provides guidance on determining the type of soil on which it is
intended to lay a foundation.

Subsoil drainage

Where the water table can rise to within 0.25 m of the lowest   C3.2
floor of the building, or where surface water could enter or

adversely affect the building, either the ground to be covered by the building should be drained by gravity, or other effective means of safeguarding the building should be taken.

If an active subsoil drain is cut during excavation and if it passes under the building it should be either: C3.3

- re-laid in pipes with sealed joints and have access points outside the building; or
- re-routed around the building; or
- re-run to another outfall (see Figure 6.4).

Where contaminants are present in the ground, consideration should be given to subsoil drainage to prevent the transportation of water-borne contaminants to the foundations or into the building or its services.   C3.7

Table 6.4 Types of subsoil

Type	Applicable field test
Rock (being stronger/ denser than sandstone, limestone or firm chalk)	Requires at least a pneumatic or other mechanically operated pick for excavation.
Compact gravel and/or sand	Requires a pick for excavation. Wooden peg 50 mm square in cross section hard to drive beyond 150 mm.
Stiff clay or sandy clay	Cannot be moulded with the fingers and requires a pick or pneumatic or other mechanically operated spade for its removal.
Firm clay or sandy clay	Can be moulded by substantial pressure with the fingers and can be excavated with a spade.
Loose sand, silty sand or clayey sand	Can be excavated with a spade. Wooden peg 50 mm square in cross section can be easily driven.
Soft silt, clay, sandy clay or silty clay	Fairly easily moulded in the fingers and readily excavated.
Very soft silt, clay, sandy clay or silty clay	Natural sample in winter conditions exudes between the fingers when squeezed in fist.

Ground movement

Known or recorded conditions of ground instability, such as that arising from landslides, disused mines or unstable strata should be taken into account in the design of the building and its foundations.   A1/2 1.9

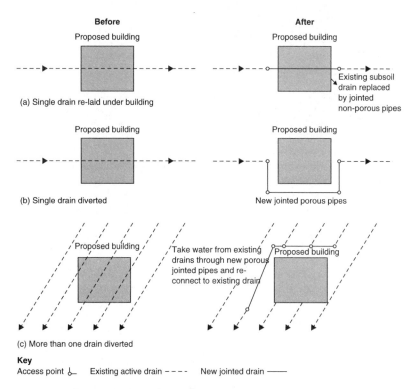

Figure 6.4 Subsoil drain cut during excavation

Foundations – plain concrete

There should **not** be:

- non-engineered fill (see BRE Digest 427) or a wide   A1/2 2E1a
  variation in ground conditions within the loaded area;
- weaker or more compressible ground at such a depth   A1/2 2E1b
  below the foundation as could impair the stability of
  the structure.

The foundations should be situated centrally under   A1/2 2E2a
the wall.

In non-aggressive soils, concrete should be composed of   A1/2 2E2b
Portland cement to BS EN 197 1 & 2 and fine and coarse
aggregate conforming to BS EN 12620 and the mix
should either be:

- 50 kg of Portland cement to not more than 200 kg
  (0.1 m³) of fine aggregate and 400 kg (0.2 m³) of
  coarse aggregate, or
- Grade ST2 or Grade GEN I concrete to BS 8500-2.

> For foundations in chemically aggressive soil conditions guidance in BS 8500-1:Part 1 and BRE Special Digest 1 should be followed.   (8.18)
>
> The minimum thickness T of concrete foundation should be 150 mm or P, whichever is the greater where P is derived using Table 6.4 and Figure 6.5   A1/2 2E2c
>
>   (8.18a)
>
> Foundations stepped on elevation should overlap by twice the height of the step, by the thickness of the foundation, or 300 mm, whichever is greater (see Figure 6.6).   A1/2 2E2d

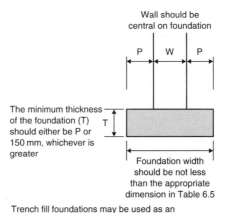

Figure 6.5 Foundation dimensions

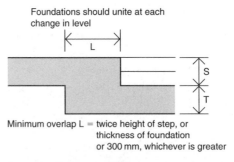

Figure 6.6 Elevation of stepped foundation

The overlap for trench fill foundations should be twice the   A1/2 2E2d
height of step or 1 metre, whichever is greater.

💡 Trench fill foundations may be used as an   A1/2 2E2c
acceptable alternative to strip foundations.

Steps in foundations should not be of greater height than
the thickness of the foundation (see Figure 6.6).

Foundations for piers, buttresses and chimneys should   A1/2 2E2f
project as shown in Figure 6.7

💣 The projection X should never be less than the
value of P where there is no local thickening of the wall.

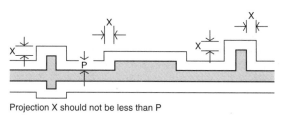

Projection X should not be less than P

Figure 6.7 Piers and chimneys

Strip foundations

The recommended minimum widths of strip foundations   A1/2 2E3
shall be as indicated in Table 6.5

Where strip foundations are founded on rock, the strip   A1/2 2E4
foundations should have a minimum depth of 0.45 m to
their underside to avoid the action of frost.

💣 This depth, however, will commonly need to be
increased in areas subject to long periods of frost or in order
to transfer the loading onto satisfactory ground.

In clay soils subject to volume change on drying (i.e.   A1/2 2E4
'shrinkable clays' with a Plasticity Index greater than or equal
to 10%), strip foundations should be taken to a depth where
anticipated ground movements (caused by vegetation and
trees on the ground) will not impair the stability of any part
of the building.

> The depth to the underside of foundations on clay soils   A1/2 2E4
> should not be less than 0.75 m.
>
> Although this depth will commonly need to be
> increased in order to transfer the loading onto
> satisfactory ground.

Table 6.5 Minimum width of strip footings

Type of ground (including engineered fill)	Condition of ground	Field test applicable	Total load of load-bearing walling not more than (kN/linear metre)					
			20	30	40	50	60	70
			Minimum width of strip foundation (mm)					
I Rock	Not inferior to sandstone, limestone or firm chalk	Requires at least a pneumatic or other mechanically operated pick for excavation	In each case equal to the width of wall					
II Gravel or sand	Medium dense	Requires pick for excavation. Wooden peg 50 mm square in cross-section hard to drive beyond 150 mm	250	300	400	500	600	650
III Clay Sandy clay	Stiff Stiff	Can be indented slightly by thumb	250	300	400	500	600	650
IV Clay Sandy clay	Firm Firm	Thumb makes impression easily	300	350	450	600	750	850
V Sand Silty sand Clayey sand	Loose Loose Loose	Can be excavated with a spade. Wooden peg 50 mm square in cross-section can be easily driven	400	600	Note: Foundations on soil types V and VI do not fall within the provisions of this section if the total load exceeds 30 kN/m.			
VI Silt Clay Sandy clay Clay or silt	Soft Soft Soft Soft	Finger pushed in up to 10 mm	450	650				
VII Silt Clay Sandy clay Clay or silt	Very soft Very soft Very soft Very soft	Finger easily pushed in up to 25 mm	Refer to specialist advice					

Disproportionate collapse

All buildings should be built so that their sensitivity to disproportionate collapse in the event of an accident is reduced.

Buildings shall remain sufficiently robust to sustain a limited   A1/2 5.1
extent of damage or failure, depending on the class of the
building, without collapse (see below).

 Notes:
(1) Buildings intended for more than one type of use should
     adopt the most onerous class.
(2) In determining the number of storeys in a building,
     basement storeys may be excluded provided that they
     meet the robustness requirements of Class 2B buildings.

Class 1

Building type and occupancy	Requirements
• Houses not exceeding 4 storeys. • Agricultural buildings. • Buildings into which people rarely go, provided no part of the building is closer to another building (or area where people go) than 1.5 times the building height.	Provided the building has been designed and constructed in accordance with Building Regulations and is in normal use, no additional measures are likely to be necessary.

Class 2A

Building type and occupancy	Requirements
• 5 storey single occupancy houses. • Hotels not exceeding 4 storeys. • Flats, apartments and other residential buildings not exceeding 4 storeys. • Offices not exceeding 4 storeys. • Industrial buildings not exceeding 3 storeys. • Retailing premises not exceeding 3 storeys of less than 2000 m² floor area in each storey.	Effective horizontal ties (or effective anchorage of suspended floors to walls) is required.

- Single storey educational buildings.
- All buildings not exceeding
  2 storeys to which members of the
  public are admitted and which
  contain floor areas not exceeding
  2000 m² at each storey.

Class 2B

Building type and occupancy	Requirements
• Hotels, flats, apartments and other residential buildings greater than 4 storeys but not exceeding 15 storeys.	
• Educational buildings greater than 1 storey but not exceeding 15 storeys.	Effective horizontal ties need to be provided.
• Retailing premises greater than 3 storeys but not exceeding 15 storeys.	
• Hospitals not exceeding 3 storeys.	Effective vertical ties need to
• Offices greater than 4 storeys but not exceeding 15 storeys.	be provided in all supporting columns and walls.
• All buildings to which members of the public are admitted which contain floor areas exceeding 2000 m² but less than 5000 m² at each storey.	
• Car parking not exceeding 6 storeys.	

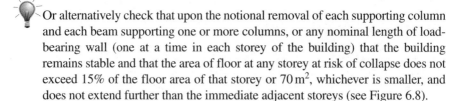

Or alternatively check that upon the notional removal of each supporting column and each beam supporting one or more columns, or any nominal length of load-bearing wall (one at a time in each storey of the building) that the building remains stable and that the area of floor at any storey at risk of collapse does not exceed 15% of the floor area of that storey or 70 m², whichever is smaller, and does not extend further than the immediate adjacent storeys (see Figure 6.8).

Where the notional removal of such columns and lengths of walls would result in damage in excess of the above limit, then such elements should be designed as a 'key element' (i.e. it should be capable of sustaining an accidental design loading of 34 kN/m²) applied in the horizontal and vertical directions

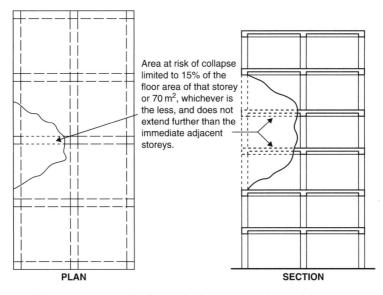

Figure 6.8 Area at risk of collapse in the event of an accident

(in one direction at a time) to the member and any attached components (e.g. cladding, etc.).

Class 3

Building type and occupancy	Requirements
• All buildings defined above as Class 2A and 2B that exceed the limits on area and/or number of storeys. • Grandstands accommodating more than 5000 spectators. • Buildings containing hazardous substances and/or processes.	A systematic risk assessment of the building should be undertaken taking into account all the normal hazards that may reasonably be foreseen, together with any abnormal hazards. Critical situations for design should be selected that reflect the conditions that can reasonably be foreseen as possible during the life of the building. Protective measures should be chosen and the detailed design of the structure and its elements

Continued over page

Continued

> undertaken in accordance with
> the following recommendations:
> - BS 5628: Part 1 – Structural
>   use of unreinforced masonry
> - BS 5950: Part 1 – Structural
>   use of steelwork in building
> - BS 8110: Parts 1 and 2 –
>   Structural use of plain,
>   reinforced and prestressed
>   concrete.

 For any building which does not fall into one of the classes listed above, or where the consequences of collapse may warrant particular examination of the risks involved, see one of the following Reports:

'Guidance on Robustness and Provision against Accidental Actions' dated July 1999, together with the accompanying BRE Report No. 200682.

'Calibration of Proposed Revised Guidance on Meeting Compliance with the Requirements of Building Regulation Part A3'.

Both of the above documents are available on the following ODPM website http://www.odpm.gov.uk.

Maximum floor area

No floor enclosed by structural walls on all sides shall exceed 70 m² (see Figure 6.9).	A1/2 (2C14)
No floor with a structural wall on one side shall exceed 36 m² (see Figure 6.9).	A1/2 (2C14)

Maximum height of buildings

The maximum height of a building shall not exceed the heights given in Table 6.6 with regard to the relevant wind speed.	A1/2 (1C17)

Heights of walls and storeys

The measured height of a wall or a storey should be in accordance with Figure 6.10.	A1/2 2C18

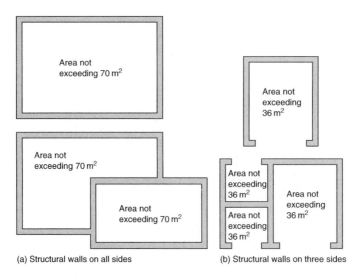

(a) Structural walls on all sides   (b) Structural walls on three sides

Figure 6.9 Maximum floor area that is enclosed by structural walls

Table 6.6 Maximum allowable building height

Factor S	Country sites			Town sites		
	Distance to the coast			Distance to the coast		
	<10 km	15–50 km	>50 km	<10 km	15–50 km	>50 km
24	15	15	15	15	15	15
25	11.5	14.5	15	15	15	15
26	8	10.5	13	15	15	15
27	6	8.5	10	15	15	15
28	4.5	6.5	8	13.5	15	15
29	3.5	5	6	11	13	14.5
30	3	4	5	9	11	12.5
31		3.5	4	8	9.5	10.5
32		3	3.5	7	8.5	9.5
33			3	6	7.5	8.5
34				5	7	8
35				4	6	7
36				3	5.5	6
37					4.5	5.5
38					4	5
39					3	4
40						3

Imposed loads on roofs, floors and ceilings

The imposed loads on roofs, floors and ceilings shall   A1/2 (2C15)
not exceed those shown in Table 6.7.

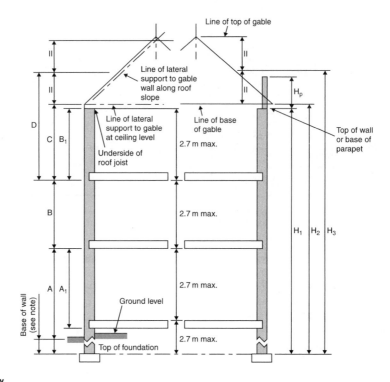

Key

(a) **Measuring Storey Heights**

A_1 is the ground storey height if the ground floor provides effective lateral support to the wall i.e. is adequately tied to the wall or is a suspended floor bearing on the wall.

A   is the ground storey height if the ground floor does not provide effective lateral support to the wall.

   Note: If the wall is supported adequately and permanently on both sides by suitable compact material, the base of the wall for the purposes of the storey height may be taken as the lower level of this support. (Not greater than 3.7 m ground storey height.)

B   is the intermediate storey height.

B_1 is the top storey height for walls which do not include a gable.

C   is the top storey height where lateral support is given to the gable at both ceiling level and along the roof slope.

D   is the top storey height for the external walls which include a gable where lateral support is given to the gable only along the roof slope.

(b) **Measuring Wall Heights**

H_1 is the height of an external wall that does no include a gable.

H_2 is the height of an internal or separating wall which is built up to the underside of the roof.

H_3 is the height of an external wall which includes a gable.

H_p is the height of a parapet. If H_p is more than 1.2 m add to H_p to H_1.

Figure 6.10 Method for measuring the heights of storeys and walls

Table 6.7 Imposed loads

Element	Distributed loads	Concentrated load
Roofs	1.00 kN/m^2 for spans not exceeding 12 m	
	1.50 kN/m^2 for spans not exceeding 6 m	
Floors	2.00 kN/m^2	
Ceilings	0.25 kN/m^2	0.9 kN/m^2

Structural safety

The safety of a structure depends on the successful combination of design and completed construction, particularly:

- the design – which should also:
  - be based on identification of the hazards (to which   A1/2 0.2a
    the structure is likely to be subjected) and an
    assessment of the risks;
  - reflect conditions that can reasonably be foreseen
    during future use;
- loading – dead load, imposed load and wind load;   A1/2 0.2b
- the properties of materials used;   A1/2 0.2c
- the detailed design and assembly of the structure;   A1/2 0.2d
- safety factors;   A1/2 0.2e
- workmanship.   A1/2 0.2f

Basic requirements for stability

Adequate provision shall be made to ensure that the   A1/2 1A2
building is stable under the likely imposed and wind
loading conditions.

The overall size and proportioning of the building shall   A1/2 1A2a
be limited according to the specific guidance for each
form of construction.

The layout of walls (both internal and external) forming   A1/2 1A2b
a robust three-dimensional box structure in plan shall be
constructed according to the specific guidance for each
form of construction.

The internal and external walls shall be adequately   A1/2 1A2c
connected by either masonry bonding or by using
mechanical connections.

The intermediate floors and roof shall be constructed
so that they:

- provide local support to the walls;   A1/2 1A2d
- act as horizontal diaphragms capable of transferring
  the wind forces to buttressing elements of the building.

> **Note:** A traditional cut timber roof (i.e. using rafters, purlins and ceiling joists) generally has sufficient built-in resistance to instability and wind forces (e.g. from either hipped ends, tiling battens, rigid sarking, or the like). However, the need for diagonal rafter bracing equivalent to that recommended in BS 5268: Part 3: 1998 or Annex H of BS 8103: Part 3: 1996 for trussed rafter roofs, should be considered especially for single-hipped and non-hipped roofs of greater than 40° pitch to detached houses.

6.2 Buildings – size

6.2.1 Classification of purpose groups

Many of the provisions in Approved Documents are related to the use of the building. The classifications 'use' are termed purpose groups and represent different levels of hazard. They can apply to a whole building, or (where a building is compartmented) to a compartment in the building and the relevant purpose group should be taken from the main use of the building or compartment. Table 6.8 sets out the purpose group classification.

Table 6.8 Classification of purpose groups

Title	Group	Purpose for which the building or compartment of a building is intended to be used
Residential[1] (dwellings)	1(a)	Flat or maisonette.
	1(b)	Dwelling house which contains a habitable storey with a floor level which is more than 4.5 m above ground level.
	1(c)	Dwelling house which does not contain a habitable storey with a floor level which is more than 4.5 m above ground level.
Residential (institutional)	2(a)	Hospital, home, school or other similar establishment used as living accommodation for, or for the treatment, care or maintenance of persons suffering from disabilities due to illness or old age or other physical or mental incapacity, or under the age of five years, or place of lawful detention, where such persons sleep on the premises.
Other	2(b)	Hotel, boarding house, residential college, hall of residence, hostel, and any other residential purpose not described above.
Office	3	Offices or premises used for the purpose of administration, clerical work (including writing, book keeping, sorting papers, filing, typing, duplicating, machine calculating, drawing and the editorial preparation of matter for publication, police and fire service work), handling

Title	Group	Purpose for which the building or compartment of a building is intended to be used
		money (including banking and building society work), and communications (including postal, telegraph and radio communications) or radio, television, film, audio or video recording, or performance (not open to the public) and their control.
Shop and commercial	4	Shops or premises used for a retail trade or business (including the sale to members of the public of food or drink for immediate consumption and retail by auction, self-selection and over-the-counter wholesale trading, the business of lending books or periodicals for gain and the business of a barber or hairdresser) and premises to which the public is invited to deliver or collect goods in connection with their hire, repair or other treatment, or (except in the case of repair of motor vehicles) where they themselves may carry out such repairs or other treatments.
Assembly and recreation	5	Place of assembly, entertainment or recreation; including bingo halls, broadcasting, recording and film studios open to the public, casinos, dance halls; entertainment, conference, exhibition and leisure centres; funfairs and amusement arcades; museums and art galleries; non-residential clubs, theatres, cinemas and concert halls; educational establishments, dancing schools, gymnasia, swimming pool buildings, riding schools, skating rinks, sports pavilions, sports stadia; law courts; churches and other buildings of worship, crematoria; libraries open to the public, non-residential day centres, clinics, health centres and surgeries; passenger stations and termini for air, rail, road or sea travel; public toilets; zoos and menageries.
Industrial	6	Factories and other premises used for manufacturing, altering, repairing, cleaning, washing, breaking-up, adapting or processing any article; generating power or slaughtering livestock.
Storage and other non-industrial[2]	7(a)	Place for the storage or deposit of goods or materials (other than described under 7(b)) and any non-residential building not within any of the purpose groups 1 to 6.
	7(b)	Car parks designed to admit and accommodate only cars, motorcycles and passenger or light goods vehicles weighing no more than 2500 kg gross.

Notes:

(1) Includes any surgeries, consulting rooms, offices or other accommodation, not exceeding 50 m² in total, forming part of a dwelling and used by an occupant of the dwelling in a professional or business capacity.

(2) A detached garage not more than 40 m² in area is included in purpose group 1(c); as is a detached open carport of not more than 40 m², or a detached building which consists of a garage and open carport where neither the garage nor open carport exceeds 40 m² in area.

(3) 'Room for residential purposes' means a room, or suite of rooms, which is not a dwellinghouse or flat and which is used by one or more persons to live and sleep in, including rooms in hotels, hostels, boarding houses, halls of residence and residential homes but not including rooms in hospitals, or other similar establishments, used for patient accommodation.

6.2.2 Requirements – size of residential buildings

The building shall be constructed so that the combined dead, imposed and wind loads are sustained and transmitted by it to the ground

- *safely;*
- *without causing such deflection or deformation of any part of the building (or such movement of the ground) as will impair the stability of any part of another building.*

(Approved Document A1)

The building shall be constructed so that ground movement caused by:

- *swelling, shrinkage or freezing of the subsoil; or*
- *landslip or subsidence (other than subsidence arising from shrinkage)*

will not impair the stability of any part of the building.

(Approved Document A2)

The maximum height of the building measured from the lowest finished ground level to the highest point of any wall or roof should be less than 15 m (see Figure 6.11).	A1/2 (2C4i)
The height of the building should not exceed twice the least width of the building (see Figure 6.11).	A1/2 (2C4ii)
The height of the wing H_2 should not be greater than twice the least width of the wing W_2 where the projection P exceeds twice the width W_2.	A1/2 (2C4iii)

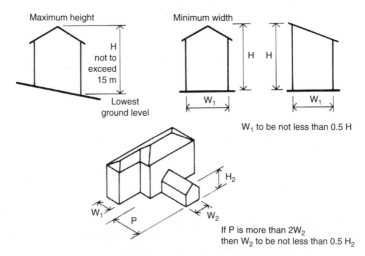

Figure 6.11 Residential buildings not more than three storeys

Small single storey non-residential buildings

The height (H) should not exceed 3 m and the width   A1/2 (2C4b)
(or greater length) should not exceed 9 m (see
Figure 6.12).

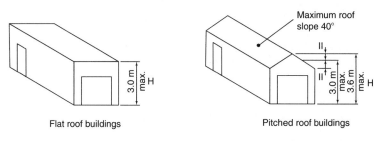

Figure 6.12 Size and proportion of non-residential buildings

Size of annexes

The height H (as variously shown in Figure 6.13)   A1/2 (2C4b)
should not exceed 3 m.

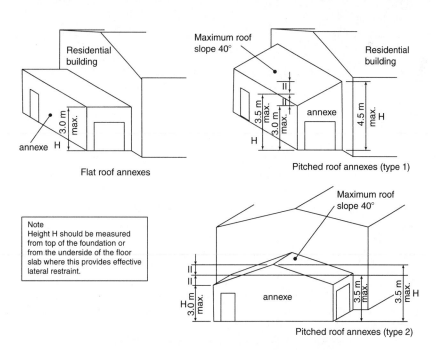

Figure 6.13 Size and proportion of non-residential annexes

6.2.3 Requirements – ventilation

Ventilation (mechanical and/or air-conditioning systems designed for domestic buildings) shall be capable of restricting the accumulation of moisture and pollutants originating within a building.

(Approved Document F1)

Ventilation

Buildings (or spaces within buildings) other than those:

- into which people do not normally go; or
- which are used solely for storage; or
- which are garages used solely in connection with a single dwelling, shall be provided with ventilation to:

• extract water vapour from non-habitable areas where it is produced in significant quantities (e.g. kitchens, utility rooms and bathrooms).	F1 (1.1–1.5)
• extract pollutants (which are a hazard to health) from areas where they are produced in significant quantities (e.g. rooms containing processes which generate harmful contaminants and rest rooms where smoking is permitted).	F1 (2.3–2.5)
• rapidly dilute (when necessary) pollutants and water vapour produced in habitable rooms, occupiable rooms and sanitary accommodation.	F1 (1.1–1.4) F1 (1.6–1.8)
• provide a minimum supply of fresh air for occupants.	F1 (2.6–2.8)

Table 6.9 Ventilation of rooms containing openable windows (i.e. located on an external wall)

Room	Rapid ventilation (e.g. opening windows)	Background ventilation	Extract ventilation fan rates or passive stack (PSV)
Habitable room	1/20th of floor area	8000 mm^2	
Kitchen	Opening window (no minimum size)	4000 mm^2	30 litres/second adjacent to a hob or 60 litres/second elsewhere or PSV
Utility room	Opening window (no minimum size)	4000 mm^2	30 litres/second or PSV
Bathroom (with or without WC)	Opening window (no minimum size)	4000 mm^2	15 litres/second or PSV
Sanitary accommodation (separate from bathroom)	1/20th of floor area or mechanical extract at 6 litres/second	4000 mm^2	

6.3 Drainage

6.3.1 The requirement (Building Act 1984 Sections 21 and 22)

All plans for building work need to show that drainage of refuse water (e.g. from sinks) and rainwater (from roofs) have been adequately catered for. Failure to do so will mean that these plans will be rejected by the local authority.

All plans for buildings must include at least one (or more) water or earth closets **unless** the local authority are satisfied that one is not required (for example in a large garage separated from the house).

 If you propose using an earth closet, the local authority cannot reject the plans unless they consider that there is insufficient water supply to that earth closet.

What are the rules about drainage? (Building Act 1984 Section 59)

The Building Act requires that all drains are connected either with a sewer (unless the sewer is more than 120 ft away or the person carrying out the building work is not entitled to have access to the intervening land) or is able to discharge into a cesspool, settlement tank or other tank designed for the reception and/or disposal of foul matter from buildings.

The local authorities view this requirement very seriously and will need to be satisfied that:

- satisfactory provision has been made for drainage;
- all cesspools, private sewers, septic tanks, drains, soil pipes, rain water pipes, spouts, sinks or other appliances are adequate for the building in question;
- all private sewers that connect directly or indirectly to the public sewer are not capable of admitting subsoil water;
- the condition of a cesspool is not detrimental to health, or does not present a nuisance;
- cesspools, private sewers and drains previously used, but now no longer in service, do not prejudice health or become a nuisance.

 This requirement can become quite a problem if it is not recognized in the early planning stages and so it is always best to seek the advice of the local authority. In certain circumstances, the local authority might even help to pay for the cost of connecting you up to the nearest sewer!

 The local authority has the authority to make the owner renew, repair or cleanse existing cesspools, sewers and drains etc.

Can two buildings share the same drainage?

Usually the local authority will require every building to be drained separately into an existing sewer but in some circumstances they may decide that it would be more cost effective if the buildings were drained in combination. On occasions, they might even recommend that a private sewer is constructed.

What about ventilation of soil pipes? (Building Act 1984 Section 60)

A major requirement of the Building Regulations is that all soil pipes from water closets shall be properly ventilated and that no use shall be made of:

- an existing or proposed pipe designed to carry rain water from a roof to convey soil and drainage from a sanitary convenience;
- an existing pipe designed to carry surface water from a premises to act as a ventilating shaft to a drain or a sewer conveying foul water.

What happens if I need to disconnect an existing drain? (Building Act 1984 Section 662)

If, in the course of your building work, you need to:

- reconstruct, renew or repair an existing drain that is joined up with a sewer or another drain;
- alter the position of an existing drain that is joined up with a sewer or another drain;
- seal off an existing drain that is joined up with a sewer or another drain,

then, provided that you give 48 hours' notice to the local authority, the person undertaking the reconstruction may break open any street for this purpose.

 You do not need to comply with this requirement if you are demolishing an existing building.

Can I repair an existing water closet or drain? (Building Act 1984 Section 63)

Repairs can be carried out to water closets, drains and soil pipes, but if that repair or construction work is prejudicial to health and/or a public nuisance, then the person who completed the installation or repair is liable, on conviction, to a heavy fine.

 In the Greater London area, a 'water closet' can **also** be taken to mean a urinal.

Can I repair an existing drain? (Building Act 1984 Section 61)

Only in extreme emergencies are you allowed to repair, reconstruct or alter the course of an underground drain that joins up with a sewer, cesspool or other drainage method (e.g. septic tank).

 If you have to carry out repairs etc. in an emergency, then make sure that you do **not** cover over the drain or sewer without notifying the local authority of your intentions!

Drains – Fire protection

Drains should also provide a degree of fire protection as shown by the following requirement:

- *all openings in fire-separating elements shall be suitably protected in order to maintain the integrity of the continuity of the fire separation,*
- *any hidden voids in the construction shall be sealed and subdivided to inhibit the unseen spread of fire and products of combustion, in order to reduce the risk of structural failure, and the spread of fire.*

(Approved Document B3)

Foul water drainage

The foul water drainage system shall:

- *convey the flow-off foul water to a foul water outfall (i.e. sewer, cesspool, septic tank or settlement (i.e. holding) tank),*
- *minimise the risk of blockage or leakage,*
- *prevent foul air from the drainage system from entering the building under working conditions,*
- *be ventilated,*
- *be accessible for clearing blockages,*
- *not increase the vulnerability of the building to flooding.*

(Approved Document H1)

Wastewater treatment systems and cesspools

Wastewater treatment systems shall:

- *have sufficient capacity to enable breakdown and settlement of solid matter in the wastewater from the buildings;*
- *be sited and constructed so as to prevent overloading of the receiving water.*

Cesspools shall have sufficient capacity to store the foul water from the building until they are emptied.

Wastewater treatment systems and cesspools shall be sited and constructed so as not to:

- *be prejudicial to health or a nuisance;*
- *adversely affect water sources or resources;*
- *pollute controlled waters;*
- *be in an area where there is a risk of flooding.*

Septic tanks and wastewater treatment systems and cesspools are constructed and sited so as to:

- *have adequate ventilation;*
- *prevent leakage of the contents and ingress of subsoil water;*
- *have regard to water table levels at any time of the year and rising groundwater levels.*

Drainage fields are sited and constructed so as to:

- *avoid overloading of the soakage capacity, and*
- *provide adequately for the availability of an aerated layer in the soil at all times.*

(Approved Document H2)

Rainwater drainage

Rainwater drainage systems shall:

- *minimise the risk of blockage or leakage;*
- *be accessible for clearing blockages;*
- *ensure that rainwater soaking into the ground is distributed sufficiently so that it does not damage foundations of the proposed building or any adjacent structure;*
- *ensure that rainwater from roofs and paved areas is carried away from the surface either by a drainage system or by other means;*
- *ensure that the rainwater drainage system carries the flow of rainwater from the roof to an outfall (e.g. a soakaway, a watercourse, a surface water or a combined sewer).*

(Approved Document H3)

Building over existing sewers

Building or extension or work involving underpinning shall:

- *be constructed or carried out in a manner which will not overload or otherwise cause damage to the drain, sewer or disposal main either during or after the construction;*
- *not obstruct reasonable access to any manhole or inspection chamber on the drain, sewer or disposal main;*
- *in the event of the drain, sewer or disposal main requiring replacement, not unduly obstruct work to replace the drain, sewer or disposal main, on its present alignment;*
- *reduce the risk of damage to the building as a result of failure of the drain, sewer or disposal main.*

(Approved Document H4)

Separate systems for drainage

Separate systems of drains and sewers shall be provided for foul water and rainwater where:

(a) *the rainwater is not contaminated; and*
(b) *the drainage is to be connected either directly or indirectly to the public sewer system and either –*
  (i) *the public sewer system in the area comprises separate systems for foul water and surface water; or*
  (ii) *a system of sewers which provides for the separate conveyance of surface water is under construction either by the sewerage undertaker or*

by some other person (where the sewer is the subject of an agreement to make a declaration of vesting pursuant to section 104 of the Water Industry Act 1991).

(Approved Document H5)

Solid waste storage shall be:

- *designed and sited so as not to be prejudicial to health,*
- *of sufficient capacity having regard to the quantity of solid waste to be removed and the frequency of removal,*
- *sited so as to be accessible for use by people in the building and of ready access from a street for emptying and removal.*

(Approved Document H6)

(Building Act 1984 Section 84)

You are required by the Building Act 1984 to ensure that all courts, yards and passageways giving access to a house, industrial or commercial building (not maintained at public expense) are capable of allowing satisfactory drainage of its surface or subsoil to a proper outfall.

 The local authority can require the owner of any of the buildings to complete such works as may be necessary to remedy the defect.

6.3.2 Meeting the requirement

Enclosures for drainage and/or water supply pipes

The enclosure should:	B3 (11.8)
<ul><li>be bounded by a compartment wall or floor, an outside wall, an intermediate floor, or a casing</li><li>have internal surfaces (except framing members) of Class 0</li><li>not have an access panel which opens into a circulation space or bedroom</li><li>be used only for drainage, or water supply, or vent pipes for a drainage system.</li></ul>	
The casing should:	B3 (11.8)
<ul><li>be imperforate except for an opening for a pipe or an access pane</li><li>not be of sheet metal</li><li>have (including any access panel) not less than 30 minutes' fire resistance.</li></ul>	
The opening for a pipe, either in the structure or the casing, should be as small as possible and fire-stopped around the pipe.	B3 (11.8)

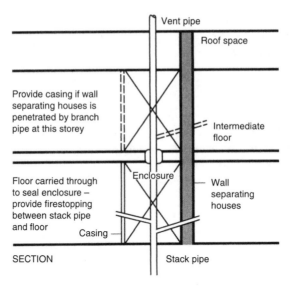

Figure 6.14 Enclosure for drainage or water supply pipes (house with any number of storeys)

Protection of openings for pipes

Pipes which pass through a compartment wall or compartment floor (unless the pipe is in a protected shaft), or through a cavity barrier, should conform to one of the following alternatives:

Proprietary seals (any pipe diameter) that maintain the fire resistance of the wall, floor or cavity barrier.	B3 (11.5–11.6)
Pipes with a restricted diameter should be used where fire-stopping is used around the pipe, keeping the opening as small as possible.	B3 (11.5 and 11.7)
Sleeving – a pipe of lead, aluminium, aluminium alloy, fibre-cement or UPVC, with a maximum nominal internal diameter of 160 mm, may be used with a sleeving of non-combustible pipe as shown in Figure 6.15.	B3 (11.5 and 11.8)

Foul water drainage

The capacity of the system should be large enough to carry the expected flow at any point (*BS 5572, BS 8301*).	H1 (0.1)

All pipes, fittings and joints should be capable of withstanding an air test of positive pressure of at least 38 mm water gauge for at least 3 minutes.	H1 (1.38)
Every trap should maintain a water seal of at least 25 mm.	H1 (1.38)

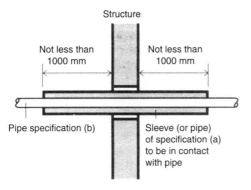

Figure 6.15 Pipes penetrating a structure

Make the opening in the structure as small as possible and provide fire-stopping between pipe and structure.

Traps

All points of discharge into the system should be fitted with a trap (e.g. a water seal) to prevent foul air from the system entering the building.	H1 (1.3–1.4)
All traps should be fitted directly over an appliance and should be removable or be fitted with a cleaning eye.	H1 (1.6)

Branch discharge pipes

Branch pipes should either discharge into another branch pipe or a discharge stack (unless the appliances discharge into a gully on the ground floor or at basement level).	H1 (1.5)
If the appliances are on the ground floor, the pipe(s) may discharge to a stub stack, discharge stack, directly to a drain, or (if the pipe carries only waste water) to a gully.	H1 (1.5–1.17) H1 (1.11) H1 (1.30)

A branch pipe from a ground floor closet should
only discharge directly to a drain if the depth from
the floor to the drain is 1.3 m or less (see Figure 6.16).

H1 (1.9)

A branch pipe serving any ground floor appliance
may discharge direct to a drain or into its own stack.

H1 (A5)

A branch pipe should not discharge into a stack
in a way which could cause cross flow into any
other branch pipe (see Figure 6.17).

H1 (1.10)

Table 6.10 Minimum trap sizes and seal depths

Appliance	Diameter of trap (mm)	Depth of seal (mm of water or equivalent)
Washbasin Bidet	32	75
Bath Shower	40	50
Food waste disposal unit Urinal bowl	40	75
Sink Washing machine Dishwashing machine		
WC pan (outlet < 80 mm) WC pan (outlet > 80 mm)	75 100	50 50

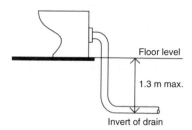

Floor level

1.3 m max.

Invert of drain

Figure 6.16 Direct connection of ground floor WC to drain

Branch discharge pipes

A branch discharge pipe should not discharge into
a stack lower than 450 mm above the invert of the
tail of the bend at the foot of the stack in single
dwellings up to 3 storeys (see Figure 6.18).

H1 (1.8)
H1 (A3, A4)
H1 (1.21)

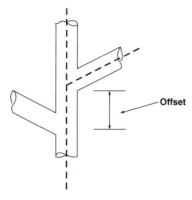

Figure 6.17 Branch connections

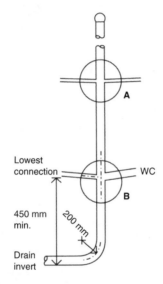

Figure 6.18 Branch discharge stack

Branch discharge pipes

Branch pipes may discharge into a stub stack.	H1 (1.12)
	H1 (1.30)
A branch pipe discharging to a gully should terminate between the grating or sealing plate and the top of the water seal.	H1 (1.13)
Bends in branch pipes should be avoided if possible.	H1 (1.16)
Junctions on branch pipes should be made with a sweep of 25 mm radius or at 45°.	H1 (1.17)

Rodding points should be provided to give access   H1 (1.25)
to any lengths of discharge pipes which cannot be   H1 (1.6)
reached by removing traps or appliances with
integral traps.

A branch pipe discharging to a gully should terminate   H1 (1.13)
between the grating or sealing plate and the top
of the water seal.

Condensate drainage from boilers may be connected   H1 (1.14)
to sanitary pipework provided:

(a)   The connection should preferably be made to an
       internal stack with a 75 mm condensate trap.
(b)   If the connection is made to a branch pipe,
       the connection should be made downstream
       of any sink waste connection.
(c)   All sanitary pipework receiving condensate should
       be made from materials resistant to a pH value
       of 6.5 and lower and be installed in accordance
       with BS 6798.

Pipes serving a single appliance should have
at least the same diameter as the appliance trap
(see Table 6.10).

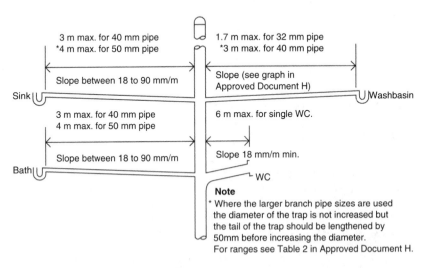

Figure 6.19 Branched connections

A separate ventilating stack is only likely to be preferred where the numbers of
sanitary appliances and their distance to a discharge stack are large.

Branch ventilation stacks

Should be connected to the discharge pipe within 750 mm of the trap and should connect to the ventilating stack or the stack vent, above the highest 'spillover' level of the appliances served.	H1 (1.22)
The ventilating pipe should have a continuous incline from the discharge pipe to the point of connection to the ventilating stack or stack vent.	H1 (1.22)
Branch ventilating pipes which run direct to outside air should finish at least 900 mm above any opening into the building nearer than 3 m (see Figure 6.21).	H1 (1.23)
A dry stack may provide ventilation for branch ventilation pipes as an alternative to carrying them to outside air or to a ventilated discharge stack (ventilated system).	H1 (A7 and 1.21)
Ventilation stacks serving buildings with not more than 10 storeys and containing only dwellings should be at least 32 mm diameter (for all other buildings see paragraph H1 (1.29)).	H1 (A8) H1 (1.21 and 1.29)
A separate ventilating stack is only likely to be preferred where the numbers of ventilating pipes and their distance to a discharge stack are large.	H1 (1.19) H1 (Table 2)

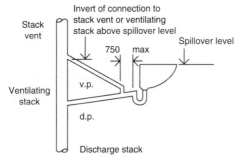

Figure 6.20 Branch ventilation pipes

Discharge stacks

All stacks should discharge to a drain.	H1 (1.26)
The bend at the foot of the stack should have as large a radius (i.e. at least 200 mm) as possible.	H1 (1.26)

Discharge stacks should be ventilated.	H1 (1.29)
Offsets in the 'wet' portion of a discharge stack should be avoided.	H1 (1.27)
Stacks serving urinals should be not less than 50 mm.	H1 (1.28)
Stacks serving closets with outlets less than 80 mm should be not less than 75 mm.	H1 (1.28)
Stacks serving closets with outlets greater than 80 mm should be not less than 100 mm.	H1 (1.28)
The internal diameter of the stack should be not less than that of the largest trap or branch discharge pipe.	H1 (1.28)

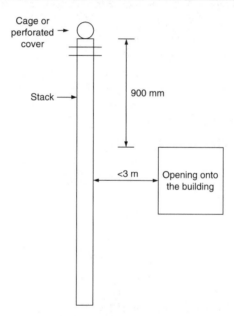

Figure 6.21 Termination of ventilation stacks

Ventilating pipes open to outside air should finish at least 900 mm above any opening into the building within 3 m and should be fitted with a perforated cover or cage (see Figure 6.21) which should be metal if rodent control is a problem.	H1 (1.31)
Ventilating pipes open to outside air should finish at least 900 mm above any opening into the building within 3 m.	H1 (1.31)

Ventilating pipes should be finished with a wire cage (metallic in areas with a rodent problem) or other perforated cover, fixed to the end of the ventilating pipe.	H1 (1.31)
Stack ventilation pipes should be not less than 75 mm.	H1 (1.32)
Ventilated discharge stacks may be terminated inside a building when fitted with air admittance valves complying with prEN 12380.	H1 (1.33)
Discharge stacks may terminate inside a building when fitted with air admittance valves.	H1 (1.29)
Rodding points should be provided to give access to any lengths of pipe that cannot be reached from any other part of the system.	H1 (1.30–1.31)
Pipes should be firmly supported without restricting thermal movement.	H1 (1.31)
Pipes, fittings and joints should be airtight.	H1 (1.32)
A stub stack may be used if it connects into a ventilated discharge stack or into a ventilated drain not subject to surcharging.	H1 (1.30)
Air admittance valves should be located in areas that have adequate ventilation.	H1 (1.33)
Air admittance valves should not be used outside buildings or in dust laden atmospheres.	H1 (1.33)
Rodding points should be provided in discharge stacks.	H1 (1.34)
Pipes should be firmly supported without restricting thermal movement.	H1 (1.35)
Sanitary pipework connected to WCs should not allow light to be visible through the pipe wall, as this is believed to encourage damage by rodents.	H1 (1.36)

 Drainage serving kitchens in commercial hot food premises should be fitted with a grease separator complying with prEN 1825-1.

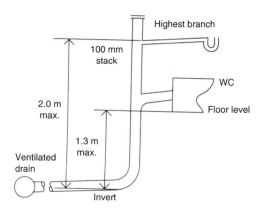

Figure 6.22 Stub stack

Foul drainage

Some public sewers may carry foul water and rainwater in the same pipe. If the drainage system is also to carry rainwater to such a sewer these combined systems should not be capable of discharging into a cesspool or septic tank.	H1 (2.1)
Foul drainage should be connected to either:	
• a public foul or combined sewer (wherever this is reasonably practicable)	H1 (2.3)
• an existing private sewer that connects with a public sewer, or	H1 (2.6)
• a wastewater treatment system or cesspool should be provided.	H1 (2.7)
Combined and rainwater sewers are designed to surcharge (i.e. the water level in the manhole rises above the top of the pipe) in heavy rainfall.	H1 (2.8)
Basements containing sanitary appliances, where the risk of flooding due to sewer surcharge of the sewer is possible should either use an anti-flooding valve (if the risk is low) or be pumped.	H1 (2.9) H1 (2.36–2.39) H1 (2.10)

☀ For other low lying sites (i.e. not basements) where the risk is considered low, a gully (at least 75 mm below the floor level) can be dug outside the building.

Anti-flooding valves should preferably be a double valve type that complies with prEN 13564.	H1 (2.11)
The layout of the drainage system should be kept simple.	H1 (2.13)
Pipes should (wherever possible) be laid in straight lines. Changes of direction and gradient should be minimized.	
Access points should be provided only if blockages could not be cleared without them.	H1 (2.13)
Connections should be made using prefabricated components.	H1 (2.15)
Connection of drains to other drains or private or public sewers and of private sewers to public sewers should be made obliquely, or in the direction of flow.	H1 (2.14)
The system should be ventilated by a flow of air.	H1 (2.18) H1 (1.27–1.29)
Ventilating pipes should not finish near openings in buildings.	H1 (2.18) H1 (1.31)
Pipes should be laid to even gradients and any change of gradient should be combined with an access point.	H1 (2.19) H1 (2.49)
Pipes should also be laid in straight lines where practicable.	H1 (2.20) H1 (2.49)

Rodent control

If the site has been previously developed, the local authority should be consulted to determine whether any special measures are necessary for control of rodents. Special measures which may be taken include the following:

Sealed drainage – should have access covers to the pipework in the inspection chamber instead of an open channel.	H1 (2.22a)
Intercepting traps – should be of the locking type that can be easily removed from the chamber surface and securely replaced.	H1 (2.22b)
Rodent barriers – including enlarged sections on discharge stacks to prevent rats climbing, flexible	H1 (2.22c)

downward facing fins in the discharge stack, or
one-way valves in underground drainage.

Metal cages on ventilator stack terminals – to discourage rats from leaving the drainage system.	H1 (2.22d) H1 (1.31)
Covers and gratings to gullies – used to discourage rats from leaving the system.	H1 (2.22e)
During construction, drains and sewers that are left open should be covered when work is not in progress to prevent entry by rats.	H1 (2.56)
Disused drains or sewers less than 1.5 m deep that are in open ground should as far as is practicable be removed. Other pipes should be sealed at both ends (and at any point of connection) and grout filled to ensure that rats cannot gain access.	H1 (B18)

Protection from settlement

• A drain may run under a building if at least 100 mm of granular or other flexible filling is provided round the pipe.	H1 (2.23)
• Where pipes are built into a structure (e.g. inspection chamber, manhole, footing, ground beam or wall) suitable measures (such as using rocker joints or a lintel) should be taken to prevent damage or misalignment (see Figures 6.23 and 6.24).	H1 (2.24)
The depth of cover will usually depend on the levels of the connections to the system, the gradients at which the pipes should be laid and the ground levels.	H1 (2.27) H1 (2.41–2.45)
All drain trenches should not be excavated lower than the foundations of any building nearby (see Figure 6.25).	H1 (2.25)

Pipe gradients and sizes

Drains should have enough capacity to carry the anticipated maximum flow (see Table 6.11).	H1 (2.29)

Sewers (i.e. a drain serving more than one property)   H1 (2.30)
should have a minimum diameter of 100 mm when serving
10 dwellings or diameter of 150 mm if more than 10.

Drains carrying foul water should have an internal diameter   H1 (2.33)
of at least 75 mm.

Drains carrying effluent from a WC or trade effluent   H1 (2.33)
should have an internal diameter of at least 100 mm.

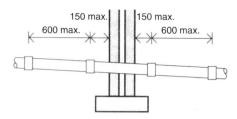

Figure 6.23 Pipe imbedded in the wall. Short length of pipe bedded in a wall with joints 150 mm of either wallface. Additional rocker pipes (max length 600 mm) with flexible joints are then added

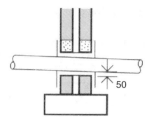

Figure 6.24 Pipe shielded by a lintel. Both sides are masked with rigid sheet material (to prevent entry of fill or vermin) and the void is filled with a compressible sealant to prevent entry of gas

Pumping installations

Where gravity drainage is impracticable, or protection   H1 (2.36)
against flooding due to surcharge in downstream sewers
is required, a pumping installation will be needed.

Where foul water drainage from a building is to be pumped,   H1 (2.39)
the effluent receiving chamber should be sized to
contain 24-hour inflow to allow for disruption in service.

The minimum daily discharge of foul drainage should be   H1 (2.39)
taken as 150 litres per head per day for domestic use.

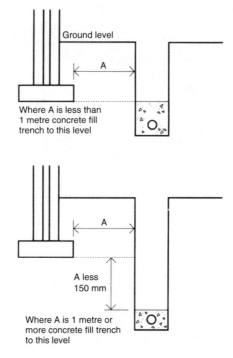

Figure 6.25 Pipe runs near buildings

Table 6.11 Flow rates from dwellings

Number of dwellings	Flow rate (litres/sec)
1	2.5
5	3.5
10	4.1
15	4.6
20	5.1
26	5.4
30	5.8

Table 6.12 Materials for below-ground gravity drainage

Material	British Standard
Rigid pipes	
Vitrified clay	BS 65, BSEN 295
Concrete	BS 5911
Grey iron	BS 437
Ductile iron	BSEN 598
Flexible pipes	
UPVC	BSEN 1401
PP	BSEN 1852
Structured walled plastic pipes	BSEN 13476

Materials for pipes and jointing

To minimize the effects of any differential settlement, pipes should have flexible joints.	H1 (2.40)
All joints should remain watertight under working and test conditions.	H1 (2.40)
Nothing in the pipes, joints or fittings should project into the pipe line or cause an obstruction.	H1 (2.40)
Different metals should be separated by non-metallic materials to prevent electrolytic corrosion.	H1 (2.40)

Bedding and backfill

The choice of bedding and backfill depends on the depth at which the pipes are to be laid and the size and strength of the pipes.	H1 (2.41)
Special precautions should be taken to take account of the effects of settlement where pipes run under or near buildings.	
The depth of the pipe cover will usually depend on the levels of the connections to the system and the gradients at which the pipes should be laid and the ground levels.	
Pipes need to be protected from damage particularly pipes which could be damaged by the weight of backfilling.	
Rigid pipes should be laid in a trench as shown in Figure 6.26.	H1 (2.42)

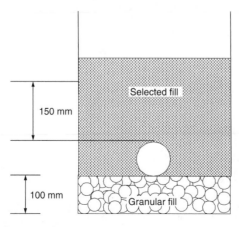

Figure 6.26 Bedding for rigid pipes

Flexible pipes shall be supported to limit deformation under load.   H1 (2.44)

Flexible pipes with very little cover shall be protected from damage by a reinforced cover slab with a flexible filler and at least 75 mm of granular material between the top of the pipe and the underside of the flexible filler below the slabs (see Figure 6.28).   H1 (2.42–2.44)

Trenches may be backfilled with concrete to protect nearby foundations. In these cases a movement joint (as shown in Figure 6.29) formed with a compressible board should be provided at each socket or sleeve joint.   H1 (2.45)

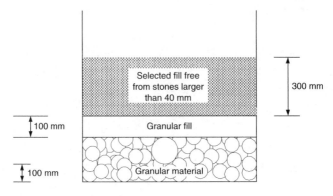

Figure 6.27 Bedding for flexible pipes

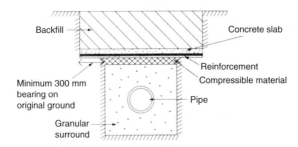

Figure 6.28 Protection of pipes laid in shallow depths

Access points

Access should be provided to long runs.   H1 (2.50)

Sufficient and suitable access points should be provided for clearing blockages from drain runs that cannot be reached by any other means.   H1 (2.46)

Access points should be provided: • on or near the head of each drain run • at a bend • at a change of gradient or pipe size • at a junction.	H1 (2.49)
Access points should be either: • rodding eyes – capped extensions of the pipes; • access fittings – small chambers on (or an extension of) the pipes but not with an open channel; • inspection chambers – chambers with working space at ground level; • manholes – deep chambers with working space at drain level.	H1 (2.48)
Access points should be constructed so as to resist the ingress of ground water or rainwater.	H1 (2.52)
Inspection chambers and manholes should have removable non-ventilating covers of durable material (such as cast iron, cast or pressed steel, precast concrete or UPVC).	H1 (2.54)
Access points to sewers (serving more than one property) should be located in places where they are accessible and apparent for use in an emergency (e.g. highways, public open space, unfenced front gardens, and shared or unfenced driveways).	H1 (2.51)
Inspection chambers and manholes in buildings should have mechanically fixed airtight covers unless the drain itself has watertight access covers.	H1 (2.54)
Manholes deeper than 1 m should have metal step irons or fixed ladders.	H1 (2.54)

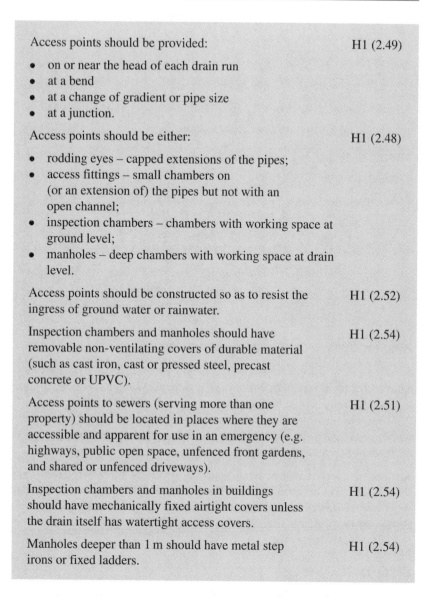

Figure 6.29 Joints for concrete encased pipes

General

Drains and sewers should be protected from damage by construction traffic and heavy machinery.	H1 (2.57)
Heavy materials should not be stored over drains or sewers.	H1 (2.57)
After laying (including any necessary concrete or other haunching or surrounding and backfiring) gravity drains and private sewers should be tested for watertightness.	H1 (2.59)
All pipework carrying greywater for reuse should be clearly marked with the word 'GREYWATER'.	H1 (A11)
Material alterations to existing drains and sewers are subject to (and covered by) the Building Regulations.	H1 (b7)
Repairs, reconstruction and alterations to existing drains and sewers should be carried out to the same standards as new drains and sewers.	H1 (B15)

Wastewater treatment systems and cesspools

A notice giving information as to the nature and frequency of maintenance required for the cesspool or wastewater treatment system to continue to function satisfactorily should be displayed within each of the buildings.

The use of non-mains foul drainage, such as wastewater treatment systems, septic tanks or cesspools, should only be considered where connection to mains drainage is **not** practicable.

 Any discharge from a wastewater treatment system is likely to require a consent from the Environment Agency. For the detailed design and installation of small sewage treatment works, specialist knowledge is advisable. Guidance is also given in BS 6297: 1983 *Code of practice for design and installation of small sewage treatment works and cesspools*.

Septic tanks

Septic tanks with some form of secondary treatment (such as from a drainage field/mound or constructed wetland such as a reed bed) will normally be the most economic means of treating wastewater from small developments (e.g. 1 to 3 dwellings). They provide suitable conditions for the settlement, storage and partial decomposition of solids which need to be removed at regular intervals.

Septic tanks should be sited at least 7 m from any habitable parts of buildings, and preferably down a slope.	H2 (1.16)

Septic tanks should only be used in conjunction with a form of secondary treatment (e.g. a drainage field, drainage mound or constructed wetland).	H2 (1.15)
Septic tanks should be sited within 30 m of a vehicle access to enable the tank to be emptied and cleaned without hazard to the building occupants and without the contents being taken through a dwelling or place of work.	H2 (1.17 and 1.64)
Septic tanks and settlement tanks should have a capacity below the level of the inlet of at least 2700 litres (2.7 m^3) for up to 4 users. This size should be increased by 180 litres for each additional user.	H2 (1.18)
Septic tanks may be constructed in brickwork or concrete (roofed with heavy concrete slabs) or factory-manufactured septic tanks (made out of glass reinforced plastics, polyethylene or steel) can be used.	H2 (1.19–20 and 1.65–66)
The brickwork should consist of engineering bricks at least 220 mm thick. The mortar should be a mix of 1:3 cement sand ratio and in-situ concrete should be at least 150 mm thick of C/25/P mix (see BS 5328).	H2 (1.20 and 1.66)
Septic tanks should be ventilated.	H2 (1.21)
Septic tanks should incorporate at least two chambers or compartments operating in series.	H2 (1.22)
Septic tanks should be provided with access for emptying and cleaning.	H2 (1.24)
A notice should be fixed within the building describing the necessary maintenance.	H2 (1.25)
Septic tanks should be inspected monthly to check they are working correctly.	H2 (A.11)
Septic tank should be emptied at least once a year.	H2 (A.13)

Cesspools

A cesspool is a watertight tank, installed underground, for the storage of sewage. No treatment is involved.

Cesspools should be sited at least 7 m from any habitable parts of buildings and preferably downslope.	H2 (1.58)
Cesspools should be provided with access for emptying and cleaning.	H2 (1.60)
Cesspools should be inspected fortnightly for overflow.	H2 (A.20)
Cesspools should be emptied on a monthly basis by	H2 (1.60)
a licensed contractor.	H2 (A.21)
A filling rate of 150 litres per person per day is assumed and if the cesspool does not fill within the estimated period, the tank should be inspected for leakage.	H2 (A.22)
Cesspools should be ventilated.	H2 (1.63)
The inlet of a cesspool should be provided with access for inspection.	H2 (1.67)
Cesspools and settlement tanks (if they are to be desludged using a tanker) should be sited within 30 m of a vehicle access.	H2 (1.64)
Cesspools and settlement tanks should prevent leakage of the contents and ingress of subsoil water.	H2 (1.63)
Cesspools should have a capacity below the level of the inlet of at least 18 000 litres (18 m^3) for 2 users increased by 6800 litres (6.8 m^3) for each additional user.	H2 (1.61)
Cesspools, septic tanks and settlement tanks may be constructed in brickwork, concrete, or glass reinforced concrete.	H2 (1.65–66)
Factory-made cesspools and septic tanks are available in glass reinforced plastic, polyethylene or steel.	
The brickwork should consist of engineering bricks at least 220 mm thick. The mortar should be a mix of 1:3 cement sand ratio and in-situ concrete should be at least 150 mm thick of C/25/P mix (see BS 5328).	H2 (1.66)
Cesspools should be covered (with heavy concrete slabs) and ventilated.	
Cesspools should have no openings except for the inlet, access for emptying and ventilation.	H2 (1.62)
Cesspools should be inspected fortnightly for overflow and emptied as required.	H2 (A.20)

Packaged treatment works

This term is applied to a range of systems designed to treat a given hydraulic and organic load using prefabricated components which can be installed with minimal site work. They are capable of treating effluent more efficiently than septic tank systems and this normally allows the product to be directly discharged to a watercourse.

The discharge from the wastewater treatment plant should be sited at least 10 m away from watercourses and any other buildings.	H2 (1.54)
Regular maintenance and inspection should be carried out in accordance with the manufacturer's instructions.	H2 (A.17)

Drainage fields and mounds

Drainage fields (or mounds) serving a wastewater treatment plant or septic tank should be located: • at least 10 m from any watercourse or permeable drain • at least 50 m from the point of abstraction of any groundwater supply • at least 15 m from any building • sufficiently far from any other drainage fields, drainage mounds or soakaways so that the overall soakage capacity of the ground is not exceeded.	H2 (1.27)
No water supply pipes or underground services other than those required by the disposal system itself should be located within the disposal area.	H2 (1.29)
No access roads, driveways or paved areas should be located within the disposal area.	H2 (1.30)
The ground water table should not rise to within 1 m of the invert level of the proposed effluent distribution pipes.	H2 (1.33)
An inspection chamber should be installed between the septic tank and the drainage field.	H2 (1.43)
Constructed wetlands should not be located in the shade of trees or buildings.	H2 (1.47)
The drainage field/mound should be checked on a monthly basis to ensure that it is not waterlogged and that the effluent is not backing up towards the septic tank.	H2 (A.15)

Under Section 50 (overflowing and leaking cesspools) of the Public Health Act 1936 action could be taken against a builder who had caused the problem, and **not** just against the owner.

 Under Section 59 (drainage of building) of the Building Act 1984, local authorities can require either the owner or the occupier to remove (or otherwise make innocuous) any disused cesspool, septic tank or settlement tank.

Greywater and rainwater tanks

Greywater and rainwater tanks should: H2 (1.70)

- prevent leakage of the contents and ingress of subsoil water;
- be ventilated;
- have an anti-backflow device;
- be provided with access for emptying and cleaning.

Rainwater drainage

The capacity of the drainage system should be large enough   H3 (0.3)
to carry the expected flow at any point in the system.

Rainwater or surface water should not be discharged to   H3 (0.6)
a cesspool or septic tank.

Gutters and rainwater pipes

Although this part of the Building Regulations only actually applies to draining the rainfall from areas of $6\,m^2$ or more (unless they receive a flow from a rainwater pipe or from paved and/or other hard surfaces). Each case should be considered separately and a decision made. This particularly applies to small roofs and balconies. Table 6.13 shows the largest effective area that should be drained into the gutter sizes most often used.

 For eaves gutters the design rainfall intensity should be 0.021 litres/second/m^2. In some cases, eaves drop systems may be used (H3 (1.13)).

Gutters should be laid with any fall towards the nearest outlet.

Gutters should be laid so that any overflow in excess of   H3 (1.7)
the design capacity (e.g. above normal rainfall) will be
discharged clear of the building.

Table 6.13 Gutter and outlet sizes

Max effective roof area (m²)	Gutter size (mm dia)	Outlet size (mm dia)	Flow capacity (litres/sec)
6.0	–	–	–
18.0	75	50	0.38
37.0	100	63	0.78
53.0	115	63	1.11
65.0	125	75	1.37
103.0	150	89	2.16

Rainwater pipes should discharge into a drain or gully (but may discharge to another gutter or onto another surface if it is drained).	H3 (1.8)
Any rainwater pipe which discharges into a combined system should do so through a trap.	H3 (1.8)
The size of a rainwater pipe should be at least the size of the outlet from the gutter.	H3 (1.10)
A down pipe which serves more than one gutter should have an area at least as large as the combined areas of the outlets.	H3 (1.10)
On flat roofs, valley gutters and parapet gutters additional outlets may be necessary.	H3 (1.7)
Where a rainwater pipe discharges onto a lower roof or paved area, a pipe shoe should be fitted to divert water away from the building.	H3 (1.9)
Gutters and rainwater pipes should be firmly supported without restricting thermal movement.	
The materials used should be of adequate strength and durability, and	H3 (1.16)

- all gutter joints should remain watertight under working conditions
- pipework in siphonic roof drainage systems should be able to resist to negative pressures in accordance with the design
- gutters and rainwater pipes should be firmly supported
- different metals should be separated by non-metallic material to prevent electrolytic corrosion.

Drainage of paved areas

Surface gradients should direct water draining from a paved area away from buildings.	H3 (2.2)
Gradients on impervious surfaces should be designed to permit the water to drain quickly from the surface. A gradient of at least 1 in 60 is recommended.	H3 (2.3)
Paths, driveways and other narrow areas of paving should be free draining to a pervious area such as grassland, provided that:	H3 (2.6)

- the water is not discharged adjacent to buildings where it could damage foundations; and
- the soakage capacity of the ground is not overloaded.

Where water is to be drained onto the adjacent ground the edge of the paving should be finished above or flush with the surrounding ground to allow the water to run off.	H3 (2.7)

- Where the surrounding ground is not sufficiently permeable to accept the flow, filter drains may be provided. — H3 (2.8 and 3.33)
- Pervious paving should not be used where excessive amounts of sediment are likely to enter the pavement and block the pores. — H3 (2.11)
- Pervious paving should not be used in oil storage areas, or where runoff may be contaminated with pollutants. — H3 (2.12)
- Gullies should be provided at low points where water would otherwise pond. — H3 (2.15)
- Gully gratings should be set approximately 5 mm below the level of the surrounding paved area in order to allow for settlement. — H3 (2.16)
- Provision should be made to prevent silt and grit entering the system, either by provision of gully pots of suitable size, or catchpits. — H3 (2.17)

Surface water drainage

Discharge to a watercourse may require a consent from the Environment Agency, who may limit the rate of discharge. Where other forms of outlet are not practicable, discharge should be made to a sewer (H3 (3.2–3.3)). For design purposes a rainfall interval of 0.014 litres/second/m² can be assumed as normal.

 Some drainage authorities have sewers that carry both foul water and rain-water (i.e. combined systems) in the same pipe. Where they do, they can allow rainwater to discharge into the system if the sewer has enough capacity to take the added flow. Some private sewers (drains serving more than one property) also carry both foul water and rainwater. If a sewer (or private sewer) operated as a combined system does not have enough capacity, the rainwater should be run in a separate system with its own outfall.

Surface water drainage should discharge to a soakaway or other infiltration system where practicable.	H3 (3.2)
Surface water drainage connected to combined sewers should have traps on all inlets.	H3 (3.7)
Drains should be at least 75 mm diameter.	H3 (3.14)
Where any materials that could cause pollution are stored or used, separate drainage systems should be provided.	H3 (3.21)
On car parks, petrol filling stations or other areas where there is likely to be leakage or spillage of oil, drainage systems should be provided with oil interceptors.	H3 (3.22) H3 (A)
Separators should be leak tight and comply with the requirements of the Environmental Agency and prEN858.	H3 (A.9–10)
Infiltration devices (including soakaways, swales, infiltration basins, and filter drains) should not be built: • within 5 m of a building or road or in areas of unstable land; • in ground where the water table reaches the bottom of the device at any time of the year; • sufficiently far from any drainage fields, drainage mounds or other soakaways; • where the presence of any contamination in the runoff could result in pollution of groundwater source or resource.	H3 (3.23–26)
Soakaways should be designed to a return period of once in ten years.	H3 (3.27)
Soakaways for areas less than 100 m² shall consist of square or circular pits, filled with rubble or lined with dry jointed masonry or perforated ring units. Soakaways serving larger areas shall be lined pits or trench type soakaways.	H3 (3.26)

> The storage volume should be calculated so that, over   H3 (3.29)
> the duration of a storm, it is sufficient to contain the
> difference between the inflow volume and the outflow
> volume.
>
> Soakaways serving larger areas should be designed in   H3 (3.30)
> accordance with BS EN 752-4.

 Under Section 85 (offences concerning the polluting of controlled waters) of the Water Resources Act 1991 it is an offence to discharge any noxious or polluting material into a watercourse, coastal water, or underground water. Most surface water sewers discharge to watercourses.

 Under Section 111 (restrictions on use of public sewers) of the Water Industry Act 1991 it is an offence to discharge petrol into any drain or sewer connected to a public sewer.

Building over existing sewers

> Where it is proposed to construct a building over or near   H4 (0.3)
> a drain or sewer shown on any map of sewers, the
> developer should consult the owner of the drain or sewer.
>
> A building constructed over or within 3 m of any   H4 (1.2)
>
> • rising main
> • drain or sewer constructed from brick or masonry
> • drain or sewer in poor condition
>
> shall not be constructed in such a position unless special
> measures are taken.
>
> Buildings or extensions should not be constructed over   H4 (1.3)
> a manhole or inspection chamber or other access
> fitting on any sewer (serving more than one property).
>
> A satisfactory diversionary route should be available so   H4 (1.4)
> that the drain or sewer could be reconstructed without
> affecting the building.
>
> The length of drain or sewer under a building should not   H4 (1.5)
> exceed 6 m except with the permission of the owners of
> the drain or sewer.
>
> Buildings or extensions should not be constructed over   H4 (1.60)
> or within 3 m of any drain or sewer more than 3 m deep,
> or greater than 225 mm in diameter except with the
> permission of the owners of the drain or sewer.

Where a drain or sewer runs under a building at least 100 mm of granular or other suitable flexible filling should be provided round the pipe.   H4 (1.9)

Where a drain or sewer running below a building is less than 2 m deep, the foundation should be extended locally so that the drain or sewer passes through the wall.   H4 (1.10)

Where the drain or sewer is more than 2 m deep to invert and passes beneath the foundations, the foundations should be designed with a lintel spanning over the line of the drain or sewer. The span of the lintel should extend at least 1.5 m either side of the pipe and should be designed so that no load is transmitted onto the drain or sewer.   H4 (1.12)

A drain trench should not be excavated lower than the foundations of any building nearby.   H4 (1.13)

Separate systems for drainage

Separate systems of drains and sewers shall be provided for foul water and rainwater where:

(a)   the rainwater is not contaminated; and
(b)   the drainage is to be connected either directly or indirectly to the public sewer system, which has separate systems for foul water and surface water.

Solid waste storage

 Although the requirements of the Building Regulations do not cover the recycling of household and other waste, H6 sets out general requirements for solid waste storage.

For domestic developments space should be provided for storage of containers for separated waste (i.e. waste that can be recycled is stored separately from waste that cannot) and having a combined capacity of 0.25 m² per dwelling.   H6 (1.1)

In low-rise domestic developments (houses, bungalows and flats up to the 4th floor) any dwelling should have, or have access to, a location with at least two movable, individual or communal waste containers.   H6 (1.2)

In multistorey domestic developments, dwellings above the 4th storey should share a container fed by a chute unless siting or operation of a chute is impracticable. In such a case a satisfactory management arrangement for conveying refuse to the storage area should be assured.   H6 (1.6)

In multistorey domestic developments, dwellings up to the 4th floor may each have their own waste container or may share a waste container.   H6 (1.5)

For waste containers up to 250 litres, steps should be avoided between the container store and collection point wherever possible.   H6 (1.10)

Containers and chutes should be sited so that householders are not required to carry refuse further than 30 m.   H6 (1.8)

Containers should be within 25 m of the vehicle access.   H6 (1.8)

Containers should be sited so that they can be collected without being taken through a building, unless it is a garage, carport or other open covered space.   H6 (1.10)

This provision applies only to new buildings.

The collection point should be reasonably accessible to the size of waste collection vehicles typically used by the waste collection authority.   H6 (1.11)

External storage areas for waste containers should be away from windows and ventilators and preferably be in the shade or under a shelter.   H6 (1.12)

Storage areas should not interfere with pedestrian or vehicle access to buildings.   H6 (1.12)

Where enclosures, compounds or storage rooms are provided they should allow room for filling and emptying and provide a clear space of 150 mm between and around the containers.   H6 (1.13)

- Enclosures, compounds or storage rooms for communal containers should be a minimum of 2 m high.   H6 (1.13)
- Enclosures for individual containers should be sufficiently high to allow the lid to be opened for filling.   H6 (1.13)
- The enclosure should be permanently ventilated at the top and bottom and should have a paved impervious floor.   H6 (1.13)
- Communal storage areas should have provision for washing down and draining the floor into a system suitable for receiving a polluted effluent.   H6 (1.14)

- Gullies should incorporate a trap that maintains a seal   H6 (1.14)
  even during prolonged periods of disuse.
- Any room (or compound) for the open storage of   H6 (1.15)
  waste should be secure to prevent access by vermin.
- Where storage rooms are provided, separate rooms   H6 (1.16)
  should be provided for the storage of waste that
  cannot be recycled, and waste that can be recycled.
- Where the location for storage is in a publicly   H6 (1.17)
  accessible area or in an open area around a building
  (e.g. a front garden) an enclosure or shelter should be
  considered.
- In high-rise domestic developments, where chutes are   H6 (1.18)
  provided they should be at least 450 mm in diameter
  and should have a smooth non-absorbent surface and
  close-fitting access doors at each storey that has a
  dwelling and be ventilated at the top and bottom.

6.4 Water supplies

6.4.1 The requirement (Building Act 1984 Sections 25 and 69)

The Building Act stipulates that plans for proposed buildings will ensure that
all occupants of the house will be provided with a supply of '*wholesome water,
sufficient for their domestic purposes*'. This can be achieved by either:

- connecting the house to water supplies from the local water authority (normally referred to as the '*statutory water undertaker*');
- by otherwise taking water into the house by means of a pipe (e.g. from a
  local recognized supply);
- by providing a supply of water within a reasonable distance from the house
  (e.g. such as from a well).

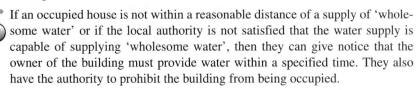

If an occupied house is not within a reasonable distance of a supply of 'wholesome water' or if the local authority is not satisfied that the water supply is
capable of supplying 'wholesome water', then they can give notice that the
owner of the building must provide water within a specified time. They also
have the authority to prohibit the building from being occupied.

What happens if there is more than one property?

Where the local authority are satisfied that two or more houses can most con-
veniently be met by means of a joint supply, they may give notice accordingly.

Can I ask the local authority to provide me with a supply of water?

If you are unable to provide a suitable supply of water, the local authority can
themselves provide, or secure the provision of, a supply of water to the house

or houses in question and then recover any expenses reasonably incurred from the owner of the house, or (where two or more houses are concerned), the owners of those houses.

 The maximum amount that a local authority can charge for providing a suitable supply of water is £3000 in respect of any one house.

Where a supply of water is provided to a house by statutory water undertakers, water rates will be included in the normal rateable value of the house.

Where two or more houses are supplied with water by a common pipe belonging to the owners or occupiers of those houses, the local authority may, when necessary, repair or renew the pipe and recover any expenses reasonably incurred by them from the owners or occupiers of the houses.

6.4.2 Meeting the requirement

Protection of openings for pipes

Pipes that pass through a compartment wall or compartment floor (unless the pipe is in a protected shaft), or through a cavity barrier, should be one of the following alternatives:

Proprietary seals (any pipe diameter) that maintain the fire resistance of the wall, floor or cavity barrier.	B3 (11.5–11.6)
Pipes with a restricted diameter where fire-stopping is used around the pipe, keeping the opening as small as possible.	B3 (11.5 and 11.7)
Sleeving – a pipe of lead, aluminium, aluminium alloy, fibre-cement or UPVC, with a maximum nominal internal diameter of 160 mm, may be used with a sleeving of non-combustible pipe as shown in Figure 6.15.	B3 (11.5 and 11.8)

6.5 Cellars

6.5.1 The requirement (Building Act 1984 Section 74)

Unless you have the consent of the local authority, you are not allowed to construct a cellar or room *in (or as part of) a house, an existing cellar, a shop, inn, hotel or office if the floor level of the cellar or room is lower than the ordinary level of the subsoil water on, under or adjacent to the site of the house, shop, inn, hotel or office.*

This does not apply to:

- the construction of a cellar or room carried out in accordance with plans deposited on an application under the Licensing Act 1964;
- the construction of a cellar or room in connection with a shop, inn, hotel or office that forms part of a railway station.

 If the owner of the house, shop, inn, hotel or office allows a cellar or room forming part of it to be used in a manner that he knows to be in contravention with the Building Regulations, he is liable, on summary conviction, to a fine.

Fire precautions

Facilities for venting for heat and smoke from basement areas shall be made available.

(Approved Document B5)

Fire precautions (construction)

Any hidden voids in the construction shall be sealed and subdivided to inhibit the unseen spread of fire and products of combustion, in order to reduce the risk of structural failure, and the spread of fire.

(Approved Document B3)

6.5.2 Meeting the requirements

Venting of heat and smoke from basements

Smoke outlets (also referred to as smoke vents) should be available so as to provide a route for heat and smoke to escape to the open air from the basement level.	B4 (19.2)
Where practicable, each basement space should have one or more smoke outlets (see Figure 6.30).	B4 (19.3)

Smoke outlets

Smoke outlets, connected directly to the open air, should be provided from every basement storey, except for a basement in a single family dwellinghouse.	B4 (19.4)
Smoke outlets should be sited at high level, either in the ceiling or in the wall of the space they serve.	B4 (19.7)
Outlet ducts or shafts, including any bulkheads over them (see Figure 6.30), should be enclosed in non-combustible construction having not less fire resistance than the element through which they pass.	B4

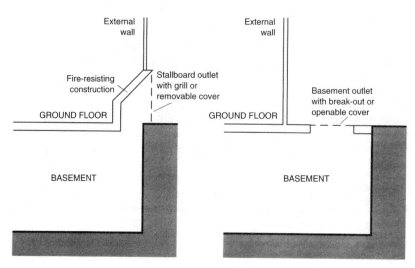

Figure 6.30 Fire resistant construction for smoke outlet shafts

6.6 Floors and ceilings

The ground floor of a building is either solid concrete or a suspended timber type. With a concrete floor, a Damp Proof Membrane (DPM) is laid between walls. With timber floors, sleeper walls of honeycomb brickwork are built on oversite concrete between the base brickwork; a timber sleeper plate rests on each wall and timber joists are supported on them. Their ends may be similarly supported, let into the brickwork or suspended on metal hangers. Floorboards are laid at right angles to joists. First-floor joists are supported by the masonry or hangers.

Similar to a brick built house, the floors in a timber-framed house are either solid concrete or suspended timber. In some cases, a concrete floor may be screeded or surfaced with timber or chipboard flooring. Suspended timber floor joists are supported on wall plates and surfaced with chipboard.

6.6.1 Requirements

The building shall be constructed so that the combined dead, imposed and wind loads are sustained and transmitted by it to the ground

- *safely;*
- *without causing such deflection or deformation of any part of the building (or such movement of the ground) as will impair the stability of any part of another building.*

(Approved Document A1)

The building shall be constructed so that ground movement caused by:

- *swelling, shrinkage or freezing of the subsoil; or*
- *landslip or subsidence (other than subsidence arising from shrinkage)*

will not impair the stability of any part of the building.

(Approved Document A2)

Fire precautions

The building shall be designed and constructed so that there are appropriate provisions for the early warning of fire, and appropriate means of escape in case of fire from the building to a place of safety outside the building capable of being safely and effectively used at all material times.

(Approved Document B1)

 For a typical one- or two-storey dwelling, the requirement is limited to the provision of smoke alarms and to the provision of openable windows for emergency exit (see B1.i).

As a fire precaution, all materials used for internal linings of a building should have a low rate of surface flame spread and (in some cases) a low rate of heat release.

(Approved Document B2)

- *all loadbearing elements of structure of the building shall be capable of withstanding the effects of fire for an appropriate period without loss of stability;*
- *ideally the building should be subdivided by elements of fire-resisting construction into compartments;*
- *all openings in fire-separating elements shall be suitably protected in order to maintain the integrity of the continuity of the fire separation;*
- *any hidden voids in the construction shall be sealed and subdivided to inhibit the unseen spread of fire and products of combustion, in order to reduce the risk of structural failure, and the spread of fire.*

(Approved Document B3)

The floors of the building shall adequately protect the building and people who use the building from harmful effects caused by:

- *ground moisture;*
- *precipitation and wind-driven spray;*
- *interstitial and surface condensation; and*
- *spillage of water from or associated with sanitary fittings or fixed appliances.*

(Approved Document C2)

Airborne and impact sound

Dwellings shall be designed so that the noise from domestic activity in an adjoining dwelling (or other parts of the building) is kept to a level that:

- *does not affect the health of the occupants of the dwelling;*
- *will allow them to sleep, rest and engage in their normal activities in satisfactory conditions.*

(Approved Document E1)

Dwellings shall be designed so that any domestic noise that is generated internally does not interfere with the occupants' ability to sleep, rest and engage in their normal activities in satisfactory conditions.

(Approved Document E2)

Domestic buildings shall be designed and constructed so as to restrict the transmission of echoes.

(Approved Document E3)

Schools shall be designed and constructed so as to reduce the level of ambient noise (particularly echoing in corridors).

(Approved Document E4)

 Note: The normal way of satisfying Requirement E4 will be to meet the values for sound insulation, reverberation time and internal ambient noise which are given in Section 1 of Building Bulletin 93 '*The Acoustic Design of Schools*', produced by DFES and published by the Stationery Office (ISBN: 0 11 271105 7).

6.6.2 Meeting the requirements

General

Floors next to the ground should:	C4.2
• resist the passage of ground moisture to the upper surface of the floor; • not be damaged by moisture from the ground; • not be damaged by groundwater; • resist the passage of ground gases.	
Floors next to the ground and floors exposed from below should be designed and constructed so that their structural and thermal performance are not adversely affected by interstitial condensation.	C4.4
All floors should be designed so they do not promote surface condensation or mould growth.	C4.5

Ground supported floors exposed to moisture from the ground

Unless subjected to water pressure, the ground of a ground supported floor should be covered with dense concrete laid on a hardcore bed and a damp-proof membrane as shown in Table 6.14.

 Note: Suitable insulation may also be incorporated.

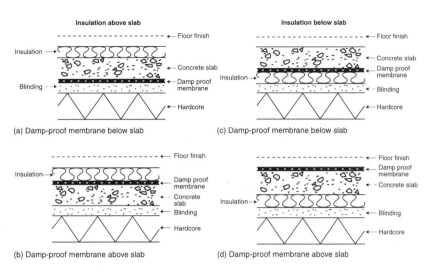

Figure 6.31 Ground supported floor – construction

Table 6.14 Ground supported floor – construction

Hardcore	Well compacted, no greater than 600 mm deep, of clean, broken brick or similar inert material, free from materials including water-soluble sulphates in quantities which could damage the concrete.	C4.7a
Concrete	At least 100 mm thick to mix ST2 in BS 8500 or (if there is embedded reinforcement) to mix ST4 in BS 8500.	C4.7b
Damp-proof membrane	Above or below the concrete which is continuous with the damp-proof courses in walls and piers etc.	C4.7c
	If below the concrete, the membrane could be formed with a sheet of polyethylene, at least 300 mm thick with sealed joints and laid on a bed of material that will not damage the sheet.	C4.8
	If laid above the concrete, the membrane may be either polyethylene (but without the bedding material) or three coats of cold applied bitumen solution or similar moisture and water vapour resisting material.	C4.9
	In each case it should be protected either by a screed or a floor finish, unless the membrane is pitchmastic or similar material which will also serve as a floor finish.	C4.9
Insulation	If placed beneath floor slabs should have sufficient strength to resist the weight of the slab, the anticipated floor loading as well as any possible overloading during construction.	C4.10
	If placed below the damp-proof membrane, then it should have low water absorption and (if considered necessary) should be resistant to contaminants in the ground.	C4.10
Timber floor finish	If laid directly on concrete, it may be bedded in a material which can also serve as a damp-proof membrane.	C4.11
	Timber fillets that are laid in the concrete as a fixing for a floor finish should be treated with an effective preservative unless they are above the damp-proof membrane.	C4.11

Suspended timber ground floors exposed to moisture from the ground

Any suspended timber floor next to the ground should:

- ensure that the ground is covered so as to resist moisture and prevent plant growth;   C4.13a
- have a ventilated air space between the ground covering and the timber;   C4.13b
- have damp-proof courses between the timber and any material which can carry moisture from the ground.   C4.13c

Unless covered with a highly vapour resistant floor finish, a suspended timber floor next to the ground may be built as shown in Figure 6.32 and as follows:

Hardcore	A bed of clean, broken brick or any other inert material free from materials including water-soluble sulphates in quantities which could damage the concrete.
Concrete	A ground covering of unreinforced concrete at least 100 mm thick to mix ST 1 in BS 8500 or C4.14a(i) laid on at least 300 μm polyethylene sheet with sealed joints (and itself laid on a bed of material which will not damage the sheet). C4.14a(ii) **Note:** To prevent water collecting on the ground covering, either the top should be entirely above the highest level of the adjoining ground or, on sloping sites, consideration should be given to installing drainage on the outside of the upslope side of the building.
Ventilation	There should be a ventilated air space at least 75 mm from the ground (and covering the underside of any wall plates) and at least 150 mm from the underside of the suspended timber floor. C4.14b Two opposing external walls should have ventilation openings placed so that the ventilating air will have a free path between opposite sides and to all parts. C4.14b Ventilating openings should be not less than either 1500 mm^2/m run of external wall or C4.14b

500 mm²/m² of floor area – whichever gives the greater opening area.

	Any pipes needed to carry ventilating air should have a diameter of at least 100 mm.	C4.14b
	Ventilation openings should incorporate suitable grilles to prevent the entry of vermin to the subfloor.	C4.14b
	If floor levels need to be nearer to the ground to provide level access, subfloor ventilation can be provided through offset (periscope) ventilators.	C4.14b
Damp-proof membrane	DPMs should be of impervious sheet material, engineering brick or slates in cement mortar or other material which will prevent the passage of moisture.	C4.14c

Insulation

Timber floor finish	In areas such as kitchens, utility rooms and bathrooms where water may be spilled, any board used as a flooring, irrespective of the storey, should be moisture resistant.	C4.15
	In the case of chipboard it should be of one of the grades with improved moisture resistance specified in BS 7331: 1990 or BS EN 312 Part 5: 1997.	C4.15
	Identification marks should be facing upwards.	C4.15
	Any softwood boarding should be at least 20 mm thick and from a durable species or treated with a suitable preservative.	C4.15

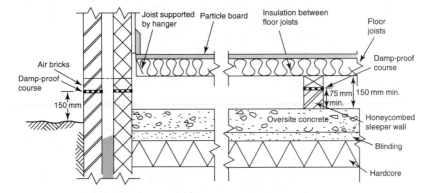

Figure 6.32 Suspended timber floor – construction

Suspended concrete ground floors exposed to moisture from the ground

Concrete suspended floors (including beam and block floors) that are next to the ground should: • adequately prevent the passage of moisture to the upper surface; • be reinforced to protect against moisture.		C4.17
There should be a facility for inspecting and clearing out the subfloor voids beneath suspended floors – particularly in localities where flooding is likely.		C4.20

Hardcore		
Concrete	In-situ concrete at least 100 mm thick containing at least 300 kg of cement for each m³ of concrete; or precast concrete construction (with or without infilling slabs).	C4.1
	Reinforcing steel should be protected by a concrete cover of at least 40 mm (if the concrete is in situ) and at least the thickness required for a moderate exposure, if the concrete is precast.	C4.1
Ventilation	There should be a ventilated air space at least 150 mm clear from the ground to the underside of the floor (or insulation if provided).	C4.19b
	Two opposing external walls should have ventilation openings placed so that the ventilating air will have a free path between opposite sides and to all parts.	C4.19b
	Ventilating openings should be not less than either 1500 mm²/m run of external wall or 500 mm²/m² of floor area – whichever gives the greater opening area.	C4.19b
	Any pipes needed to carry ventilating air should have a diameter of at least 100 mm.	C4.19b
	Ventilation openings should incorporate suitable grilles to prevent the entry of vermin to the subfloor.	C4.19b
Damp-proof membrane	A suspended concrete floor should contain a damp-proof membrane (if the ground below the floor has been excavated below the lowest level of the surrounding ground and will not be effectively drained).	C4.19a

Ground floors and floors exposed from below (resistance to damage from interstitial condensation)

A ground floor (or floor exposed from below such as above an open parking space or passageway – see Figure 6.33) shall be designed in accordance with Clause 8.5 and Appendix D of BS 5250: 2002.

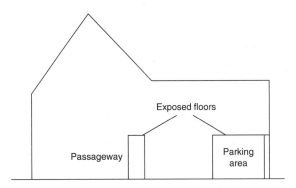

Figure 6.33 Typical floors exposed from below

Floors (resistance to surface condensation and mould growth)

Ground floors should be designed and constructed so that the thermal transmittance (U-value) does not exceed 0.7 W/m²K at any point.	C4.22a
Junctions between elements should be designed in accordance with robust construction recommendations.	C4.22b
Junctions between elements should be designed in accordance with robust construction recommendations.	C4.22b

Means of escape

Floors more than 7.5 m above ground level (where, in the case of a fire, the risk of the stairway becoming impassable before occupants of the upper parts of the house have escaped is appreciable) will be provided with an alternative route.	B1 (2.1)
In the event of fire, suitable means shall be provided for emergency egress from each storey.	B1 (2.1)

Floors more than 7.5 m above ground level (where,   B1 (2.1)
in the case of a fire, the risk that the stairway will
become impassable before occupants of the upper
parts of the house have escaped is appreciable) will
be provided with an alternative route.

Except for kitchens, all habitable rooms in the ground   B1 (2.8)
storey should either open directly onto a hall leading
to the entrance or other suitable exit, or be provided
with a window (or door).

Where a sleeping gallery is provided: B1 (2.9a–c)

- the gallery should be not more than 4.5 m above
  ground level;
- the distance between the foot of the access stair to the
  gallery and the door to the room containing the gallery
  should not exceed 3 m;
- galleries longer than 7.5 m should be provided
  with a separate alternative exit.

Houses with floors above 4.5 m above ground level

The top storey should be separated from the lower storeys   B1 (2.13b)
by fire-resisting construction and be provided with an
alternative escape route leading to its own final exit.

If a house has two or more storeys with floors more   B1 (2.14)
than 4.5 m above ground level (typically a house of
four or more storeys), then an alternative escape route
should be provided from each storey or level situated
7.5 m or more above ground level.

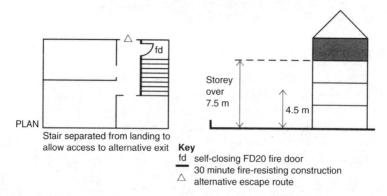

PLAN

Stair separated from landing to
allow access to alternative exit   **Key**
   fd   self-closing FD20 fire door
   ——   30 minute fire-resisting construction
   △   alternative escape route

Storey
over
7.5 m

4.5 m

Figure 6.34 Fire separation in houses with more than one floor over 4.5 m
above ground level

Lateral support by floors

Floors should:

- act to transfer lateral forces from walls to buttressing walls, piers or chimneys;   A1/2 2C33a
- be secured to the supported wall by connections (see Figure 6.35 and Table 6.15).   A1/2 2C33b

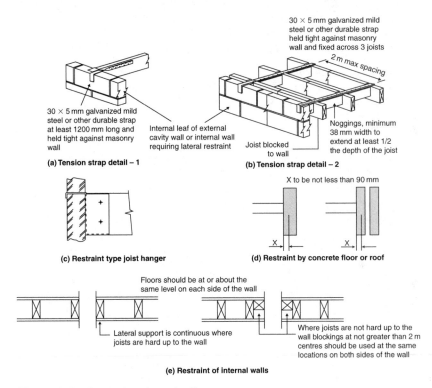

Figure 6.35 Lateral support by floors

Table 6.15 Lateral support by walls

Wall type	Wall length	Lateral support required
Solid or cavity: external compartment separating	Any length	Roof lateral support by every roof forming a junction with the supported wall
	Greater than 3 m	Floor lateral support by every floor forming a junction with the supported wall
Internal load-bearing wall (not being a compartment or separating wall)	Any length	Roof or floor lateral support at the top of each storey

Intermediate floors and roof shall be constructed so that they provide local support to the walls and act as horizontal diaphragms capable of transferring the wind forces to buttressing elements of the building.   A1/2 1A2d

A wall in each storey of a building should:

- extend to the full height of that storey;   A1/2 2C32
- have horizontal lateral supports to restrict movement of the wall at right angles to its plane.

Walls should be strapped to floors above ground level, at intervals not exceeding 2 m and as shown in Figure 6.35 by tension straps conforming to BS EN 845-1.   A1/2 2C35

Where an opening in a floor for a stairway or the like adjoins a supported wall and interrupts the continuity of lateral support:

- the maximum permitted length of the opening is to be 3 m, measured parallel to the supported wall;   A1/2 2C37a
- connections (if provided by means other than by anchor) should be throughout the length of each portion of the wall situated on each side of the opening;   A1/2 2C37b
- connections via mild steel anchors should be spaced closer than 2 m on each side of the opening to provide the same number of anchors as if there were no opening;   A1/2 2C37c
- there should be no other interruption of lateral support.   A1/2 2C37d

The maximum span for any floor supported by a wall is 6 m where the span is measured centre to centre of bearing (see Figure 6.36).   A1/2 2C23

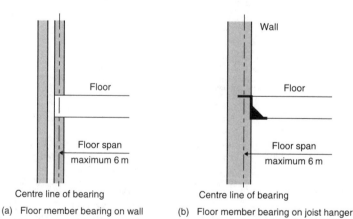

(a) Floor member bearing on wall   (b) Floor member bearing on joist hanger

Figure 6.36 Maximum span of floors

No openings should be provided in walls below ground floor except for small holes for services and ventilation etc. which should be limited to a maximum area of 0.1 m² at not less than 2 m centres (see Figure 6.37 and Table 6.16).   A1/2 2C29

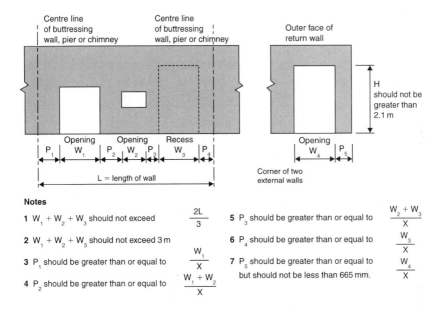

Notes

1 $W_1 + W_2 + W_3$ should not exceed $\dfrac{2L}{3}$

2 $W_1 + W_2 + W_3$ should not exceed 3 m

3 P_1 should be greater than or equal to $\dfrac{W_1}{X}$

4 P_2 should be greater than or equal to $\dfrac{W_1 + W_2}{X}$

5 P_3 should be greater than or equal to $\dfrac{W_2 + W_3}{X}$

6 P_4 should be greater than or equal to $\dfrac{W_3}{X}$

7 P_5 should be greater than or equal to $\dfrac{W_4}{X}$ but should not be less than 665 mm.

Figure 6.37 Sizes or openings and recesses

Table 6.16 Value of factor 'X' (see Figure 6.37)

Nature of roof span	Maximum roof span (m)	Minimum thickness of wall inner (mm)	Span of floor is parallel to wall	Span of timber floor into wall		Span of concrete floor into wall	
				Max 4.5 m	Max 6.0 m	Max 4.5 m	Max 6.0 m
			Value of factor 'X'				
Roof spans parallel to wall	Not applicable	100	6	6	6	6	6
		90	6	6	6	6	5
Timber roof spans into wall	9	100	6	6	5	4	3
		90	6	4	4	3	3

Small single-storey non-residential buildings and annexes

The floor area of the building or annexe shall not exceed 36 m².	A1/2 2C38(1)a
Where the floor area of the building or annexe exceeds 10 m², the walls shall have a mass of not less than 130 kg/m².	A1/2 2C38(1)c

Tension straps

Tension straps (conforming to BS EN 845-1) should be used to strap walls to floors above ground level, at intervals not exceeding 2 m.	A1/2 2C35
For corrosion resistance purposes, the tension straps should be material reference 14 or 16.1 or 16.2 (galvanized steel) or other more resistant specifications including material references 1 or 3 (austenitic stainless steel).	A1/2 2C35

The declared tensile strength of tension straps should not be less than 8 kN.

Tension straps need **not** be provided:

- in the longitudinal direction of joists in houses of not more than 2 storeys if the joists:
  - are at not more than 1.2 m centres; A1/2 2C35a
  - have at least 90 mm bearing on the supported walls or 75 mm bearing on a timber wallplate at each end; or A1/2 2C35a
  - are carried on the supported wall by joist hangers (in accordance with BS EN 845-1 and BS 5628 – see Figure 6.35(c)); A1/2 2C35b
  - and are incorporated at not more than 2 m centres; A1/2 2C35b
- when a concrete floor has at least 90 mm bearing on the supported wall (see Figure 6.35(d)); and A1/2 2C35c
- where floors are at or about the same level on each side of a supported wall, and contact between the floors and wall is either continuous or at intervals not exceeding 2 m. Where contact is intermittent, the points of contact should be in line or nearly in line on plan (see Figure 6.35(e)). A1/2 2C35d

Interruption of lateral support

Where an opening in a floor or roof for a stairway or the like adjoins a supported wall and interrupts the continuity of lateral support:

- the maximum permitted length of the opening is to be 3 m, measured parallel to the supported wall;   A1/2 2C37a
- connections (if provided by means other than by anchor) should be throughout the length of each portion of the wall situated on each side of the opening;   A1/2 2C37b
- connections via mild steel anchors should be spaced closer than 2 m on each side of the opening to provide the same number of anchors as if there were no opening;   A1/2 2C37c
- there should be no other interruption of lateral support.   A1/2 2C37d

Ceiling joists

Softwood timber used for roof construction or fixed in the roof space (including ceiling joists within the void spaces of the roof) should be adequately treated to prevent infestation by the house longhorn beetle (*Hylotrupes bajulus* L.), particularly in the following areas: A1/2 2B2

- the Borough of Bracknell Forest, in the parishes of Sandhurst and Crowthorne;
- the Borough of Elmbridge;
- the District of Hart, in the parishes of Hawley and Yateley;
- the District of Runnymede;
- the Borough of Spelthorne;
- the Borough of Surrey Heath;
- the Borough of Rushmoor, in the area of the former district of Farnborough;
- the Borough of Woking.

📖 **Note:** Guidance on suitable preservative treatments is given within the *British Wood Preserving and Damp-Proofing Association's Manual* (2000 revision), available from 1 Gleneagles House, Vernongate, South Street, Derby DE1 1UP.

Note: Guidance on the sizing of certain members in floors and roofs is given in BS 5268: Part 2: 2002 and Part 3: 1998 as 'Span tables for solid timber members in floors, ceilings and roofs (excluding trussed rafter roofs) for dwellings', published by TRADA, available from Chiltern House, Stocking Lane, Hughenden Valley, High Wycombe, HP14 4ND, Bucks.

Internal fire spread (structure)

Loadbearing elements of structure

All loadbearing elements of a structure shall have a minimum standard of fire resistance.	B3 (8.1)
Structural frames, beams, floor structures and gallery structures should have at least the fire resistance given in Appendix A of Approved Document B.	B3 (8.2)
When altering an existing two-storey, single-family dwellinghouse to provide additional storeys, the floor(s), both old and new, shall have the full 30 minute standard of fire resistance.	B3 (8.7)

Fire resistance – compartmentation

To prevent the spread of fire within a building, whenever possible, the building should be sub-divided into compartments separated from one another by walls and/or floors of fire-resisting construction.	B3 (9.1)
Parts of a building that are occupied mainly for different purposes, should be separated from one another by compartment walls and/or compartment floors.	B3 (9.11)
The wall and any floor between the garage and the house shall have a 30 minute fire resistance. Any opening in the wall to be at least 100 mm above the garage floor level with an FD30 door.	B3
In buildings containing flats or maisonettes compartment walls or compartment floors shall be constructed between: • every floor (unless it is within a maisonette) • one storey and another within one dwelling • every wall separating a flat or maisonette from any other part of the building • every wall enclosing a refuse storage chamber.	B3 (9.15)

Every compartment floor should: B3 (9.22)

- form a complete barrier to fire between the compartments they separate; and
- have the appropriate fire resistance as indicated in Appendix A of Approved Document B, Tables A1 and A2.

Where a compartment wall or compartment floor meets   B3 (9.27)
another compartment wall, or an external wall, the junction
should maintain the fire resistance of the compartmentation.

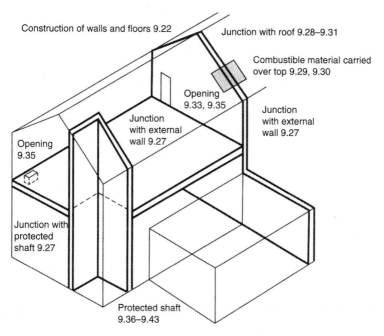

Figure 6.38 Compartment walls and compartment floors with reference to the relevant paragraphs in Approved Document B

Concrete

With a concrete intermediate floor:

The ground floor may be a solid slab, laid on the ground, or a suspended concrete floor.	E2.51 E2.88 E2.126
A concrete slab floor on the ground may be continuous	E2.51

under a solid separating wall but may not be continuous under the cavity masonry core of the separating wall.	E2.88 E2.127 E2.130

An internal concrete floor slab may only be carried through a separating wall if the floor base has a mass of at least 365 kg/m².	E2.46 E2.121

Note: Internal concrete floors should generally be built into a separating wall and carried through to the cavity face of the leaf.

The cavity should not be bridged (E2.85, E2.122). Internal hollow-core concrete plank floors (and concrete beams with infilling block floors) should **not** be continuous through or under a separating wall (E2.47, E2.53, E2.129).

A suspended concrete floor may only pass under a separating wall if the floor has a mass of at least 365 kg/m².	E2.52 E2.89

Note: A suspended concrete floor should **not** be carried through to the cavity face of the leaf and the cavity should not be bridged (E2.89, E2.132).	E2.128 E2.131

Enclosures for drainage and/or water supply pipes

The enclosure should:	B3 (11.8)

- be bounded by a compartment wall or floor, an outside wall, an intermediate floor, or a casing
- have internal surfaces (except framing members) of Class 0
- not have an access panel that opens into a circulation space or bedroom
- be used only for drainage, or water supply, or vent pipes for a drainage system.

The casing should:	B3 (11.8)

- be imperforate except for an opening for a pipe or an access pane
- not be of sheet metal
- have (including any access panel) not less than 30 minutes fire resistance.

Suspended ceilings

Table 6.17 sets out criteria appropriate to the suspended ceilings that can be accepted as contributing to the fire resistance of a floor.

Table 6.17 Limitations on fire protected suspended ceilings

Height of building or separated part	Type of floor	Provision for fire resistance of floor	Description of suspended ceiling
<18 m	Not compartment	60 mins or less	Type A, B, C or D
	Compartment	<60 mins	Type A, B, C or D
		60 mins	Type B, C or D
18 m or more	Any	60 mins or less	Type C or D
No limit	Any	60 mins	Type D

Requirements – ceilings

General

The resistance to airborne and impact sound depends on the independence and isolation of the ceiling and the type of material used.	E
Three ceiling treatments (which are ranked in order of sound insulation) may be used:	E3

- ceiling treatment A – independent ceiling with absorbent material;
- ceiling treatment B – plasterboard on proprietary resilient bars with absorbent material;
- ceiling treatment C – plasterboard on timber battens (or proprietary resilient channels) with absorbent material.

Note: These ceiling treatments are described in more detail in p. 263.

If the roof or loft space is not a habitable room (and there is a ceiling with a minimum mass per unit area of 10 kg/m^2 with sealed joints) then:	E

- the mass per unit area of the separating wall above the ceiling may be reduced to 150 kg/m^2;
- the independent panels may be omitted in the roof space but the cavity masonry core should be x maintained to the underside of the roof;
- the linings on each frame may be reduced to two layers of plasterboard or the cavity may be closed at ceiling level without connecting the two frames rigidly together.

All junctions between ceilings and independent panels (and joints between casings and ceiling) should be sealed with tape or caulked with sealant.	E
At junctions with external cavity walls (with masonry inner leaf) the ceiling should be taken through to the masonry.	E3
The ceiling void and roof space detail can only be used where the Requirements of Building Regulation Part B – Fire safety can also be satisfied.	E3
If there is an existing lath and plaster ceiling it should be retained as long as it satisfies Building Regulation Part B – Fire safety.	E3
If the existing ceiling is **not** lath and plaster, it should be upgraded to provide at least two layers of plasterboard with joints staggered, total mass per unit area 20 kg/m².	E4
Care should be taken at the design stage to ensure that adequate ceiling height is available in all rooms to be treated.	
The ceiling should be supported by either:	E2

- independent joists fixed only to the surrounding walls; or
- independent joists fixed to the surrounding walls with additional support provided by resilient hangers attached directly to the existing floor base.

Where a window head is near to the existing ceiling, the new independent ceiling may be raised to form a pelmet recess.	E4
A rigid or direct connection should not be created between an independent ceiling and the floor base.	E4

💡 For more details about ceilings see Section 6.8.

Floors – general

Floors that separate a dwelling from another dwelling (or part of the same building) shall resist the transmission of airborne sounds.	E

Floors above a dwelling that separate it from another dwelling (or another part of the same building) shall resist:	E

- the transmission of impact sound (such as speech, musical instruments and loudspeakers and impact sources such as footsteps and furniture moving);

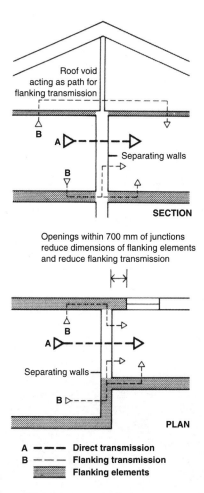

Figure 6.39 Direct and flanking transmission
For clarity not all flanking paths have been shown.

- the flow of sound energy through walls and floors;
- the level of airborne sound.

Air paths, including those due to shrinkage, must be avoided – E
porous materials and gaps at joints in the structure must be sealed.

The possibility of resonance in parts of the structure (such as E
a dry lining) should be avoided.

Flanking transmission (i.e. the indirect transmission of sound from E
one side of a wall or floor to the other side) should be minimized.

Requirement E1

Figure 6.40 illustrates the relevant parts of the building that should be protected from airborne and impact sound in order to satisfy Requirement E1.

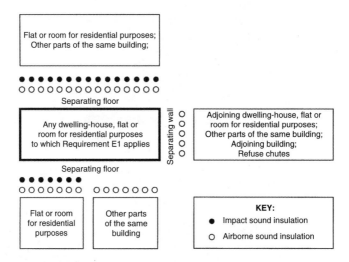

Figure 6.40 Requirement E1 – resistance to sound

All new floors constructed within a dwelling-house (flat or   E0.9
room used for residential purposes) – whether purpose built
or formed by a material change of use – shall meet the
laboratory sound insulation values set out in Table 6.18.

Floors that have a separating function should achieve   E0.1
the sound insulation values:

- for rooms for residential purposes as set out in Table 6.18;
- for dwelling-houses and flats as set out in Table 6.18.

Table 6.18 Dwelling-houses and flats – performance standards for separating floors and stairs that have a separating function

	Airborne sound insulation $D_{nT,w} + C_{tr}$ dB (minimum values)	Impact sound insulation $L_{nT,w}$ dB (maximum values)
Purpose built rooms for residential purposes	45	62
Purpose built dwelling houses and flats	45	62
Rooms for residential purposes formed by material change of use	43	64
Dwelling-houses and flats formed by material change of use	43	64

📖 **Note:**

(1)   The sound insulation values in this table include a built-in allowance for 'measurement uncertainty' and so if any these test values are not met, then that particular test will be considered as failed.

(2)   Occasionally a higher standard of sound insulation may be required between spaces used for normal domestic purposes and noise generated in and to an adjoining communal or non-domestic space. In these cases it would be best to seek specialist advice before committing yourself.

Flanking transmission from walls and floors connected to the separating wall shall be controlled.	E2
Tests should be carried out between rooms or spaces that share a common area formed by a separating wall or separating floor.	E1
Impact sound insulation tests should be carried out without a soft covering (e.g. carpet, foam backed vinyl etc.) on the floor.	E1
When a separating floor is used, a minimum mass per unit area of $120\,\text{kg/m}^2$ (excluding finish) shall always apply, irrespective of the presence or absence of openings.	E2
Care should be taken to correctly detail the junctions between the separating floor and other elements such as external walls, separating walls and floor penetrations.	E3
Spaces between floor joists should be sealed with full depth timber blocking.	E2
If the floor joists are to be supported on the separating wall then they should be supported on hangers and should not be built in.	E2
If the joists are at right angles to the wall, spaces between the floor joists should be sealed with full depth timber blocking.	E3
The floor base (excluding any screed) should be built into a cavity masonry external wall and carried through to the cavity face of the inner leaf.	E
The floor base should be continuous or above an internal masonry wall.	E3
All pipes and ducts that penetrate a floor separating habitable rooms in different flats should:	E3
• be enclosed for their full height in each flat; • have fire protection to satisfy Building Regulation Part B – Fire safety.	

Notes:

(1) Where any building element functions as a separating element (e.g. a ground floor that is also a separating floor for a basement flat) then the separating element requirements should take precedence.

(2) In some circumstances (for example, when a historic building is undergoing a material change of use) it may not be practical to improve the sound insulation to the standards set out in Approved Document E1 particularly if the special characteristics of such a building need to be recognized. In these circumstances the aim should be to improve sound insulation to the 'extent that it is practically possible'.

(3) BS 7913:1998 *The principles of the conservation of historic buildings* provides guidance on the principles that should be applied when proposing work on historic buildings.

Requirement E2

Constructions for new floors within a dwelling-house (flat or room for residential purposes) – whether purpose built or formed by a material change of use – shall meet the laboratory sound insulation values set out in Table 6.19.	E0.9

Table 6.19 Laboratory values for new internal walls within dwelling-houses, flats and rooms for residential purposes – whether purpose built or formed by a material change of use

	Airborne sound insulation R_W dB (minimum values)
Purpose built dwelling houses and flats	
Floors	40

Figure 6.41 illustrates the relevant parts of the building that should be protected from airborne and impact sound in order to satisfy Requirement E2.

Requirement E3

Sound absorption measures described in Section 7 of Approved Document N shall be applied.	E0.11

Requirement E4

The values for sound insulation, reverberation time and indoor ambient noise as described in Section 1 of Building	E0.12

Bulletin 93 *'The Acoustic Design of Schools'* (produced
by DFES and published by the Stationery Office
(ISBN 0 11 271105 7)) shall be satisfied.

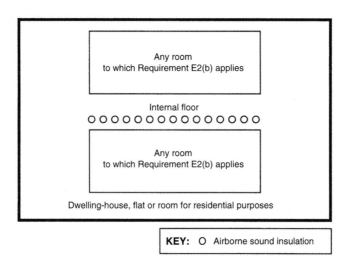

KEY: O Airborne sound insulation

Figure 6.41 Requirement E2(b) – internal floors

Separating floors and associated flanking constructions for new buildings

There are three groups of floors as shown below:

Floor type 1 Concrete base with ceiling and soft floor covering	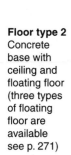	The resistance to airborne sound depends mainly on the mass per unit area of the concrete base and partly on the mass per unit area of the ceiling. The soft floor covering reduces impact sound at source.
Floor type 2 Concrete base with ceiling and floating floor (three types of floating floor are available see p. 271)	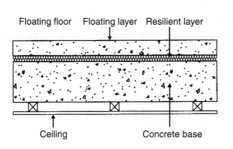	The resistance to airborne and impact sound depends on the mass per unit area of the concrete as well as the mass per unit area and isolation of the floating layer and ceiling. The floating floor reduces impact sound at source.

Floor type 3
Timber frame
base with
ceiling and
platform floor

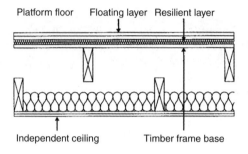

Platform floor   Floating layer Resilient layer

Independent ceiling   Timber frame base

The resistance
to airborne and
impact sound
depends on the
structural floor
base and the
isolation of the
platform floor
and the ceiling.
The platform
floor reduces
impact sound
at source.

General requirements

Floor types should be capable of achieving the performance standards shown in Table 6.18 on p. 258	E3.1
Care should be taken to correctly detail the junctions between the separating floor and other elements such as external walls, separating walls and floor penetrations.	E3.10

 Note: Where any building element functions as a separating element (e.g. a ground floor that is also a separating floor for a basement flat) then the separating element requirements should take precedence.

Ceiling treatments

Each floor type should use one of the following three ceiling treatments which are ranked in order of sound insulation performance from A to C. **Note:** Use of a better performing ceiling than that described in the guidance should improve the sound insulation of the floor provided there is no significant flanking transmission.	E3.17 to E3.18

Ceiling treatment A
Independent ceiling
with absorbent
material

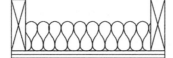

- at least 2 layers of
  plasterboard with
  staggered joints;
- minimum total mass
  per unit area of
  plasterboard 20 kg/m^2;
- an absorbent layer
  of mineral wool
  (minimum thickness
  100 mm, minimum
  density 10 kg/m^3) laid
  in the cavity formed
  above the ceiling.

The type of ceiling
support depends on
the floor type

For floor types 1, 2 and 3
Use independent joists fixed
only to the surrounding walls

For floor type 3
Use independent joists fixed
to the surrounding walls
with additional support
provided by resilient
hangers attached
directly to the floor

Always ensure
- that you seal the
  perimeter of the
  independent ceiling
  with tape or
  sealant;
- you do not
  create a rigid or
  direct connection
  between the
  independent ceiling
  and the floor base.

Ceiling treatment B
Plasterboard on
proprietary resilient
bars with absorbent
material

- single layer of
  plasterboard,
  minimum mass per
  unit area of
  plasterboard
  10 kg/m^2;
- fixed using
  proprietary resilient
  metal bars;
- an absorbent layer
  of mineral wool
  (minimum density
  10 kg/m^3) should
  fill the ceiling void.

Ceiling treatment C
Plasterboard on
timber battens or
proprietary resilient
channels with
absorbent material

If timber battens are used

If resilient channels are used

- single layer of
  plasterboard,
  minimum mass per
  unit area 10 kg/m^2;
- fixed using timber
  battens or proprietary
  resilient channels;
- if resilient channels
  are used, incorporate
  an absorbent
  layer of mineral wool
  (minimum density
  10 kg/m^3) that fills
  the ceiling void.

Notes: (1) Electrical cables give off heat when in use and special precautions may be required when they are covered by thermal insulating materials. See BRE BR 262, Thermal Insulation: avoiding risks, Section 2.3. (2) Installing recessed light fittings in ceiling treatments A to C can reduce their resistance to the passage of airborne and impact sound.

Floor type 1 Concrete base with ceiling and soft floor covering

A floor type 1 consists of a concrete floor base with a soft floor covering and a ceiling. Its resistance to airborne sound mainly depends on:

- the mass per unit area of the concrete base;
- the mass per unit area of the ceiling;
- the soft floor covering (which helps to reduce the source of the impact sound).

General requirements

To allow for future replacements, the soft floor covering should be fixed or glued to the floor.	E3.27a
To avoid air paths all joints between parts of the floor should be filled.	E3.27b
To reduce flanking transmission, air paths should be avoided at all points where a pipe or duct penetrates the floor.	E3.27c
A separating concrete floor should be built into the walls (around its entire perimeter) if the walls are masonry.	E3.27d
All gaps between the head of a masonry wall and the underside of the concrete floor should be filled with masonry.	E3.27e
Flanking transmission from walls connected to the separating floor should be controlled.	E3.27f
The floor base shall not bridge the cavity in a cavity masonry wall.	E3.27a2
Non-resilient floor finishes (such as ceramic floor tiles and wood block floors that are rigidly connected to the floor base) shall not be used.	E3.27b2
Any soft floor covering that is used should be of a resilient material with an overall uncompressed thickness of at least 4.5 mm (also see BS EN ISO 140-8:1998).	E3.28

Two floor types (see below) will meet these requirements.

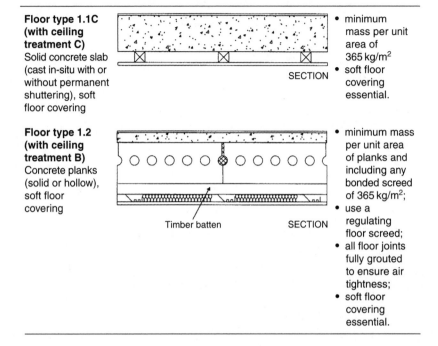

| **Floor type 1.1C (with ceiling treatment C)** Solid concrete slab (cast in-situ with or without permanent shuttering), soft floor covering | SECTION | • minimum mass per unit area of 365 kg/m² • soft floor covering essential. |
| **Floor type 1.2 (with ceiling treatment B)** Concrete planks (solid or hollow), soft floor covering | Timber batten   SECTION | • minimum mass per unit area of planks and including any bonded screed of 365 kg/m²; • use a regulating floor screed; • all floor joints fully grouted to ensure air tightness; • soft floor covering essential. |

Junction requirements for floor type 1

Junctions with an external cavity wall with masonry inner leaf

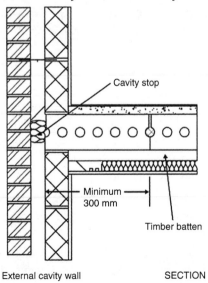

Cavity stop

Minimum 300 mm

Timber batten

External cavity wall   SECTION

Figure 6.42 Junctions with an external cavity wall with masonry inner leaf

If the external wall is a cavity wall:

- the outer leaf of the wall may be of any construction;   E3.31a
- the cavity should be stopped with a flexible closer to   E3.31b
  ensure adequate drainage.

The masonry inner leaf of an external cavity wall should have   E3.32
a mass per unit area of at least 120 kg/m^2 excluding finish.

The floor base (excluding any screed) should be built into   E3.33
a cavity masonry external and carried through to the cavity
face of the inner leaf.

Junctions with an external cavity wall with timber frame inner leaf

Where the external wall is a cavity wall:

- the outer leaf of the wall may be of any construction;   E3.36a
- the cavity should be stopped with a flexible closer;   E3.36b
- the wall finish of the inner leaf of the external wall should   E3.36c
  be two layers of plasterboard, each sheet of plasterboard
  to be a minimum mass per unit area 10 kg/m^2;
- all joints should be sealed with tape or caulked unenclosed.   E3.36c

Junctions with an external solid masonry wall

No official guidance currently available about junctions   E3.37
with a solid masonry wall. Best to seek specialist advice.

Junctions with internal framed walls

There are no restrictions on internal walls meeting a   E3.38
type 1 separating floor.

Junctions with internal masonry walls

The floor base should be continuous through or above   E3.39
an internal masonry wall.

The mass per unit area of any load bearing internal wall   E3.40
(or any internal wall rigidly connected to a separating floor)
should be at least 120 kg/m^2 excluding finish.

Junctions with floor penetrations (excluding gas pipes)

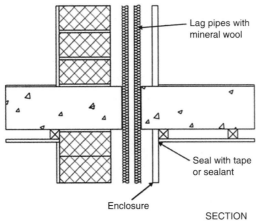

Lag pipes with
mineral wool

Seal with tape
or sealant

Enclosure

SECTION

Figure 6.43 Floor type 1 floor penetrations

Pipes and ducts should be in an enclosure (both above and below the floor). In all cases:

The enclosure should be constructed of material having a mass per unit area of at least $15\,kg/m^2$.	E3.32
The enclosure should either be lined or the duct (or pipe) within the enclosure wrapped with 25 mm unfaced mineral fibre.	E3.42
Penetrations through a separating floor by ducts and pipes should have fire protection to satisfy Building Regulation Part B – Fire safety.	E3.43
Fire stopping should be flexible to prevent a rigid contact between the pipe and the floor.	E3.43
Gas pipes may be contained in a separate (ventilated) duct or can remain unenclosed.	E3.43
If a gas service is installed it shall comply with the Gas Safety (Installation and Use) Regulations 1998, S1 1998 No 2451.	E3.43
If the pipes and ducts penetrate a floor separating habitable rooms in different flats, then they should be enclosed for their full height in each flat.	E3.41

Junctions with separating wall type 1 – solid masonry

For floor types 1.1C and 1.2C, two possibilities exist:

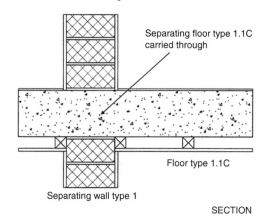

Figure 6.44 Floor type 1.1C – wall type 1

Requirements – floors and ceilings

A separating floor type 1.1 C base (excluding any screed)   E3.44
should pass through a separating wall type 1 (for flats where
there are separating walls, guidance on p. 276 may also apply).

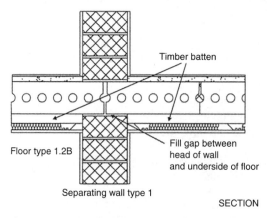

Figure 6.45 Floor type 1.1C – wall type 1

A separating floor type 1.2B base (excluding any screed) should   E3.44
not pass through a separating wall type 1 (for flats where there
are separating walls, guidance on p. 276 may also apply).

Junctions with separating wall type 2 cavity masonry

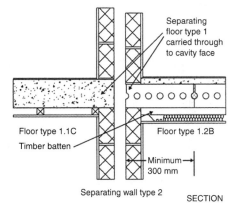

Figure 6.46 Floor types 1.1C and 1.2B – wall type 2

The mass per unit area of any leaf that is supporting or adjoining the floor should be at least 15 kg/m² excluding finish.	E3.46
The floor base (excluding any screed) should be carried through to the cavity face of the leaf.	E3.47
The wall cavity should not be bridged.	E3.47
Where floor type 1.2B is used (and the planks are parallel to the separating wall) the first joint should be a minimum of 300 mm from the inner face of the adjacent cavity leaf.	E3.48

Junctions with separating wall types 3.1 and 3.2 (solid masonry core)

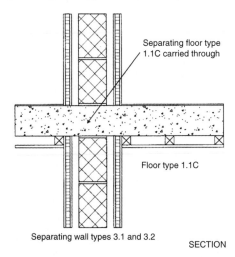

Figure 6.47 Floor type 1.1C – wall types 3.1 and 3.2

A separating floor type 1.1C base (excluding any screed)   E3.49
should pass through separating wall types 3.1 and 3.2.

A separating floor type 1.2B base (excluding any screed)   E3.50
should not be continuous through a separating wall type 3.

Where a separating wall type 3.2 is used with floor type   E3.51
1.2B (and the planks are parallel to the separating wall)
the first joint should be a minimum of 300 mm from the
centreline of the masonry core.

Junctions with separating wall type 3.3 (cavity masonry core)

The mass per unit area of any leaf that is supporting or   E3.52
adjoining the floor should be at least 120 kg/m^2 excluding
finish.

The floor base (excluding any screed) should be carried   E3.53
through to the cavity face of the leaf of the core.

The cavity should not be bridged.   E3.53

Where floor type 1.2B is used (and the planks   E3.54
are parallel to the separating wall) the first
joint should be a minimum of 300 mm from the
inner face of the adjacent cavity leaf of the
masonry core.

Junctions with separating wall type 4 timber frames with absorbent material

No official guidance currently available. Best to seek   E3.55
specialist advice.

Floor type 2: Concrete base with ceiling and floating floor

A floor type 2 consists of a concrete floor base with a floating floor (which in
turn consists of a floating layer and a resilient layer) and a ceiling. Its resistance
to airborne and impact sound depends on:

- the mass per unit area of the concrete base;
- the mass per unit area and isolation of the floating layer and the ceiling;
- the floating floor (which reduces impact sound at source).

General requirements

All joints between parts of the floor should be filled to avoid air paths.	E3.61a
To reduce flanking transmission, air paths should be avoided at all points where a pipe or duct penetrates the floor.	E3.61b
A separating concrete floor should be built into the walls (around its entire perimeter) if the walls are masonry.	E3.61c
All gaps between the head of a masonry wall and the underside of the concrete floor should be filled with masonry.	E3.61d
Flanking transmission from walls connected to the separating floor should be controlled.	E3.61e
The floor base shall not bridge a cavity in a cavity masonry wall.	E3.61

Two floor types (consisting of a floating layer and resilient layer – see below) will meet these requirements. A performance-based approach (type C) is also available.

Floating floor (a)
Timber raft floating layer with resilient layer

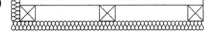

- timber raft of board material (with bonded edges, e.g. tongued and grooved);
- minimum mass per unit area 12 kg/m^2;
- fixed to 45 mm × 45 mm battens laid loose on the resilient layer (but **not** along any joints in the resilient layer);
- resilient layer of mineral wool (which may be paper faced on the underside) with density 36 kg/m^{-3} and minimum thickness 25 mm.

Floating floor (b)
Sand cement screed floating layer with resilient layer

Floating layer
- of 65 mm sand cement screed with a mass per unit area of at least 80 kg/m^2.

Resilient layer
- protected while the screed is being

Floating floor (b) – *continued*

laid (e.g. by a 20–50 mm wire mesh) and consisting of either:

- a layer of mineral wool of minimum thickness 25 mm with density 36 kg/m^{-3} (paper faced on the upper side);
- an alternative type of resilient layer with maximum dynamic stiffness of 15 kg/m^3;
- an alternative type of resilient layer with minimum thickness of 5 mm (see BS EN ISO 29052-1:1992).

Floating floor (c)
Performance-based approach

—

- rigid boarding above a resilient and/or damping layer;
- weighted reduction in impact sound pressure level of not less than 29 dB (see BS EN ISO 717-2:1997 and BS EN ISO 140-8:1998).

General requirements

A small gap filled with a flexible sealant should be left between the floating layer and wall at all room edges.	E3.63a
A small gap (approx. 5 mm and filled with a flexible sealant) should be left between skirting and floating layer.	E3.63b
Resilient materials should be laid in rolls or sheets either with lapped joints or with joints tightly butted and taped.	E3.63c
Paper facing should be used on the upper side of fibrous materials to prevent screed entering the resilient layer.	E3.63d
The floating layer and the base or surrounding walls shall not be bridged (e.g. with services or fixings that penetrate the resilient layer).	E3.63a2
The floating screed shall create a bridge (for example, through a gap in the resilient layer) to the concrete floor base or surrounding walls.	E3.63b2

Depending on the type of ceiling two options can be used:

Floor type 2.1C(a) (with ceiling treatment C and floating floor (a))

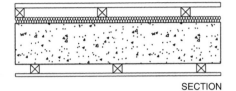

SECTION

Floor type a

- minimum mass per unit area of 300 kg/m²;

Solid concrete slab (cast in-situ with or without permanent shuttering), floating floor, ceiling treatment

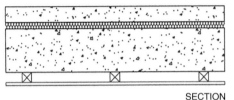

SECTION

Floor type b

- regulating floor screed optional;
- floating floor (a), (b) or (c) essential;
- ceiling treatment C (or better) essential.

Floor type 2.1C(b) (with ceiling treatment C and floating floor (b))

Concrete planks (solid or hollow), floating floor, ceiling treatment B

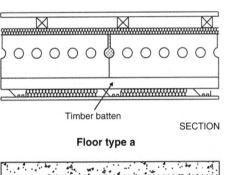

Timber batten

SECTION

Floor type a

Timber batten

SECTION

Floor type b

- minimum mass per unit area of planks and any bonded screed of 300 kg/m²;
- use a regulating floor screed;
- all floor joints fully grouted to ensure air tightness;
- floating floor (a), (b) or (c) essential;
- ceiling treatment B (or better) essential.

Junction requirements for floor type 2

Junctions with an external cavity wall with type 2 timber frame inner leaf

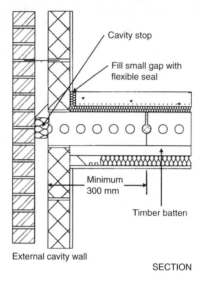

Figure 6.48 Floor type 2 – external cavity wall with masonry internal leaf

Where the external wall is a cavity wall:	
• the outer leaf of the wall may be of any construction;	E3.69a
• the cavity should be stopped with a flexible closer.	E3.69b
The masonry inner leaf of an external cavity wall should have a mass per unit area of at least $120\,kg/m^2$.	E3.70
The floor base (excluding any screed) should be built into a cavity masonry external wall and carried through to the cavity face of the inner leaf.	E3.71
The cavity should not be bridged.	E3.71
If a floor 2.2B is used (and the planks through, or above, an internal masonry wall are parallel to the external wall) the first joint should be a minimum of 300 mm from the cavity face of the inner leaf.	E3.72

Junctions with an external cavity wall with timber frame inner leaf

Where the external wall is a cavity wall:	
• the outlet leaf of the wall may be of any construction;	E3.74a
• the cavity should be stopped with a flexible closer;	E3.74b

• the wall face of the inner leaf of the external wall should be two layers of plasterboard;	E3.74c
• each sheet of plasterboard to be of minimum mass per unit area 10 kg/m²;	E3.74c
• all joints should sealed or caulked with sealant.	E3.74c

Junctions with an external solid masonry wall

No official guidance currently available. Best to seek specialist advice.	E3.75

Junctions with internal framed walls

There are no restrictions on internal walls meeting a type 4 separating wall.	E3.76

Junctions with internal masonry walls

The floor base should be continuous or above an internal masonry wall.	E3.77
The mass per unit area of any load bearing internal wall or any internal wall rigidly connected to a separating floor should be at least 120 kg/m² excluding finish.	E3.78

Junctions with floor penetrations (excluding gas pipes)

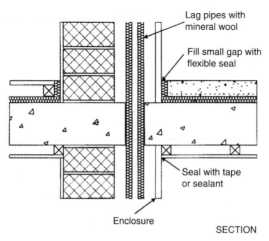

Figure 6.49 Floor type 2 – floor penetrations

Pipes and ducts that penetrate a floor separating habitable rooms in different flats should be enclosed for their full height in each flat.	E3.79
The enclosure should be constructed of material having a mass per unit area of at least 15 kg/m².	E3.80
Either line the enclosure, or wrap the duct or pipe within the enclosure, with 25 mm unfaced mineral wool.	E3.80
A small gap (sealed with sealant or neoprene) of about 5 mm should be left between the enclosure and the floating.	E3.81
Where floating floor (a) or (b) is used the enclosure may go down to the floor base (provided that the enclosure is isolated from the floating layer).	E3.81
Penetrations through a separating floor by ducts and pipes should have fire protection to satisfy Building Regulation Part B – Fire safety.	E3.82
Gas pipes may be contained in a separate (ventilated) duct or can remain unenclosed.	E3.82
If a gas service is installed it shall comply with the Gas Safety (Installation and Use) Regulations 1998, S1 1998 No 2451.	E3.82

Junctions with a separating wall type 1 – solid masonry

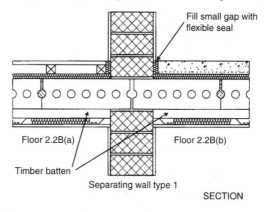

Figure 6.50 Floor type 2.1C – wall types 3.1 and 3.2

A separating floor type 2.1C base (excluding any screed) should pass through a separating wall type 1.	E3.84
A separating floor type 2.2B base (excluding any screed) should **not** be continuous through a separating wall type 1.	E3.84

Junctions with a separating wall type 2 cavity masonry

The floor base (excluding any screed) should be carried through to the cavity face of the leaf.	E3.85
The cavity should not be bridged.	E3.85
If a floor type 2.2B is used (and the planks are parallel to the separating wall) the first joint should be a minimum of 300 mm from the cavity face of the leaf.	E3.86

Junctions with separating wall type 3.1 and 3.2 (solid masonry core)

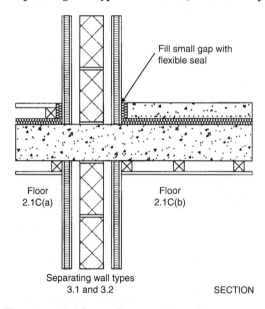

Figure 6.51 Floor type 2.1C – wall types 3.1 and 3.2

A separating floor type 2.1C base (excluding any screed) should pass through separating wall types 3.1 and 3.2.	E3.87
A separating floor type 2.2B base (excluding any screed) should not be continuous through a separating wall type 3.	E3.88
If a separating wall type 3.2 is used with floor type 2.2B (and the planks are parallel to the separating wall) the first joint should be a minimum of 300 mm from the centreline of the masonry core.	E3.89

Junctions with separating wall type 3.3 (cavity masonry core)

The mass per unit area of any leaf that is supporting or adjoining the floor should be at least $120\,kg/m^2$ excluding finish.	E3.90
The floor base (excluding any screed) should be carried through to the cavity face of the leaf of the core.	E3.91
The cavity should not be bridged.	E3.91
If a floor type 2.2B is used (and the planks are parallel to the separating wall) the first joint should be a minimum of 300 mm from the inner face of the adjacent cavity leaf of the masonry core.	E3.92

Junctions with separating wall type 4 timber frames with absorbent material

No official guidance currently available. Best to seek specialist advice.	E3.93

Floor type 3: Timber frame base with ceiling and platform floor

A floor type 3 consists of a timber frame structural floor base with a deck, platform floor (consisting of a floating layer and a resilient layer) and ceiling treatment. Its resistance to airborne and impact sound depends on:

- the structural floor base;
- the isolation of the platform floor and the ceiling;
- the platform floor (which reduces impact sound at source).

General requirements

To reduce flanking transmission, air paths should be avoided at all points where the floor is penetrated.	E3.99a
Flanking transmission from walls connected to the separating floor should be as described in the following junction requirements for floor type 3	E3.99b
There should be no bridge (e.g. formed by services or fixings that penetrate the resilient layer) between the floating layer and the base or surrounding walls.	E3.99

For the platform floor, ensure that:

• the correct density of resilient layer is used;	E3.99c
• the layer can carry the anticipated load;	E3.99c
• during construction a gap is maintained between the wall and the floating layer (filled with a flexible sealant, expanded or extruded polystyrene strip);	E3.99d
• resilient materials are laid in sheets with all joints tightly butted and taped.	E3.99e

The following floor type (floor type 3.1A) will meet these requirements.

Floor type 3.1A

Timber frame base with ceiling treatment A and platform floor

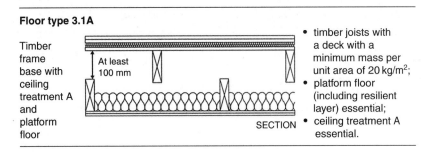

At least 100 mm

SECTION

• timber joists with a deck with a minimum mass per unit area of 20 kg/m²;
• platform floor (including resilient layer) essential;
• ceiling treatment A essential.

Platform floor

The floating layer should: E3.101

- be a minimum of two layers of board material;
- be minimum total mass per unit area 25 kg/m²;
- have layers of minimum thickness 8 mm;
- be fixed together with joints staggered;
- be laid loose on a resilient layer.

Resilient layer

The resilient layer should be of mineral wool: E3.102

- minimum thickness 25 mm;
- density 60 to 100 kg/m³;
- paper faced on the underside.

Junction requirements for floor type 3

Junctions with an external cavity wall with masonry inner leaf

Where the external wall is a cavity wall:	
• the outer leaf of the wall may be of any construction;	E3.103a
• the cavity should be stopped with a flexible closer.	E3.103b
The masonry inner leaf of a cavity wall should be lined with an independent panel.	E3.104
The ceiling should be taken through to the masonry.	E3.105
The junction between the ceiling and the independent panel should be sealed with tape or caulked with sealant.	E3.105
Air paths between floor and wall cavities should be blocked.	E3.106

Note:

(1)   Any normal method of connecting floor base to wall may be used.

(2)   Independent panels are not required if the mass per unit area of the inner leaf is greater than $375 \, \text{kg/m}^2$.

Junctions with an external cavity wall with timber frame inner leaf

Where the external wall is a cavity wall:	
• the outer leaf of the wall may be of any construction;	E3.109a
• the cavity should be stopped with a flexible closer.	E3.109b
The wall finish of the inner leaf of the external wall should:	
• be two layers of plasterboard;	E3.110a
• be each sheet of plasterboard of minimum mass per unit area $10 \, \text{kg/m}^2$;	E3.110b
• have all joints sealed with tape or caulked with sealant.	E3.110c
Any normal method of connecting floor base to wall may be used.	E3.111
If the joists are at right angles to the wall, spaces between the floor joists should be sealed with full depth timber blocking.	E3.112
The junction between the ceiling and wall lining should be sealed with tape or caulked with sealant.	E3.113

Junctions with an external solid masonry wall

No official guidance currently available. Best to seek specialist advice.	E3.113

Junctions with internal framed walls

The spaces between joists are at right angles and should be sealed with full depth timber blocking.	E3.114
The junction between the ceiling and the internal framed wall should be sealed with tape or caulked with sealant.	E3.115

Junctions with internal masonry walls

No official guidance currently available. Best to seek specialist advice.	E3.116

Junctions with floor penetrations (excluding gas pipes)

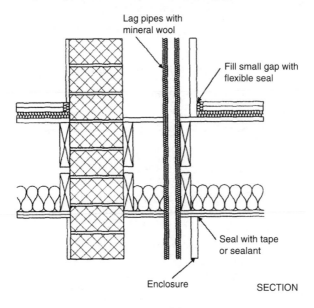

Figure 6.52 Floor type 3 – floor penetrations

Pipes and ducts that penetrate a floor separating   E3.117
habitable rooms in different flats should be enclosed
for their full height in each flat.

The enclosure should:

- be constructed of material having a mass per unit area of   E3.118
  at least 15 kg/m^2;
- have a small, sealed (with sealant or neoprene) 5 mm   E3.119
  gap between the enclosure and floating layer;
- go down to the floor base;   E3.119
- be isolated from the floating layer.   E3.119

The duct or pipe within the enclosure should be lined or   E3.118
wrapped with 25 mm unfaced mineral wool.

Penetrations through a separating floor by ducts and pipes   E3.120
should have fire protection to satisfy Building Regulation
Part B – Fire safety.

Fire stopping should be flexible and also prevent rigid   E3.121
contact between the pipe and floor.

Gas pipes may be contained in a separate (ventilated)   E3.120
duct or can remain unenclosed.

If a gas service is installed it shall comply with the Gas Safety   E3.120
(Installation and Use) Regulations 1998, S1 1998 No 2451.

Junctions with a separating wall type 1 solid masonry

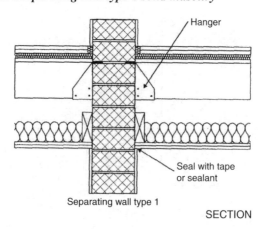

Hanger

Seal with tape
or sealant

Separating wall type 1

SECTION

Figure 6.53 Floor type 3 – wall type 1

Floor joists supported on a separating wall should be supported on hangers as opposed to being built in.	E3.121
The junction between the ceiling and wall should be sealed with tape or caulked with sealant.	E3.122

 Note: The above is particularly relevant for flats where there are separating walls.

Junctions with a separating wall type 2 – cavity masonry

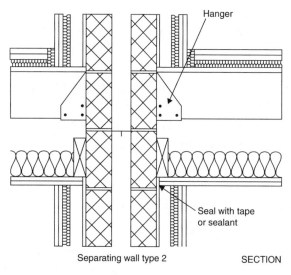

Figure 6.54 Floor type 3 – wall type 2

Floor joists that are supported on a separating wall should be supported on hangers and not built in.	E3.123
The adjacent leaf of a cavity separating wall should be lined with an independent panel.	E3.124
The ceiling should be taken through to the masonry.	E3.125
The junction between the ceiling and the independent panel should be sealed with tape or caulked with sealant.	E3.125

 Note: Independent panels are not required if the mass per unit area of the inner leaf is greater than 375 kg/m^2.

Junctions with a separating wall type 3 – masonry between independent panels

Floor joists that are supported on a separating wall should be supported on hangers and not built in.	E3.127
The ceiling should be taken through to the masonry.	E3.128
The junction between the ceiling and the independent panel should be sealed with tape or caulked with sealant.	E3.128

Junctions with a separating wall type 4 – timber frames with absorbent material

Spaces between the floor joists that are at right angles to the wall should be sealed with full depth timber blocking.	E3.129
The junction of the ceiling and wall lining should be sealed with tape or caulked with sealant.	E3.130

6.6.3 The use of Robust Standards

Background

The 2003 edition of Part E of the Building Regulations (i.e. 'Resistance to the passage of sound') involves Pre-Completion Sound Testing (PCT) for certain types of homes. In an attempt to eliminate the risk of any remedial work being required to completed floor and/or wall constructions (together with the potential for delays in completing the property) the House Builders Federation (HBF) suggested that a series of construction solutions (called Robust Details) should be developed as an alternative to PCT.

This approach was agreed and in May 2003, the Office of the Deputy Prime Minister (ODPM) published the first batch of Robust Detail proposals for public consultation. At the same time, the introduction of PCT in new homes was postponed until July 2004 – on the assumption that the Robust Details scheme would eventually receive ministerial approval.

In January 2004 the Minister responsible for Building Regulations announced that he would allow Robust Details to be used as an alternative to PCT and that it would take effect from 1st July 2004 (i.e. so as to coincide with the introduction of PCT). Under a Memorandum of Understanding with the OPDM, a Limited Company (Robust Details Ltd) was set up to approve, manage, monitor and promote the use of Robust Details as a method of satisfying Building Regulations.

What is a Robust Detail?

Robust Details provide builders with a choice of possible construction solutions that have been proved to outperform the standards of Part E, thus eliminating the need for routine pre-completion sound testing!

A Robust Detail is only used in connection with Part E and is defined as *'a separating wall or floor (of concrete, masonry, timber, steel or steel-concrete composite) construction, which has been assessed and approved by Robust Details Limited'*.

How are Robust Details approved?

In order to be approved, each Robust Detail must:

- be capable of consistently exceeding the performance standards given in Approved Document E to the Building Regulations;
- be practical to construct on site;
- be reasonably tolerant to workmanship.

How can Robust Details be used?

Builders are only permitted to use Robust Details instead of PCT **if** the plots concerned have been registered in advance with Robust Details Limited.

Once a plot has been registered, Robust Details Ltd will provide the relevant registration documentation (which will be accepted by all building control bodies as evidence that the builder is entitled to use Robust Details instead of PCT). The builder will then need to select the Robust Detail specific to the walls and/or floors they wish to build from the Robust Details Handbook (available from Robust Details Ltd) and produce a sitework checklist to show how they are going to ensure that building work is carried out exactly in accordance with the Robust Detail specifications.

Will there be more Robust Details?

Trade associations, manufacturers or other interested parties may submit applications for new robust details which will be evaluated and if found acceptable, approved and published.

Where can I obtain more information?

Robust Details Limited
PO Box 7289
Milton Keynes
MK14 6ZQ
Telephone/Fax:
Business line: 0870 240 8210
Technical support line: 0870 240 8209

Fax: 0870 240 8203
e-mail Support:
Technical email support (technical@robustdetails.com)
Administrative email support (administration@robustdetails.com)
Other support (customerservice@robustdetails.com)

6.7 Walls

In a brick built house, the external walls are loadbearing elements that support
the roof, floors and internal walls. These walls are normally cavity walls comprising of two leaves braced with metal ties but older houses will have solid
walls, at least 225 mm (9″) thick. Bricks are laid with mortar in overlapping
bonding patterns to give the wall rigidity and a Damp-Proof Course (DPC) is
laid just above ground level to prevent the moisture rising. Window and door
openings are spanned above with rigid supporting beams called lintels. The
internal walls of a brick built house are either non-loadbearing divisions (made
from lightweight blocks, manufactured boards or timber studding) or load-
bearing structures made of brick or block.

Modern timber-framed house walls are constructed of vertical timber studs
with horizontal top and bottom plates nailed to them. The frames, which are
erected on a concrete slab or a suspended timber platform supported by cavity
brick walls, are faced on the outside with plywood sheathing to stiffen the
structure. Breather paper is fixed over the top to act as a moisture barrier.
Insulation quilt is used between studs. Rigid timber lintels at openings carry
the weight of the upper floor and roof. Brick cladding is typically used to
cover the exterior of the frame. It is attached to the frame with metal ties.
Weatherboarding often replaces the brick cladding on upper floors.

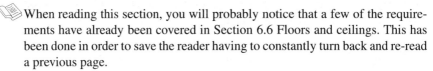When reading this section, you will probably notice that a few of the require-
ments have already been covered in Section 6.6 Floors and ceilings. This has
been done in order to save the reader having to constantly turn back and re-read
a previous page.

6.7.1 Requirements

*As a fire precaution, all materials used for internal linings of a building should
have a low rate of surface flame spread and (in some cases) a low rate of heat
release.*

(Approved Document B2)

Fire precautions

- *all loadbearing elements of structure of the building shall be capable of withstanding the effects of fire for an appropriate period without loss of stability;*
- *ideally the building should be subdivided by elements of fire-resisting
  construction into compartments;*

- *all openings in fire-separating elements shall be suitably protected in order to maintain the integrity of the continuity of the fire separation;*
- *any hidden voids in the construction shall be sealed and subdivided to inhibit the unseen spread of fire and products of combustion, in order to reduce the risk of structural failure and the spread of fire.*

(Approved Document B3)

- *external walls shall be constructed so that the risk of ignition from an external source and the spread of fire over their surfaces, is restricted*
- *the amount of unprotected area in the side of the building shall be restricted so as to limit the amount of thermal radiation that can pass through the wall*
- *the roof shall be constructed so that the risk of spread of flame and/or fire penetration from an external fire source is restricted*
- *the risk of a fire spreading from the building to a building beyond the boundary, or vice versa shall be limited.*

(Approved Document B4)

The walls of the building shall adequately protect the building and people who use the building from harmful effects caused by:

- *ground moisture;*
- *precipitation and wind-driven spray;*
- *interstitial and surface condensation; and*
- *spillage of water from or associated with sanitary fittings or fixed appliances.*

(Approved Document C2)

Cavity insulation

Fumes given off by insulating materials such as by Urea Formaldehyde (UF) foams should not be allowed to penetrate occupied parts of buildings to an extent where it could become a health risk to persons in the building by becoming an irritant concentration.

(Approved Document D)

Airborne and impact sound

Dwellings shall be designed so that the noise from domestic activity in an adjoining dwelling (or other parts of the building) is kept to a level that:

- *does not affect the health of the occupants of the dwelling;*
- *will allow them to sleep, rest and engage in their normal activities in satisfactory conditions.*

(Approved Document E1)

Dwellings shall be designed so that any domestic noise that is generated internally does not interfere with the occupants' ability to sleep, rest and engage in their normal activities in satisfactory conditions.

(Approved Document E2)

Domestic buildings shall be designed and constructed so as to restrict the transmission of echoes.

(Approved Document E3)

Schools shall be designed and constructed so as to reduce the level of ambient noise (particularly echoing in corridors).

(Approved Document E4)

6.7.2 Meeting the requirements

General

Walls should comply with the relevant requirements of BS 5628: Part 3:2001.	A1/2 2C2c

Basic requirements for stability

The layout of walls (both internal and external) shall:	A1/2 1A2b
• form a robust three-dimensional box structure in plan; • be constructed according to the specific guidance for each form of construction.	
Internal and external walls shall be adequately connected by either masonry bonding or by using mechanical connections.	A1/2 1A2c

Building height

For residential buildings, the maximum height of the building measured from the lowest finished ground level adjoining the building to the highest point of any wall or roof should not be greater than 15 m.	A1/2 2C4i
Types of wall shown in Table 6.20 must extend to the full storey height.	A1/2 2C2

Table 6.20 Wall types considered in this section

Residential buildings of up to three storeys	Small single storey non-residential buildings and annexes
External walls Internal load-bearing walls Compartment walls Separating walls	External walls Internal load-bearing walls

Thickness of walls

The thickness of the wall depends on the general conditions relating to the building of which the wall forms a part (e.g. floor area, roof loading, wind speed, etc.) and the design conditions relating to the wall (e.g. type of materials, loading, end restraints, openings, recesses, overhangs and lateral floor support requirements, etc.).

 Note: Where walls are constructed of bricks or blocks, they shall be in accordance with BS 6649: 1985.

Masonry units

Walls should be properly bonded and solidly put together with mortar and constructed of masonry units conforming to:

• clay bricks or blocks conforming to BS 3921: 1985 or BS 6649: 1985 or BS EN 771-1;	A1/2 2C20a
• calcium silicate bricks conforming to BS 187: 1978 or BS 6649: 1985 or BS EN 771-2;	A1/2 2C20b
• concrete bricks or blocks conforming to BS 6073: Part 1: 1981 or BS EN 771-3 or 4;	A1/2 2C20c
• square dressed natural stone conforming to the appropriate requirements described in BS EN 771-6 or BS 5628: Part 3: 2001;	A1/2 2C20d
• Manufactured stone complying with BS 6457: 1984 and BS EN 771-5.	A1/2 2C20e

 Note: See BS 3921, BS 6073-1, BS 187, BS 5390 and BS 6649 for further details about the minimum compressive strength requirements for masonry units.

Mortar

Mortar should be equivalent to (or of greater strength and durability) than:

• mortar designation (iii) according to BS 5628: Part 3: 2001;	A1/2 2C22a
• strength class M4 according to BS EN 998-2;	A1/2 2C22b
• 1:1: 5 or 6 CEM 1, lime and fine aggregate measured by volume of dry materials.	A1/2 2C22c

Tension straps

Tension straps (conforming to BS EN 845-1) should be used to strap walls to floors above ground level, at intervals not exceeding 2 m and as shown in Figure 6.55.   A1/2 2C35

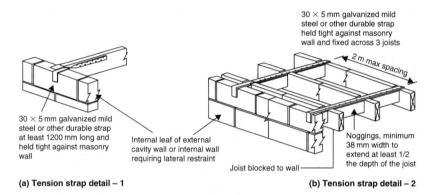

30 × 5 mm galvanized mild steel or other durable strap held tight against masonry wall and fixed across 3 joists

2 m max spacing

30 × 5 mm galvanized mild steel or other durable strap at least 1200 mm long and held tight against masonry wall

Internal leaf of external cavity wall or internal wall requiring lateral restraint

Joist blocked to wall

Noggings, minimum 38 mm width to extend at least 1/2 the depth of the joist

(a) Tension strap detail – 1

(b) Tension strap detail – 2

Figure 6.55 Lateral support by floors

Gable walls should be strapped to roofs as shown in Figure 6.56(a) and (b) by tension straps.   A1/2 2C36

For corrosion resistance purposes, tension straps should be material reference 14 or 16.1 or 16.2 (galvanized steel) or other more resistant specifications including material references 1 or 3 (austenitic stainless steel).   A1/2 2C35

The declared tensile strength of tension straps should not be less than 8 kN.

Tension straps need **not** be provided:

- in the longitudinal direction of joists in houses of not more than two storeys if the joists:
  - are at not more than 1.2 m centres;   A1/2 2C35a
  - have at least 90 mm bearing on the supported walls or 75 mm bearing on a timber wall-plate at each end; or   A1/2 2C35a
  - are carried on the supported wall by joist hangers (in accordance with BS EN 845-1 and BS 5628 – see Figure 6.57);   A1/2 2C35b
  - and are incorporated at not more than 2 m centres;   A1/2 2C35b
- when a concrete floor has at least 90 mm bearing on the supported wall (see Figure 6.58); and   A1/2 2C35c

● where floors are at or about the same level on each   A1/2 2C35d
  side of a supported wall, and contact between the
  floors and wall is either continuous or at intervals not
  exceeding 2 m. Where contact is intermittent, the
  points of contact should be in line or nearly in line
  on plan (see Figure 6.59).

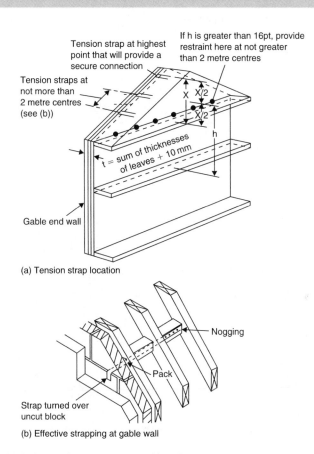

(a) Tension strap location

Tension strap at highest point that will provide a secure connection

If h is greater than 16pt, provide restraint here at not greater than 2 metre centres

Tension straps at not more than 2 metre centres (see (b))

t = sum of thicknesses of leaves + 10 mm

Gable end wall

Nogging

Pack

Strap turned over uncut block

(b) Effective strapping at gable wall

Figure 6.56 Lateral support at roof level

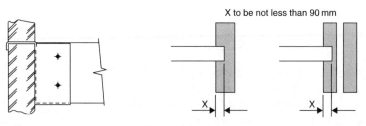

X to be not less than 90 mm

Figure 6.57 Restraint type joist hanger

Figure 6.58 Restraint by concrete floor or roof

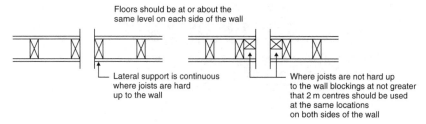

Floors should be at or about the
same level on each side of the wall

Lateral support is continuous
where joists are hard
up to the wall

Where joists are not hard up
to the wall blockings at not greater
that 2 m centres should be used
at the same locations
on both sides of the wall

Figure 6.59 Restraint of internal walls

Internal load-bearing walls in brickwork or blockwork

The maximum span for any floor supported by a wall is 6 m where the span is measured centre to centre of bearing (see Figure 6.60).	A1/2 2C23

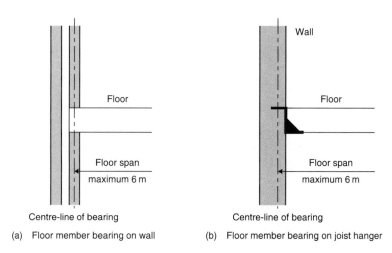

Wall

Floor

Floor

Floor span
maximum 6 m

Floor span
maximum 6 m

Centre-line of bearing

Centre-line of bearing

(a) Floor member bearing on wall

(b) Floor member bearing on joist hanger

Figure 6.60 Maximum span of floors

Vertical loading on walls should be distributed.	A1/2 2C23a
Differences in level of ground or other solid construction between one side of the wall and the other should be less than four times the thickness of the wall as shown in Figure 6.61.	A1/2 2C23b

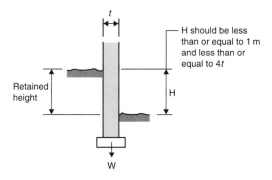

Figure 6.61 Maximum difference in permitted level

Dead load, imposed load and wind load should be in accordance with current codes of practice.	A1/2 0.2b
Loads used in calculations should allow for possible dynamic, concentrated and peak load effects that may occur.	A1/2 0.2
All walls (except compartment and/or separating walls) should have a thickness not less than:	A1/2 2C10

$$\frac{\text{Specified thickness from Table 6.21}}{2} - 5\,mm$$

Note: Except for a wall in the lowest storey of a three storey building, carrying load from both upper storeys, which should have a thickness as determined by the equation or 140 mm whichever is the greatest.

Solid external walls, compartment walls and separating walls in coursed brickwork or blockwork

Solid walls constructed of coursed brickwork or blockwork should be at least as thick as 1/16 of the storey height.	A1/2 2C6

Solid external walls, compartment walls and separating walls in uncoursed stone, flints, etc.

The thickness of walls constructed in uncoursed stone, flints, clunches, of bricks or other burnt or vitrified material should not be less than 1.33 times the thickness of the storey height.	A1/2 2C7

Cavity walls in coursed brickwork or blockwork

All cavity walls should have leaves at least 90 mm thick and cavities at least 50 mm wide.	A1/2 2C8
The combined thickness of the two leaves plus 10 mm should be at least 1/16 of the storey height (see as per Table 6.21).	A1/2 2C8

Table 6.21 Minimum thickness of certain external walls, compartment walls and separating walls

Height of wall	Length of wall	Minimum thickness of wall
not exceeding 3.5 m	not exceeding 12 m	190 mm for whole of its height
exceeding 3.5 m but not exceeding 9 m	not exceeding 9 m	190 mm for whole of its height
	exceeding 9 m	290 mm from the base for the height of one storey and 190 mm for the rest of its height
exceeding 9 m but not exceeding 12 m	not exceeding 9 m	290 mm from the base for the height of one storey and 190 mm for the rest of its height
	exceeding 9 m but not exceeding 12 m	290 mm from the base for the height of two storeys and 190 mm for the rest of its height

Wall ties should either comply with BS 1243, DD 140 or BS EN 845-1.	A1/2 2C19
Wall ties should have a horizontal spacing of 900 mm and a vertical spacing of 450 mm.	A1/2 2C8
💡 Equivalent to 2.5 ties per square metre.	
Wall ties should also be provided, spaced not more than 300 mm apart vertically, within a distance of 225 mm from the vertical edges of all openings, movement joints and roof verges.	A1/2 2C8
For external walls, compartment walls and separating walls in cavity construction, the combined thickness of the two leaves plus 10 mm should be at least as thick as 1/16 of the storey height.	A1/2 2C8

Walls providing vertical support to other walls

Irrespective of the material used in the construction, a wall should not be less than the thickness of any part of the wall to which it gives vertical support.	A1/2 2C9

Parapet walls

The minimum thickness and maximum height of parapet walls should be as shown in Figure 6.62.	A1/2 2C11

Single leaves of certain external walls

The single leaf of external walls of small single storey non-residential buildings and of annexes need be only 90 mm thick.	A1/2 2C12

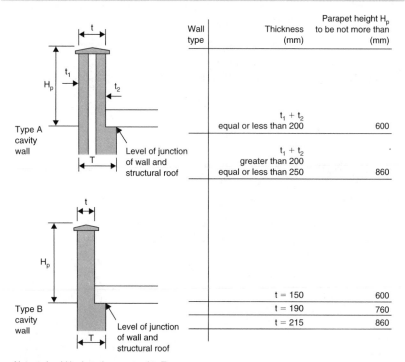

Wall type	Thickness (mm)	Parapet height H_p to be not more than (mm)
Type A cavity wall	$t_1 + t_2$ equal or less than 200	600
	$t_1 + t_2$ greater than 200 equal or less than 250	860
Type B cavity wall	$t = 150$	600
	$t = 190$	760
	$t = 215$	860

Note: t should be less than or equal to T

Figure 6.62 Height of parapet walls

The combined dead and imposed load should not exceed 70 kN/m at base of wall (see Figure 6.63).	A1/2 2C23c
Walls should not be subjected to lateral load other than from wind.	A1/2 2C23c

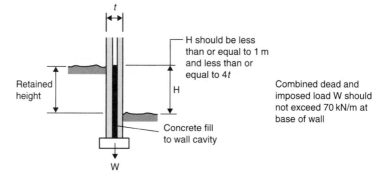

Figure 6.63 Combined and imposed dead load

Vertical lateral restraint to walls

The ends of every wall should be bonded or otherwise securely tied throughout their full height to a buttressing wall, pier or chimney.	A1/2 2C25

Long walls may be provided with intermediate buttressing walls, piers or chimneys dividing the wall into distinct lengths within each storey.	A1/2 2C25
Note: Each distinct length is considered to be a supported wall for the purposes of the Building Regulations.	
Intermediate buttressing walls, piers or chimneys should provide lateral restraint to the full height of the supported wall.	A1/2 2C25
They may be staggered at each storey.	
A wall in each storey of a building should:	A1/2 2C32
• extend to the full height of that storey;	

- have horizontal lateral supports to restrict movement of the wall at right angles to its plane.

The requirements for lateral restraint are shown in Table 6.22.

A1/2 2C34

Table 6.22 Lateral support for walls

Wall type	Wall length	Lateral support required
Solid or cavity: external compartment separating	Any length	Roof lateral support by every roof forming a junction with the supported wall
	Greater than 3 m	Floor lateral support by every floor forming a junction with the supported wall
Internal load-bearing wall (not being a compartment or separating wall)	Any length	Roof or floor lateral support at the top of each storey

Walls should be strapped to floors above ground level, at intervals not exceeding 2 m and as shown in Figure 6.65 by tension straps conforming to BS EN 845-1.

A1/2 2C35

Buttressing walls

If the buttressing wall is not itself a supported wall its thickness T_2 should not be less than:

- half the thickness required for an external or separating wall of similar height and length less 5 mm; or
- 75 mm if the wall forms part of a dwelling-house and does not exceed 6 m in total height and 10 m in length; and
- 90 mm in other cases.
- The length of the buttressing wall should be:
  - at least 1/6 of the overall height of the supported wall
  - be bonded or securely tied to the supporting wall and at the other end to a buttressing wall, pier or chimney.
- The size of any opening in the buttressing wall should be restricted as shown in Figure 6.64.

A1/2 2C26a

A1/2 2C26b

A1/2 2C26c

A1/2 2C26c

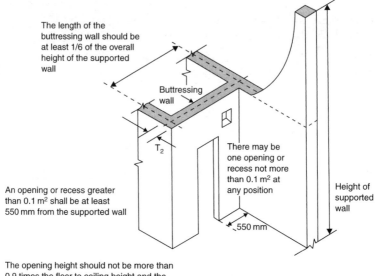

The length of the
buttressing wall should be
at least 1/6 of the overall
height of the supported
wall

Buttressing
wall

T_2

There may be
one opening or
recess not more
than 0.1 m² at
any position

An opening or recess greater
than 0.1 m² shall be at least
550 mm from the supported wall

Height of
supported
wall

550 mm

The opening height should not be more than
0.9 times the floor to ceiling height and the
depth of the lintel including any masonry
over the opening should be not less than 150 mm

Figure 6.64 Openings in a buttressing wall

Gable walls

Gable walls should be strapped to roofs as shown in Figure 6.65(a) and (b) by tension straps.	A1/2 2C36
Vertical strapping at least 1 m in length should be provided at eaves level at intervals not exceeding 2 m as shown in Figure 6.65(c) and (d).	A1/2 2C36
Vertical strapping may be omitted if the roof:	A1/2 2C36a–d

- has a pitch of 15° or more; and
- is tiled or slated; and
- is of a type known by local experience to be resistant to wind gusts; and
- has main timber members spanning onto the supported wall at not more than 1.2 m centres.

Piers

Piers should have a minimum width of 190 mm (see Figure 6.66).	A1/2 2C27a
Piers should measure at least three times the thickness of the supported wall.	A1/2 2C27a

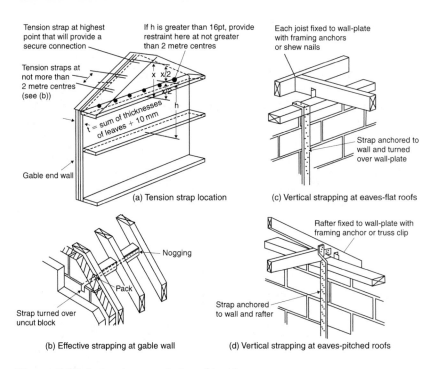

Tension strap at highest point that will provide a secure connection

Tension straps at not more than 2 metre centres (see (b))

If h is greater than 16pt, provide restraint here at not greater than 2 metre centres

x   x/2

$t = $ sum of thicknesses of leaves + 10 mm

h

Gable end wall

(a) Tension strap location

Each joist fixed to wall-plate with framing anchors or shew nails

Strap anchored to wall and turned over wall-plate

(c) Vertical strapping at eaves-flat roofs

Nogging

Pack

Strap turned over uncut block

(b) Effective strapping at gable wall

Rafter fixed to wall-plate with framing anchor or truss clip

Strap anchored to wall and rafter

(d) Vertical strapping at eaves-pitched roofs

Figure 6.65 Lateral support at roof level

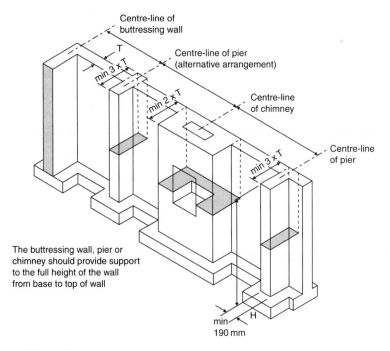

Centre-line of buttressing wall

T

min 3 × T

Centre-line of pier (alternative arrangement)

min 2 × T

Centre-line of chimney

min 3 × T

Centre-line of pier

The buttressing wall, pier or chimney should provide support to the full height of the wall from base to top of wall

min 190 mm

H

Figure 6.66 Buttressing

Chimneys

Chimneys should measure at least twice the thickness, measured at right angles to the wall (see Figure 6.66).   A1/2 2C27a

The sectional area on plan of chimneys (excluding openings for fireplaces and flues):   A1/2 2C27b

- should be not less than the area required for a pier in the same wall; and
- the overall thickness should not be less than twice the required thickness of the supported wall (see Figure 6.66).

Openings and recesses

The number, size and position of openings and recesses should not impair the stability of a wall or the lateral restraint afforded by a buttressing wall to a supported wall.   A1/2 2C28

Construction over openings and recesses should be adequately supported.   A1/2 2C28

No openings should be provided in walls below ground floor except for small holes for services and ventilation, etc. which should be limited to a maximum area of $0.1\,m^2$ at not less than 2 m centres (see Figure 6.67 and Table 6.23).   A1/2 2C29

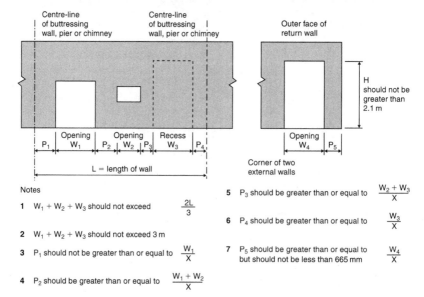

Notes

1 $W_1 + W_2 + W_3$ should not exceed $\dfrac{2L}{3}$

2 $W_1 + W_2 + W_3$ should not exceed 3 m

3 P_1 should not be greater than or equal to $\dfrac{W_1}{X}$

4 P_2 should be greater than or equal to $\dfrac{W_1 + W_2}{X}$

5 P_3 should be greater than or equal to $\dfrac{W_2 + W_3}{X}$

6 P_4 should be greater than or equal to $\dfrac{W_3}{X}$

7 P_5 should be greater than or equal to but should not be less than 665 mm $\dfrac{W_4}{X}$

Figure 6.67 Sizes of openings and recesses

Table 6.23 Value of 'X' factor for Figure 6.67

Nature of roof span	Maximum roof span (m)	Minimum thickness of wall inner (mm)	Span of floor is parallel to wall	Span of timber floor into wall		Span of concrete floor into wall	
				Max 4.5 m	Max 6.0 m	Max 4.5 m	Max 6.0 m
			Value of factor 'X'				
Roof spans parallel to wall	Non-applicable	100	6	6	6	6	6
		90	6	6	6	6	5
Timber roof spans into wall	9	100	6	6	5	4	3
		90	6	4	4	3	3

Overhangs

The amount of any projection should not impair the stability of the wall.	A1/2 2C31

Chases

Vertical chases should not be deeper than 1/3 of the wall thickness.	A1/2 2C30a
Note: Or, in cavity walls, 1/3 of the thickness of the leaf.	
Horizontal chases should not be deeper than 1/6 of the thickness of the leaf of the wall.	A1/2 2C30b
Chases should not be so positioned as to impair the stability of the wall (particularly where hollow blocks are used).	A1/2 2C30c

Small single-storey non-residential buildings and annexes

General

The walls shall be solidly constructed in brickwork or blockwork.	A1/2 2C38(i)b
Where the floor area of the building or annexe exceeds $10\,m^2$, the walls shall have a mass of not less than $130\,kg/m^2$.	A1/2 2C38(i)c

The only lateral loads are wind loads.	A1/2 2C38(i)e
The maximum length or width of the building or annexe shall not exceed 9 m.	A1/2 2C38(i)f
The height of the building or annexe shall not exceed the lower value derived from Figure 6.68.	A1/2 2C38(i)g

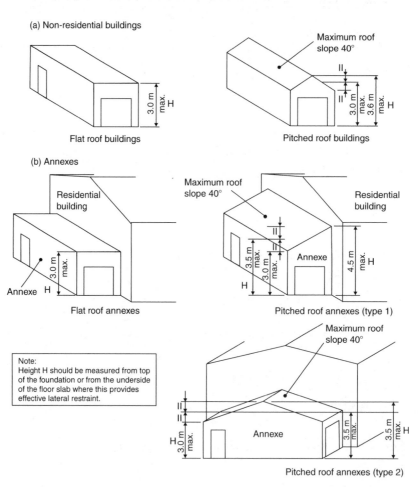

Figure 6.68 Size and proportions of non-residential buildings and annexes

| Walls shall be tied to the roof structure vertically and horizontally and have a horizontal lateral restraint at roof level. | A1/2 2C38(i)i |

Size and location of openings

One or two major openings not more than 2.1 m in height are permitted in one wall of the building or annexe only.	A1/2 2C38(ii)
The width of a single opening or the combined width of two openings should not exceed 5 m.	A1/2 2C38(ii)
The only other openings permitted in a building or annexe are for windows and a single leaf door.	A1/2 2C38(ii)
The size and location of these openings should be in accordance with Figure 6.69.	A1/2 2C38(ii)
Major openings should be restricted to one wall only. Their aggregate width should not exceed 0.5 m and their height should not be greater than 2.1 m.	A1/2 2C38(ii)
There should be no openings within 2.0 m of a wall containing a major opening.	A1/2 2C38(ii)
The aggregate size of the openings in a wall not containing a major opening should not exceed 2.4 m.	A1/2 2C38(ii)
There should not be more than one opening between piers.	A1/2 2C38(ii)
Unless there is a corner pier the distance from a window or a door to a corner should not be less than 390 mm.	A1/2 2C38(ii)

Wall thicknesses and recommendations for piers

The walls should have a minimum thickness of 90 mm.	A1/2 2C38(iii)
Walls which do not contain a major opening but exceed 2.5 m in length or height should be bonded or tied to piers for their full height at not more than 3 m centres as shown in Figure 6.70.	A1/2 2C38(iii)
Walls which contain one or two major openings should in addition have piers as shown in Figure 6.70(b) and (c).	A1/2 2C38(iii)

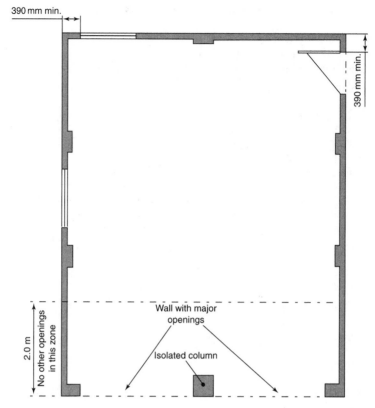

Figure 6.69 Size and location of openings

Where ties are used to connect piers to walls they   A1/2 2C38(iii)
should be:

- flat;
- 20 mm × 3 mm in cross-section;
- stainless steel;
- placed in pairs;
- spaced at not more than 300 mm centre vertically.

Walls should be tied horizontally at no more than 2 m   A1/2 2C38(iv)
centres to the roof structure at eaves level, base of
gables and along roof slopes (as shown in Figure 6.71)
with straps.

Where straps cannot pass through a wall, they should be   A1/2 2C38(iv)
adequately secured to the masonry using suitable fixings.

Isolated columns should also be tied to the roof   A1/2 2C38(iv)
structure (see Figure 6.71).

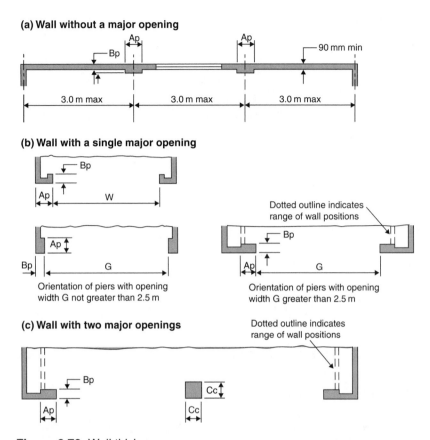

(a) Wall without a major opening

(b) Wall with a single major opening

Orientation of piers with opening
width G not greater than 2.5 m

Orientation of piers with opening
width G greater than 2.5 m

(c) Wall with two major openings

Dotted outline indicates
range of wall positions

Figure 6.70 Wall thicknesses

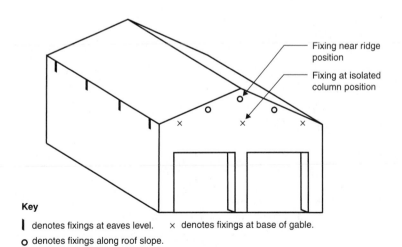

Key

❙ denotes fixings at eaves level.   × denotes fixings at base of gable.

○ denotes fixings along roof slope.

Figure 6.71 Lateral restraint at roof level

Foundations

A wall shall be erected to prevent undue moisture from the   C3
ground reaching the inside of the building and (if it is an
outside wall) adequately resisting the penetration of rain
and snow to the inside of the building (see Figure 6.72).

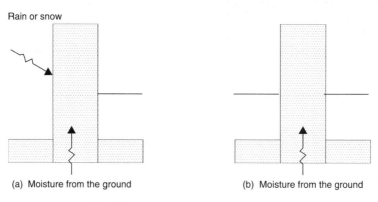

Rain or snow

(a) Moisture from the ground   (b) Moisture from the ground

Figure 6.72 Wall, resistance to moisture. (a) External wall. (b) Internal wall

Resistance to the passage of moisture

Walls should:

- resist the passage of moisture from the ground   C5.2a
  to the inside of the building*;
- not be damaged by moisture from the ground;   C5.2b
- not carry moisture from the ground to any part   C5.2b
  which would be damaged by moisture.

*For buildings used wholly for storing goods or where provisions   C5.3
put in place do not increase the health and safety of persons
employed in that building, this requirement may not apply.

External walls should:

- resist rain penetrating components of the   C5.2c
  structure that might be damaged by moisture;
- resist rain penetrating to the inside of the building*;   C5.2d
- be designed and constructed so that their   C5.2e
  structural and thermal performance are not
  adversely affected by interstitial condensation;
- not promote surface condensation or mould growth.   C5.2f

*For buildings used wholly for storing goods or where provisions   C5.3
put in place do not increase the health and safety of persons
employed in that building, this requirement may not apply.

Internal and external walls exposed to moisture from the ground

Internal and external walls (subject to moisture from the ground)   C5.5a
shall have a damp-proof course of bituminous material,
polyethylene, engineering bricks or slates in cement mortar or
any other material that will prevent the passage of moisture.

The damp-proof course should be continuous with   C5.5a
any damp-proof membrane in the floors.

If the wall is an external wall, the damp-proof course should   C5.5b
be at least 150 mm above the level of the adjoining ground
(see Figure 6.73) unless the design is such that a part of
the building will protect the wall.

If the wall is an external cavity wall (see Figure 6.74) the
cavity should either:

- be taken down at least 225 mm below the level   C5.5c
  of the lowest damp-proof course; or
- a damp-proof tray should be provided so as to prevent   C5.5c
  precipitation passing into the inner leaf (see Figure 6.75),
  with weep holes every 900 mm to assist in the transfer
  of moisture through the external leaf.

Where the damp-proof tray does not extend the full length   C5.5c
of the exposed wall (i.e. above an opening) stop ends and at
least two weep holes should be provided.

As well as giving protection against moisture from the ground,   C5.7
an external wall should give protection against precipitation.

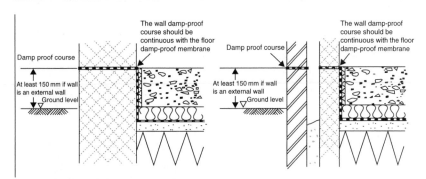

Figure 6.73 Damp proof courses

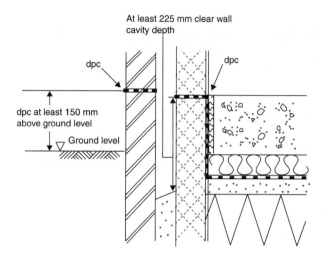

Figure 6.74 Cavity carried down

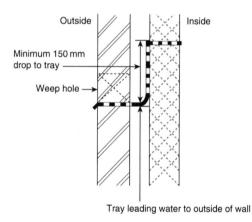

Figure 6.75 Damp proof (cavity) tray

Solid external walls

Solid walls shall hold moisture arising from rain and snow until it can be released in a dry period without penetrating to the inside of the building, or causing damage to the building.	C5.8
Solid external walls exposed to **very severe** conditions should be protected by external impervious cladding.	C5.9
Solid external walls exposed to **severe** conditions may be built with:	C5.9a

- brickwork (or stonework) at least 328 mm thick;
- dense aggregate concrete blockwork at least 250 mm thick; or

- lightweight aggregate (aerated autoclaved concrete blockwork) at least 215 mm thick.

Solid external walls exposed to **severe** conditions may be built, providing:

- the rendering is in two coats with a total thickness   C5.9b of at least 20 mm and has a scraped or textured finish;
- the strength of the mortar is compatible with the   C5.9b strength of the bricks or blocks;
- the joints (if the wall is to be rendered) are   C5.9b raked out to a depth of at least 10 mm;
- the rendering mix is 1 part of cement, 1 part of lime and   C5.9b 6 parts of well-graded sharp sand (nominal mix 1:1:6) unless the blocks are of dense concrete aggregate, in which case the mix may be 1:½.

Adequate protection should be provided at the top   C5.9c of walls, etc. (see Figure 6.76).

Unless the protection and joints are a complete barrier to   C5.9c moisture, a damp-proof course should also be provided.

Damp-proof courses, cavity trays and closers should be provided and designed to ensure that water drains outwards:

- where the downward flow will be interrupted   C5.9d(i) by an obstruction (e.g. from some types of lintel);
- under openings – unless there is a sill and the   C5.9d(ii) sill and its joints will form a complete barrier;
- at abutments between walls and roofs.   C5.9d(iii)

A solid external wall may be insulated on the   C5.10 inside or on the outside.

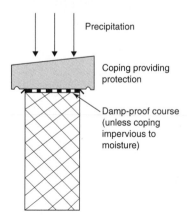

Figure 6.76 Projection of wall head from precipitation

Where the insulation is on the inside, a cavity should be provided to give a break in the path for moisture.	C5.10
Where the insulation is on the outside, it should provide some resistance to the ingress of moisture to ensure the wall remains relatively dry (see Figure 6.77).	C5.10

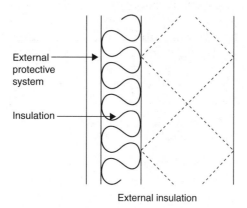

External protective system

Insulation

External insulation

Figure 6.77 Insulated (solid) external wall

Cavity external walls

The outer leaf shall be separated from the inner leaf by a drained air space (or in any other way which will prevent precipitation from being carried to the inner leaf).	C5.12
The construction of a cavity external wall could include:	
• outer leaf masonry (bricks, blocks, stone or manufactured stone);	C5.13a
• a cavity at least 50 mm wide;	C5.13b
• inner leaf masonry or frame with lining.	C5.13b
Masonry units should be laid on a full bed of mortar with the cross joints substantially and continuously filled to ensure structural robustness and weather resistance.	C5.13c
Where a cavity is to be partially filled, the residual cavity should not be less than 50 mm wide (see Figure 6.78).	C5.13c

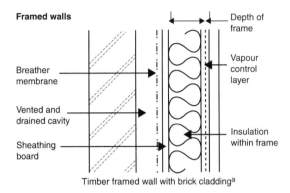

Figure 6.78 Insulated framed wall

Cavity insulation

The suitability of the wall for installing insulation material(s) is to be assessed before the work is carried out.	C5.15a and d
When the cavity of an existing house is being filled, attention should be given to the condition of the external leaf of the wall, e.g. its state of repair and type of pointing.	C5.15e
A full or partial fill insulating material may be placed in the cavity between the outer leaf and an inner leaf of masonry subject to the suitability of a wall for installing insulation into the cavity (see Table 6.24).	C5.15a
The insulating material should be the subject of current certification from an appropriate body or a European Technical Approval.	C5.15c
When partial fill materials are used, the residual cavity should not be less than 50 mm nominal.	C5.15b
Rigid (board or batt) thermal insulating material built into the wall must be certified as being in conformance by an approved installer.	C5.15b
Urea-formaldehyde foam inserted into the cavity should be: • in accordance with BS 5617: 1985; • installed in accordance with BS 5618: 1985.	C5.15d
The person undertaking installation work should operate under an Approved Installer Scheme.	C5.15c

Table 6.24 Maximum recommended exposure zones for insulated masonry walls

| Wall construction | | Maximum recommended exposure zone for each construction | | | | | | |
| Insulation method | Min. width of filled or clear cavity (mm) | Impervious cladding | | Rendered finish | | Facing masonry | | Flush skills and copings |
		Full height of wall	Above facing masonry	Full height of wall	Above facing masonry	Tooled flush joints	Recessed mortar joints	
Built-in full fill	50	4	3	3	3	2	1	1
	75	4	3	4	3	3	1	1
	100	4	4	4	3	3	1	2
	125	4	4	4	3	3	1	2
	150	4	4	4	4	4	1	2
Injected fill *not* UF foam	50	4	2	3	2	2	1	1
	75	4	3	4	3	3	1	1
	100	4	3	4	3	3	1	1
	125	4	4	4	3	3	1	2
	150	4	4	4	4	4	1	2
Injected fill UF foam	50	4	2	3	2	1	1	1
	75	4	2	3	2	2	1	1
	100	4	2	3	2	2	1	1
Partial fill								
Residual 50 mm cavity	50	4	4	4	4	3	1	1
Residual 75 mm cavity	75	4	4	4	4	4	1	1
Residual 100 mm cavity	100	4	4	4	4	4	2	1
Internal insulation								
Clear cavity 50 mm	50	4	3	4	3	3	1	1
Clear cavity 100 mm	100	4	4	4	4	4	2	2
Fully filled								
Cavity 50 mm	50	4	3	3	3	2	1	1
Cavity 100 mm	100	4	4	4	3	3	1	2

Framed external walls

The cladding shall be separated from the insulation or sheathing by a vented and drained cavity with a membrane that is vapour open, but resists the passage of liquid water, on the inside of the cavity (see Figure 6.78).	C5.17

Cracking of external walls

The possibility of severe rain penetration occurring through cracks in masonry external walls should be taken into account when designing a building.	C5.18

Impervious cladding systems for walls

Cladding systems for walls should:

- resist the penetration of precipitation to the inside of the building; C5.19a
- not be damaged by precipitation; C5.19b
- not carry precipitation to any part of the building which would be damaged by it. C5.19b

Cladding that is designed to protect a building from precipitation shall be:

- jointless or have sealed joints; C5.21a
- impervious to moisture. C5.21a

If the cladding has overlapping dry joints it shall be:

- impervious or weather resisting; C5.21b
- backed by a material which will direct precipitation which C5.21b
  enters the cladding towards the outer face.

Materials that can deteriorate rapidly without special care C5.22
should only be used as the weather-resisting part of a
cladding system.

Cladding may be:

- impervious – e.g. metal, plastic, glass and bituminous C5.23a
  products;
- weather-resisting – e.g. natural stone or slate, C5.23b
  cement-based products, fired clay and wood;
- moisture-resisting – e.g. bituminous and plastic C5.23c
  products lapped at the joints;

- jointless materials and sealed joints – i.e. to   C5.23d
  allow for structural and thermal movement.

Dry joints between cladding units should be designed so that:

- precipitation will not pass through them;   C5.24
- precipitation which enters the joints will be directed C5.24
  towards the exposed face without it penetrating beyond
  the back of the cladding.

Note: Whether dry joints are suitable will depend
on the design of the joint or the design of the cladding
and the severity of the exposure to wind and rain.

Each sheet, tile and section of cladding should be securely   C5.25
fixed (as per guidance contained in BS 8000-6: 1990).

Particular care should be taken with detailing and   C5.25
workmanship at the junctions between cladding and window
and door openings as they are vulnerable to moisture ingress.

Insulation may be incorporated into the construction provided C5.26
it is either protected from moisture or is unaffected by it.

Where cladding is supported by timber components (or is on   C5.27
the façade of a timber framed building) the space between the
cladding and the building should be ventilated to ensure rapid
drying of any water that penetrates the cladding.

Joint between walls and doors/window frames

The joint between walls and doors and window frames should:

- resist the penetration of precipitation to the   C5.29a
  inside of the building;
- not be damaged by precipitation;   C5.29b
- not permit precipitation to reach any part of   C5.29c
  the building which would be damaged by it.

Damp-proof courses should be provided to direct
moisture towards the outside, particularly:

- where the downward flow of moisture would be   C5.30a
  interrupted at an obstruction, e.g. at a lintel;
- where sill elements (including joints) do not   C5.30b
  form a complete barrier to the transfer of precipitation,
  e.g. under openings, windows and doors;

- where reveal elements, including joints, do not   C5.30c
  form a complete barrier to the transfer of rain
  and snow, e.g. at openings, windows and doors.

Direct plastering of the internal reveal of any   C5.31
window frame should only be used with a backing
of expanded metal lathing or similar.

In areas of the country that are exposed to very
severe driving rain:

- checked rebates should be used in all window and   C5.32
  door reveals;
- the frame should be set back behind the outer   C5.32
  leaf of masonry as shown in Figure 6.79;
- alternatively an insulated finned cavity closer may be used.   C5.32

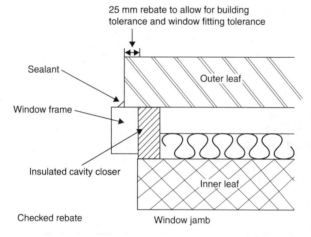

Figure 6.79 Window reveals for use in areas subject to very severe driving rain

Door thresholds

Where an accessible threshold is provided to allow
unimpeded access (as specified in Part M):

- the external landing (see Figure 6.80) should   C5.33a
  be laid to a fall between 1 in 40 and 1 in 60 in a single
  direction away from the doorway;
- the sill leading up to the door threshold has a   C5.33b
  maximum slope of 15.

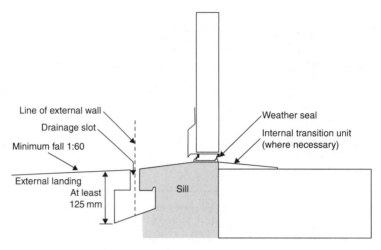

Figure 6.80 Accessible threshold for use in exposed areas

Interstitial condensation (external doors)

External walls shall be designed and constructed in accordance with Clause 8.3 of BS 5250: 2002.	C5.34
💣 Specialist advice should be sought when designing swimming pools and other buildings where interstitial condensation in the walls (caused by high internal temperatures and humidities) can cause high levels of moisture being generated.	C5.35

Surface condensation and mould growth (external doors)

External walls shall be designed and constructed so that the:

- thermal transmittance (U-value) does not exceed   C5.36a
  $0.7 \, W/m^2 K$ at any point;
- junctions between elements and details of openings (such   C5.36b
  as doors and windows) meet with the recommendations
  in the report on robust construction details.

Wall cladding

Wall cladding presents a hazard if it becomes detached from the building. An acceptable level of safety can be achieved depending on the type and location of the cladding.

Note: The guidance given below relates to all forms of cladding, including curtain walling and glass façades.

Cladding shall be capable of safely sustaining and transmitting (to the supporting structure of the building) all dead, imposed and wind loads.	A1/2 3.2a
Provision shall be made, where necessary, to accommodate differential movement of the cladding and the supporting structure of the building.	A1/2 3.2c
Wind loading on the cladding should be derived from BS 6399, Part 2: 2001.	A1/2 3.3
Due consideration shall be given to local increases in wind suction arising from funnelling of the wind through gaps between buildings.	A1/2 3.3
Note: Guidance on funnelling effects is given in BRE Digest 436 '*Wind loading on buildings – Brief guidance for using BS 6399-2: 1997*' available from BRE, Bucknalls Lane, Garston, Watford, Herts WD2 7JR.	
The cladding shall be securely fixed to, and supported by, the structure of the building using both vertical support and horizontal restraint.	A1/2 3.2b
The cladding and its fixings (including any support components) shall be of durable materials.	A1/2 3.2d
The design life of the fixings shall not be less than that of the cladding.	A1/2 3.2d
Fixings shall be corrosion resistant and of a material type appropriate for the local environment.	A1/2 3.2d
Fixings for supporting cladding should be determined from a consideration of the proven performance of the fixing and the risks associated with the particular application.	A1/2 3.7
The strength of fixings should be derived from tests using materials representative of the material into which the fixing is to be anchored, taking account of any inherent weaknesses that may affect the strength of the fixing, e.g. cracks in concrete due to shrinkage and flexure, or voids in masonry construction.	A1/2 3.8
Where the cladding is required to support other fixtures (e.g. handrails or fittings such as antennae and signboards)	A1/2 3.4

account should be taken of the loads and forces arising
from such fixtures and fittings.

Where the wall cladding is required to function as pedestrian   A1/2 3.5
guarding to stairs, ramps, vertical drops of 600 mm or
greater or as a vehicle barrier, account should be taken of
the additional imposed loading as stipulated in Part K.

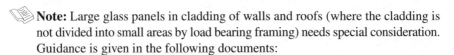 Where wall cladding is required to safely withstand
lateral pressures from crowds, an appropriate design loading
is given in BS 6399 Part 1 and the 'Guide to Safety at
Sports Grounds' (4th Edition, 1997).

Applications should be designated as being either   A1/2 3.7
non-redundant (where the failure of a single fixing could
lead to the detachment of the cladding) or redundant
(where failure or excessive movement of one fixing results
in load sharing by adjacent fixings) and the required
reliability of the fixing determined accordingly.

All cladding (used to protect the building from rain   C4 (5.1–5.6)
or snow) shall be jointless or have sealed joints.

Note: Large glass panels in cladding of walls and roofs (where the cladding is
not divided into small areas by load bearing framing) needs special consideration.
Guidance is given in the following documents:

The Institution of Structural Engineers' Report on 'Structural use of glass in build-
ings' dated 1999, available from 11 Upper Belgrave Street, London SW1X 8BH.
'Nickel sulfide in toughened glass' published by the Centre for Window Cladding
and Technology, dated 2000.
Further guidance on cladding is given in the following documents:
The Institution of Structural Engineers' Report on 'Aspects of cladding', dated
1995.
The Institution of Structural Engineers' Report on 'Guide to the structural use
of adhesives', dated 1999.
BS 8297 'Code of practice for the design and installation of non-load bearing
precast concrete cladding'.
BS 8298 'Code of practice for the design and installation of natural stone
cladding and lining'.

Internal fire spread (linings)

The choice of materials for walls and ceilings can significantly affect the
spread of a fire and its rate of growth, even though they are not likely to be the
materials first ignited. Although furniture and fittings can have a major effect
on fire spread it is not possible to control them through Building Regulations.

The surface linings of walls should meet the classifications shown in Table 6.25.

Table 6.25 Classification of linings

Location	Class*
Small rooms with an area of not more than 4 m² (in residential accommodation) or 30 m² (in non-residential accommodation)	3
Domestic garages not more than 40 m²	3
Other rooms (including garages)	1
Circulation spaces within buildings	1
Other circulation spaces (including the common area of flats and maisonettes)	0

* Classifications are based on tests as per BS 476 and as described in Appendix A of Approved Document B.

Air supported structures should comply with the recommendations given in BS 6661.	B2 (7.8)
Any flexible membrane covering a structure (other than an air supported structure) should comply with the recommendations given in Appendix A of BS 7157.	B2 (7.9)
The wall and any floor between the garage and the house shall have a 30 minute standard of fire resistance.	B2

Loadbearing elements of structure

All loadbearing elements of a structure shall have a minimum standard of fire resistance.	B3 (8.1)
Structural frames, beams, columns, loadbearing walls (internal and external), floor structures and gallery structures, should have at least the fire resistance given in Appendix A of Approved Document B.	B3 (8.2)

Compartmentation

To prevent the spread of fire within a building, whenever possible, the building should be sub-divided into compartments separated from one another by walls and/or floors of fire-resisting construction.	B3 (9.1)
A wall common to two or more buildings should be constructed as a compartment wall.	B3 (9.10)

Parts of a building that are occupied mainly for different purposes, should be separated from one another by compartment walls and/or compartment floors.	B3 (9.11)
Structural frames, beams, columns, loadbearing walls (internal and external), floor structures and gallery structures, should have at least the fire resistance given in Appendix A of Approved Document B.	B3 (8.2)
When altering an existing two-storey single-family dwellinghouse to provide additional storeys, the floor(s), both old and new, shall have the full 30 minute standard of fire resistance.	B3 (8.7)
If an existing house or other building is converted, the means of escape shall be adequately protected and there shall be a 30 minute fire resistance standard.	B3 (8.10–8.11)
There should be continuity at the junctions of the fire resisting elements enclosing a compartment.	B3 (9.6)
Spaces that connect compartments, such as stairways and service shafts, need to be protected to restrict fire spread between the compartments.	B3 (9.7)
Every place that is a potential fire hazard should be enclosed with fire-resisting construction.	B3 (9.12)
Every wall separating semi-detached houses, or houses in terraces, should be constructed as a compartment wall, and the houses should be considered as separate buildings.	B3 (9.13)
If a domestic garage is attached to (or forms an integral part of) a house, the garage should be separated from the rest of the house, as shown in Figure 6.81.	B3 (9.14)

The wall and any floor between the garage and the house shall have a 30 minute standard of fire resistance. Any opening in the wall should be at least 100 mm above the garage floor level with an FD30 door.

In buildings containing flats or maisonettes, compartment walls (or compartment floors) shall be constructed between:	B3 (9.15)

- every floor (unless it is within a maisonette);
- one storey and another within one dwelling;
- every wall separating a flat or maisonette from any other part of the building;

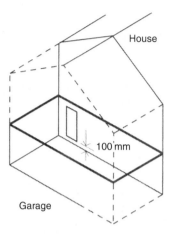

Figure 6.81 Separation between garage and dwellinghouse

- every wall enclosing a refuse storage chamber.

Every compartment wall and compartment floor should: B3 (9.22)

- form a complete barrier to fire between the compartments they separate; and
- have the appropriate fire resistance as indicated in Appendix A, Tables A6.1 and A6.2.

Timber beams, joists, purlins and rafters may be built into B3 (9.22)
or carried through a masonry or concrete compartment
wall if the openings for them are kept as small as practicable
and then fire-stopped.

If trussed rafters bridge the wall, they should be designed B3 (9.22)
so that failure of any part of the truss due to a fire in one
compartment will not cause failure of any part of the truss
in another compartment.

Compartment walls that are common to two or more B3 (9.23)
buildings should run the full height of the building
in a continuous vertical plane so that the adjoining
buildings are separated by walls, not floors.

Compartment walls (used to form a separated part of B3 (9.24)
a building) should run the full height of the building
in a continuous vertical plane.

Compartment walls in a top storey beneath a roof B3 (9.26)
should be continued through the roof space.

Where a compartment wall or compartment floor meets   B3 (9.27)
another compartment wall, or an external wall, the junction
should maintain the fire resistance of the compartmentation.

When a compartment wall meets the underside of the   B3 (9.28)
roof covering or deck, the wall/roof junction shall
maintain continuity of fire resistance.

Double skinned insulated roof sheeting should   B3 (9.29)
incorporate a band of material of limited combustibility.

Any openings in a compartment wall which is common   B3 (9.33)
to two or more buildings should be provided with
an escape door in case of fire.

Openings in compartment walls should have the   B3 (9.35)
appropriate fire resistance and be limited to those for:

- doors that have the appropriate fire resistance;
- the passage of pipes, ventilation ducts, chimneys,
  appliance ventilation ducts or ducts encasing one or
  more flue pipes;
- refuse chutes of non-combustible construction;
- protected shafts.

Any stairway or other shaft passing directly from   B3 (9.36)
one compartment to another should be enclosed in
a protected shaft so as to delay or prevent the spread
of fire between compartments.

Protection of openings for pipes

Pipes that pass through a compartment wall or compartment floor (unless the pipe is in a protected shaft), or through a cavity barrier, should conform to one of the alternatives shown in Figure 6.82.

Flue walls

Flue walls should have a fire resistance of at least   B3 (11.11)
one half of that required for the compartment wall
or floor and be of non-combustible construction.

If a flue, or duct containing flues or appliance ventilation B3 (11.11)
duct(s), passes through a compartment wall or
compartment floor, or is built into a compartment wall,
each wall of the flue or duct should have a fire resistance
of at least half that of the wall or floor in order to prevent
the by-passing of the compartmentation (see Figure 6.83).

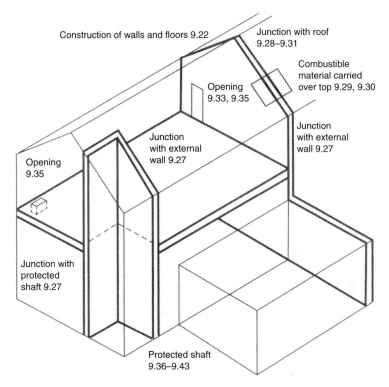

Figure 6.82 Compartment walls and compartment floors with reference to the relevant paragraphs in Approved Document B

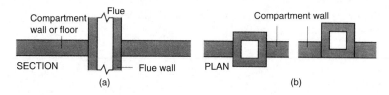

Figure 6.83 Flues penetrating compartment walls or floors. (a) Flue passing through compartment wall or floor. (b) Flue built into compartment wall

Fire resistance

Proprietary fire-stopping and sealing systems (including those designed for service penetrations) which have been shown by test to maintain the fire resistance of the wall or other element, are available and may be used. Other fire-stopping materials include:

- cement mortar,
- gypsum-based plaster,

- cement or gypsum-based vermiculite/perlite mixes,
- glass fibre, crushed rock, blast furnace slag or ceramic-based products (with or without resin binders), and
- intumescent mastics (B3 11.14).

Joints between fire separating elements should be fire-stopped.	B3 (11.12a)
All openings for pipes, ducts, conduits or cables to pass through any part of a fire separating element should be:	B3 (11.12b)

- kept as few in number as possible;
- kept as small as practicable;
- fire-stopped (which in the case of a pipe or duct, should allow thermal movement).

To prevent displacement, materials used for fire-stopping should be reinforced with (or supported by) materials of limited combustibility.	B3 (11.13)

Construction of an external wall

Where a portal framed building is near a relevant boundary, the external wall near the boundary may need fire resistance to restrict the spread of fire between buildings.	B4 (13.4)
In cases where the external wall of the building cannot be wholly unprotected, the rafter members of the frame, as well as the column members, may need to be fire protected.	B4 (13.4)
The external surfaces of walls should meet the provisions shown in Figure 6.84.	B4 (13.5)

It should be noted that the use of combustible materials for cladding frame-work, or the use of combustible thermal insulation as an overcladding may be risky in tall buildings, even though the provisions for external surfaces in Figure 6.84 may have been satisfied.

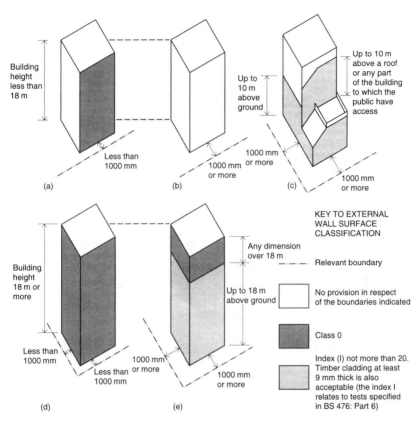

Figure 6.84 Provisions for external surfaces of walls. (a), (d), (e) Any building. (b) Any building other than (c). (c) Assembly or recreation building of more than one storey

| The external envelope of a building should not provide a medium for fire spread if it is likely to be a risk to health or safety. | B4 (13.7) |
| In a building with a storey 18 m or more above ground level, insulation material used in ventilated cavities in the external wall construction should be of limited combustibility (this restriction does not apply to masonry cavity wall construction). | B4 (13.7) |

Combustible material should not be placed in or exposed to the cavity, except for:

- timber lintels, window or door frames, or the end stairway of timber joists;
- pipes, conduits or cables;
- DPC, flashing, cavity closer or wall ties;
- fire-resisting thermal insulating material;
- a domestic meter cupboard.

Masonry wall construction

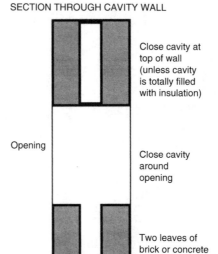

SECTION THROUGH CAVITY WALL

Close cavity at
top of wall
(unless cavity
is totally filled
with insulation)

Opening

Close cavity
around
opening

Two leaves of
brick or concrete
each at least
75 mm thick

Figure 6.85 Masonry cavity walls excluded from the previous for cavity barriers

Cavity insulation

The outer leaf of the wall should be built of masonry or concrete.	D1 (1.1–1.2)
The inner leaf of the wall should be built of masonry (bricks or blocks).	D1 (1.1–1.2)
The wall being insulated with UF (area formaldehyde) shall be assessed (in accordance with BS 8208) for suitability before any work commences.	D1 (1.1–1.2)
The person carrying out the work needs to hold (or operate under) a current BSI Certificate of Registration of Assessed Capability for the work he is doing.	D1 (1.1–1.2)
The installation shall be in accordance with BS 5618: 1985.	D1 (1.1–1.2)
The material shall be in accordance with the relevant recommendations of BS 5617: 1985.	D1 (1.1–1.2)

Airborne sound

The flow of sound energy through walls should be restricted.	E
Walls should reduce the level of airborne sound.	E
Walls that separate a dwelling from another building (oranother dwelling) shall resist the transmission of airborne sound.	E
Habitable rooms (or kitchens) within a dwelling shall resist the transmission of airborne sound.	E
Air paths, including those due to shrinkage, must be avoided.	E
Porous materials and gaps at joints in the structure must be sealed.	E
Flanking transmission (i.e. the indirect transmission of sound from one side of a wall to the other side) should be minimized.	E
The possibility of resonance in parts of the structure (such as a dry lining) should be avoided.	E

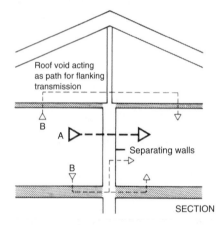

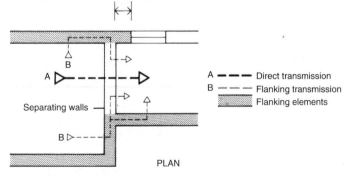

Figure 6.86 Direct and flanking transmission

 For clarity not all flanking paths have been shown.

Separating walls (new buildings)

Walls – general

All new walls constructed within a dwelling-house (flat or room used for residential purposes) – whether purpose built or formed by a material change of use – shall meet the laboratory sound insulation values set out in Table 6.26.	E0.9
Walls that have a separating function should achieve the sound insulation values:	E0.1

- for rooms for residential purposes as set out in Table 6.26;
- dwelling-houses and flats as set out in Table 6.26.

Table 6.26 Dwelling houses and flats – performance standards for separating walls that have a separating function

	Airborne sound insulation $D_{nT,w} + C_{tr}$ dB (minimum values)
Purpose built rooms for residential purposes	43
Purpose built dwelling-houses and flats	45
Rooms for residential purposes formed by material change of use	43
Dwelling-houses and flats formed by material change of use	43

 Notes:

(1)   The sound insulation values in this table include a built-in allowance for 'measurement uncertainty' and so if any of these test values are not met, then that particular test will be considered as failed.

(2)   Occasionally a higher standard of sound insulation may be required between spaces used for normal domestic purposes and noise generated in and to an adjoining communal or non-domestic space. In these cases it would be best to seek specialist advice before committing yourself.

Flanking transmission from walls connected to the separating wall shall be controlled.	E2
Tests should be carried out between rooms or spaces that share a common area formed by a separating wall or separating floor.	E1
Impact sound insulation tests should be carried out without a soft covering (e.g. carpet, foam backed vinyl etc.) on the floor.	E1

If the floor joists are to be supported on the separating wall then they should be supported on hangers and should not be built in.	E2
If the joists are at right angles to the wall, spaces between the floor joists should be sealed with full depth timber blocking.	E3
The floor base (excluding any screed) should be built into a cavity masonry external wall and carried through to the cavity face of the inner leaf.	E

Walls that separate a dwelling from another dwelling (or part of the same building) shall resist: • the level (and transmission) of airborne sounds; • the transmission of impact sound (such as speech, musical instruments and loudspeakers and impact sources such as footsteps and furniture moving); • the flow of sound energy through walls and floors.	E

Requirements

Requirement E1

Figure 6.87 illustrates the relevant parts of the building that should be protected from airborne and impact sound in order to satisfy Requirement E1.

In some circumstances (for example, when a historic building is undergoing a material change of use) it may not be practical to improve the sound insulation to the standards set out in Approved Document E1 particularly if the special characteristics of such a building need to be recognized. In these circumstances the aim should be to improve sound insulation to the '*extent that it is practically possible*'.

 Note: BS 7913:1998 *The principles of the conservation of historic buildings* provides guidance on the principles that should be applied when proposing work on historic buildings.

Requirement E2

Constructions for new walls within a dwelling-house (flat or room for residential purposes) – whether purpose built or formed by a material change of use – shall meet the laboratory sound insulation values set out in Table 6.27.	E0.9

Figure 6.88 illustrates the relevant parts of the building that should be protected from airborne and impact sound in order to satisfy Requirement E2.

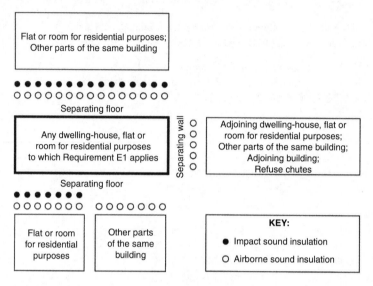

Figure 6.87 Requirement E1 – resistance to sound

Table 6.27 Laboratory values for new internal walls within dwelling-houses, flats and rooms for residential purposes – whether purpose built or formed by a material change of use

	Airborne sound insulation R_W dB (minimum values)
Purpose built dwelling-houses and flats	40

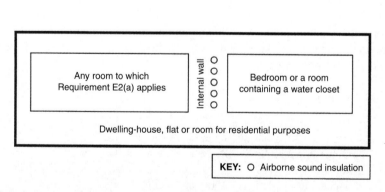

Figure 6.88 Requirement E2a – internal walls

Requirement E3

Sound absorption measures described in Section 7 of Approved Document N shall be applied.	E0.11

Requirement E4

The values for sound insulation, reverberation time and indoor ambient noise as described in Section 1 of Building Bulletin 93 *'The Acoustic Design of Schools'* (produced by DFES and published by the Stationery Office (ISBN 0 11 271105 7)) shall be satisfied.	E0.12

Types of wall

As shown in Figure 6.89 there are four main types of separating walls that can be used in order to achieve the required performance standards shown in Table 6.28.

Solid masonry
(Wall type 1)

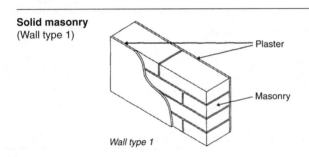

Plaster

Masonry

Wall type 1

The resistance to airborne sound depends mainly on the mass per unit area of the wall.

Cavity masonry
(Wall type 2)

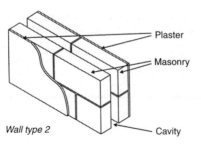

Plaster

Masonry

Wall type 2

Cavity

The resistance to airborne sound depends on the mass per unit area of the leaves and on the degree of isolation achieved. The isolation is affected by connections (such as wall ties and foundations) between the wall leaves and by the cavity width.

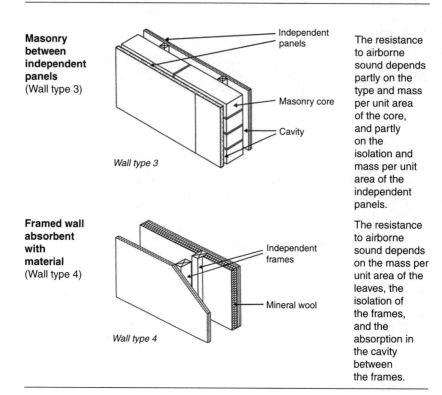

Masonry between independent panels (Wall type 3)

Wall type 3

Independent panels

Masonry core

Cavity

The resistance to airborne sound depends partly on the type and mass per unit area of the core, and partly on the isolation and mass per unit area of the independent panels.

Framed wall absorbent with material (Wall type 4)

Wall type 4

Independent frames

Mineral wool

The resistance to airborne sound depends on the mass per unit area of the leaves, the isolation of the frames, and the absorption in the cavity between the frames.

Figure 6.89 Types of separating walls

 Other designs, materials and/or products may also be available and so it is always worthwhile talking to the manufacturers and/or suppliers first.

 The resistance to airborne sound depends mainly on the mass of the wall.

Table 6.28 Dwelling-houses and flats – performance standards for separating walls, separating floors and stairs that have a separating function

	Airborne sound insulation $D_{nT,w} + C_{tr}$ dB (minimum values)	Impact sound insulation $L'_{nT,w}$ dB (maximum values)
Purpose built dwelling-houses and flats		
Walls	45	–
Floors and stairs	45	62
Dwelling-houses and flats formed by material change of use		
Walls	43	–
Floors and stairs	43	64

Junctions between separating walls and other building elements

Care should be taken to correctly detail the junctions between the separating wall and other elements, such as floors, roofs, external walls and internal walls.	E2.9

 Note: Where any building element functions as a separating element (e.g. a ground floor that is also a separating floor for a basement flat) then the separating element requirements should take precedence.

Mass per unit area of walls

The mass per unit area of a wall is expressed in kilograms per square metre (kg/m^2) and is equivalent to:

$$\text{mass per unit area of a wall} = \frac{\text{mass of co-ordinating area}}{\text{co-ordinating area}} \qquad (6.1)$$

Mass per unit area of a wall can be calculated as follows:

$$\text{mass per unit area of a wall} = \frac{M_B + \rho_m\,[Td(1 + h - d) + V]}{LH}\,\text{kg/m}^2 \quad (6.2)$$

Where:
M_B = brick/block mass (kg) at appropriate moisture content
ρ_m = density of mortar (kg/m^3) at appropriate mortar content
T = the brick/block finish without surface finish (m)
d = mortar thickness (m)
L = co-ordinating length (m)
H = co-ordinating height (m)
V = volume of any frog/void filled with mortar (m^3)

 Note: The method for calculating mass per unit area is provided in Annex A to Part E of the Regulations together with some examples.

Density of the materials

The density of the materials used (and on which the mass per unit area of the wall depends) is expressed in kilograms per cubic metre (kg/m^3).

Plasterboard linings on separating and external masonry walls

Wherever plasterboard is recommended (or the finish is not specified) a drylining laminate of plasterboard with mineral wool may be used.	E2.15
Plasterboard linings should be fixed according to manufacturer's instructions.	E2.16

Note: Recommended cavity widths in separating cavity masonry walls are minimum values.

Wall ties in separating and external cavity masonry walls

There are two types of wall ties that can be used in masonry cavity walls, type A (butterfly ties), which are normal and type B (double triangle ties), which are used only in external masonry cavity walls where tie type A does not satisfy the requirements of Building Regulation Part A – Structure.

 Notes:
(1)   Recommended cavity widths in separating cavity masonry walls are minimum values.
(2)   In external cavity masonry walls, tie type B may decrease the airborne sound insulation due to flanking transmission via the external wall leaf compared to tie type A.

Stainless steel cavity wall ties are specified for all houses regardless of their location.	A1/2
Wall ties should have a horizontal spacing of 900 mm and a vertical spacing of 450 mm.	A1/2 2C8
💡 Equivalent to 2.5 ties per square metre.	
Wall ties should be spaced not more than 300 mm apart vertically, within a distance of 225 mm from the vertical edges of all openings, movement joints and roof verges.	A1/2 2C8
Wall ties should either comply with BS 1243, DD 140, or BS EN 845-1.	A1/2 2C19
Wall ties should be selected in accordance with Table 6.	A1/2 2C19
The leaves of a cavity masonry wall construction should be connected by either butterfly ties or double-triangle ties spaced as per BS 5628-3:2001 which limits this tie type and spacing to cavity widths of 50 mm to 75 mm with a minimum masonry leaf thickness of 90 mm.	E2.19
Note: Wall ties may be used provided that they have the measured dynamic stiffness for the cavity width (see E2.20 and E2.21 for details of the relevant formula for measuring the dynamic stiffness).	
In conditions of severe exposure, austenitic stainless steel or suitable non-ferrous ties should be used.	A1/2 (1C20)

The number of ties per square metre, n, shall be calculated from the horizontal (S_x) and vertical (S_y) tie spacing distances (in metres) using the formula $n = 1/(S_x . S_y)$.	E2.22
All wall ties and spacings specified using the dynamic stiffness parameter should also satisfy the requirements of Building Regulation Part A – Structure.	E2.24

Corridor walls and doors

Separating walls should be used between corridors and rooms in flats, in order to control flanking transmission and to provide the required sound insulation. **Note:** It is highly likely that the amount of sound insulation gained by using a separating wall will be reduced by the presence of a door.	E2.25
Noisy parts of the building should preferably have a lobby, double door or high performance doorset to contain the noise.	E2.27
All corridor doors shall have a good perimeter sealing (including the threshold where practical).	E2.26
All corridor doors shall have a minimum mass per unit area of 25 kg/m².	E2.26
All corridor doors shall have a minimum sound reduction index of 29 dB Rw (measured according to BS EN ISO 140-3:1995 and rated according to BS EN ISO 717-1:1997).	E2.26
All corridor doors shall meet the requirements for fire safety (see Building Regulations Part B – Fire Safety)	E2.26

Refuse chutes

A wall separating a habitable room (or kitchen) from a refuse chute should have a mass per unit area (including any finishes) of at least 1320 kg/m².	E2.28
A wall separating a non-habitable room from a refuse chute should have a mass per unit area (including any finishes) of at least 220 kg/m².	E2.28

Wall type 1 (solid masonry)

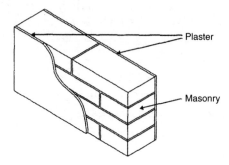

When using a solid masonry wall, the resistance to airborne sound depends mainly on the mass per unit area of the wall. As shown below, there are three different categories of solid masonry walls:

Table 6.29　Wall type 1 – categories

Wall type 1		Minimum mass per unit area (including plaster) 415 kg/m²
Category 1.1 **Solid masonry** Dense aggregate concrete block, plaster on both room faces		Plaster on both room faces Blocks laid flat to the full thickness of the wall For example: Size　　215 mm laid flat Density　1840 kg/m³ Coursing 110 mm Plaster　13 mm lightweight
Wall type 1 **Category 1.2** **Dense aggregate concrete** Dense aggregate concrete cast in-situ, plaster on both room faces		Minimum mass per unit area (including plaster) 415 kg/m² Plaster on both room faces For example: Concrete　190 mm Density　2200 kg/m³ Plaster　13 mm lightweight

Wall type 1

Category 1.3
Brick

Brick, plaster on
both room faces

Minimum mass per unit
area (including plaster)
375 kg/m²

Bricks to be laid frog
up, coursed with
headers

For example:
Size   215 mm laid
          flat
Density   1610 kg/m³
Coursing 75 mm
Plaster   13 mm
lightweight

General requirements

Fill and seal all masonry joints with mortar.	E2.32a
Lay bricks frog up to achieve the required mass per unit area and avoid air paths.	E2.32b
Use bricks/blocks that extend to the full thickness of the wall.	E2.32c
Ensure that an external cavity wall is stopped with a flexible closer at the junction with a separating wall.	E2.32d
Unless the cavity is fully filled with mineral wool or expanded polystyrene beads.	
Control flanking transmission from walls and floors connected to the separating wall (see guidance on junctions).	E2.32e
Deep sockets and chases should **not** be used in separating walls.	E2.32
Stagger the position of sockets on opposite sides of the separating wall.	E2.32f
Ensure flue blocks: • will not adversely affect the sound insulation; • use a suitable finish.	E2.32g
A cavity separating wall may **not** be changed into a solid masonry (i.e. type 1) wall by filling in the cavity with mortar and/or concrete.	E2.32
When the cavity wall is bridged by the solid wall, ensure that there is no junction between the solid masonry wall and a cavity wall.	E2.32

Wall type 1 – Junction requirements

Junctions with an external cavity wall with masonry inner leaf

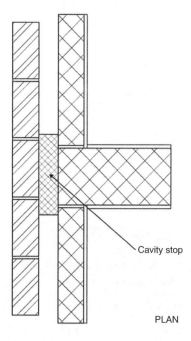

Cavity stop

PLAN

Where the external wall is a cavity wall:

- the outer leaf of the wall may be of any construction;   E2.36a
- the cavity should be stopped with a flexible closer.   E2.36b

The masonry inner leaf should have a mass per unit   E2.38a
area of at least 120 kg/m² excluding finish unless
there are openings in the external wall (see Figure 6.90)
that are:

- not less than 1 metre high;   E2.38b
- on both sides of the separating wall at every storey;   E2.38c
- not more than 700 mm from the face of the   E2.39
  separating wall on both sides.

Note: If there is also a separating floor, then the minimum mass
per unit area of 120 kg/m² (excluding finish) will always apply,
irrespective of the presence or absence of openings.

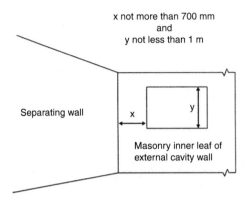

Figure 6.90 Wall type 1 – position of openings in a masonry inner leaf of an external cavity wall

The separating wall should be joined to the inner leaf of the external cavity wall by one of the following methods:

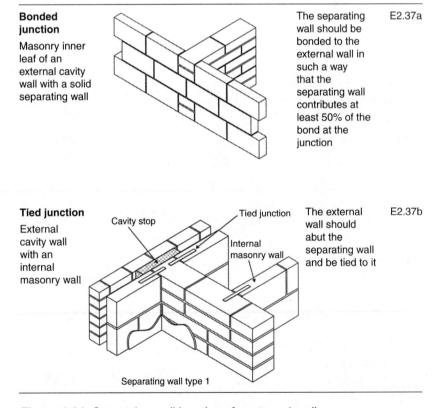

Bonded junction Masonry inner leaf of an external cavity wall with a solid separating wall		The separating wall should be bonded to the external wall in such a way that the separating wall contributes at least 50% of the bond at the junction	E2.37a
Tied junction External cavity wall with an internal masonry wall		The external wall should abut the separating wall and be tied to it	E2.37b

Figure 6.91 Separating wall junctions for a type 1 wall

Junctions with an external cavity wall with timber frame inner leaf

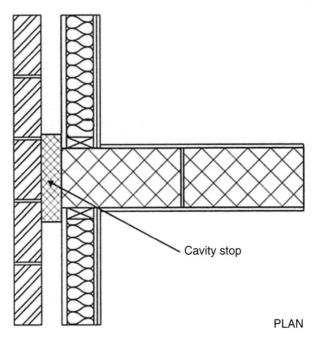

Cavity stop

PLAN

Where the external wall is a cavity wall:

- the outer leaf of the wall may be of any construction;   E2.40a
- the cavity should be stopped with a flexible closer.   E2.40b

Where the inner leaf of an external cavity wall is of framed
construction, the framed inner leaf should:

- abut the separating wall;   E2.41a1
- be tied to it with ties at no more than 300 mm centres   E2.41b1
  vertically.

The wall finish of the framed inner leaf of the external
wall should be:

- one layer of plasterboard; or   E2.41a2
- two layers of plasterboard where there is a   E2.41b2
  separating floor;
- each sheet of plasterboard should be of minimum   E2.41c
  mass per unit area 10 kg/m^2;
- all joints should be sealed with tape or caulked   E2.41d
  with sealant.

Junctions with internal timber floors

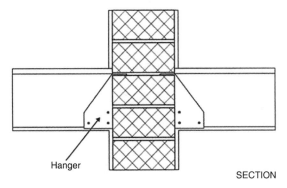

SECTION

Hanger

If the floor joists are to be supported on a type 1 separating wall then they should be supported on hangers as opposed to being built in.	E2.45

Junctions with internal concrete floors

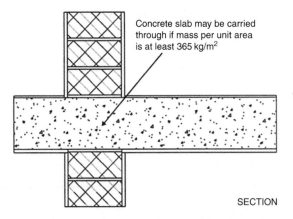

Concrete slab may be carried through if mass per unit area is at least 365 kg/m^2

SECTION

An internal concrete floor slab may only be carried through a type 1 separating wall if the floor base has a mass per unit area of at least 365 kg/m^2.	E2.46
Internal hollow-core concrete plank floors and concrete beams with infilling block floors should not be continuous through a type 1 separating wall.	E2.47

Note: For internal floors of concrete beams with infilling blocks, avoid beams built into the separating wall unless the blocks in the floor fill the space between the beams where they penetrate the wall.

Junctions with concrete ground floors

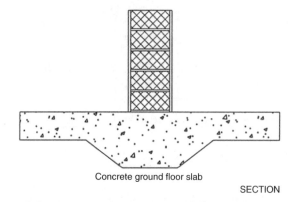

Concrete ground floor slab

SECTION

The ground floor may be a solid slab, laid on the ground, or a suspended concrete floor.	E2.51
A concrete slab floor on the ground may be continuous under a type 1 separating wall.	E2.51
A suspended concrete floor may only pass under a type 1 separating wall if the floor has a mass of at least 365 kg/m².	E2.52
Hollow core concrete plank and concrete beams with infilling block floors should not be continuous under a type 1 separating wall.	E2.53

Note: See also Building Regulation Part C – Site preparation and resistance to moisture, and Building Regulation Part L – Conservation of fuel and power.

Junctions with ceiling and roof

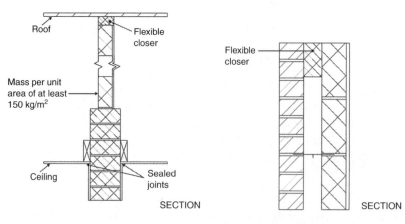

Ceiling and roof junction   External cavity at roof level

Where a type 1 separating wall is used it should be continuous to the underside of the roof.	E2.55
The junction between the separating wall and the roof should be filled with a flexible closer which is also suitable as a fire stop.	E2.56
Where the roof or loft space is not a habitable room (and there is a ceiling with a minimum mass per unit area of $10\,kg/m^2$ with sealed joints) then the mass per unit area of the separating wall above the ceiling may be reduced to $150\,kg/m^2$.	E2.57
If lightweight aggregate blocks of density less than $1200\,kg/m^3$ are used above ceiling level, then one side should be sealed with cement paint or plaster skim.	E2.58
Where there is an external cavity wall, the cavity should be closed at eaves level with a suitable flexible material (e.g. mineral wool).	E2.59
A rigid connection between the inner and external wall leaves should be avoided.	Ep23
If a rigid material is used, then it should only be rigidly bonded to one leaf.	Ep23

Guidance for other types of wall type 1 junctions

Junctions with an external solid masonry wall	No guidance available (seek specialist advice).	E2.42
Junctions with internal framed walls	There are no restrictions on internal framed walls meeting a type 1 separating wall.	E2.43
Junctions with internal masonry walls	Internal masonry walls that abut a type 1 separating wall should have a mass per unit area of at least $120\,kg/m^2$ excluding finish.	E2.44
Junctions with timber ground floors	If the floor joists are to be supported on a type 1 separating wall then they should be supported on hangers and should not be built in.	E2.49

Wall type 2 (cavity masonry)

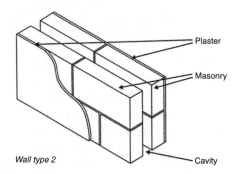

Wall type 2

When using a cavity masonry wall, the resistance to airborne sound depends on the mass per unit area of the leaves and on the degree of isolation achieved. The isolation is affected by connections (such as wall ties and foundations) between the wall leaves and by the cavity width.

As shown below, there are four different categories of cavity masonry walls:

Table 6.30 Wall type 2 – categories

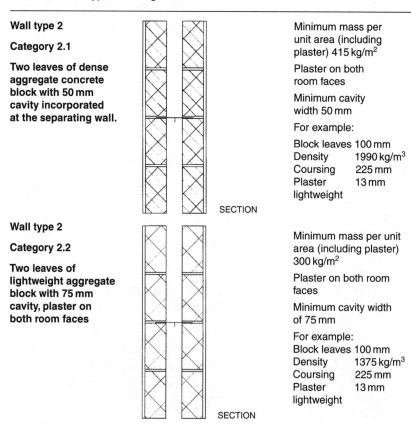

Wall type 2 **Category 2.1** **Two leaves of dense aggregate concrete block with 50 mm cavity incorporated at the separating wall.**	SECTION	Minimum mass per unit area (including plaster) 415 kg/m^2 Plaster on both room faces Minimum cavity width 50 mm For example: Block leaves 100 mm Density 1990 kg/m^3 Coursing 225 mm Plaster 13 mm lightweight
Wall type 2 **Category 2.2** **Two leaves of lightweight aggregate block with 75 mm cavity, plaster on both room faces**	SECTION	Minimum mass per unit area (including plaster) 300 kg/m^2 Plaster on both room faces Minimum cavity width of 75 mm For example: Block leaves 100 mm Density 1375 kg/m^3 Coursing 225 mm Plaster 13 mm lightweight

Wall type 2

Category 2.3

Two leaves of lightweight aggregate block with 75 mm cavity and step/stagger plasterboard on both room faces

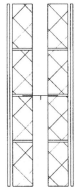

SECTION

Minimum mass per unit area (including plaster) 290 kg/m²

Lightweight aggregate blocks should have a density in the range 1350 to 1600 m³

Minimum cavity width of 75 mm

Wall type 2.3 should **only** be used where there is a step and/or stagger of at least 300 mm

Increasing the size of the step or stagger in the separating wall tends to increase the airborne sound insulation

Plasterboard (lightweight) each sheet of minimum mass per unit area 10 kg/m² on both room faces

For example:
Block leaves   100 mm
Density   1375 kg/m³
Coursing   225 mm
Lightweight plasterboard (minimum mass per unit area 10 kg/m²) on both room faces

Wall type 2

Category 2.4

Two leaves of Aircrete block with 75 mm cavity and step/stagger plasterboard or plaster on both room faces

Wall type 2.4 should **only** be used where there is a step and/or stagger of at least 300 mm

SECTION

Minimum mass per unit area (including plaster) 150 kg/m²

Lightweight aggregate blocks should have a density in the range 1350 to 1600 kg/m³

Minimum cavity width of 75 mm

Plasterboard (lightweight) minimum mass per unit area 10 kg/m² on both room faces or 13 mm plasterboard on both faces

Increasing the size of the step or stagger in the separating wall tends to increase the airborne sound insulation

For example:
Aircrete block leaves 100 mm
Density   650 kg/m³
Coursing   225 mm
Plaster (lightweight) minimum mass per unit area 10 kg/m² on both room faces

General requirements

Fill and seal all masonry joints with mortar.	E2.65a
Keep the cavity leaves separate below ground floor level.	E2.65b
Ensure that any external cavity wall is stopped with a flexible closer at the junction with the separating wall.	E2.65c
Control flanking transmission from walls and floors connected to the separating wall.	E2.65d
Stagger the position of sockets on opposite sides of the separating wall.	E2.65e
Ensure that flue blocks will not adversely affect the sound insulation and that a suitable finish is used over the flue blocks.	E2.65f
The cavity separating wall should **not** be converted to a type 1 (solid masonry) separating wall by inserting mortar or concrete into the cavity between the two leaves.	E2.65a2
A solid wall construction in the roof space should **not** be changed.	E2.65b2
Cavity walls should **not** be built off a continuous solid concrete slab floor.	E2.65c2
Deep sockets and chases should **not** be used in a separating wall.	E2.65d2
Deep sockets and chases in a separating wall should **not** be placed back to back.	E2.65d2
Wall ties used to connect the leaves of a cavity masonry wall should be tie type A.	E2.66

Wall type 2 – Junction requirements

Junctions with an external cavity wall with masonry inner leaf

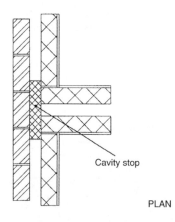

Wall types 2.1 and 2.2 – external cavity wall with masonry inner leaf

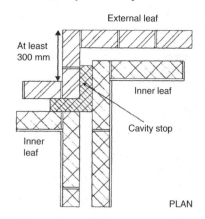

Wall types 2.3 and 2.4 – external cavity wall with masonry inner leaf – stagger

Where the external wall is a cavity wall:

- the outer leaf of the wall may be of any construction;   E2.73a
- the cavity should be stopped with a flexible closer.   E2.73b

The separating wall should be joined to the inner leaf of   E2.74
the external cavity wall by one of the following methods:

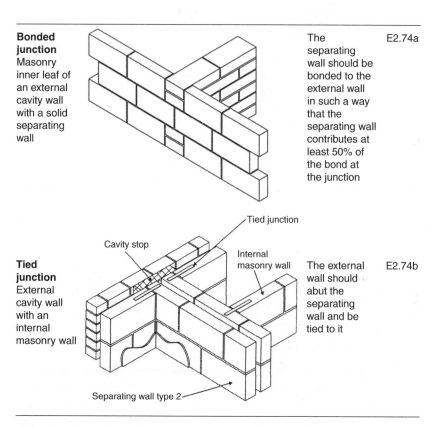

Bonded junction
Masonry inner leaf of an external cavity wall with a solid separating wall

The separating wall should be bonded to the external wall in such a way that the separating wall contributes at least 50% of the bond at the junction   E2.74a

Tied junction
External cavity wall with an internal masonry wall

Tied junction
Cavity stop
Internal masonry wall
Separating wall type 2

The external wall should abut the separating wall and be tied to it   E2.74b

Figure 6.92 Separating wall junctions for a type 2 wall

The masonry inner leaf should have a mass per unit area   E2.75
of at least 120 kg/m² excluding finish.

There is no minimum mass requirement where   E2.76
separating wall type 2.1, 2.3 or 2.4 is used **unless** there is
also a separating floor.

Junctions with an external cavity wall with timber frame inner leaf

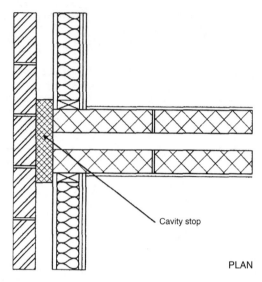

PLAN

Figure 6.93 Wall type 2 – external cavity wall with timber frame inner leaf

Where the external wall is a cavity wall:

• the outer leaf of the wall may be of any construction;	E2.77a
• the cavity should be stopped with a flexible closer (see Figure 6.93).	E2.77b

Where the inner leaf of an external cavity wall is of framed construction, the framed inner leaf should:

• abut the separating wall;	E2.78a
• be tied to it with ties at no more than 300 mm centres vertically.	E2.78b

The wall finish of the inner leaf of the external wall should be:

• one layer of plasterboard;	E2.98a2
• two layers of plasterboard where there is a separating floor;	E2.78b2
• each sheet of plasterboard to be of minimum mass per unit area 10 kg/m²;	E2.78c2
• all joints should be sealed with tape or caulked with sealant.	E2.78d2

Junctions with internal timber floors

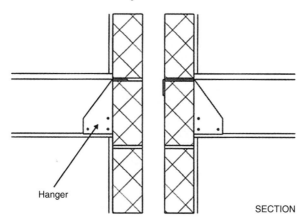

Hanger

SECTION

Figure 6.94 Wall type 2 – internal timber floor

If the floor joists are to be supported on the separating wall then they should be supported on hangers as opposed to being built in.	E2.84

Junctions with internal concrete floors

Internal concrete floors should generally be built into a type 2 separating wall and carried through to the cavity face of the leaf. The cavity should not be bridged.	E2.85

Junctions with concrete ground floors (see Figure 6.95)

The ground floor may be a solid slab, laid on the ground, or a suspended concrete floor.	E2.88
A concrete slab floor on the ground should not be continuous under a type 2 separating wall.	E2.88
A suspended concrete floor should not be continuous under a type 2 separating wall.	E2.89
A suspended concrete floor should be carried through to the cavity face of the leaf. The cavity should not be bridged.	E2.89

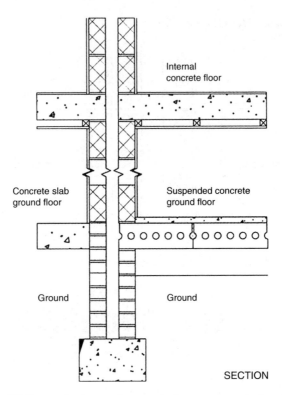

Figure 6.95 Wall type 2 – internal concrete floor and concrete ground floor

Junctions with ceiling and roof space

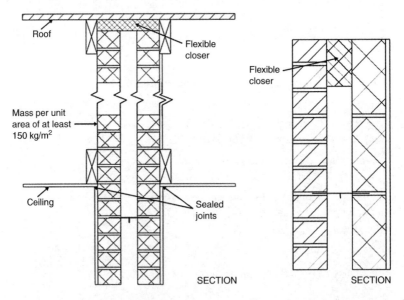

Wall type 2 – ceiling and roof junction   External cavity wall at eaves level

A type 2 separating wall should be continuous to the underside of the roof.	E2.91
The junction between the separating wall and the roof should be filled with a flexible closer which is also suitable as a fire stop.	E2.92
If lightweight aggregate blocks (with a density less than $1200\,kg/m^3$ are used) above ceiling level, then one side should be sealed with cement paint or plaster skim.	E2.94
The cavity of an external cavity wall should be closed at eaves level with a suitable flexible material (e.g. mineral wool).	E2.95
A rigid connection between the inner and external wall leaves should be avoided.	E2.95
If a rigid material has to be used, then it should **only** be rigidly bonded to one leaf.	E2.95

Note: If the roof or loft space is not a habitable room (and there is a ceiling with a minimum mass per unit area of $10\,kg/m^2$ with sealed joints) then the mass per unit area of the separating wall above the ceiling may be reduced to $150\,kg/m^2$ – **but** it should still be a cavity wall.

Guidance for other wall type 2 junctions

Junctions with internal masonry walls	• Internal masonry walls that abut a type 2 separating wall should have a mass per unit area of at least $120\,kg/m^2$ excluding finish;	E2.81
	• When there is a separating floor, the internal masonry walls should have a mass per unit area of at least $120\,kg/m^2$ excluding finish.	E2.82
Junctions with internal framed walls	There are no restrictions on internal framed walls meeting a type 2 separating wall.	E2.80
Junctions with an external solid masonry wall	No guidance available (seek specialist advice).	E2.79
Junctions with timber ground floors	If the floor joists are to be supported on a type 1 separating wall then they should be supported on hangers and should not be built in.	E2.49

Wall type 3 (masonry between independent panels)

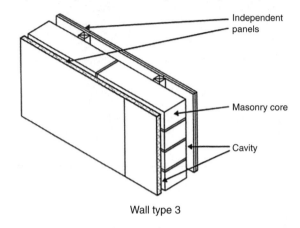

Wall type 3

Wall type 3 provides a high resistance to the transmission of both airborne sound and impact sound on the wall. As shown below there are three different categories of Wall type 3 which comprise either a solid or a cavity masonry core wall with independent panels on both sides. Their resistance to sound depends partly on the type (and mass) of the core and partly on the isolation and mass of the panels.

General requirements

Fill and seal all masonry joints with mortar.	E2.101a
Control flanking transmission from walls and floors connected to the separating wall.	E2.101b
The panels and any frame should not be in contact with the core wall.	E2.99
The panels and/or supporting frames should be fixed to the ceiling and floor only.	E101c
All joints should be taped and sealed.	E2.101d
Flue blocks shall not adversely affect the sound insulation.	E2.101e
A suitable finish is used over the flue blocks (see BS 1289-1:1986).	E2.101e
Free-standing panels and/or the frame should not be fixed, tied or connected to the masonry core.	E2.101
Wall ties in cavity masonry cores used to connect the leaves of a cavity masonry core together should be tie type A.	E2.102

Table 6.31 Wall type 3 – categories

Wall type 3

Category 3.1

Solid masonry core (dense aggregate concrete block), independent panels on both room faces

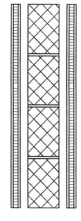

SECTION

- Minimum mass per unit area (including plaster) 300 kg/m²;
- Independent panels on both room faces;
- Minimum core width determined by structural requirements.

For example:
Size   140 mm core block
Density   2200 kg/m³
Coursing 110 mm
Independent panels – each panel of mass per unit area 20 kg/m², to be two sheets of plasterboard with joints staggered

Wall type 3

Category 3.2

Solid masonry core (lightweight concrete block), independent panels on both room faces

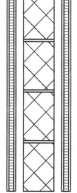

SECTION

- Minimum mass per unit area (including plaster) 150 kg/m²;
- Independent panels on both room faces;
- Minimum core width determined by structural requirements.

For example:
Size   140 mm core block
Density   1400 kg/m³
Coursing 225 mm
Independent panels – each panel of mass per unit area 20 kg/m², to be two sheets of plasterboard with joints staggered

Wall type 3

Category 3.3

Cavity masonry core (brickwork or block work), 50 mm cavity, independent panels on both room faces

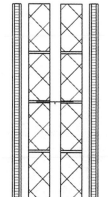

SECTION

- Core mass – unrestricted;
- Minimum cavity width of 50 mm;
- Independent panels on both room faces;
- Minimum core width determined by structural requirements.

For example:
Concrete block – two leaves (each leaf at least 100 mm thick)
Minimum cavity width – 50 mm
Independent panels – each panel of mass per unit area 20 kg/m², to be two sheets of plasterboard with joints staggered

The minimum mass per unit area of independent panel (excluding any supporting framework) should be 20 kg/m².	E2.104
Panels should be either at least two layers of plasterboard with staggered joints or a composite panel consisting of two sheets of plasterboard separated by a cellular core.	E2.104
Panels that are not supported on a frame should be at least 35 mm from the masonry core.	E2.104
Panels which are supported on a frame should have a gap of at least 10 mm between the frame and the masonry core.	E2.104

Junction requirements for wall type 3

Junctions with an external cavity wall with masonry inner leaf

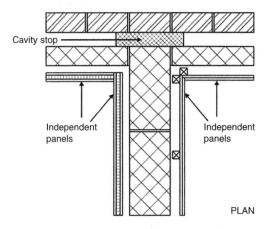

Figure 6.96 Wall type 3 – external cavity wall with masonry inner leaf

If the external wall is a cavity wall:	E2.108
• the outer leaf of the wall may be of any construction; • the cavity should be stopped with a flexible closer.	
If the inner leaf of an external cavity wall is masonry:	E2.109
• the inner leaf of the external wall should be bonded or tied to the masonry core; • the inner leaf of the external wall should be lined with independent panels.	

If there is a separating floor, the masonry inner leaf (of   E2.110
the external wall) should have a minimum mass per unit
area of at least 120 kg/m² excluding finish.

If there is no separating floor:

- the external wall may be finished with plaster or   E2.111
  plasterboard of minimum mass per unit area 10 kg/m²
  (provided the masonry inner leaf of the external wall
  has a mass of at least 120 kg/m² excluding finish);
- there is no minimum mass requirement on the masonry   E2.112
  inner leaf (provided that the masonry inner leaf of the
  external wall is lined with independent panels in the
  same manner as the separating walls);

Junctions with internal framed walls

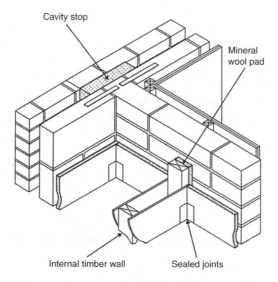

Figure 6.97 Wall type 3 – external cavity wall with internal timber wall

Load bearing (framed) internal walls should be fixed to   E2.115
the masonry core through a continuous pad of mineral wool.

Non-load bearing internal walls should be butted to the   E2.116
independent panels.

All joints between internal walls and panels should be   E2.117
sealed with tape or caulked with sealant.

Junctions with internal timber floors

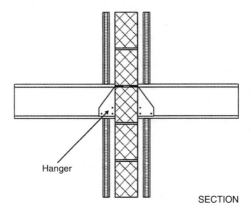

Figure 6.98 Wall type 3 – internal timber floor

Junctions with internal masonry walls

If the floor joists are to be supported on the separating wall then they should be supported on hangers as opposed to being built in.	E2.119
Spaces between the floor joists should be sealed with full depth timber blocking.	E2.120

Junctions with internal concrete floors

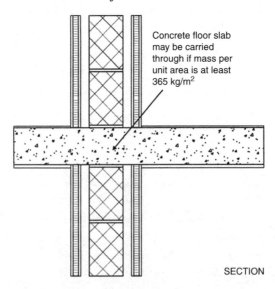

Figure 6.99 Wall types 3.1 and 3.2 – internal concrete floor

> For wall types 3.1 and 3.2 (i.e. those with solid masonry   E2.121
> cores) internal concrete floor slabs may only be carried
> through a solid masonry core if the floor base has a
> mass per unit area of at least 365 kg/m².
>
> For wall type 3.3 (cavity masonry core):   E2.122
>
> • internal concrete floors should generally be built into
>   a cavity masonry core and carried through to the
>   cavity face of the leaf;
> • the cavity should not be bridged.

Junctions with ceiling and roof space

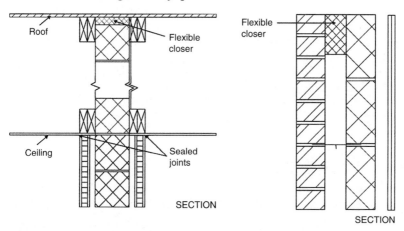

Wall types 3.1 and 3.2 – ceiling and roof junction   External cavity wall at eaves level

> The masonry core should be continuous to the underside   E2.133
> of the roof.
>
> The junction between the separating wall and the roof   E2.134
> should be filled with a flexible closer which is also
> suitable as a fire stop.
>
> The junction between the ceiling and independent panels   E2.135
> should be sealed with tape or caulked with sealant.
>
> If there is an external cavity wall, the cavity should be   E2.136
> closed at eaves level with a suitable flexible material
> (e.g. mineral wool).

Rigid connections between the inner and external wall leaves should be avoided where possible.	E2.136
If a rigid material is used, then it should only be rigidly bonded to one leaf.	E2.136

For wall types 3.1 and 3.2 (solid masonry core):

- if the roof or loft space is not a habitable room (and there is a ceiling with a minimum mass per unit area $10\,kg/m^2$ and it has sealed joints) the independent panels may be omitted in the roof space and the mass per unit area of the separating wall above the ceiling may be a minimum of $150\,kg/m^2$;   E2.137
- if lightweight aggregate blocks with a density less than $1200\,kg/m^3$ are used above ceiling level, then one side should be sealed with cement paint or plaster skim.   E2.138

For wall type 3.3 (cavity masonry core) if the roof or loft space is not a habitable room (and there is a ceiling with a minimum mass per unit area $10\,kg/m^2$ and it has sealed joints) the independent panels may be omitted in the roof space, but the cavity masonry core should be maintained to the underside of the roof.   E2.139

Junctions with internal masonry floors

Internal walls that abut a type 2 separating wall should not be of masonry construction.	E2.118

Junctions with timber ground floors

Floor joists supported on a separating wall should be supported on hangers as opposed to being built in.	E2.123
The spaces between floor joists should be sealed with full depth timber blocking.	E2.124

Junctions with an external cavity wall with timber frame inner leaf

No official guidance currently available. Best to seek specialist advice.	E2.113

Junctions with an external solid masonry wall

No official guidance currently available. Best to seek specialist advice.	E2.114

Junctions with concrete ground floors

The ground floor may be a solid slab, laid on the ground, or a suspended concrete floor.   E2.126

For wall types 3.1 and 3.2 (solid masonry core):

- a concrete slab floor on the ground may be continuous under the solid masonry core of the separating wall;   E2.127
- a suspended concrete floor may only pass under the solid masonry core if the floor has a mass per unit area of at least 365 kg/m^2;   E2.128
- hollow core concrete plank (and concrete beams with infilling block floors) should **not** be continuous under the solid masonry core of the separating wall.   E2.129

For wall type 3.3 (cavity masonry core):

- a concrete slab floor on the ground should **not** be continuous under the cavity masonry core of the separating wall;   E2.130
- a suspended concrete floor should **not** be continuous under the cavity masonry core of a Type 3.3 separating wall;   E2.131
- a suspended concrete floor **should** be carried through to the cavity face of the lea but the cavity should not be bridged.   E2.132

Junctions with internal masonry walls

Internal walls that abut a type 3 separating wall should not be of masonry construction.	E2.118

Wall type 4 (framed walls with absorbent material)

A wall type 4 consists of a timber frame with a plasterboard lining on the room surface with an absorbent material between the frames. Its resistance to airborne sound depends on:

- the mass per unit area of the leaves;
- the isolation of the frames;
- the absorption in the cavity between the frames.

General requirements

If a fire stop is required in the cavity between frames, then it should either be flexible or only be fixed to one frame.	E2.146a
Layers of plasterboard should:	
• be independently fixed to the stud frame;	E2.146c
• not be chased.	E2.146b2
If two leaves have to be connected together for structural reasons, then:	
• the cross-section of the ties shall be less than 40 mm × 3 mm;	E2.146a2
• ties should be fixed to the studwork at or just below ceiling level;	E2.146a2
• ties should not be set closer than 1.2 m centres.	E2.146a2
Sockets should:	
• be positioned on opposite sides of a separating wall;	E2.146b
• not be connected back to back;	E2.146b2
• be staggered a minimum of 150 mm edge to edge.	E2.146b2
The flanking transmission from walls and floors connected to a separating wall should be controlled (see guidance on junctions).	E2.146d

Wall type 4.1 (double leaf frames with absorbent material)

General requirements

The lining shall be two or more layers of plasterboard with a minimum sheet mass per unit area 10 kg/m^2 and with staggered joints.	E2.147
If a masonry core is used for structural purposes, then the core should only be connected to one frame.	E2.147
The minimum distance between inside lining faces shall be 200 mm.	E2.147
Plywood sheathing may be used in the cavity if required for structural reasons.	E2.147
Absorbent material:	
• shall have a minimum density of 10 kg/m^3;	E2.147
• shall be unlaced mineral wool batts (or quilt);	

- may be wire reinforced;
- shall have a minimum thickness of between 25 and 50 mm as shown in Figure 6.100.

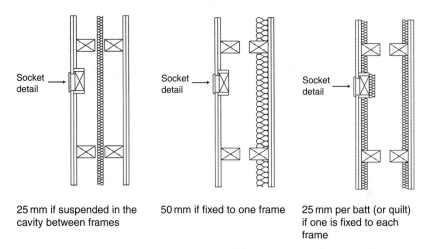

25 mm if suspended in the cavity between frames

50 mm if fixed to one frame

25 mm per batt (or quilt) if one is fixed to each frame

Figure 6.100 Wall type 4.1 – minimum thickness of absorbent material

Junction requirements for wall type 4

Junctions with an external cavity wall with timber frame inner leaf

If the external wall is a cavity wall:	
• the outer leaf of the wall may be of any construction;	E2.149
• the cavity should be stopped between the ends of the separating wall and the outer leaf with a flexible closer.	E2.149
The wall finish of the inner leaf of the external wall should be one layer of plasterboard (or two layers of plasterboard if there is a separating floor).	E2.150a and b
Each sheet of plasterboard to be of minimum mass per unit area 10 kg/m².	E2.150c
All joints should be sealed with tape or caulked with sealant.	E2.150

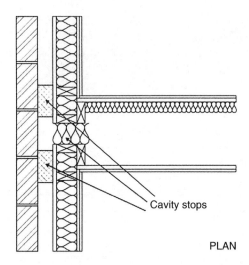

Figure 6.101 Junctions with an external solid masonry wall

Junction with ceiling and roof space

The wall should preferably be continuous to the underside of the roof.	E2.160
The junction between the separating wall and the roof should be filled with a flexible closer.	E2.161
The junction between the ceiling and the wall linings should be sealed with tape or caulked with sealant.	E2.162
If the roof or loft space is not a habitable room (and there is a ceiling with a minimum mass per unit area of $10\,kg/m^2$ with sealed joints), then:	
• either the linings on each frame may be reduced to two layers of plasterboard, each sheet with a minimum mass per unit area of $10\,kg/m^2$; or	E2.162a
• the cavity may be closed at ceiling level without connecting the two frames rigidly together.	E2.162b
Note: In which case there need only be one frame in the roof space provided there is a lining of two layers of plasterboard, each sheet of minimum mass per unit area of $10\,kg/m^2$, on both sides of the frame.	
External wall cavities should be closed at eaves level with a suitable material.	E2.163

Junctions with timber ground floors

Air paths through the wall into the cavity shall be blocked using solid timber blockings, continuous ring beam or joists.	E2.156

 See also Building Regulation Part C – Site preparation and resistance to moisture, and Building Regulation Part L – Conservation of fuel and power.

Junctions with concrete ground floors

If the ground floor is a concrete slab laid on the ground, it may be continuous under a type 4 separating wall.	E2.158
If the ground floor is a suspended concrete floor, it may only pass under a wall type 4 if the floor has a mass per unit area of at least $365 \, \text{kg/m}^2$.	

 See also Building Regulation Part C – Site preparation and resistance to moisture, and Building Regulation Part L – Conservation of fuel and power.

Junctions with internal timber floors

Air paths through the wall into the cavity shall be blocked using solid timber blockings, continuous ring beam or joists.	E2.154

Junctions with internal concrete floors

No official guidance currently available. Best to seek specialist advice.	E2.155

Junctions with internal framed walls

There are no restrictions on internal framed walls meeting a type 4 separating wall.	E2.152

Junctions with internal masonry walls

There are no restrictions on internal masonry walls meeting a type 4 separating wall.	E2.153

Junctions with an external solid masonry wall

> No official guidance currently available. Best to seek   E2.151
> specialist advice.

Junctions with an external cavity wall with masonry inner leaf

> No official guidance currently available. Best to seek   E2.148
> specialist advice.

Walls adjacent to hearths

> Walls that are not part of a fireplace recess or a prefabricated   J (2.31)
> appliance chamber but are adjacent to hearths or appliances
> also need to protect the building from catching fire. A way of
> achieving the requirement is shown in Figure 6.102.

See also p. 284, Appendix A (The Use of Robust Standards)

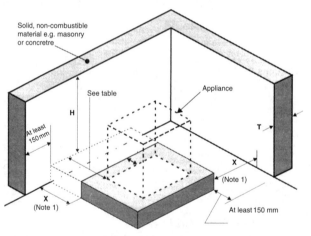

Location of hearth or appliance	Solid, non-combustible material	
	Thickness (T)	Height (H)
Where the hearth abuts a wall and the appliance is not more than 50 mm from the wall	200 mm	At least 300 mm above the appliance and 1.2 m above the hearth
Where the hearth abuts a wall and the appliance is more than 50 mm but not more than 300 mm from the wall	75 mm	At least 300 mm above the appliance and 1.2 m above the hearth
Where the hearth does not abut a wall and is no more than 150 mm from the wall (see Note 1)	75 mm	At least 1.2 m above the hearth
Note 1: There is no requirement for protection of the wall where X is more than 150 mm		

Figure 6.102 Walls adjacent to hearths

6.8 Ceilings

6.8.1 The requirement

As a fire precaution, all materials used for internal linings of a building should have a low rate of surface flame spread and (in some cases) a low rate of heat release.

(Approved Document B2)

Dwellings shall be designed so that the noise from domestic activity in an adjoining dwelling (or other parts of the building) is kept to a level that:

- *does not affect the health of the occupants of the dwelling;*
- *will allow them to sleep, rest and engage in their normal activities in satisfactory conditions.*

(Approved Document E1)

Dwellings shall be designed so that any domestic noise that is generated internally does not interfere with the occupants' ability to sleep, rest and engage in their normal activities in satisfactory conditions.

(Approved Document E2)

Domestic buildings shall be designed and constructed so as to restrict the transmission of echoes.

(Approved Document E3)

Schools shall be designed and constructed so as to reduce the level of ambient noise (particularly echoing in corridors).

(Approved Document E4)

6.8.2 Meeting the requirements

Suspended ceilings

Table 6.32 sets out criteria appropriate to the suspended ceilings that can be accepted as contributing to the fire resistance of a floor.

Table 6.32 Limitations on fire protected suspended ceilings

Height of building or separated part	Type of floor	Provision for fire resistance of floor	Description of suspended ceiling
<18 m	Not compartment	60 mins or less	Type A, B, C or D
	Compartment	<60 mins	Type A, B, C or D
		60 mins	Type B, C or D
18 m or more	Any	60 mins or less	Type C or D
No limit	Any	60 mins	Type D

Requirements – floors and ceilings

Ceilings – General

The resistance to airborne and impact sound depends on the independence and isolation of the ceiling and the type of material used.   E

Three ceiling treatments (which are ranked in order of sound insulation) may be used: E3

- Ceiling treatment A – independent ceiling with absorbent material;
- Ceiling treatment B – plasterboard on proprietary resilient bars with absorbent material;
- Ceiling treatment C – plasterboard on timber battens (or proprietary resilient channels) with absorbent material.

Note: These ceiling treatments are described in more detail starting on p. 262.

If the roof or loft space is not a habitable room (and provided that there is a ceiling with a minimum mass per unit area of $10 \, kg/m^2$ with sealed joints and the cavity masonry core is maintained to the underside of the roof) then: E

- the mass per unit area of the separating wall above the ceiling may be reduced to $150 \, kg/m^2$;
- the independent panels may be omitted in the roof space;
- the linings on each frame may be reduced to two layers of plasterboard or the cavity may be closed at ceiling level without connecting the two frames rigidly together.

All junctions between ceilings and independent panels (and joints between casings and ceiling) should be sealed with tape or caulked with sealant.   E

At junctions with external cavity walls (with masonry inner leaf) the ceiling should be taken through to the masonry.   E3

The ceiling void and roof space detail can only be used where the requirements of Building Regulation Part B – Fire safety can also be satisfied.   E3

If there is an existing lath and plaster ceiling it should be retained as long as it satisfies Building Regulation Part B – Fire safety.   E3

If the existing ceiling is not lath and plaster, it should be   E4
upgraded to provide:

- at least two layers of plasterboard with staggered joints;
- a minimum total mass per unit area $20\,kg/m^2$;
- an absorbent layer of mineral wool laid on the ceiling
  (minimum thickness 100 mm, minimum density $10\,kg/m^3$);
- plasterboard with joints staggered, total mass per unit
  area $20\,kg/m^2$.

Care should be taken at the design stage to ensure that
adequate ceiling height is available in all rooms to be treated.

The ceiling should be supported by either: E2

- independent joists fixed only to the surrounding walls; or
- independent joists fixed to the surrounding walls with
  additional support provided by resilient hangers attached
  directly to the existing floor base.

Note: A clearance of at least 25 mm should be left
between the top of the independent ceiling joists and the
underside of the existing floor construction.

Where a window head is near to the existing ceiling, the new   E4
independent ceiling may be raised to form a pelmet recess.

A rigid or direct connection should not be created between an   E4
independent ceiling and the floor base.

Where the roof or loft space is not a habitable room   E2.57
(and there is a ceiling with a minimum mass per unit area of
$10\,kg/m^2$ with sealed joints) then the mass per unit area of the
separating wall above the ceiling may be reduced to $150\,kg/m^2$.

If lightweight aggregate blocks of density less than   E2.58
$1200\,kg/m^3$ are used above ceiling level, then one side   E2.94
should be sealed with cement paint or plaster skim.   E2.138

Where the external wall is a cavity wall with a masonry   E3.105
inner leaf (or a simple cavity masonry wall or masonry   E3.125
between independent panels), the ceiling should be taken   E3.125
through to the masonry.

Where a window head is near to the existing ceiling, the new   E4.29
independent ceiling may be raised to form a pelmet recess.

A rigid or direct connection should not be created between the   E4.30
independent ceiling and the floor base.

6.9 Roofs

The roof of a brick-built house is normally an aitched (sloping) roof comprising rafters fixed to a ridge board, braced by purlins, struts and ties and fixed to wall plates bedded on top of the walls. They are then usually clad with slates or tiles to keep the rain out.

Timber-framed houses usually have trussed roofs – prefabricated triangulated frames that combine the rafters and ceiling joists – which are lifted into place and supported by the ails. The trusses are joined together with horizontal and diagonal ties. A ridge board is not fitted, nor are purlins required. Roofing felt battens and tiling are applied in the usual way.

6.9.1 Requirements

As a fire precaution, all materials used for internal linings of a building should have a low rate of surface flame spread and (in some cases) a low rate of heat release.

(Approved Document B2)

 It is not necessary, however, to ventilate warm deck roofs or inverted roofs, i.e. those roofs where the moisture from the building cannot permeate the insulation.

External fire spread

- *The roof shall be constructed so that the risk of spread of flame and/or fire penetration from an external fire source is restricted.*
- *The risk of a fire spreading from the building to a building beyond the boundary, or vice versa shall be limited.*

(Approved Document B4)

Internal fire spread (structure)

- *Ideally the building should be sub-divided by elements of fire-resisting construction into compartments.*
- *All openings in fire-separating elements shall be suitably protected in order to maintain the integrity of the continuity of the fire separation.*
- *Any hidden voids in the construction shall be sealed and sub-divided to inhibit the unseen spread of fire and products of combustion, in order to reduce the risk of structural failure, and the spread of fire.*

(Approved Document B3)

Dwellings shall be designed so that the noise from domestic activity in an adjoining dwelling (or other parts of the building) is kept to a level that:

- *does not affect the health of the occupants of the dwelling;*

- *will allow them to sleep, rest and engage in their normal activities in satis-factory conditions.*

(Approved Document E1)

Dwellings shall be designed so that any domestic noise that is generated intern-ally does not interfere with the occupants' ability to sleep, rest and engage in their normal activities in satisfactory conditions.

(Approved Document E2)

Domestic buildings shall be designed and constructed so as to restrict the transmission of echoes.

(Approved Document E3)

Schools shall be designed and constructed so as to reduce the level of ambient noise (particularly echoing in corridors).

(Approved Document E4)

6.9.2 Meeting the requirements

Precipitation

Roofs should:	
• resist the penetration of precipitation to the inside of the building;	C6.2a
• not be damaged by precipitation;	C6.2b
• not carry precipitation to any part of the building which would be damaged by it;	C6.2b
• be designed and constructed so that their structural and thermal performance are not adversely affected by interstitial condensation.	C6.2c

Resistance to moisture from the outside

Roofs should be designed so as to protect the building from precipitation either by holding the precipitation at the face of the roof or by stopping it from penetrating beyond the back of the roofing system.	C6.3
Roofs that are jointless or have sealed joints should be impervious to moisture.	C6.4a
Roofs that have overlapping dry joints should be weather resistant and backed by a material (such as	C6.4b

roofing felt) to direct any precipitation that does
enter the roof towards the outer face.

Materials that can deteriorate rapidly without special   C6.5
care should only be used as the weather-resisting
part of a roof.

Weather-resistant parts of a roofing system shall not   C6.5
include paint or include any coating, surfacing or
rendering which will not itself provide all the weather
resistance.

Roofing systems may be:
- impervious – such as metal, plastic and bituminous   C6.6a
  products;
- weather-resistant – such as natural stone or slate,   C6.6b
  cement-based products, fired clay and wood;
- moisture-resisting – such as bituminous and plastic   C6.6c
  products lapped at the joints;
- jointless materials and sealed joints – that would allow   C6.6d
  for structural and thermal movement.

Dry joints between roofing sheets should be designed   C6.7
so that precipitation will not pass through them.

Any precipitation that does enter a joint shall be drained   C6.7
away without penetrating beyond the back of the
roofing system.

Each sheet, tile and section of roof should be fixed in   C6.8
accordance with the guidance contained in
BS 8000-6: 1990.

Resistance to damage from interstitial condensation

Roofs shall be designed and constructed in accordance   C6.10
with Clause 8.4 of BS 5250: 2002 and BS EN ISO
13788: 2001.

Cold deck roofs (i.e. those roofs where the moisture   C6.11
from the building can permeate the insulation) shall
be ventilated.

Any parts of a roof which have a pitch of 70° or more   C6.11
shall be insulated as though it were a wall.

All gaps and penetrations for pipes and electrical wiring should be filled and sealed to avoid excessive moisture transfer to roof voids.	C6.12
An effective draught seal should be provided to loft hatches to reduce inflow of warm air and moisture.	C6.12
Specialist advice should be sought when designing swimming pools and other buildings where interstitial condensation in the walls (caused by high internal temperatures and humidities) can cause high levels of moisture being generated.	C6.13

Resistance to surface condensation and mould growth

Roofs shall be designed and constructed so that the:	
• thermal transmittance (U-value) does not exceed 0.35 W/m^{2}K at any point;	C6.14a
• junctions between elements and the details of openings (such as windows) are in accordance with the recommendations in the report on robust construction details.	C6.14b

Building height

For residential buildings, the maximum height of the building measured from the lowest finished ground level adjoining the building to the highest point of any roof should not be greater than 15 m.	A1/2 2C4i

General

Roofs shall be constructed so that they:	A1/2 1A2d
• provide local support to the walls;	
• act as horizontal diaphragms capable of transferring the wind forces to buttressing elements of the building.	

Note: A traditional cut timber roof (i.e. using rafters, purlins and ceiling joists) generally has sufficient built-in resistance to instability and wind forces. However, the need for diagonal rafter bracing equivalent to that recommended in BS 5268: Part 3: 1998 or Annex H of BS 8103: Part 3: 1996 for trussed rafter roofs, should be considered especially for single-hipped and non-hipped roofs of greater than 40° pitch to detached houses.

Roofs should:

- act to transfer lateral forces from walls to buttressing walls, piers or chimneys;   A1/2 2C33a
- be secured to the supported wall.   A1/2 2C33b

The roof shall be braced (in accordance with   A1/2 2C38(i)h
BS 5268: Part 3):

- at rafter level;
- horizontally at eaves level;
- at the base of any gable by roof decking, rigid sarking or diagonal timber bracing (as appropriate).

Vertical strapping may be omitted if the roof: A1/2 2C36a–d

- has a pitch of 15° or more; and
- is tiled or slated; and
- is of a type known by local experience to be resistant to wind gusts; and
- has main timber members spanning onto the supported wall at not more than 1.2 m centres.

Gable walls should be strapped to roofs as shown   A1/2 2C36
in Figure 6.103(a) and (b) by tension straps.

Walls shall be tied to the roof structure vertically   A1/2 2C38(i)i
and horizontally and have a horizontal lateral
restraint at roof level.

Wall ties should also be provided, spaced not   A1/2 2C8
more than 300 mm apart vertically, within a
distance of 225 mm from the vertical edges of
all roof verges.

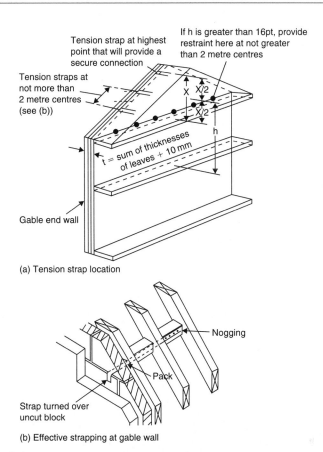

(a) Tension strap location

(b) Effective strapping at gable wall

Figure 6.103 Lateral support at roof level

Walls shall be tied to the roof structure vertically and horizontally and have a horizontal lateral restraint at roof level.	A1/2 2C38(i)i
Walls should be tied horizontally at no more than 2 m centres to the roof structure at eaves level, base of gables and along roof slopes (as shown in Figure 6.104) with straps.	A1/2 2C38(iv)
Isolated columns should also be tied to the roof structure (see Figure 6.104).	A1/2 2C38(iv)
The roof structure of an annexe shall be secured to the structure of the main building at both rafter and eaves level.	A1/2 2C38(1)j

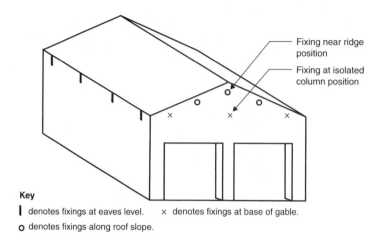

Key

I denotes fixings at eaves level.   × denotes fixings at base of gable.

o denotes fixings along roof slope.

Figure 6.104 Lateral restraint at roof level

Access to the roof shall only be for the purposes   A1/2 2C38(1)d
of maintenance and repair.

Timber

Softwood timber used for roof construction or fixed   A1/2 2B2
in the roof space (including ceiling joists within the
void spaces of the roof), should be adequately treated
to prevent infestation by the house longhorn beetle
(*Hylotrupes bajulus* L.), particularly in the following
areas:

- The Borough of Bracknell Forest, in the parishes
  of Sandhurst and Crowthorne
- The Borough of Elmbridge
- The District of Hart, in the parishes of Hawley and
  Yateley
- The District of Runnymede
- The Borough of Spelthorne
- The Borough of Surrey Heath
- The Borough of Rushmoor, in the area of the
  former district of Farnborough
- The Borough of Woking.

Note: Guidance on suitable preservative treatments
is given within the *British Wood Preserving*

and Damp-Proofing Association's Manual (2000 revision), available from 1 Gleneagles House, Vernongate, South Street, Derby DE1 1UP.

📖 **Note:** Guidance on the sizing of roof members is given in BS 5268: Part 2: 2002 & Part 3: 1998 as well as 'Span tables for solid timber members in floors, ceilings and roofs (excluding trussed rafter roofs) for dwellings', published by TRADA (available from Chiltern House, Stocking Lane, Hughenden Valley, High Wycombe, Bucks HP14 4ND).

Openings

Where an opening in a roof for a stairway adjoins a supported wall and interrupts the continuity of lateral support:

- the maximum permitted length of the opening is to be 3 m, measured parallel to the supported wall;   A1/2 2C37a
- connections (if provided by means other than by anchor) should be throughout the length of each portion of the wall situated on each side of the opening;   A1/2 2C37b
- connections via mild steel anchors should be spaced closer than 2 m on each side of the opening to provide the same number of anchors as if there were no opening;   A1/2 2C37c
- there should be no other interruption of lateral support.   A1/2 2C37d

Means of escape

A flat roof being used as a means of escape should: C1

- be part of the same building from which escape is being made;
- should lead to a storey exit or external escape route;
- should provide 30 minutes' fire resistance.

💡 Where a balcony or flat roof is provided for escape purposes guarding may be needed (see also Approved Document K, Protection from falling, collision and impact).

Compartmentation

To prevent the spread of fire within a building, whenever possible, the building should be sub-divided into compartments separated from one another by walls and/or floors of fire-resisting construction.	B3 (9.1)
Compartment walls in a top storey beneath a roof should be continued through the roof space.	B3 (9.26)
When a compartment wall meets the underside of the roof covering or deck, the wall/roof junction shall maintain continuity of fire resistance.	B3 (9.28)
Double skinned insulated roof sheeting should incorporate a band of material of limited combustibility.	B3 (9.29)

Roof covering

The re-covering of roofs is commonly undertaken to extend the useful life of buildings; however, roof structures may be required to carry underdrawing or insulation provided at a time later than their initial construction.

All materials used to cover roofs (including transparent or translucent materials, but excluding windows of glass in residential buildings with roof pitches of not less than 15°) shall be capable of safely withstanding the concentrated imposed loads upon roofs specified in BS 6399 Pt 3.	A1/2 4.1
Where the work involves a significant change in the applied loading the structural integrity of the roof structure and the supporting structure should be checked to ensure that upon completion of the work the building is not less compliant.	A1/2 4.3
Note: Re-covering roofs is commonly undertaken to extend the useful life of buildings (for example, roof structures may be required to be insulated at a later date).	
Where such checking of the existing roof structure indicates that the construction is unable to sustain any proposed increase in loading (e.g. due to overstressed members or unacceptable deflection leading to ponding), appropriate strengthening work or replacement of roofing members should be undertaken.	A1/2 4.5

This is classified as a material alteration.

Where work will significantly decrease the roof dead   A1/2 4.7
loading, the roof structure and its anchorage to the
supporting structure should be checked to ensure
that an adequate factor of safety is maintained
against uplift of the roof under imposed wind loading.

Note: A significant change in roof loading is when the
loading upon the roof is increased by more than 15%.

Plastic rooflights should have a minimum   B4 (15.6)
of class 3 lower surface.

When used in rooflights, unwired glass shall be at least   B4 (15.8)
4 mm thick and shall be AA designated (see Table 6.32).

Thatch and wood shingles should be regarded as having   B4 (15.9)
an AD/BD/CD designation (see Table 6.33).

Table 6.33 Limitations on roof coverings

Designation of roof covering	Minimum distance from any point on relevant boundary			
	Less than 6 m	At least 6 m	At least 12 m	At least 20 m
AA, AB or AC	Acceptable	Acceptable	Acceptable	Acceptable
BA, BB or BC	Not acceptable	Acceptable	Acceptable	Acceptable
CA, CB or CC	Not acceptable	Acceptable	Acceptable	Acceptable
AD, BD or CD	Not acceptable	Acceptable	Acceptable	Acceptable
DA, DB, DC or DD	Not acceptable	Not acceptable	Not acceptable	Acceptable

Pitched roofs covered with slates or tiles

Covering material	Supporting structure	Designation
1. Natural slates 2. Fibre reinforced cement slates 3. Clay tiles 4. Concrete tiles	Timber rafters with or without underfelt, sarking, boarding, woodwool slabs, compressed straw slabs, plywood, wood chipboard, or fibre insulating board.	AA

Although the table does not include guidance for pitched roofs covered with bitumen felt, it should be noted that there is a wide range of materials on the market
and information on specific products is readily available from manufacturers.

Pitched roofs covered with self-supporting sheet

Roof covering material	Construction	Supporting structure	Designation
Profiled sheet of galvanized steel, aluminium, fibre reinforced cement, or pre-painted (coil coated) steel or aluminium with a PVC or PVF2 coating	Single skin without underlay, or with underlay or plasterboard, or woodwool slab	Structure of timber, steel or concrete	AA
Profiled sheet of galvanized steel, aluminium, fibre reinforced cement, or pre-painted (coil coated) steel or aluminium with a PVC or PVF2 coating	Double skin without interlayer, or with interlayer of resin bonded or concrete glass fibre, mineral wool slab, polystyrene, or polyurethane	Structure of timber, steel or concrete	AA

Flat roofs with bitumen felt

A flat roof consisting of bitumen felt should (irrespective of the felt specification) be deemed to be of designation AA if the felt is laid on a deck constructed of 6 mm plywood, 12.5 mm wood chipboard, 16 mm (finished) plain-edged timber boarding, compressed straw slab, screeded woodwool slab, profiled fibre reinforced cement or steel deck (single or double skin) with or without fibre insulating board overlay, profiled aluminium deck (single or double skin) with or without fibre insulating board overlay, or concrete or clay pot slab (in situ or pre-cast), and has a surface finish of:

(a)   bitumen-bedded stone chippings covering the whole surface to a depth of at least 12.5 mm;

(b)   bitumen-bedded tiles of a non-combustible material;

(c)   sand and cement screed; or

(d)   tarmacadam.

Pitched or flat roofs covered with fully supported materials

Covering material	Supporting structure	Designation
1. Aluminium sheet 2. Copper sheet 3. Zinc sheet	Timber joists and tongued and grooved boarding, or plain-edged boarding	AA
4. Lead sheet 5. Mastic asphalt 6. Vitreous enamelled steel 7. Lead/tin alloy-coated steel sheet	Steel or timber joists with deck of woodwool slabs, compressed straw slab, wood chipboard, fibre insulating	AA

Covering material	Supporting structure	Designation
8. Zinc/aluminium alloy-coated steel sheet	board, or 9.5 mm plywood	
9. Pre-painted (coil coated) steel sheet including liquid-applied PVC coatings	Concrete or clay pot slab (in situ or pre-cast) or non-combustible deck of steel, aluminium, or fibre cement (with or without insulation)	AA

 Lead sheet supported by timber joists and plain edged boarding should be regarded as having a BA designation.

Rating of material and products

Table 6.34 Typical performance ratings of some generic materials and products

Rating	Material or product
Class 0	1. Any non-combustible material or material of limited combustibility.
	2. Brickwork, blockwork, concrete and ceramic tiles.
	3. Plasterboard (painted or not with a PVC facing not more than 0.5 mm thick) with or without an air gap or fibrous or cellular insulating material behind.
	4. Woodwool cement slabs.
	5. Mineral fibre tiles or sheets with cement or resin binding.
Class 3	6. Timber or plywood with a density more than 400 kg/ml, painted or unpainted.
	7. Wood particle board or hardboard, either untreated or painted.
	8. Standard glass reinforced polyesters.

Roof with a pitch of 15° or more

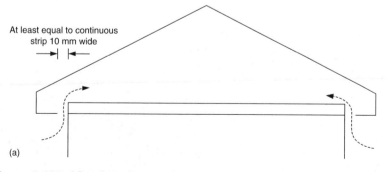

Figure 6.105 (*Continued*)

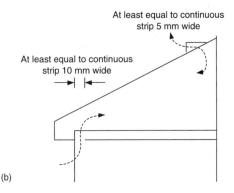

Figure 6.105 Ventilating roof voids. (a) Pitched roof. (b) Lean-to roof

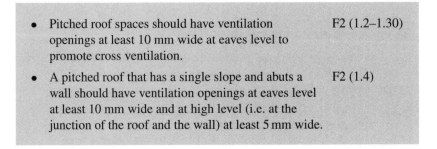

- Pitched roof spaces should have ventilation openings at least 10 mm wide at eaves level to promote cross ventilation.   F2 (1.2–1.30)

- A pitched roof that has a single slope and abuts a wall should have ventilation openings at eaves level at least 10 mm wide and at high level (i.e. at the junction of the roof and the wall) at least 5 mm wide.   F2 (1.4)

Roof with a pitch of less than 15°

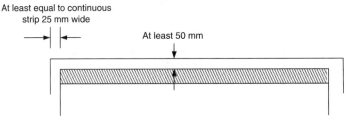

Figure 6.106 Ventilating roof void – flat roof

Roof spaces should have ventilation openings at least 25 mm wide in two opposite sides to promote cross ventilation.	F2 (2.2)
The void should have a free air space of at least 50 mm between the roof deck and the insulation.	F2 (2.4)
Pitched roofs where the insulation follows the pitch of the roof need ventilation at the ridge at least 5 mm wide.	F2 (2.5)
Where the edges of the roof abut a wall or other obstruction in such a way that free air paths cannot be formed to	F2 (2.6)

promote cross ventilation or the movement of air outside
any ventilation openings would be restricted, an alternative
form of roof construction should be adopted.

Roofs with a span exceeding 10 m may require more ventilation, totalling 0.6%
of the roof area.

Ventilation openings may be continuous or distributed along the full length and
may be fitted with a screen, facia, baffle, etc.

Where necessary (i.e. for the purposes of health and safety), ventilation to
small roofs such as those over porches and bay windows should always be
provided and a roof which has a pitch of 70° or more shall be insulated as
though it were a wall.

If the ceiling of a room follows the pitch of the roof, ventilation should be provided as if it were a flat roof.

Ceiling and roof junctions

General

Where a type 1 separating wall is used it should be continuous to the underside of the roof.	E2.55 E2.91 E2.133
The junction between the separating wall and the roof should be filled with a flexible closer which is also suitable as a fire stop.	E2.56 E2.92 E2.134

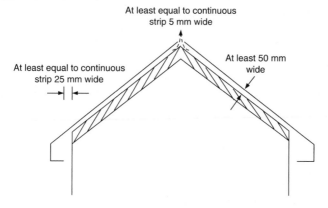

Figure 6.107 Ventilating roof void – ceiling following pitch of roof

Wall type 1 – solid masonry

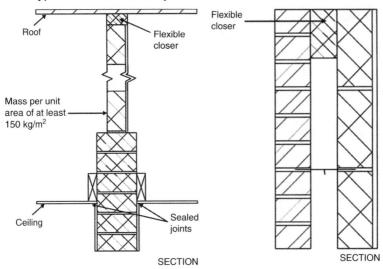

SECTION   SECTION

Ceiling and roof junction   External cavity at roof level

If lightweight aggregate blocks of density less than 1200 kg/m^3 are used above ceiling level, then one side should be sealed with cement, paint or plaster skim.	E2.58 E2.94 E2.138
Where the roof or loft space is not a habitable room (and there is a ceiling with a minimum mass per unit area of 10 kg/m^2 with sealed joints) then the mass per unit area of the separating wall above the ceiling may be reduced to 150 kg/m^2.	E2.57
Where there is an external cavity wall, the cavity should be closed at eaves level with a suitable flexible material (e.g. mineral wool).	E2.59
A rigid connection between the inner and external wall leaves should be avoided.	Ep23
If a rigid material is used, then it should only be rigidly bonded to one leaf.	Ep23

Wall type 2 – cavity masonry

Where a type 2 separating wall is used, it should be continuous to the underside of the roof.	E2.91

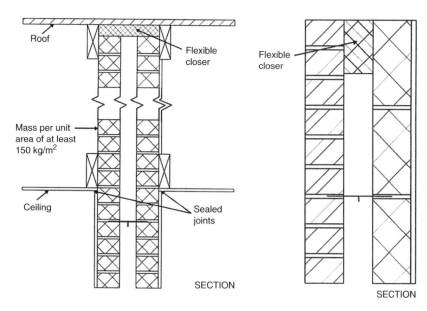

Wall type 2 – ceiling and roof junction   External cavity wall at eaves level

The junction between the separating wall and the roof should be filled with a flexible closer which is also suitable as a fire stop.	E2.92
If lightweight aggregate blocks (with a density less than 1200 kg/m³ are used) above ceiling level, then one side should be sealed with cement paint or plaster skim.	E2.94
The cavity of an external cavity wall should be closed at eaves level with a suitable flexible material (e.g. mineral wool).	E2.95
A rigid connection between the inner and external wall leaves should be avoided.	E2.95
If a rigid material has to be used, then it should **only** be rigidly bonded to one leaf.	E2.95

 Note: If the roof or loft space is not a habitable room (and there is a ceiling with a minimum mass per unit area of $10 \, \text{kg/m}^2$ with sealed joints) then the mass per unit area of the separating wall above the ceiling may be reduced to $150 \, \text{kg/m}^2$ – **but** it should still be a cavity wall.

Wall type 3 – masonry between independent panels

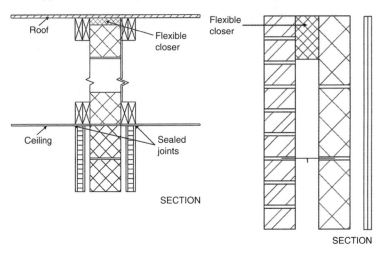

Wall types 3.1 and 3.2 – ceiling and roof External cavity wall at eaves level junction

Where a type 3 separating wall is used, the masonry core should be continuous to the underside of the roof.	E2.133
The junction between the separating wall and the roof should be filled with a flexible closer which is also suitable as a fire stop.	E2.134
The junction between the ceiling and independent panels should be sealed with tape or caulked with sealant.	E2.135
If there is an external cavity wall, the cavity should be closed at eaves level with a suitable flexible material (e.g. mineral wool).	E2.136
Rigid connections between the inner and external wall leaves should be avoided where possible.	E2.136
If a rigid material is used, then it should only be rigidly bonded to one leaf.	E2.136

For wall types 3.1 and 3.2 (solid masonry core):

- if the roof or loft space is not a habitable room (and there   E2.137
  is a ceiling with a minimum mass per unit area of $10\,\text{kg/m}^2$
  and it has of sealed joints) the independent panels may be
  omitted in the roof space and the mass per unit area of the
  separating wall above the ceiling may be a minimum of
  $150\,\text{kg/m}^2$;

- if lightweight aggregate blocks with a density less   E2.138
  than 1200 kg/m^3 are used above ceiling level, then one
  side should be sealed with cement paint or plaster skim.

For wall type 3.3 (cavity masonry core):

- if the roof or loft space is not a habitable room (and   E2.139
  there is a ceiling with a minimum mass per unit area
  of 10 kg/m^2 and it has sealed joints) the independent
  panels may be omitted in the roof space, but the
  cavity masonry core should be maintained to the
  underside of the roof.

Wall type 4 – framed walls with absorbent material

Where a type 4 separating wall is used, the wall should preferably be continuous to the underside of the roof.	E2.160
The junction between the separating wall and the roof should be filled with a flexible closer.	E2.161
The junction between the ceiling and the wall linings should be sealed with tape or caulked with sealant.	E2.162
If the roof or loft space is not a habitable room (and there is a ceiling with a minimum mass per unit area of 10 kg/m^2 with sealed joints), then:	E2.162a

- either the linings on each frame may be reduced   E2.162a
  to two layers of plasterboard, each sheet with a
  minimum mass per unit area of 10 kg/m^2; or
- the cavity may be closed at ceiling level without   E2.162b
  connecting the two frames rigidly together.

📖 **Note:** In which case there need only be one frame in
the roof space provided there is a lining of two layers of
plasterboard, each sheet of minimum mass per unit area
10 kg/m^2, on both sides of the frame.

External wall cavities should be closed at eaves level with a suitable material.	E2.163

6.10 Chimneys

6.10.1 The requirement (Building Act 1984 Section 73)

If a person erects or raises a building that is (or is going to be) taller than the chimneys and/or flues from an adjoining building that is either joined by a party wall or less than six feet away from the taller building, then the local authority may:

- if reasonably practical, require that person to build up those chimneys and flues, so that their top is of the same height as the top of the chimneys of the taller building or the top of the taller building, whichever is the higher;
- require the owner or occupier of the adjoining building to allow the person erecting or raising the building, access to the adjacent building so that he can carry out such work as may be necessary to comply with the notice served on him.

 The owner or occupier of the adjacent building is entitled to complete the work himself by (within fourteen days) serving a 'counter-notice' that he has elected to carry out the work himself.

The building shall be constructed so that the combined dead, imposed and wind loads are sustained and transmitted by it to the ground

- *safely;*
- *without causing such deflection or deformation of any part of the building (or such movement of the ground) as will impair the stability of any part of another building.*

(Approved Document A1)

The building shall be constructed so that ground movement caused by:

- *swelling, shrinkage or freezing of the subsoil; or*
- *landslip or subsidence (other than subsidence arising from shrinkage)*

will not impair the stability of any part of the building.

(Approved Document A2)

Fire precautions (construction)

Any hidden voids in the construction shall be sealed and subdivided to inhibit the unseen spread of fire and products of combustion, in order to reduce the risk of structural failure, and the spread of fire.

(Approved Document B3)

Protection of building

Combustion appliances and fluepipes shall be so installed, and fireplaces and chimneys shall be so constructed and installed, as to reduce to a reasonable level the risk of people suffering burns or the building catching fire in consequence of their use.

(Approved Document J3)

Provision of information

Where a hearth, fireplace, flue or chimney is provided or extended, a durable notice containing information on the performance capabilities of the hearth, fireplace, flue or chimney shall be affixed in a suitable place in the building for the purpose of enabling combustion appliances to be safely installed.

(Approved Document J4)

6.10.2 Meeting the requirement

End restraints

The ends of every wall (except single leaf walls less than 2.5 m in storey height and length) in small single storey non-residential buildings and annexes should be bonded or otherwise securely tied throughout their full height to a buttressing wall, pier or chimney.	A1/2 (2C25)
Long walls may be provided with intermediate support, dividing the wall into distinct lengths; each distinct length is a supported wall for the purposes of this section.	A1/2 (2C25)
The buttressing wall, pier or chimney should provide support from the base to the full height of the wall.	A1/2 (2C25)
The sectional area on plan of chimneys (excluding openings for fireplaces and flues) should be not less than the area required for a pier in the same wall, and the overall thickness should not be less than twice the required thickness of the supported wall (see Figure 6.70).	A1/2 (2C27b)
Floors and roofs should act to transfer lateral forces (see Table 6.15) from walls to buttressing walls, piers or chimneys and be secured to the supported wall as shown.	A1/2 (2C33a)

Masonry chimneys

Where a chimney is not adequately supported by ties or   1D1
securely restrained in any way, its height H (measured from the
highest point of any chimney pot or other flue terminal) should
not exceed 4.5 times the width W (the least horizontal dimension
of the chimney measured at the same point of intersection) –
provided that the density of the masonry is greater than 1500 kg/m³.

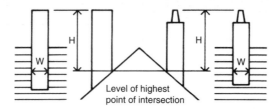

Figure 6.108 Proportions for masonry chimneys

The foundation of piers, buttresses and chimneys should project as indicated
in Figure 6.109 and the projection X should never be less than P.

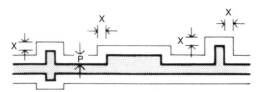

Projection X should not be less than P

Figure 6.109 Piers and chimneys

Flues, etc.

Flue walls should have a fire resistance of at least one   B3 (11.11)
half of that required for the compartment wall or floor
and be of non-combustible construction.

If a flue, or duct containing flues or appliance ventilation   B3 (11.11)
duct(s), passes through a compartment wall or compartment
floor (or is built into a compartment wall) each wall of the
flue or duct should have a fire resistance of at least half
that of the wall or floor in order to prevent the by-passing of
the compartmentation.

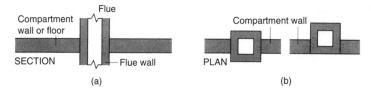

Figure 6.110 Flues penetrating compartment walls or floors. (a) Flue passing through compartment wall or floor. (b) Flue built into compartment wall

Fire stopping

Proprietary fire-stopping and sealing systems (including those designed for service penetrations) which have been shown by test to maintain the fire resistance of the wall or other element, are available and may be used. Other fire-stopping materials include:

- cement mortar,
- gypsum-based plaster,
- cement or gypsum-based vermiculite/perlite mixes,
- glass fibre, crushed rock, blast furnace slag or ceramic based products (with or without resin binders), and
- intumescent mastics (B3 11.14).

Joints between fire separating elements should be fire-stopped.	B3 (11.12a)
All openings for pipes, ducts, conduits or cables to pass through any part of a fire separating element should be: • kept as few in number as possible; • kept as small as practicable; • fire-stopped (which in the case of a pipe or duct, should allow thermal movement).	B3 (11.12b)
Cables concealed in floors and walls (in certain circumstances) are required to have an earthed metal covering, be enclosed in steel conduit, or have additional mechanical protection (see BS 7671 for more information).	P AppA 2d
To prevent displacement, materials used for fire-stopping should be reinforced with (or supported by) materials of limited combustibility.	B3 (11.13)

Chimney construction

> Chimneys shall consist of a wall or walls enclosing one   J (0.4–7)
> or more flues (see Figure 6.111).
>
> 📖 In the gas industry, the chimney for a gas appliance is
> commonly called the flue.
>
> Down-draughts that could interfere with the combustion   J (0.4–11)
> performance of an open-flued appliance (see Figure 6.112)
> shall be minimized.

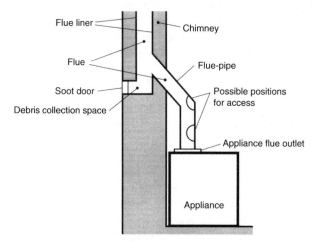

Figure 6.111 Chimneys and flues

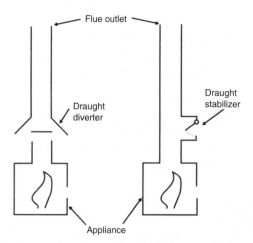

Figure 6.112 Draught diverters and draught stabilizers

Fireplaces

Fireplace recesses (sometimes called a builder's opening)   J (0.4–16)
shall be formed in a wall or in a chimney breast, from
which a chimney leads and which has a hearth at its base.

Simple recesses are suitable for closed appliances
such as roomheaters, stoves, cookers or boilers.
They are **not** suitable for an open fire without a
canopy.

Fireplace recesses are used for accommodating
open fires and freestanding fire baskets.

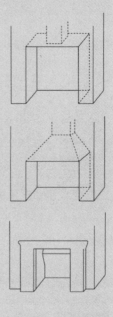

Fireplace recesses are often lined with firebacks
to accommodate inset open fires.

 Lining components and decorative
treatments fitted around openings reduce the
opening area. It is the finished fireplace opening
area that determines the size of flue required for
an open fire in such a recess.

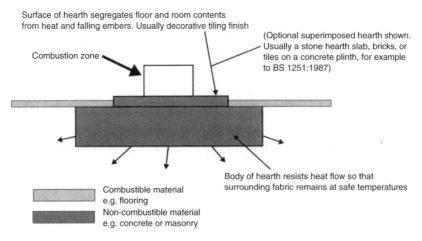

Surface of hearth segregates floor and room contents
from heat and falling embers. Usually decorative tiling finish

(Optional superimposed hearth shown.
Usually a stone hearth slab, bricks, or
tiles on a concrete plinth, for example
to BS 1251:1987)

Combustion zone

Body of hearth resists heat flow so that
surrounding fabric remains at safe temperatures

Combustible material
e.g. flooring

Non-combustible material
e.g. concrete or masonry

Figure 6.113 The functions of a hearth

Hearths

A hearth shall safely isolate a combustion appliance from people, combustible parts of the building fabric and soft furnishings.   J (0.4–26)

Flueblock chimneys

Flueblock chimneys should be constructed of factory-made components suitable for the intended application installed in accordance with manufacturer's instructions.   J (1.29)

Joints should be sealed in accordance with the flueblock manufacturer's instructions.   J (1.30)

Bends and offsets should only be formed with matching factory-made components.   J (1.30)

Masonry chimneys (change of use)

Where a building is to be altered for different use (e.g. it is being converted into flats) the fire resistance of walls of existing masonry chimneys may need to be improved.   J (1.31)

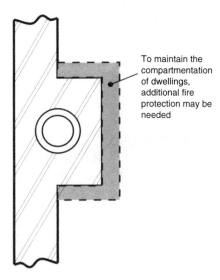

To maintain the compartmentation of dwellings, additional fire protection may be needed

Figure 6.114 Fire protection of chimneys passing through other dwellings

Connecting fluepipes

Whenever possible, fluepipes should be manufactured from: J (1.32)

- cast iron (BS41: 1973 (1998));
- mild steel fluepipes (BS1449, Part 1: 1991, with a flue wall thickness of at least 3 mm);
- stainless steel (BS EN 10088-1: 1995 grades 1.4401, 1.4404, 1.4432 or 1.4436 with a flue wall thickness of at least 1 mm);
- vitreous enamelled steel (BS 6999: 1989 (1996)).

Fluepipes with spigot and socket joints should be fitted with the socket facing upwards to contain moisture and other condensates in the flue.   J (1.33)

Joints should be made gas-tight.   J (1.33)

Repair of flues

If renovation, refurbishment or repair amounts to or involves the provision of a new or replacement flue liner, it is considered 'building work' within the meaning of Regulation 3 of the Building Regulations and must, therefore, **not** be undertaken without prior notification to the local authority. Examples of work that would need to be notified include: J (1.34–1.35)

- relining work comprising the creation of new flue walls by the insertion of new linings such as rigid or flexible prefabricated components;
- a cast in situ liner that significantly alters the flue's internal dimensions.

If you are in doubt you should consult the building control department of your local authority, or an approved inspector.

Re-use of existing flues

Where it is proposed to bring a flue in an existing chimney back into use (or to re-use a flue with a different type or rating of appliance) the flue and the chimney should be checked and, if necessary, altered to ensure that they satisfy the requirements for the proposed use. — J (1.36)

Oversize flues can be unsafe. A flue may, however, be lined to reduce the flue area to suit the intended appliance. — J (1.38)

Relining

If a chimney has been relined in the past using a metal lining system and the appliance is being replaced, the metal liner should also be replaced unless the metal liner can be proven to be recently installed and can be seen to be in good condition. — J (1.39)

In certain circumstances, relining is considered 'building work' within the meaning of Regulation 3 of the Building Regulations and must, therefore, **not** be undertaken without prior notification to the local authority. If you are in doubt you should consult the building control department of your local authority, or an approved inspector.

Flexible flue liners should only be used to reline a chimney and should not be used as the primary liner of a new chimney. — J (1.40)

Plastic fluepipe systems can be acceptable in some cases, for example with condensing boiler installations, where the fluepipes are supplied by or specified by the appliance manufacturer. — J (1.41)

Factory-made metal chimneys

Where a factory-made metal chimney passes through a wall, sleeves should be provided to prevent damage to the flue or building through thermal expansion. — J (1.43)

To facilitate the checking of gas-tightness, joints between chimney sections should not be concealed within ceiling joist spaces or within the thicknesses of walls. — J (1.43)

When installing a factory-made metal chimney, provision   J (1.44)
should be made to withdraw the appliance without the need to
dismantle the chimney.

Factory-made metal chimneys should be kept a suitable   J (1.45)
distance away from combustible materials.

📖 One way of meeting this requirement is by locating the
chimney not less than distance 'X' from combustible material,
where 'X' is defined in BS 4543-1: 1990 (1996) as shown
in Figure 6.115.

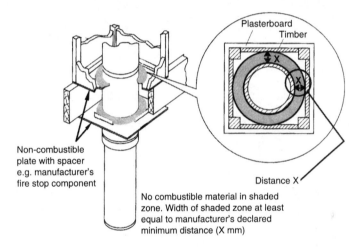

Figure 6.115 The separation of combustible material from a factory-made metal chimney meeting BS 4543, Part 1 (1990)

Flue systems

Flue systems should offer least resistance to the passage of flue   J (1.47)
gases by minimizing changes in direction or horizontal length.

Wherever possible flues should be built so that they are   J (1.47)
straight and vertical except for the connections to combustion
appliances with rear outlets where the horizontal section
should not exceed 150 mm. Where bends are essential, they
should be angled at no more than 45° to the vertical.

Provisions should be made to enable flues to be swept and   J (1.48)
inspected (see Figure 6.116).

A flue should **not** have openings into more than one room or   J (1.49)
space except for the purposes of:

• inspection or cleaning; or
• fitting an explosion door, draught break, draught stabilizer
  or draught diverter.

Openings for inspection and cleaning should be formed using   J (1.50)
purpose factory-made components compatible with the flue
system, having an access cover that has the same level of
gas-tightness as the flue system and an equal level of thermal
insulation.

After the appliance has been installed, it should be possible   J (1.50)
to sweep the whole flue.

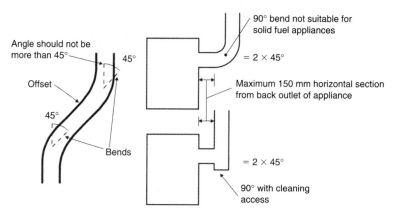

Figure 6.116 Bends in flues

Dry lining around fireplace openings

Where a decorative treatment, such as a fireplace surround,   J (1.52)
masonry cladding or dry lining is provided around a
fireplace opening, any gaps that could allow flue gases to
escape from the fireplace opening into the void behind the
decorative treatment, should be sealed to prevent such leakage.

The sealing material should be capable of remaining in place   J (1.53)
despite any relative movement between the decorative
treatment and the fireplace recess.

Notice plates for hearths and flues (Requirement J4)

Where a hearth, fireplace (including a flue box), flue or chimney is provided or extended (including cases where a flue is provided as part of the refurbishment work), information essential to the correct application and use of these facilities should be permanently posted in the building. A way of meeting this requirement would be to provide a notice plate conveying the following information:

- the location of the hearth, fireplace (or flue box) or the location of the beginning of the flue;
- the category of the flue and generic types of appliances that can be safely accommodated;
- the type and size of the flue (or its liner if it has been relined) and the manufacturer's name;
- the installation date.

Additional provisions for appliances burning solid fuel (with a rated output up to 50 kW)

> Any room or space containing an appliance should burning   J (2.2)
> solid fuel (with a rated output up to 50 kW) have a permanent
> air vent opening of at least the size shown in Figure 6.117.

Open fire with no throat (e.g. a fire under a canopy)

Permanently open air vent(s) should have a total free area of at least 50% of the crosssectional area of the flue.

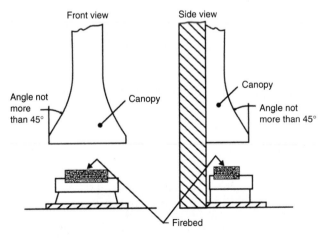

Figure 6.117 Canopy for an open solid fuel fire

Open fire with a throat and gather

Permanently open air vent(s) should have a total free area of at least 50% of the throat opening area.

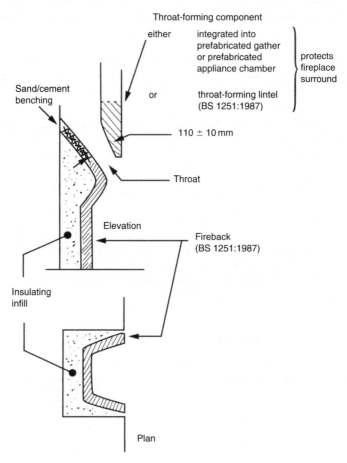

Figure 6.118 Open fireplaces – throat and fireplace components

Other appliance (such as a stove, cooker or boiler)

Permanently open air vent(s).

Size of flues

Fluepipes should have the same diameter or equivalent cross-sectional area as that of the appliance flue outlet.	J (2.4)
Flues should be not less than the size of the appliance flue outlet or that recommended by the appliance manufacturer.	J (2.5)

Flues in chimneys

Table 6.35 Size of flues in chimneys (see Figure 6.119)

Fireplace with an opening up to 500 mm × 550 mm.	200 mm diameter or rectangular/square flues having the same cross-sectional dimension not less than 175 mm.
Fireplace with an opening in excess of 500 mm × 550 mm or a fireplace exposed on two or more sides.	If rectangular/square flues are used the minimum dimension should not be less than 200 mm.
Closed appliance up to 20 kW rated output which: • burns smokeless or low volatile fuel; or • is an appliance that meets the requirements of the Clean Air Act when burning an appropriate bitumous coal (these appliances are known as 'exempted fireplaces').	125 mm diameter or rectangular/square flues having the same cross-sectional area and a minimum dimension not less than 100 mm for straight flues or 125 mm for flues with bends or offsets.
Other closed appliance of up to 35 kW rated output burning any fuel.	150 mm diameter or rectangular/square flues having the same cross-sectional area and a minimum dimension not less than 125 mm.
Closed appliance of up to 30 kW and up to 50 kW rated output burning any fuel.	175 mm diameter or rectangular/square flues having the same cross-sectional area and a minimum dimension not less than 150 mm.
For fireplaces with openings larger than 500 mm × 550 mm or fireplaces exposed on two or more sides (such as a fireplace under a canopy or open on both sides of a central chimney breast) a way of showing compliance would be to provide a flue with a cross-sectional area equal to 15% of the total face area of the fireplace opening(s) using the formula: Fireplace opening area (mm²) × Total horizontal length of fireplace opening L (mm) × Height of fireplace opening H (mm) **Examples of L and H for large and unusual fireplace openings are shown in Figure 6.119**	J (2.7)

Height of flues (see Figure 6.119)

Flues should be high enough (normally 4.5 m is sufficient) to ensure sufficient draught to clear the products of combustion.	J (2.8)
The outlet from a flue should be above the roof of the building in a position where the products of combustion can discharge freely and will not present a fire hazard, whatever the wind conditions (see Figure 6.120 and Table 6.36).	J (2.10)

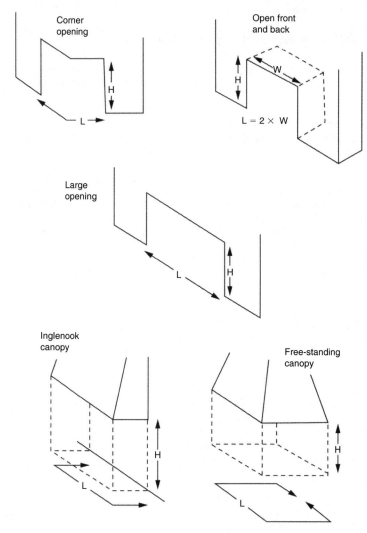

Figure 6.119 Large or unusual fireplace openings

Flue outlet clearances – thatched or shingled roof

The clearances to flue outlets which discharge on, or are in   J (2.12)
close proximity to, roofs with surfaces which are readily
ignitable (e.g. covered in thatch or shingles) should be
increased to those shown in Figure 6.121.

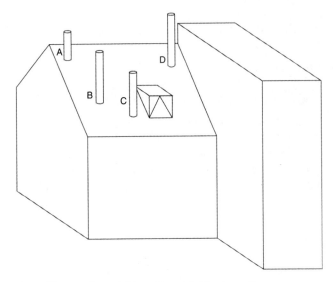

Figure 6.120 Flue outlet positions for solid fuel appliances

Table 6.36 Flue outlet positions

Point where flue passes through weather surfaces (e.g. roof, tiles or external walls)	Clearance to flue outlet
A	At or within 600 mm of the ridge.	At least 600 mm above the ridge.
B	Elsewhere on a roof (whether pitched or flat).	At least 2300 mm horizontally from the nearest point on the weather surface and: • at least 1000 mm above the highest point of intersection of the chimney and the weather surface; or • at least as high as the ridge.
C	Below (on a pitched roof) or within 2300 mm horizontally to an openable rooflight, dormer window or other opening.	At least 1000 mm above the top of the opening.
D	Within 2300 mm of an adjoining or adjacent building whether or not beyond the measurements is the boundary.	At least 600 mm above the adjacent building.

Connecting fluepipes

> Connecting fluepipes should not pass through   J (2.14)
> any roof space, partition, internal wall or floor,
> unless they pass directly into a chimney through

either a wall of the chimney or a floor supporting
the chimney.

Connecting fluepipes should be guarded if they   J (2.14)
are likely to be damaged or if the burn hazard
they present to people is not immediately apparent.

Connecting fluepipes should be located so as   J (2.15 and 1.45)
to avoid igniting combustible material by
minimizing horizontal and sloping runs and
separation of the fluepipe from combustible
material.

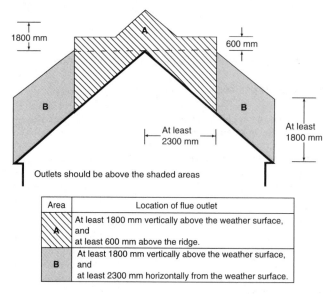

Area	Location of flue outlet
A	At least 1800 mm vertically above the weather surface, and at least 600 mm above the ridge.
B	At least 1800 mm vertically above the weather surface, and at least 2300 mm horizontally from the weather surface.

Figure 6.121 Flue outlet positions for solid fuel appliances – discharging
near easily ignited roof coverings

Masonry and flueblock chimneys

The thickness of the walls around the flues, excluding the   J (2.17)
thickness of any flue liners, should be in accordance with
Figure 6.123.

Combustible material should not be located where it could   J (2.18)
be ignited by the heat dissipating through the walls of
fireplaces or flues.

Construction of fireplace gathers

To minimize resistance to the proper working of flues, tapered gathers should be provided in fireplaces for open fires J (2.21), or corbelling of masonry, as

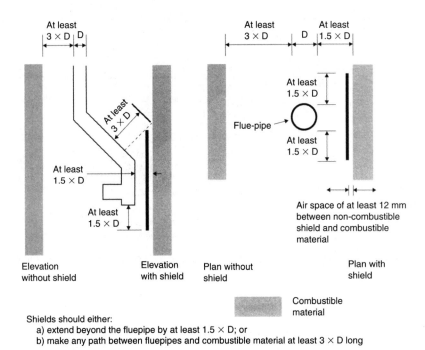

Shields should either:
  a) extend beyond the fluepipe by at least 1.5 × D; or
  b) make any path between fluepipes and combustible material at least 3 × D long

Figure 6.122 Protecting combustible material from uninsulated fluepipes for solid fuel appliances

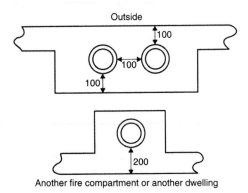

Figure 6.123 Wall thickness for masonry and flueblock chimneys. Dimensions in mm

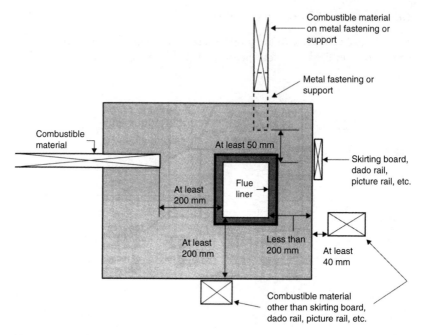

Figure 6.124 Minimum separation distances for combustible material in or near a chimney

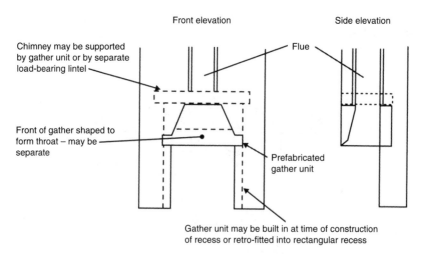

Figure 6.125 Construction of fireplace gathers – using prefabricated components

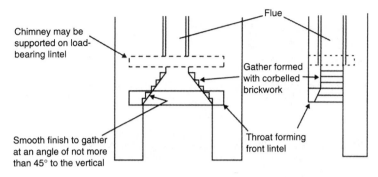

Figure 6.126 Construction of fireplace gathers – using masonry

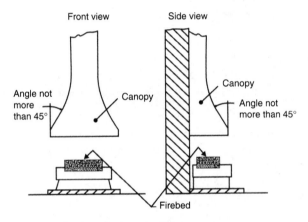

Figure 6.127 Canopy for an open fuel fire

shown in Figure 6.126. Alternatively a suitable canopy (as shown in Figure 6.127) or a prefabricated appliance chamber incorporating a gather may be used.

This can be achieved by using prefabricated gather components built into a fireplace recess, as shown in Figure 6.125.

Construction of hearths

Hearths should be constructed of suitably robust materials   J (2.22)
and to appropriate dimensions such that, in normal use, they
prevent combustion appliances setting fire to the building
fabric and furnishings, and they limit the risk of people being
accidentally burnt.

The hearth should be able to accommodate the weight of the appliance and its chimney if the chimney is not independently supported.   J (2.22)

Appliances should stand wholly above either hearths made of non-combustible board/sheet material, tiles at least 12 mm thick or constructional hearths.   J (2.23)

Constructional hearths should have plan dimensions as shown in Figure 6.128.   J (2.24a)

Constructional hearths should be made of solid, non-combustible material, such as concrete or masonry, at least 125 mm thick, including the thickness of any non-combustible floor and/or decorative surface.   J (2.24b)

Combustible material should not be placed beneath constructional hearths unless there is an air-space of at least 50 mm between the underside of the hearth and the combustible material, or the combustible material is at least 250 mm below the top of the hearth (see Figure 6.129).   J (2.25)

An appliance should be located on a hearth so that it is surrounded by a surface free of combustible material (as shown in Figure 6.130) or it may be the surface of a superimposed hearth laid wholly or partly upon a constructional hearth.   J (2.26)

The edges of this surface should be marked to provide a warning to the building occupants and to discourage combustible floor finishes such as carpet from being laid too close to the appliance. A way of achieving this would be to provide a change in level.   J (2.26)

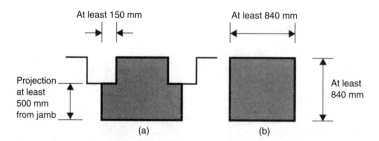

Figure 6.128 Constructional hearth suitable for solid fuel appliances (including open fires) – plan. (a) Fireplace recess. (b) Freestanding

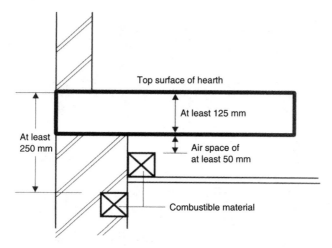

Figure 6.129 Constructional hearth suitable for solid fuel appliances (including open fires) – section

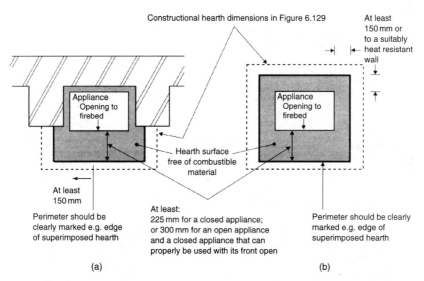

Figure 6.130 Non-combustible hearth surface surrounding a solid fuel appliance. (a) Fireplace recess. (b) Freestanding

Fireplace recesses

Fireplaces need to be constructed such that they adequately protect the building fabric from catching fire.	J (2.29)
Fireplace recesses can be from masonry or concrete as shown in Figure 6.130.	J (2.29a)

Fireplace recesses can also be prefabricated factory-made   J (2.29b)
appliance chambers using components that are made of
insulating concrete having a density of between 1200 and
1700 kg/m³ and with the minimum thickness as shown in
Table 6.37.

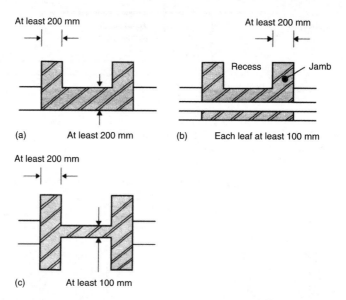

Figure 6.131 Fireplace recesses. (a) Solid wall. (b) Cavity wall. (c) Back-to-back (within the same dwelling)

Fireplace lining components

Fireplace recesses containing inset open fires, need to   J (2.30)
be heat protected and should either be lined with suitable
firebricks or lining components as shown in Table 6.37.

Table 6.37 Prefabricated appliance chambers: minimum thickness

Component	Minimum thickness (mm)
Base	50
Side section, forming wall on either side of chamber	75
Back section, forming rear of chamber	100
Top slab, lintel or gather, forming top of chamber	100

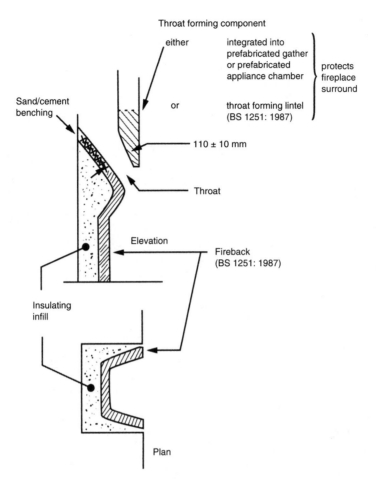

Figure 6.132 Open fireplaces – throat and fireplace components

Walls adjacent to hearths

Walls that are not part of a fireplace recess or a prefabricated   J (2.31)
appliance chamber but are adjacent to hearths or appliances
also need to protect the building from catching fire. A way
of achieving the requirement is shown in Figure 6.99.

Additional provisions for gas burning devices

The Gas Safety (Installation and Use) Regulations require that (a) gas fittings,
appliances and gas storage vessels must only be installed by a person with the
required competence and (b) any person having control to any extent of gas
work must ensure that the person carrying out that work has the required

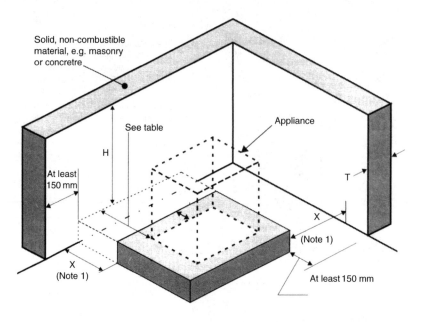

Location of hearth or appliance	Solid, non-combustible material	
	Thickness (T)	Height (H)
Where the hearth abuts a wall and the appliance is not more than 50 mm from the wall	200 mm	At least 300 mm above the appliance and 1.2 m above the hearth
Where the hearth abuts a wall and the appliance is more than 50 mm but not more than 300 mm from the wall	75 mm	At least 300 mm above the appliance and 1.2 m above the hearth
Where the hearth does not abut a wall and is no more than 150 mm from the wall (see Note 1)	75 mm	At least 1.2 m above the hearth
Note 1: There is no requirement for protection of the wall where X is more than 150 mm		

Figure 6.133 Walls adjacent to hearths

competence and (c) any person carrying out gas installation, whether an employee or self-employed, must be a member of a class of persons approved by the HSE; for the time being this means they must be registered with CORGI, the Council for Registered Gas Installers.

Important elements of the Regulations include that:

(a) any appliance installed in a room used or intended   J (3.5a)
      to be used as a bath or shower room must be of the
      room-sealed type
(b) a gas fire, other gas space heater or gas water heater   J (3.5b)
      of more than 14 kW (gross) heat input (12.7 kW (net)

heat input) must not be installed in a room used or
intended to be used as sleeping accommodation unless
the appliance is room-sealed

(c)   a gas fire, other space heater or gas water heater of up   J (3.5c)
to 14 kW (gross) heat input (12.7 kW (net) heat input)
must not be installed in a room used or intended to be
used as sleeping accommodation unless it is
room-sealed or equipped with a device designed to shut
down the appliance before there is a build-up of a
dangerous quantity of the products of combustion in
the room concerned

The restrictions in (a)–(c) above also apply in respect of any   J (3.5d)
cupboard or compartment within the rooms concerned and
to any cupboard, compartment or space adjacent to and with
an air vent into such a room.

Instantaneous water heaters (installed in any room) must be   J (3.5e)
room-sealed or have fitted a safety device to shut down the
appliance as in (c) above.

Precautions must be taken to ensure that all installation   J (3.5f)
pipework, gas fittings, appliances and flues are installed
safely. When any gas appliance is installed, checks are
required for ensuring compliance with the Regulations,
including the effectiveness of the flue, the supply of
combustion air, the operating pressure or heat input (or
where necessary both), and the operation of the appliance
to ensure its safe functioning.

All flues must be installed in a safe position.   J (3.5g)

No alteration is allowed to any premises in which a gas   J (3.5h)
fitting or gas storage vessel is fitted that would adversely
affect the safety of that fitting or vessel, causing it no longer
to comply with the Regulations.

LPG storage vessels and LPG-fired appliances fitted with   J (3.5i)
automatic ignition devices or pilot lights must not be
installed in cellars or basements.

Outlets from flues should be situated externally so as to
allow the products of combustion to dispel, and, if a
balanced flue, the intake of air – see Figure 6.134.

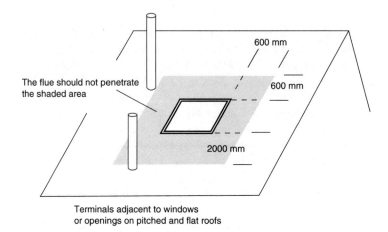

Terminals adjacent to windows
or openings on pitched and flat roofs

Figure 6.134 Location of outlets near roof windows from flues serving gas appliances

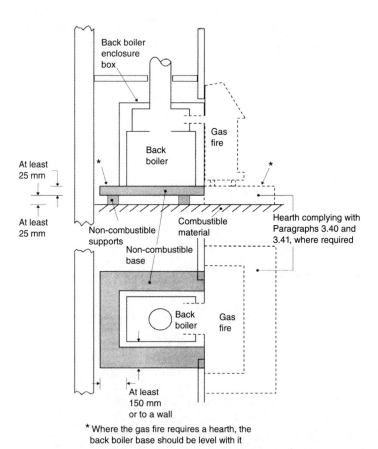

* Where the gas fire requires a hearth, the
back boiler base should be level with it

Figure 6.135 Bases for back boilers

Fireplaces – gas fires

Provided it can be shown to be safe, gas fires may be installed in fireplaces that have flues designed to serve solid fuel appliances.	J (3.7)

Bases for back boilers

Back boilers should adequately protect the fabric of the building from heat (see example at Figure 6.135).	J (3.39)

Kerosene and gas oil burning appliances

Kerosene (Class C2) and gas oil (Class D) appliances have the following, additional, requirements:

Open-fired oil appliances should not be installed in rooms such as bedrooms and bathrooms where there is an increased risk of carbon monoxide poisoning.	J (4.2)
The outlet from a flue should be so situated externally to ensure: • the correct operation of a natural draft flue; • the intake of air if a balanced flue; • dispersal of the products of combustion. Figure 6.136 (and Table 6.38) indicates typical positioning to meet this requirement.	J (4.6)

Flueblock chimneys

Flueblock chimneys should be installed with sealed joints in accordance with the flueblock manufacturer's installation instructions.	J (4.16)
Flueblocks that are not intended to be bonded into surrounding masonry should be supported and restrained in accordance with the manufacturer's installation instructions.	J (4.16)
Where a fluepipe or chimney penetrates a fire compartment wall or floor, it must not breach the fire separation requirements.	J (4.18) Approved Doc. B

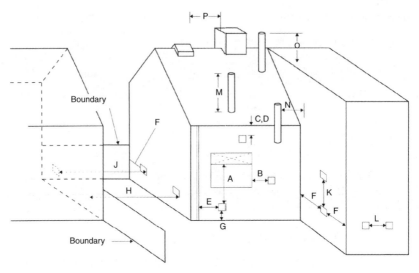

Figure 6.136 Location of outlets from flues serving oil-fired appliances

Relining chimney flues (for oil appliances)

Flexible metal flue liners should be installed in one complete length without joints within the chimney.   J (4.22)

Other than for sealing at the top and the bottom, the space between the chimney and the liner should be left empty (unless this is contrary to the manufacturer's instructions).   J (4.22)

Flues that may be expected to serve appliances burning Class D oil (i.e. gas oil) should be made of materials that are resistant to acids.   J (4.23)

Hearths for oil appliances

Oil appliance hearths are needed to prevent the building catching fire and, whilst it is not a health and safety provision, it is customary to top them with a tray for collecting spilled fuel.   J (4.24)

Table 6.38 Location of outlets from flues serving oil fired appliances

Minimum separation distances for terminals in mm

Location of outlet[1]		Appliance with pressure jet burner	Appliance with vaporizing burner
A	Below an opening[2,3]	600	should not be used
B	Horizontally to an opening[2,3]	600	should not be used
C	Below a plastic/painted gutter, drainage pipe or eaves if combustible material protected[4]	75	should not be used
D	Below a balcony or a plastic/painted gutter, drainage pipe or eaves without protection to combustible material	600	should not be used
E	From vertical sanitary pipework	300	should not be used
F	From an external or internal corner or from a surface or boundary alongside the terminal	300	should not be used
G	Above ground or balcony level	300	should not be used
H	From a surface or boundary facing the terminal	600	should not be used
J	From a terminal facing the terminal	1200	should not be used
K	Vertically from a terminal on the same wall	1500	should not be used
L	Horizontally from a terminal on the same wall	750	should not be used
M	Above the highest point of an intersection with the roof	600[6]	1000[5]
N	From a vertical structure to the side of the terminal	750[6]	2300
O	Above a vertical structure which is less than 750 mm (pressure jet burner) or 2300 mm (vaporizing burner) horizontally from the side of the terminal	600[6]	1000[5]
P	From a ridge terminal to a vertical structure on the roof	1500	should not be used

Notes:
1. Terminals should only be positioned on walls where appliances have been approved for such configurations when tested in accordance with BS EN 303-1: 1999 or OFTEC standards OFS A100 or OFS A101.
2. An opening means an openable element, such as an openable window, or a permanent opening such as a permanently open air vent.
3. Notwithstanding the dimensions above, a terminal should be at least 300 mm from combustible material, e.g. a window frame.
4. A way of providing protection of combustible material would be to fit a heat shield at least 750 mm wide.
5. Where a terminal is used with a vaporizing burner, the terminal should be at least 2300 mm horizontally from the roof.
6. Outlets for vertical balanced flues in locations M, N and O should be in accordance with manufacturer's instructions.

6.11 Stairs

6.11.1 Requirements

The building shall be constructed so that the combined dead, imposed and wind loads are sustained and transmitted by it to the ground

- *safely;*
- *without causing such deflection or deformation of any part of the building (or such movement of the ground) as will impair the stability of any part of another building.*

(Approved Document A1)

The building shall be constructed so that ground movement caused by:

- *swelling, shrinkage or freezing of the subsoil; or*
- *landslip or subsidence (other than subsidence arising from shrinkage)*

will not impair the stability of any part of the building.

(Approved Document A2)

The building shall be designed and constructed so that there are appropriate provisions for the early warning of fire, and appropriate means of escape in case of fire from the building to a place of safety outside the building capable of being safely and effectively used at all material times.

(Approved Document B)

 For a typical one- or two-storey dwelling, the requirement is limited to the provision of smoke alarms and to the provision of openable windows for emergency exit (see B1.i).

Airborne and impact sound

Dwellings shall be designed so that the noise from domestic activity in an adjoining dwelling (or other parts of the building) is kept to a level that:

- *does not affect the health of the occupants of the dwelling;*
- *will allow them to sleep, rest and engage in their normal activities in satisfactory conditions.*

(Approved Document E1)

Dwellings shall be designed so that any domestic noise that is generated internally does not interfere with the occupants' ability to sleep, rest and engage in their normal activities in satisfactory conditions.

(Approved Document E2)

Domestic buildings shall be designed and constructed so as to restrict the transmission of echoes.

(Approved Document E3)

Schools shall be designed and constructed so as to reduce the level of ambient noise (particularly echoing in corridors).

(Approved Document E4)

Stairs, ladders and ramps

All stairs, steps and ladders shall provide reasonable safety between levels in a building.

(Approved Document K1)

 In a public building the standard of stair, ladder or ramp may be higher than in a dwelling, to reflect the lesser familiarity and greater number of users.

 This requirement only applies to stairs, ladders and ramps that form part of the building.

Pedestrian guarding should be provided for any part of a floor, gallery, balcony, roof, or any other place to which people have access and any light well, basement area or similar sunken area next to a building.

(Approved Document K2)

 Requirement K2 (a) applies only to stairs and ramps that form part of the building.

Access and facilities for disabled people

In addition to the requirements of the Disability Discrimination Act 1995 precautions need to be taken to ensure that:

- *new non-domestic buildings and/or dwellings (e.g. houses and flats used for student living accommodation etc.);*
- *extensions to existing non-domestic buildings;*
- *non-domestic buildings that have been subject to a material change of use (e.g. so that they become a hotel, boarding house, institution, public building or shop);*

are capable of allowing people, regardless of their disability, age or gender, to:

- *gain access to buildings;*
- *gain access within buildings;*
- *be able to use the facilities of the buildings (both as visitors and as people who live or work in them);*
- *use sanitary conveniences in the principal storey of any new dwelling.*

(Approved Document M)

 Note: See Annex A for guidance on access and facilities for disabled people.

6.11.2 Meeting the requirements

Except for kitchens, all habitable rooms in the upper storey(s)   B1 (2.7)
of a house served by only one stair should be provided
with a window (or external door) that could be used as an
emergency exit.

Stairs

Where an opening in a floor or roof for a stairway or the like adjoins a
supported wall and interrupts the continuity of lateral support:

- the maximum permitted length of the opening   A1/2 2C37a
  is to be 3 m, measured parallel to the
  supported wall;
- connections (if provided by means other than by   A1/2 2C37b
  anchors) should be throughout the length of each
  portion of the wall situated on each side of the
  opening;
- connections via mild steel anchors should be   A1/2 2C37c
  spaced closer than 2 m on each side of the opening
  to provide the same number of anchors as if there
  were no opening;
- there should be no other interruption of lateral   A1/2 2C37d
  support.

Stairs that separate a dwelling from another dwelling (or part   E
of the same building) shall resist:

- the transmission of impact sound (such as footsteps and
  furniture moving);
- the flow of sound energy through walls and floors;
- the level of airborne sound;
- flanking transmission from stairs connected to the   E2
  separating wall

All new stairs constructed within a dwelling-house (flat or   E0.9
room used for residential purposes) – whether purpose built or
formed by a material change of use – shall meet the laboratory
sound insulation values set out in Table 6.39.

Table 6.39 Dwelling-houses and flats – performance standards for separating floors and stairs that have a separating function

	Airborne sound insulation $D_{nT,w} + C_{tr}$ dB (minimum values)	Impact sound insulation $L'_{nT,w}$ dB (maximum values)
Purpose built rooms for residential purposes	45	62
Purpose built dwelling houses and flats	45	62
Rooms for residential purposes formed by material change of use	43	64
Dwelling-houses and flats formed by material change of use	43	64

 Notes:

(1)   The sound insulation values in this table include a built-in allowance for 'measurement uncertainty' and so if any these test values are not met, then that particular test will be considered as failed.

(2)   Occasionally a higher standard of sound insulation may be required between spaces used for normal domestic purposes and noise generated in and to an adjoining communal or non-domestic space. In these cases it would be best to seek specialist advice before committing yourself.

(3)   If the stair is not enclosed, then the potential sound insulation of the internal floor will not be achieved, nevertheless, the internal floor should still satisfy Requirement E2.

(4)   In some cases it may be that an existing wall, floor or stair in a building will achieve these performance standards without the need for remedial work, for example if the existing construction was already compliant.

Figure 6.137 illustrates the relevant parts of the building that should be protected from airborne and impact sound in order to satisfy Requirement E2.

Sound insulation testing

The person carrying out the building work should arrange for sound insulation testing to be carried out (by a test body with appropriate third party accreditation) in accordance with the procedure described in Annex B of this Approved Document E.	E0.3 E0.4
Impact sound insulation tests should be carried out without a soft covering (e.g. carpet, foam backed vinyl etc.) on the stair floor.	E1.10

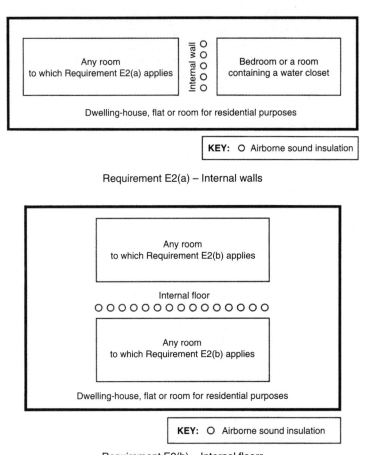

Requirement E2(a) – Internal walls

Requirement E2(b) – Internal floors

Figure 6.137 Airborne and impact sound requirements

Testing should not be carried out between living spaces, corridors, stairwells or hallways.	E1.8
Test bodies conducting testing should preferably have UKAS accreditation (or a European equivalent) for field measurements.	E0.4

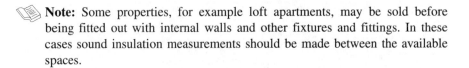 **Note:** Some properties, for example loft apartments, may be sold before being fitted out with internal walls and other fixtures and fittings. In these cases sound insulation measurements should be made between the available spaces.

If stairs form a separating function then they are subject to the same sound insulation requirements as floors. In this case, the resistance to airborne sound depends mainly on:

- the mass of the stair;
- the mass and isolation of any independent ceiling;
- the air tightness of any cupboard or enclosure under the stairs;
- the stair covering (which reduces impact sound at source).

Stair treatment 1

Stair treatment 1 consists of a stair covering and independent ceiling with absorbent material.

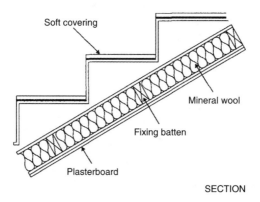

SECTION

Figure 6.138 Stair covering and independent ceiling with absorbent material

---

The soft covering should be:

- at least 6 mm thickness;   E4.37
- laid over the stair treads;
- be securely fixed (e.g. glued) so it does not become a safety hazard.

If there is a cupboard under all, or part, of the stair: E4.37

- the underside of the stair within the cupboard should be lined with plasterboard (minimum mass per unit area 10 kg/m^2) together with an absorbent layer of mineral wool (minimum density 10 kg/m^3);
- the cupboard walls should be built from two layers of plasterboard (or equivalent), each sheet with a minimum mass per unit area of 10 kg/m^2;
- a small, heavy, well fitted door should be fitted to the cupboard.

If there is no cupboard under the stair, an independent ceiling   E4.37
should be constructed below the stair (see Floor treatment 1).

Where a staircase performs a separating function it shall   E4.38
conform to Building Regulation Part B – Fire safety.

Reverberation

Requirement E3 requires that *'domestic buildings shall be designed and constructed so as to restrict the transmission of echoes'*. The guidance notes provided in Part E cover two methods (Method A and Method B) which can be used in determining the amount of additional absorption to be used in corridors, hallways, stairwells and entrance halls that give access to flats and rooms for residential purposes. Method A is applicable to stairs and requires the following to be observed:

Method A

Cover the ceiling area with the additional absorption.   E7.10

Cover the underside of intermediate landings, the underside of   E7.11
the other landings, and the ceiling area on the top floor.

The absorptive material should be equally distributed between   E7.12
all floor levels.

For stairwells (or a stair enclosure), calculate the combined   E7.11
area of the stair treads, the upper surface of the intermediate
landings, the upper surface of the landings (excluding ground
floor) and the ceiling area on the top floor. Either cover an
area equal to this calculated area with a Class D absorber, or
cover an area equal to at least 50% of this calculated area with
a Class C absorber or better.

 Note: Method A can generally be satisfied by the use of proprietary acoustic ceilings.

Piped services

Piped services (excluding gas pipes) and ducts that pass through separating floors should be surrounded with sound absorbent material for their full height and enclosed in a duct above and below the floor.

Junctions with floor penetrations (excluding gas pipes)

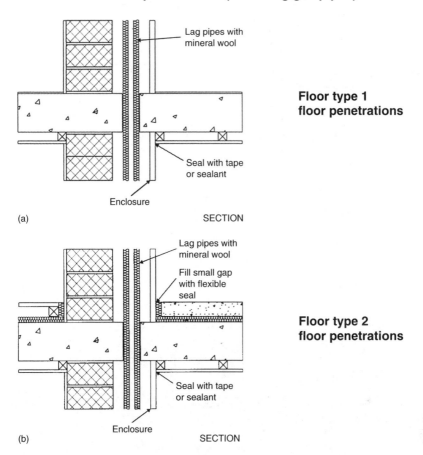

(a)

SECTION

**Floor type 1
floor penetrations**

Lag pipes with
mineral wool

Seal with tape
or sealant

Enclosure

(b)

SECTION

**Floor type 2
floor penetrations**

Lag pipes with
mineral wool

Fill small gap
with flexible
seal

Seal with tape
or sealant

Enclosure

Figure 6.139a,b,c Junctions with floor penetrations (excluding gas pipes)

Pipes and ducts that penetrate a floor separating habitable rooms in different flats should be enclosed for their full height in each flat.	E3.41 E3.79 E3.117
The enclosure should be constructed of material having a mass per unit area of at least 15 kg/m².	E3.32 E3.80 E3.118
The enclosure should either be lined or the duct (or pipe) within the enclosure wrapped with 25 mm unfaced mineral fibre.	E3.42 E3.80 E3.118

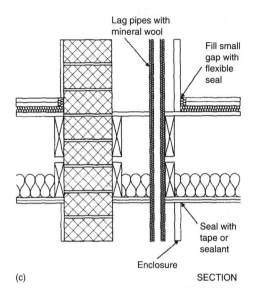

**Floor type 3
floor penetrations**

(c)   SECTION

Figure 6.139 (*Continued*)

Penetrations through a separating floor by ducts and pipes should have fire protection to satisfy Building Regulation Part B – Fire safety.	E3.43 E3.82 E3.120
Fire stopping should be flexible to prevent a rigid contact between the pipe and the floor.	E3.43 E3.121
A small gap (sealed with sealant or neoprene) of about 5 mm should be left between the enclosure and the floating floor.	E3.81 E3.119
Where floating floor (a) or (b) is used the enclosure may go down to the floor base (provided that the enclosure is isolated from the floating layer).	E3.81 E3.119

Junctions with floor penetrations (including gas pipes)

Gas pipes may be contained in a separate (ventilated) duct or can remain unenclosed.	E3.43 E3.120
If a gas service is installed it shall comply with the Gas Safety (Installation and Use) Regulations 1998, SI 1998 No 2451.	E3.43 E3.120

In the Gas Safety Regulations there are requirements for ventilation of ducts at each floor where they contain gas pipes. Gas pipes may be contained in a separate ventilated duct or they can remain unducted.

Stairs, ladders and ramps

The rise of a stair shall be between 155 mm and 220 mm with any going between 245 mm and 260 mm and a maximum pitch of 42°.	K1 (1.1–1.4)

The normal relationship between the dimensions of the rise and going is that twice the rise plus the going (2R + G) should be between 550 mm and 700 mm.

Stairs with open risers that are likely to be used by children under 5 years should be constructed so that a 100 mm diameter sphere cannot pass through the open risers.	K1 (1.9)
Stairs which have more than 36 risers in consecutive flights should make at least one change of direction, between flights, of at least 30°.	K1 (1.14)
If a stair has straight and tapered treads, then the going of the tapered treads should not be less than the going of the straight tread.	K1 (1.20)
The going of tapered treads should measure at least 50 mm at the narrow end.	K1 (1.18)
The going should be uniform for consecutive tapered treads.	K1 (1.19) K1 (1.22–1.24)
Stairs should have a handrail on both sides if they are wider than 1 m and on at least one side if they are less than 1 m wide.	K1 (1.27)
Handrail heights should be between 900 mm and 1000 mm measured to the top of the handrail from the pitch line or floor.	K1 (1.27)
Spiral and helical stairs should be designed in accordance with BS 5395.	K1 (1.21)

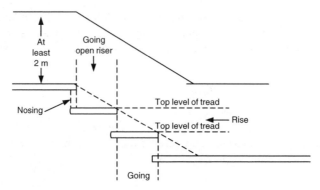

Figure 6.140 Rise and going plus headroom

Steps

Steps should have level treads.	K1 (1.8)
Steps may have open risers, but treads should then overlap each other by at least 16 mm.	K1 (1.8)
Steps should be uniform with parallel nosings, the stair should have handrails on both sides and the treads should have slip resistant surfaces.	
The headroom on the access between levels should be no less than 2 m.	K1 (1.10)
Landings should be provided at the top and bottom of every flight.	K1 (1.15)
The width and length of every landing should be the same (or greater than) the smallest width of the flight.	K1 (1.15)
Landings should be clear of any permanent obstruction.	K1 (1.16)
Landings should be level.	K1 (1.17)
Any door (entrance, cupboard or duct) that swings across a landing at the top or bottom of a flight of stairs must leave a clear space of at least 400 mm across the full width of the flight.	K1 (1.16)

| Flights and landings should be guarded at the sides when there is a drop of more than 600 mm. | K1 (1.28–1.29) |

 For stairs that are likely to be used by children under 5 years the construction of the guarding shall be such that a 100 mm sphere cannot pass through any openings in the guarding and children will not easily climb the guarding.

- For loft conversions, a fixed ladder should have   K1 (1.25)
  fixed handrails on both sides.

Whilst there are no recommendations for minimum stair widths, designers should bear in mind the requirements of Approved Documents B (means of escape) and M (access and facilities for disabled people).

Ramps

All ramps shall provide reasonable safety between levels in a building (where the difference in level is more than 600 mm) and other buildings where the change of level is more than 380 mm.

Ramps should be clear of permanent obstructions.	K1 (2.4)
The slope of a ramp shall be no more than 1:12.	K1 (2.1)
Ramps should have a handrail on both sides if they are wider than 1 m and on at least one side if they are less than 1 m wide.	K1 (2.5) M
Handrail heights should be between 900 mm and 1000 mm measured to the top of the handrail from the pitch line or floor.	K1 (2.5) M
All ramps should have landings.	K1 (2.6)
All ramps (and associated landings) should have a clear headroom throughout of at least 2 m.	K1 (2.2)
Ramps and landings should be guarded at the sides when there is a drop of more than 600 mm.	K1 (2.7)

For stairs that are likely to be used by children under 5 years the construction of the guarding shall be such that a 100 mm sphere cannot pass through any openings in the guarding and children will not easily climb the guarding.

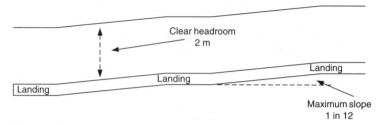

Figure 6.141 The recommended design of a ramp

Protection from falling

All stairs, landings, ramps and edges of internal floors shall have a wall, parapet, balustrade or similar guard at least 900 mm high.	K3 (3.2)
All guarding should be capable of resisting at least the horizontal force given in BS 6399: Part 1: 1996.	K3 (3.2)
If glazing is used as (or part of) the pedestrian guarding, see Approved Document N: Glazing – safety in relation to impact, opening and cleaning.	N
If a building is likely to be used by children under 5 years, the guarding should not have horizontal rails, should stop children from easily climbing it, and the construction should prevent a 100 mm sphere being able to pass through any opening of that guarding.	K3 (3.3)
All external balconies and edges of roofs shall have a wall, parapet, balustrade or similar guard at least 1100 mm high.	K3 (3.2)

Wall cladding

Where wall cladding is required to function as pedestrian guarding to stairs, ramps, vertical drops of 600 mm or greater or as a vehicle barrier, account should be taken of the additional imposed loading as stipulated in Part K.	A1/2 3.5
Where wall cladding is required to safely withstand lateral pressures from crowds, an appropriate design loading is given in BS 6399 Part 1 and the *Guide to Safety at Sports Grounds* (4th Edition, 1997).	

Access and facilities for disabled people

Internal steps, stairs and ramps

Stepped access

A stepped access should:

- have a level landing at the top and bottom of each flight; M (3.51a)
- be 1200 mm long each landing and be unobstructed.

Doors should not swing across landings.   M (3.51a)

The surface width of flights between enclosing walls,   M (3.51a)
strings or upstands should not be less than 1.2 m.

There should be no single steps.   M (3.51a)

Nosings for the tread and the riser should be 55 mm wide   M (3.51a)
and of a contrasting material.

Step nosings should not project over the tread below by   M (3.51a)
more than 25 mm (see Figure 6.142).

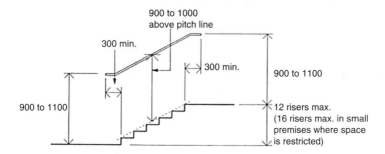

Figure 6.142 Internal stairs – key dimensions

The rise and going of each step should be consistent   M (3.51a)
throughout a flight.

The rise of each step should be between 150 mm and   M (3.51a)
170 mm.

The going of each step should be between 280 mm and   M (3.51a)
425 mm.

Rises should not be open.   M (3.51a)

There should be a continuous handrail on each side of a   M (3.51a)
flight and landings.

If additional handrails are used to divide the flight into channels, then they should not be less than 1 m wide or more than 1.8 m wide.   M (3.51a)

Flights between landings should contain no more than 12 risers.   M (3.51b)

The rise of each step should be between 150 mm and 170 mm.   M (3.51c)

The going of each step should be at least 250 mm.   M (3.51d)

For mobility-impaired people, a going of at least 300 mm is preferred.

Materials for treads should not present a slip hazard.   M (3.50)

Areas below stairs or ramps with a soffit less than 2.1 m above ground level should be protected by guarding and low level cane detection.   M (3.51e)

Any feature projecting more than 100 mm onto an access route should be protected by guarding that includes a kerb (or other solid barrier) that can be detected using a cane (see Figure 6.143).   M (3.51e)

Note: For school buildings, the rise should not exceed 170 mm, with a preferred going of 280 mm.

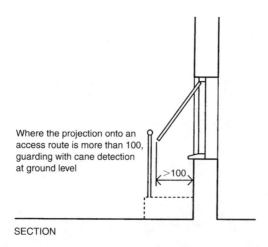

Where the projection onto an access route is more than 100, guarding with cane detection at ground level

>100

SECTION

Figure 6.143 Avoiding hazards on access routes

Internal ramps

Where an internal ramp is provided:	M (3.53)

- the approach should be clearly signposted;
- the going should be no greater than 10 m;
- the rise should be no more than 500 mm;
- if the total rise is greater than 2 m then an alternative means of access (e.g. a lift) should be provided for wheelchair users;
- the ramp surface should be slip resistant;
- the ramp surface should be of a contrasting colour with that of the landings;
- frictional characteristics of ramp and landing surfaces should be similar;
- landings at the foot and head of a ramp should be at least 1.2 m long and clear of any obstructions;
- intermediate landings should be at least 1.5 m long and clear of obstructions;
- all landings should be:
  - level;
  - have a maximum gradient of 1:60 along their length;
  - have a maximum cross fall gradient of 1:40;
- there should be a handrail on both sides;
- in addition to the guarding requirements of Park K, there should be a visually contrasting kerb on the open side of the ramp (or landing) at least 100 mm high.

Where the change in level is 300 mm or more, two or more clearly signposted steps should be provided (i.e. in addition to the ramp).	M (3.53b)
If the change in level is no greater than 300 mm, a ramp should be provided instead of a single step.	M (3.53c)
All landings should be level and a maximum gradient of 1:60 along their entire length.	M (3.53d)
Areas below stairs or ramps with a soffit less than 2.1 m above ground level should be protected by guarding and low level cane detection.	M (3.53e)
Any feature projecting more than 100 mm onto an access route should be protected by guarding that includes a kerb (or other solid barrier) that can be detected using a cane (see Figure 6.144).	M (3.53e)
Gradients should be as shallow as practicable.	M (3.52)

Handrails to internal steps, stairs and ramps

Handrails to external stepped or ramped access should be positioned as per Figure 6.144. M (1.37a)

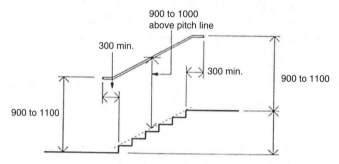

Figure 6.144 Handrails to internal steps, stairs and ramps – key dimensions

Handrails to internal steps, stairs and ramps should: M (3.55)

- be continuous across flights and landings;
- extend at least 300 mm horizontally beyond the top and bottom of a ramped access;
- not project into an access route;
- contrast visually with the background;
- have a slip resistant surface which is not cold to the touch;
- terminate in such a way that reduces the risk of clothing being caught;
- either be circular (with a diameter of between 40 and 45 mm) or oval with a width of 50 mm (see Figure 6.145).

Handrails to external stepped or ramped access should: M (3.55)

- not protrude more than 100 mm into the surface width of the ramped or stepped access where this would impinge on the stair width requirement of Part B1;
- have a clearance of between 60 and 75 mm between the handrail and any adjacent wall surface;
- have a clearance of at least 50 mm between a cranked support and the underside of the handrail;
- ensure that its inner face is located no more than 50 mm beyond the surface width of the ramped or stepped access;

- should be spaced away from the wall and rigidly supported in a way that avoids impeding finger grip;
- should be set at heights that are convenient for all users of the building.

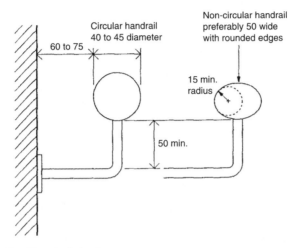

Figure 6.145 Handrail designs

Common stairs in blocks of flats

The aim for all buildings containing flats should be to make reasonable provision for disabled people to visit occupants who live on any storey of the building, via a common staircase or a lift.

Common stairs

If there is no passenger lift to provide access between storeys, a stair (designed to suit the needs of ambulant disabled people, people with impaired sight and people with sensory impairments) should be provided.	M (9.3 and 9.4)

If a passenger lift is not installed, a common stair should be provided which has:

• step nosings with contrasting brightness;	M (9.5a)
• top and bottom landings whose lengths are in accordance with Part K1;	M (9.5b)
• steps with suitable tread nosing profiles (see Figure 6.146) with a uniform rise not more than 170 mm;	M (9.5c)

- a uniform going of each step not less than 250 mm;   M (9.5d)
- risers which are not open;   M (9.5e)
- a continuous handrail on each side of flights and   M (9.5f)
  landings (if the rise of the stair comprises two or
  more rises).

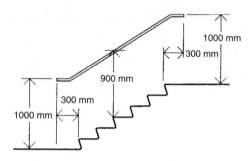

Figure 6.146 Common stairs in blocks of flats

6.12 Windows

6.12.1 Requirements

Protection from falling

Pedestrian guarding should be provided for any part of a floor (including the edge below an opening window) gallery, balcony, roof (including rooflight and other openings), any other place to which people have access and any light well, basement area or similar sunken area next to a building.

(Approved Document K2)

Conservation of fuel and power

Energy efficiency measures shall be provided which:

(a) • *limit the heat loss through the roof, wall, floor, windows and doors, etc. by suitable means of insulation;*
     • *where appropriate permit the benefits of solar heat gains and more efficient heating systems to be taken into account;*
     • *limit unnecessary ventilation heat loss by providing building fabric which is reasonably airtight;*
     • *limit the heat loss from hot water pipes and hot air ducts used for space heating;*
     • *limit the heat loss from hot water vessels and their primary and secondary hot water connections by applying suitable thicknesses of*

insulation (where such heat does not make an efficient contribution to the space heating).

(b) provide space heating and hot water systems with reasonably efficient equipment such as heating appliances and hot water vessels where relevant, such that the heating and hot water systems can be operated effectively as regards the conservation of fuel and power.

(c) provide lighting systems that utilize energy-efficient lamps with manual switching controls or, in the case of external lighting fixed to the building, automatic switching, or both manual and automatic switching controls as appropriate, such that the lighting systems can be operated effectively as regards the conservation of fuel and power.

(d) provide information, in a suitably concise and understandable form (including results of performance tests carried out during the works), that shows building occupiers how the heating and hot water services can be operated and maintained.

(Approved Document L1)

 Responsibility for achieving compliance with the requirements of Part L rests with the person carrying out the work. That person may be, for example, a developer, a main (or sub-) contractor, or a specialist firm directly engaged by a private client.

The person responsible for achieving compliance should either themselves provide a certificate, or obtain a certificate from the sub-contractor, that commissioning has been successfully carried out. The certificate should be made available to the client and the building control body.

Protection against impact

Glazing with which people are likely to come into contact whilst moving in or about the building, shall:

- *if broken on impact, break in a way which is unlikely to cause injury; or*
- *resist impact without breaking; or*
- *be shielded or protected from impact.*

(Approved Document N)

6.12.2 Meeting the requirement

Protection from falling

All stairs, landings, ramps and edges of internal floors shall have a wall, parapet, balustrade or similar guard at least 900 mm high.	K3 (3.2)

All guarding should be capable of resisting at least the horizontal force given in BS 6399: Part 1: 1996.	K3 (3.2)
If glazing is used as (or part of) the pedestrian guarding, see Approved Document N: Glazing – safety in relation to impact, opening and cleaning.	N
If a building is likely to be used by children under 5 years, the guarding should not have horizontal rails, should stop children from easily climbing it, and the construction should prevent a 100 mm sphere being able to pass through any opening of that guarding.	K3 (3.3)
All external balconies and edges of roofs shall have a wall, parapet, balustrade or similar guard at least 1100 mm high.	K3 (3.2)
All windows, skylights, and ventilators shall be capable of being left open without danger of people colliding with them by:	K

- installing windows, etc. so that projecting parts are kept away from people moving in and around the building; or
- installing features that guide people moving in or about the building away from any open window, skylight or ventilator.

Parts of windows (skylights and ventilators) that project either internally or externally more than about 1000 mm horizontally into spaces used by people moving in or about the building should not present a safety hazard.	K

Conservation of fuel and energy

Single-glazed panels can be acceptable in external doors provided that the heat loss through all the windows, doors and rooflights matches the relevant figure in Table 6.40 and the area of the windows, doors and rooflights together does not exceed 25% of the total floor area.	L1 (1.5)
The average U-value of windows, doors and rooflights should match the relevant figure in Table 6.40 and the area of the windows, doors and rooflights together should not exceed 25% of the total floor area.	L1 (1.17)

Examples of how the average U-value is calculated are given in Appendix D to Approved Document L1.

For dwellings whose windows have metal frames (including thermally broken frames) the target U-value can be increased by multiplying by a factor of 1.03, to take account of the additional solar gain due to the greater glazed proportion. L1

Table 6.40 Elemental method, U-values (W/m^2 K) for construction elements

Exposed element	U-value
Windows, doors and rooflights (area-weighted average), glazing in metal frames[1]	2.2
Windows, doors and (area-weighted average) rooflights[1], glazing in wood or PVC frames[2]	2.0

Notes:
1. Rooflights include roof windows.
2. The higher U-value for metal-framed windows allows for additional solar gain due to the greater glazed proportion.

Replacement of controlled services or fittings

When windows, etc. are being upgraded they shall be replaced with new draught-proofed ones either with an average U-value not exceeding the appropriate entry in Table 6.40, or with a centre-pane U-value not exceeding $1.2\,W/m^2\,K$. L1 (2.3a)

This requirement does not apply to repair work on parts of these elements, such as replacing broken glass or sealed double-glazing units or replacing rotten framing members.

What about glazing?

Although the installation of replacement windows or glazing (e.g. by way of repair), is not considered as building work under Regulation 3 of the Building Regulations, on the other hand, glazing that:

- is installed in a location where there was none previously;
- is installed as part of an erection;
- is installed as part extension or material alteration of a building;

is subject to these requirements.

The existence of large uninterrupted areas of transparent glazing represents a significant risk of injury through collision. The risk is at its most severe between areas of a building or its surroundings that are essentially at the same level and where a person might reasonably assume direct access between locations that are separated by glazing.

The most likely places where people can sustain injuries are due to impacts with doors, door side panels (especially between waist and shoulder level) when initial impact can be followed by a fall through the glazing resulting in additional injury to the face and body. Hands, wrists and arms are particularly vulnerable.

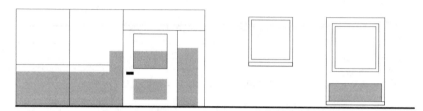

Figure 6.147 Shaded areas show critical locations in internal and external walls

Apart from doors, walls and partitions are a low-level, high-risk area, particularly where children are concerned.

The existence of large uninterrupted areas of transparent glazing represents a significant risk of injury through collision. The risk is at its most severe between areas of a building or its surroundings that are essentially at the same level and where a person might reasonably assume direct access between locations that are separated by glazing.

 Approved Document B: Fire safety includes guidance on fire-resisting glazing and the reaction of glass to fire.

 Approved Document K: Protection from falling, collision and impact covers glazing that forms part of the protection from falling from one level to another, and that needs to ensure containment as well as limiting the risk of sustaining injury through contact.

Some glazing materials, such as annealed glass, gain strength through thickness; others such as polycarbonates or glass blocks are inherently strong. Some annealed glass is considered suitable for use in large areas forming fronts to shops, showrooms, offices, factories, and public buildings.

Protection against impact

Measures shall be taken to limit the risk of sustaining   N1 (0.1)
cutting and piercing injuries.

In critical locations, if glazing is damaged the breakage should only result in small, relatively harmless particles.	N1 (0.2)
Glazing should be sufficiently robust to ensure that the risk of breakage is low.	N1 (0.4)
Steps should be taken to limit the risk of contact with the glazing.	N1 (0.5)
Glazing in critical locations should either be permanently protected, be in small panes or if it breaks, break safely (see BS 6206).	N1 (1.2)
Small panes should not exceed 250 mm and an area of 0.5 m².	N1 (1.6)

Transparent glazing

Transparent glazing with which people are likely to come into contact while moving in or about the building shall incorporate features that make it apparent.

The presence of glazing should be made more apparent or visible to people using the building.	N2 (0.8)
The presence of large uninterrupted areas of transparent glazing should be clearly indicated.	N2 (0.6, 2.1, 2.2)
In critical locations (i.e. large areas where the glazing forms part of internal or external walls and doors of shops, showrooms, transoms, offices, factories, public or other non-domestic buildings) the presence of large uninterrupted areas of transparent glazing should be clearly indicated by the use of broken or solid lines, patterns or company logos at appropriate heights and intervals.	N2 (2.4–2.5)

Safe opening and closing of windows

Windows, skylights and ventilators that can be opened by people should be capable of being opened, closed or adjusted safely.

Where controls can be reached without leaning over an obstruction they should not be more than 1.9 m above the floor. Where there is an obstruction, the control should be lower (e.g. not more than 1.7 m where there is a 600 mm deep obstruction).	N3 (3.2)

Where controls cannot be positioned within safe reach from a permanent stable surface, a safe means of remote operation, such as a manual or electrical system should be provided.	N3 (3.2)
Where there is a danger of the operator or other person falling through a window above ground floor level, suitable opening limiters should be fitted or guarding provided.	N3 (3.3)

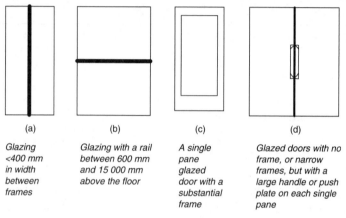

(a)	(b)	(c)	(d)
Glazing <400 mm in width between frames	*Glazing with a rail between 600 mm and 15 000 mm above the floor*	*A single pane glazed door with a substantial frame*	*Glazed doors with no frame, or narrow frames, but with a large handle or push plate on each single pane*

Figure 6.148 Examples of door height glazing not requiring identification further

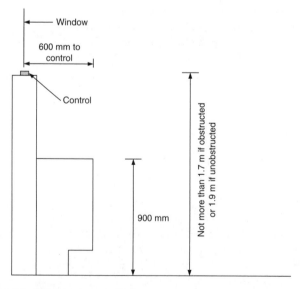

Figure 6.149 Height of controls

Safe access for cleaning

All windows, skylights (or any transparent or translucent walls, ceilings or roofs) of a dwelling should be safely accessible for cleaning.	N

Where glazed surfaces cannot be cleaned safely by a person standing on the ground, the requirement for a floor, or other permanent stable surface, could be satisfied by provisions such as the following:

Safe means of access shall be provided for cleaning both sides of glazed surfaces where there is danger of falling more than 2 m.	N4
Where possible windows should be of a size and design that allow the outside surface to be cleaned safely from inside the building.	N4 (4.2)
Windows that reverse for cleaning should be fitted with a mechanism that holds the window in the reversed position (see BS 8213).	N4 (4.2)
For large buildings (e.g. office blocks) a firm, level surface shall be provided to enable portable ladders (not more than 9 m long) to be used and the use of suspended cradles, travelling ladders, abseiling equipment should also be considered.	N4 (4.2)

6.13 Doors

6.13.1 Requirements

Conservation of fuel and power

Energy efficiency measures shall be provided that limit the heat loss through the doors, etc. by suitable means of insulation.

Responsibility for achieving compliance with the requirements of Part L rests with the person carrying out the work. That person may be, for example, a developer, a main (or sub-) contractor, or a specialist firm directly engaged by a private client.

The person responsible for achieving compliance should either themselves provide a certificate, or obtain a certificate from the sub-contractor, that commissioning has been successfully carried out. The certificate should be made available to the client and the building control body.

Access and facilities for disabled people

In addition to the requirements of the Disability Discrimination Act 1995 precautions need to be taken to ensure that:

- *new non-domestic buildings and/or dwellings (e.g. houses and flats used for student living accommodation etc.);*
- *extensions to existing non-domestic buildings;*
- *non-domestic buildings that have been subject to a material change of use (e.g. so that they become a hotel, boarding house, institution, public building or shop);*

are capable of allowing people, regardless of their disability, age or gender, to:

- *gain access to buildings;*
- *gain access within buildings;*
- *be able to use the facilities of the buildings (both as visitors and as people who live or work in them);*
- *use sanitary conveniences in the principal storey of any new dwelling.*

(Approved Document M)

 Note: See Annex A for guidance on access and facilities for disabled people.

6.13.2 Meeting the requirement

Doors and gates on main traffic routes (and those that can be pushed open from either side) should have vision panels unless they are low enough (e.g. 900 mm) to see over.	K5 (5.2a)
Sliding doors and gates should have a retaining rail to prevent them falling should the suspension system fail or the rollers leave the track.	K5 (5.2b)
Upward opening doors and gates should be fitted with a device to stop them falling in a way that could cause injury.	K5 (5.2c)
Power operated doors and gates should have safety features to prevent injury to people who are struck or trapped (such as a pressure sensitive door edge that operates the power switch).	K5 (5.2d)
Power operated doors and gates should have a readily identifiable and accessible stop switch.	K5 (5.2d)

Power operated doors and gates should be provided with   K5 (5.2d)
a manual or automatic opening device in the event of a
power failure where and when necessary for health
or safety.

Single-glazed panels can be acceptable in external doors   L1 (1.5)
provided that the heat loss through all the windows, doors
and rooflights matches the relevant figure in Table 6.37
and the area of the windows, doors and rooflights together
does not exceed 25% of the total floor area.

The average U-value of windows, doors and rooflights   L1 (1.7)
should match the relevant figure in Table 6.37 and the
area of the windows, doors and rooflights together should
not exceed 25% of the total floor area.

Examples of how the average U-value is calculated
are given in Appendix D to Approved Document L1.

Replacement of controlled services or fittings

When doors, etc. are being upgraded they shall be   L1 (2.3a)
replaced with new draught-proofed ones either with
an average U-value not exceeding the appropriate
entry in Table 6.37, or with a centre-pane U-value
not exceeding $1.2\,W/m^2\,K$.

This requirement does not apply to repair work on
parts of these elements, such as replacing broken glass or
sealed double-glazing units or replacing rotten framing
members.

Access and facilities for disabled people

Internal doors

The opening force for a manual operating door should not   M (3.10a)
exceed 20 N.

The effective clear width through a single leaf door (or one   M (3.10b)
leaf of a double leaf door) should be in accordance with
Table 6.41 and Figure 6.150.

Table 6.41 Minimum effective clear widths of doors

Direction and width of approach	New buildings (mm)	Existing buildings (mm)
Straight-on (without a turn or oblique approach)	800	750
At right angles to an access route at least 1500 mm wide	800	750
At right angles to an access route at least 1200 mm wide	825	775
External doors to buildings used by the general public	1000	775

There should be an unobstructed space of at least 300 mm on the pull side of the door between the leading edge of the door and any return wall (unless the door is a powered entrance door) (see Figure 6.150).   M (3.10c)

A space alongside the leading edge of a door should be provided to enable a wheelchair user to reach and grip the door handle.

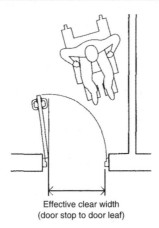

Effective clear width
(door stop to door leaf)

Figure 6.150 Effective clear width and visibility requirements of doors

Door opening furniture should:

- be easy to operate by people with limited manual dexterity;
- be capable of being operated with one hand using a closed fist (e.g. a lever handle);   M (3.10d)
- contrast visually with the surface of the door.   M (3.10e)

Door frames should contrast visually with the surrounding wall.   M (3.10f)

The surface of the leading edge of a non self-closing door should contrast visually with the other door surfaces and its surroundings.   M (3.10g)

Door leaves or side panels should be wider than 450 mm.   M (3.10h)

Vision panels towards the leading edge of the door should include a visibility zone (or zones) between 500 mm and 1500 mm from the floor. If interrupted (e.g. to accommodate an intermediate horizontal rail – see Figure 6.151) then this should be 800 mm and 1150 mm above the floor.   M (3.10h)

🔔 Glass entrance doors and glazed screen should be clearly marked (i.e. with a logo or sign) on the glass at two levels, 850 to 1000 mm and 1400 to 1600 mm above the floor.   M (3.10i)

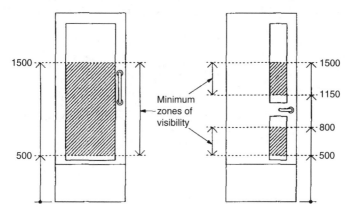

Figure 6.151 Door vision panels

It should be possible to tell between a fully glazed door and any adjacent glazed wall/partition by providing a high-contrast strip at the top and on both sides.   M (3.10j)

Fire doors (particularly those in corridors) should be held open with an electromagnetic device that is capable of self closing when: M (3.10k)

- the power supply fails;
- activated by smoke detectors;
- activated by a hand-operated switch.

Fire doors (particularly to individual rooms) should be fitted with swing-free devices that close when:	M (3.10l)

- activated by smoke detectors;
- the building's fire alarm system is activated;
- when the power supply fails.

Low energy powered door systems may be used in locations not subject to frequent use or heavy traffic.	M (3.7)

Low energy powered swing door systems should be capable of being operated:	M (3.10m)

- in manual mode;
- in powered mode; or
- in power-assisted mode.

The use of self-closing devices should be minimized as they disadvantage many people (e.g. those pushing prams or carrying heavy objects).	M (3.7)

If closing devices are needed for fire control:	M (3.7)

- they should be electrically powered hold-open devices or swing-free closing devices;
- their closing mechanism should only be activated in case of emergency.

The presence of doors, whether open or closed, should be apparent to visually impaired people.	M (3.8)

 Note: See BS 8300 for guidance on:

- electrically powered hold-open devices;
- swing-free systems;
- low energy powered door systems.

6.14 Vertical circulation within the building

6.14.1 The requirement

In addition to the requirements of the Disability Discrimination Act 1995 precautions need to be taken to ensure that:

- *new non-domestic buildings and/or dwellings (e.g. houses and flats used for student living accommodation etc.);*

- *extensions to existing non-domestic buildings;*
- *non-domestic buildings that have been subject to a material change of use (e.g. so that they become a hotel, boarding house, institution, public building or shop);*

are capable of allowing people, regardless of their disability, age or gender, to:

- *gain access to buildings;*
- *gain access within buildings;*
- *be able to use the facilities of the buildings (both as visitors and as people who live or work in them);*
- *use sanitary conveniences in the principal storey of any new dwelling.*

(Approved Document M)

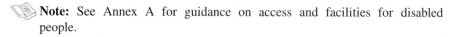 **Note:** See Annex A for guidance on access and facilities for disabled people.

6.14.2 Meeting the requirement

Lifting devices

For all buildings, a passenger lift is considered the most suitable form of access for people moving from one storey to another.

Wherever possible a passenger lift (or a lifting platform) serving all storeys should be provided in: • new developments; • existing buildings. **Note:** In exceptional circumstances (e.g. a listed building, or an infill site in a historic town centre) where a passenger lift cannot be accommodated, a wheelchair platform stair lift serving an intermediate level or a single storey may be used.	M (3.17, 3.24a to d)
The location of lifting devices that are accessible by mobility impaired people should be clearly visible from the building entrance.	M (3.18)
Signs should be available at each landing to identify the floor reached by the lifting device.	M (3.18)
In addition to the lifting device, internal stairs (designed to suit ambulant disabled people and those with impaired sight) should always be provided.	M (3.19)

General requirements for lifting devices

There should be an unobstructed manoeuvring space of 1500 mm × 1500 mm, or a straight access route 900 mm wide, in front of each lifting device.	M (3.28a)
The landing call buttons should be located between 900 mm and 1100 mm from the floor and at least 500 mm from any return wall.	M (3.28b)
Landing call buttons and lifting device control button symbols:	M (3.248c and d)
• should contrast visually with the surrounding face plate; • should be raised to facilitate tactile reading; • should be accessible by wheelchair users.	M (3.27)
The floor of the lifting device should not be of a dark colour.	M (3.24e)
A handrail (at 900 mm nominal) should be provided on at least one wall of the lifting device.	M (3.24f)
A suitable emergency communication system should be fitted.	M (3.24g)

 Note: See also:

- Lift Regulations 1997 SI 1997/831;
- Lifting Operations and Lifting Equipment Regulations 1998 SI 1998/2307;
- Provision and Use of Work Equipment Regulations 1998 SI 1998/2306;
- Management of Health and Safety at Work Regulations 1999 SI 1999/3242;
- BS 8300.

Passenger lifts

Lift sizes should be chosen to suit the anticipated density of use of the building and the needs of disabled people.

Passenger lifts should conform to the requirements of:	
• the Lift Regulations 1997, SI 1997/831; • the relevant British Standards, EN 81 series of standards; • BS EN 81-70: 2003 (*Safety rules for the construction and installation of lifts*).	M (3.34a)

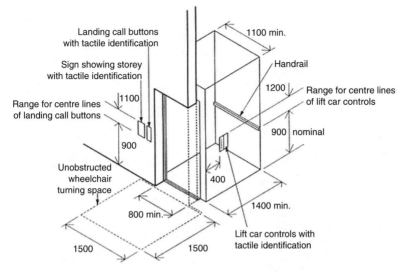

Figure 6.152 Key dimensions associated with passenger lifts

Passenger lifts should:

- be accessible from the remainder of the storey;   M (3.34b)
- have power-operated horizontal sliding doors   M (3.34e)
  which provide an effective clear width of at least
  800 mm (nominal);
- have doors fitted with timing devices (and   M (3.34f)
  re-opening activators) to allow enough time for
  people and any assistance dogs to enter or leave;
- locate the car controls between 900 mm and   M (3.34g)
  1200 mm (preferably 1100 mm) from the car
  floor and at least 400 mm from any return wall;
- locate all landing call buttons between 900 mm and   M (3.34h)
  1100 mm from the floor of the landing and at least
  500 mm from any return wall;
- be fitted with lift landing and car doors that are   M (3.34i)
  visually distinguishable from adjoining walls;
- include (in the lift car and the lift lobby) audible   M (3.31 and
  and visual information to tell them that a lift has   3.34j)
  arrived, which floor it has reached and where in a
  bank of lifts it is located;
- ensure that all glass areas can be easily identified   M (3.34k)
  by people with impaired vision;
- conform to BS 5588-8 if the lift is to be used to   M (3.34l)
  evacuate disabled people in an emergency.

If the lift is not large enough to allow a   M (3.34d)
wheelchair user to turn around within the lift car,
then a mirror should be provided that enables the
wheelchair user to see behind the wheelchair.

The minimum dimensions of the lift cars should be   M (3.34c)
1100 mm wide and 1400 mm deep (see Figure 6.152);

Visually and acoustically reflective wall surfaces   M (3.32)
should not be used.

Where possible, lift cars (used for access   M (3.33)
between two levels only) may be provided with
opposing doors to allow a wheelchair user to leave
without reversing out.

Lifting platforms

A lifting platform should only be provided to transfer wheelchair users, people with impaired mobility and their companions vertically between levels or storeys (M3.35).

Lifting platforms should:

- conform to the requirements of the Supply of   M (3.43a)
  Machinery (Safety) Regulations 1992, SI
  1992/3073, the relevant British Standards and the
  EN81 series of standards;
- restrict the vertical travel distance to: M (3.43b)
  – not more than 2 m if there is no liftway enclosure
    and/or floor penetration;
  – more than 2 m, where there is a liftway enclosure;
- restrict the rated speed of the platform so that it   M (3.43c)
  does not exceed 0.15 m/s
- locate their controls between 80 mm and 1100 mm   M (3.43d)
  from the floor of the lifting platform and at least
  400 mm from any return wall;
- locate all landing call buttons between 900 mm   M (3.43f)
  and 1100 mm from the floor of the landing and
  at least 500 mm from any return wall;
- have continuous pressure controls (e.g. push buttons); M (3.43e)

- have doors with an effective clear width of at least: M (3.43h)
  - 900 mm for an 1100 mm wide and 1400 mm deep
    lifting platform;
  - 800 mm in other cases;
- be fitted with clear instructions for use;   M (3.43i)
- have their entrances accessible from the remainder   M (3.43j)
  of the storey;
- should have doors visually distinguishable from   M (3.43k)
  adjoining walls;
- have an audible and visual announcement of platform   M (3.43l)
  arrival and level reached;
- have areas of glass that are identifiable by people   M (3.43m)
  with impaired vision.

The minimum clear dimensions of the platform should be: M (3.43g)

- 800 mm wide by 1250 mm deep (where the lifting
  platform is not enclosed and provision has been
  made for an unaccompanied wheelchair user);
- 900 mm wide by 1400 mm deep (where the lifting
  platform is enclosed and provision has been made
  for an unaccompanied wheelchair user);
- 1100 mm wide by 1400 mm deep (where two doors
  are located at 90° relative to each other, the lifting
  platform is enclosed or where provision is being made
  for an accompanied wheelchair user).

All users including wheelchair users should be able   M (3.36)
to reach and use the controls that summon and direct the
lifting platform.

Where possible, lifting platforms (used for access   M (3.41)
between two levels only) may be provided with opposing
doors to allow a wheelchair user to leave without
reversing out.

Visually and acoustically reflective wall surfaces should   M (3.42)
not be used.

Wheelchair platform stairlifts

Wheelchair platform stairlifts are only intended for the transportation of wheelchair users and should only be considered for conversions and alterations where it is not practicable to install a conventional passenger lift or a lifting platform (3.44).

Wheelchair platform stairlifts should conform to the requirements of the Supply of Machinery (Safety) Regulations 1992, SI 1992/3073 the relevant British Standards, EN81 series of standards.   M (3.49a)

Buildings with single stairways shall maintain the required clear width.   M (3.49b)

The speed of the platform should not exceed 0.15 m/s.   M (3.49c)

Continuous pressure controls (e.g. joystick) should be provided.   M (3.47 and 3.49d)

The platform should have minimum clear dimensions of 800 mm wide and 1250 mm deep.   M (3.49e)

Wheelchair platform stairlifts should:

- be fitted with clear instructions for use;   M (3.49f)
- provide an effective clear width of at least 800 mm;   M (3.49g)
- not be installed where their operation restricts the safe use of the stair by other people.   M (3.44)

Passenger lifts in blocks of flats

If a passenger lift access is installed, it should:

- be suitable for an unaccompanied wheelchair user;   M (9.4)
- have a minimum load capacity of 400 kg;   M (9.6)
- have a clear landing at least 1500 mm wide and 1500 mm long in front of its entrance;   M (9.7a)
- have a door (or doors) with a clear opening width of at least 800 mm;   M (9.7b)
- have a car at least 900 mm wide and 1250 mm long;   M (9.7c)
- have landing and car controls between 900 mm and 1200 mm above the landing and the car floor and at least 400 mm from the front wall;   M (9.7d)
- have suitable tactile indication (on the landing and adjacent to the lift call button) to identify the storey;   M (9.7e)
- have suitable tactile indication (on or adjacent to lift buttons within the car) to confirm the floor selected;   M (9.7f)
- incorporate a signalling system that provides visual notification that the lift is answering a landing call;   M (9.7g)

- have a 'dwell time' of five seconds before its doors   M (9.7g)
  begin to close after they are fully open;
- provide a visual and audible indication of the   M (9.7h)
  floor reached (when the lift serves more than
  three storeys).

Vertical circulation within the entrance storey of a dwelling

A stair providing vertical circulation within the   M (7.7a to c)
entrance/principal storey of a dwelling should have:

- flights with clear widths of at least 900 mm;
- a suitable continuous handrail on each side of the
  flight (and any intermediate landings where the rise
  of the flight comprises three or more rises);
- the rise and going in accordance with the guidance
  in Approved Document K for private stairs in
  dwelling.

6.15 Corridors and passageways

6.15.1 The requirement

In addition to the requirements of the Disability Discrimination Act 1995 precautions need to be taken to ensure that:

- *new non-domestic buildings and/or dwellings (e.g. houses and flats used for student living accommodation etc.);*
- *extensions to existing non-domestic buildings;*
- *non-domestic buildings that have been subject to a material change of use (e.g. so that they become a hotel, boarding house, institution, public building or shop);*

are capable of allowing people, regardless of their disability, age or gender, to:

- *gain access to buildings;*
- *gain access within buildings;*
- *be able to use the facilities of the buildings (both as visitors and as people who live or work in them);*
- *use sanitary conveniences in the principal storey of any new dwelling.*

(Approved Document M)

 Note: See Annex A for guidance on access and facilities for disabled people.

6.15.2 Meeting the requirement

Corridors and passageways should:

- be wide enough to allow wheelchair users, people   M (3.11)
  with buggies, people carrying cases and/or people
  on crutches to pass others on the access route;
- **not** have projecting elements such as columns,   M (3.14a)
  radiators and fire hoses;
- have an unobstructed width of at least 1200 mm.   M (3.14b)

For school buildings, the preferred corridor
width dimension is 2700 mm where there are lockers
within the corridor.

- have passing places at least 1800 mm long and   M (3.14c)
  at least 1800 mm wide at corridor junctions;
- have a floor level no steeper than 1:60;   M (3.14d)
- have an internal ramp in accordance with Table 6.42 and   M (3.14d)
  Figure 6.153 for floors with a gradient of 1:20 or steeper.

Table 6.42 Limits for ramp gradients

Going of a flight (m)	Maximum gradient	Maximum rise (mm)
10	1:20	500
5	1:15	333
2	1:12	16

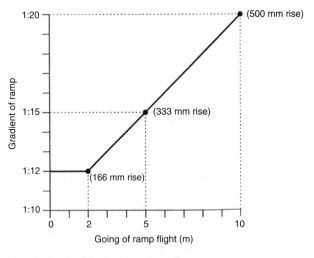

Figure 6.153 Relationship for ramp gradients

Corridors should be at least 1800 mm wide where doors from a unisex wheelchair-accessible toilet project open into that corridor.	M (3.14h)
If a section of the floor has a gradient steeper than 1:60 (but less than 1:20) it should rise no more than 500 mm without a level rest area at least 1500 mm long.	M (3.14e)
Sloping sections should extend the full width of the corridor or have the exposed edge protected by guarding.	M (3.14f)
Doors opening towards a corridor, which is a major access route or an escape route, should be recessed.	M (3.14g)
On a major access route (or an escape route) the wider leaves of double doors should be on the same side of the corridor.	M (3.14i)
The use of floor surface finishes which have patterns that could be mistaken for steps or changes of level should be avoided.	M (3.14j)
Floor finishes should be slip resistant.	M (3.14k)
Glazed screens alongside a corridor should be clearly marked (i.e. with a logo or sign) on the glass at two levels, 850 to 1000 mm and 1400 to 1600 mm above the floor.	M (3.14l)
The acoustic design should be neither too reverberant nor too absorbent.	M (3.13)

Corridors, passageways and internal doors within the entrance storey of a dwelling

The objective is to make it easy for wheelchair users, ambulant disabled people, people of either sex with babies and small children, or people with luggage to gain access to an entrance and/or principal storey of a dwelling, into habitable rooms and/or a room containing a WC on that level.

| Corridors and passageways in the entrance storey should be sufficiently wide enough for a wheelchair user to circumnavigate. | M (7.2) |
| Internal doors should be wide enough for wheelchairs to go through with ease. | M (7.4) |

Permanent obstructions in a corridor (such as a radiator) should be no longer than 2 m (provided that the width of the corridor is not less than 750 mm and the obstruction is not opposite a door to a room). M (7.2 and 7.5b)

Corridors and/or other access routes in the entrance storey should have an unobstructed width in accordance with Table 6.43 and Figure 6.154. M (7.5a)

Doors to habitable rooms and/or rooms containing a WC should have minimum clear opening widths shown in Table 6.43 and Figure 6.153. M (7.5c)

Table 6.43 Minimum widths of corridors and passageways for a range of doorway widths

Doorway clear opening width (mm)	Corridor/passageway width (mm)
750 or wider	900 (when approached head on)
750	1200 (when not approached head on)
775	1050 (when not approached head on)
800	900 (when not approached head on)

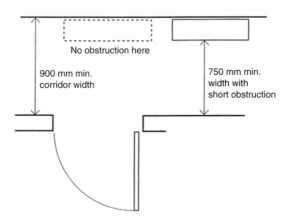

Figure 6.154 Corridors, passages and internal doors

6.16 Facilities in buildings other than dwellings

The overall aim should be that **all** people can have access to (and be able to use) **all** of the facilities provided within a building. Everyone (no matter their disability) should be able to fully participate in lecture/conference facilities as well as be able to enjoy entertainment, leisure and social venues – not just as spectators, but also as participants and/or staff. To achieve these aims:

All floor areas (even when located at different levels) should be accessible. M (4.3)

In hotels, motels and student accommodation: M (4.4)

- a proportion of the sleeping accommodation should be designed for wheelchair users;
- the remainder should include facilities suitable for people with sensory, dexterity or learning difficulties.

If there is a reception point: M (3.2 to 3.5)

- it should be easily accessible and convenient to use;
- information about the building should be clearly available from notice boards and signs;
- the floor surface should be slip resistant.

Disabled people should be able to have: M (4.2)

- a choice of seating location at spectator events;
- a clear view of the activity taking place (whilst not obstructing the view of others).

Bars and counters in refreshment areas should be at a suitable level for wheelchair users. M (4.3)

6.16.1 Audience and spectator facilities

Audience and spectator facilities fall primarily into three categories:

- lecture/conference facilities;
- entertainment facilities (e.g. theatres/cinemas);
- sports facilities (e.g. stadia).

Audience facilities generally

Wheelchair users (as well as those with mobility and/or sensory problems) may need to see or listen from a particular side, or sit at the front to lip read or read sign interpreters.

For this reason they should be provided with a selection of spaces into which they can manoeuvre easily and which offer them a clear view of an event – taking particular care that these do not become segregated into 'special areas'.

For audience seating generally

The route to wheelchair spaces should be accessible to users.	M (12a)
Stepped access routes to audience seating should be provided with fixed handrails.	M (12b)
Handrails to external stepped or ramped access should be positioned as per Figure 6.155.	M (4.12a)

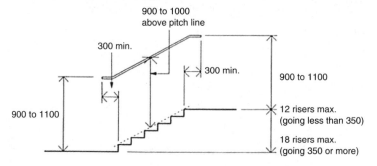

Figure 6.155 Handrails to external stepped and ramped access – key dimensions

Handrails to external stepped or ramped access should:

- be continuous across flights and landings;   M (4.12b)
- extend at least 300 mm horizontally beyond the top and bottom of a ramped access;
- not project into an access route;
- contrast visually with the background;
- have a slip resistant surface which is not cold to the touch;
- terminate in such a way that reduces the risk of clothing being caught;
- either be circular (with a diameter of between 40 and 45 mm) or oval with a width of 50 mm (see Figure 6.156).

Handrails to external stepped or ramped access should:

- not protrude more than 100 mm into the surface width   M (4.12b) of the ramped or stepped access where this would impinge on the stair width requirement of Part B1;

- have a clearance of between 60 and 75 mm between the handrail and any adjacent wall surface;
- have a clearance of at least 50 mm between a cranked support and the underside of the handrail;
- ensure that its inner face is located no more than 50 mm beyond the surface width of the ramped or stepped access;
- be spaced away from the wall and rigidly supported in a way that avoids impeding finger grip;
- be set at heights that are convenient for all users of the building.

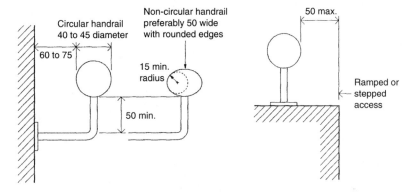

Figure 6.156 Handrail – key dimensions

Seating

The minimum number of permanent and removable spaces provided for wheelchair users is as per Table 6.44.   M (4.12c)

Table 6.44 Provision of wheelchair spaces for audience seating

Seating capacity	Minimum provision of spaces for wheelchairs	
	Permanent	Removable
Up to 600	1% of total seating capacity (rounded up)	Remainder to make a total of 6
Over 600 but less than 10 000	1% of total seating capacity (rounded up)	Additional provision if desired

Some wheelchair spaces should be provided in pairs, with   M (4.12d)
standard seating on at least one side (see Figure 6.157).

If more than two wheelchair spaces are provided, they   M (4.12e)
should be located so as to give a range of views of the event.

The minimum clear space for:

- access to wheelchair spaces should be 900 mm;   M (4.12f)
- wheelchair spaces in a parked position should be   M (4.12g)
  900 mm wide by 1400 mm deep.

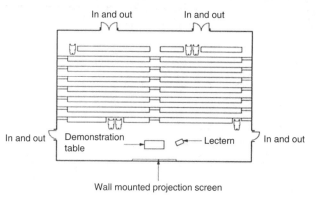

Figure 6.157 An example of wheelchair spaces in a lecture theatre

The floor of each wheelchair space should be horizontal.   M (4.12h)

Seats at the ends of rows **and** next to wheelchair spaces M (4.12j)
should have detachable (or lift-up) arms.

For seating on a stepped terraced floor

Wheelchair spaces at the back of a stepped terraced floor   M (4.12k)
should be in accordance with Figure 6.158.

Lecture/conference facilities

All people should be able to use presentation facilities.   M (4.9)

People with hearing impairments should be able to participate M (4.9)
fully in conferences, committee meetings and study groups.

The design of the acoustic environment should ensure that audible information can be heard clearly.   M (4.9)

Artificial lighting should be designed to give good colour rendering of all surfaces.   M (4.9)

Glare and reflections from shiny surfaces, and large repeating patterns, should be avoided in spaces where visual accuracy is critical.   M (4.9)

Uplighters mounted at low or floor level should be avoided as they can disorientate some visually impaired people.   M (4.9)

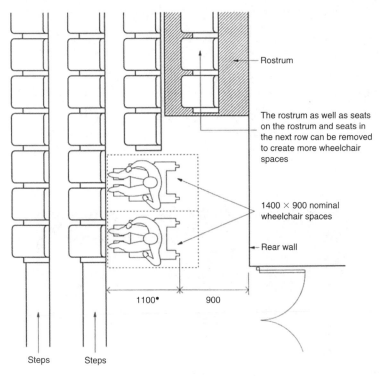

Rostrum

The rostrum as well as seats on the rostrum and seats in the next row can be removed to create more wheelchair spaces

1400 × 900 nominal wheelchair spaces

Rear wall

1100•   900

Steps   Steps

• Dimension derived from BS 8300

Figure 6.158 Possible location of wheelchair spaces in front of a rear isle

For lecture/conference facilities

Where a podium or stage is provided, wheelchair users should have access to it by means of a ramp or lifting platform.   M (4.12.1)

A clearly audible public address system should be
supplemented by visual information.

Hearing enhancement systems should be installed: M (4.12.1)

- in rooms and spaces designed for meetings, lectures,
  classes, performances, spectator sports or films;
- at service or reception counters (especially when
  situated in noisy areas or behind glazed screens).

The availability of an induction loop or infrared   M (4.12.1)
hearing enhancement system should be indicated
by the standard symbol.

Telephones suitable for hearing aid users should: M (4.12.1)

- be clearly indicated by the standard ear symbol;
- incorporate an inductive coupler and volume control.

Text telephones for deaf and hard of hearing people   M (4.12.1)
should be clearly indicated by the standard symbol.

Artificial lighting should be designed to be compatible   M (4.12.1)
with other electronic and radio frequency installations.

| Toilets are available that have been adapted for people with mobility impairments | Changing rooms are available for people with mobility impairments | Assistance dogs are welcome on the premises |

Figure 6.159 Examples of facility signs

Entertainment, leisure and social facilities

In theatres and cinemas (where seating is normally closely   M (4.10)
packed together) special care should be given to the design
and location of wheelchair spaces.

Sports facilities

 Note: See *Guide to Safety at Sports Grounds.*

Refreshment facilities

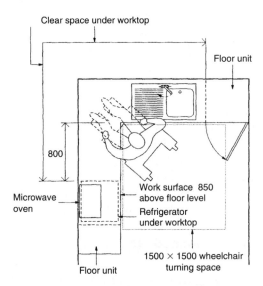

Figure 6.160 An example of a typical shared refreshment facility

Restaurants and bars should be designed so that they can be reached and used by all people independently or with companions.

All people should have access to:

• all parts of the facility;	M (4.13)
• staff areas;	M (4.13)
• public areas (e.g. lavatory accommodation, public telephones and external terraces);	M (4.14)
• self-service facilities (when provided).	M (4.14)

💡 Changes of floor level are permitted, provided that all of the different levels are accessible and raised thresholds are avoided.

Part of the working surface of a bar or serving counter should:

• be permanently accessible to wheelchair users;	M (4.16b)
• be at a level of not more than 850 mm above the floor.	M (4.16b)

Note: If unavoidable, then the total height should not be more than 15 mm, with a minimum number of upstands and slopes and with any upstands higher than 5 mm chamfered or rounded.

In addition:

- the worktop of a shared refreshment facility (e.g. for tea making) should be 850 mm above the floor with a clear space beneath at least 700 mm above the floor (see Figure 6.158);   M (4.16c)
- basin taps should either be controlled automatically or capable of being operated using a closed fist, e.g. by lever action;   M (4.16c)
- all terminal fittings should comply with Guidance Note G18.5 of the Guidance Document relating to Schedule 2: Requirements for Water Fittings, of the Water Supply (Water Fittings) Regulations 1999, SI 1999/1148.   M (4.16c)

6.16.2 Sleeping accommodation

Sleeping accommodation in hotels, motels and student accommodation, should be convenient for all types of people.   M (4.17)

A proportion of rooms should be available for wheelchair users.   M (4.17)

Wheelchair users should be able to:

- reach all the facilities available within the building;   M (4.18)
- manoeuvre around and use the facilities in the room, and operate switches and controls.   M (4.19)

En-suite sanitary facilities are the preferred option for wheelchair-accessible bedrooms.   M (4.19)

There should be at least as many en-suite shower rooms as en-suite bathrooms (i.e. some mobility-impaired people may find it easier to use a shower than a bath).   M (4.19)

In all bedrooms, built-in wardrobes and shelving should be accessible and convenient to use.   M (4.20)

Bedrooms not designed for independent use by a person in a wheelchair should have an outer door wide enough to be accessible to a wheelchair user.　　M (4.21)

For all bedrooms:

- built-in wardrobe swing doors should open through 180°;　　M (4.24b)
- handles on hinged and sliding doors:　　M (4.24c)
  - should be easy to grip and operate;
  - should contrast visually with the surface of the door;
- windows and window controls should be:　　M (4.24d)
  - located between 800 and 1000 mm above the floor;
  - easy to operate without using both hands simultaneously;
- a visual fire alarm signal should be provided in addition to the requirements of Part B;　　M (4.24e)
- room numbers should be embossed characters;　　M (4.24f)
- the effective clear width of the door from the access corridor should comply with Table 6.45.　　M (4.24a)

Table 6.45 Minimum effective clear widths of doors

Direction and width of approach	New buildings (mm)	Existing buildings (mm)
Straight-on (without a turn or oblique approach)	800	750
At right angles to an access route at least 1500 mm wide	800	750
At right angles to an access route at least 1200 mm wide	825	775
External doors to buildings used by the general public	1000	775

 For wheelchair-accessible bedrooms at least one wheelchair-accessible bedroom should be provided for every 20 bedrooms – M (4.24 g).

Wheelchair-accessible bedrooms should:

- be located on accessible routes;　　M (4.24h)
- be designed to provide a choice of location;　　M (4.24i)
- have a standard of amenity equivalent to that of other bedrooms;　　M (4.24i)
- (for en-suite bathroom and shower room doors) have an effective clear width complying with Table 6.45;　　M (4.24k)

- be large enough (see Figure 6.161) to enable a wheelchair user to manoeuvre with ease.   M (4.24l)

In wheelchair-accessible bedrooms:

- if wide angle viewers are provided in the entrance door, they should be located at 1050 mm and 1500 mm above floor level;   M (4.24n)
- if a balcony is provided, it should: M (4.24o)
  - comply with Table 6.46;
  - have a level threshold;
  - have no horizontal transoms between 900 mm and 1200 mm above the floor;
- there should be no permanent obstructions in a zone 1500 mm back from any balcony doors;   M (4.24p)

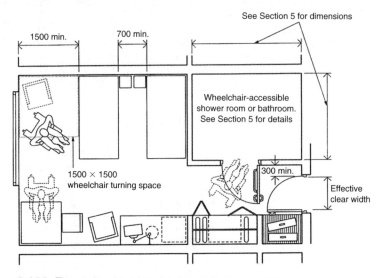

Figure 6.161 Example of a wheelchair accessible hotel bedroom

- emergency assistance alarms should be provided;   M (4.24q)
- the door from the access corridor to a wheelchair-accessible bedroom should: M (4.24j)
  - not require more than 20 N opening force;
  - have an effective clear width through a single leaf door (or one leaf of a double leaf door) in accordance with Table 6.45 and Figure 6.162;
  - have an unobstructed space of at least 300 mm on the pull side of the door – see Figure 6.162.

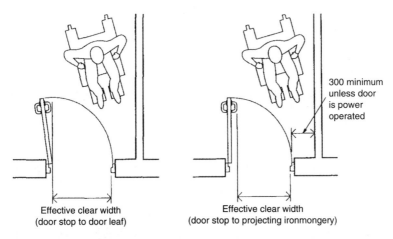

Effective clear width
(door stop to door leaf)

Effective clear width
(door stop to projecting ironmongery)

300 minimum
unless door
is power
operated

Figure 6.162 Effective clear width of doors

Sanitary facilities, en-suite to a wheelchair-accessible bedroom, should comply with the provisions of M5.15 to M5.21 for 'Wheelchair-accessible bathrooms' or 'Wheelchair-accessible shower facilities' (see pages 484 to 491).	M (4.24m)

6.16.3 Switches, outlets and controls

The aim should be to ensure that all switches, outlets and controls are easy to operate, visible and are free from obstruction.

Light switches should: • have large push pads; • align horizontally with door handles; • be within 900 to 1100 mm from the entrance door opening.	M (4.30h and I)
Switches and controls should be located between 750 mm and 1200 mm above the floor.	M (4.30c and d)
The operation of all switches, outlets and controls should not require the simultaneous use of both hands (unless necessary for safety reasons).	M (4.30j)
Switched socket outlets should indicate whether they are 'on'.	M (4.30k)
Mains and circuit isolator switches should clearly indicate whether they are 'on' or 'off'.	M (4.30l)

Individual switches on panels and on multiple socket outlets should be well separated.	M (4.29)
All socket outlets should be wall mounted.	M (4.30a and b)
All telephone points and TV sockets should be located between 400 mm and 1000 mm above the floor (or 400 mm and 1200 mm above the floor for permanently wired appliances).	M (4.30a and b)
Socket outlets should be located no nearer than 350 mm from room corners.	M (4.30g)
Controls that need close vision (e.g. thermostats) should be located between 1200 mm and 1400 mm above the floor.	M (4.30f)
Emergency alarm pull cords should be:	M (4.30e)

- coloured red;
- located as close to a wall as possible;
- have two red 50 mm diameter bangles.

Front plates should contrast visually with their backgrounds.	M (4.30m)
The colours red and green should **not** be used in combination as indicators of 'on' and 'off' for switches and controls.	M (4.28)

6.16.4 Aids to communication

The design of the acoustic environment should ensure that audible information can be heard clearly.	M (4.33)
A clearly audible public address system should be supplemented by visual information.	
Note: To assist people with impaired hearing to fully participate in public discussions (meetings and performances) may require an advanced sound level system (e.g. induction loop, infrared or radio system) to be installed.	M (4.36a)
Hearing enhancement systems should be installed:	M (4.36b)

- in all rooms and spaces designed for meetings, lectures, classes, performances, spectator sport or films;
- at service or reception counters (especially when they are are situated in noisy areas or they are behind glazed screens.

| The availability of an induction loop or infrared hearing enhancement system should be indicated by the standard symbol. | M (4.36c) |

| Telephones suitable for hearing aid users should: | M (4.36d) |

- be clearly indicated by the standard ear and 'T' symbol;
- incorporate an inductive coupler and volume control.

| Text telephones for deaf and hard of hearing people should be clearly indicated by the standard symbol. | M (4.36e) |

| Artificial lighting should be designed to be compatible with other electronic and radio frequency installations. | M (4.36f) |

| Artificial lighting should be designed to give good colour rendering of all surfaces. | M (4.34) |

| Glare and reflections from shiny surfaces, and large repeating patterns, should be avoided in spaces where visual acuity is critical. | M (4.32) |

 Uplighters mounted at low or floor level should be avoided as they can disorientate some visually impaired people. | M (4.34) |

| A hearing system is available in certain locations | Mini-com and/or text phone facility available | Staff have received disability awareness training |

Figure 6.163 Examples of facility signs

 Note: Detailed guidance on surface finishes, visual, audible and tactile signs, as well as the characteristics and appropriate choice and use of hearing enhancement systems, is available in BS 8300.

6.17 Water (and earth) closets

6.17.1 The requirement

Sanitary conveniences (provided in separate rooms or in bathrooms) shall be:

- *separated from places where food is prepared;*

- *provided with washbasins plumbed with hot and cold water;*
- *be designed and installed so as to allow effective cleaning.*

(Building Act 1984 Section 26)

All plans for buildings must include at least one (or more) water or earth closets **unless** the local authority are satisfied in the case of a particular building that one is not required (for example in a large garage separated from the house).

 If you propose using an earth closet, the local authority cannot reject the plans unless they consider that there is insufficient water supply to that earth closet.

Ventilation

Ventilation (mechanical and/or air-conditioning systems designed for domestic buildings) shall be capable of restricting the accumulation of moisture and pollutants originating within a building.

(Approved Document F)

Sanitary conveniences

All dwellings (houses, flats or maisonettes) should have at least one closet and one washbasin which should:

- *be separated by a door from any space used for food preparation or where washing-up is done in washbasins;*
- *ideally, be located in the room containing the closet;*
- *have smooth, non-absorbent surfaces and be capable of being easily cleaned;*
- *be capable of being flushed effectively;*
- *only be connected to a flush pipe or discharge pipe;*
- *washbasins should have a supply of hot and cold water.*

(Approved Document G1)

Access and facilities for disabled people

In addition to the requirements of the Disability Discrimination Act 1995 precautions need to be taken to ensure that:

- *new non-domestic buildings and/or dwellings (e.g. houses and flats used for student living accommodation etc.);*
- *extensions to existing non-domestic buildings;*
- *non-domestic buildings that have been subject to a material change of use (e.g. so that they become a hotel, boarding house, institution, public building or shop);*

are capable of allowing people, regardless of their disability, age or gender, to:

- *gain access to buildings;*
- *gain access within buildings;*
- *be able to use the facilities of the buildings (both as visitors and as people who live or work in them);*

- *use sanitary conveniences in the principal storey of any new dwelling.*

  (Approved Document M)

 Note: See Annex A for guidance on access and facilities for disabled people.

 If the proposed building is going to be used as a workplace or a factory in which persons of both sexes are going to be employed, then separate closet accommodation **must** be provided unless the local authority approve otherwise.

The Building Act 1984 Sections 64–68

Under existing regulations, all buildings (except factories and buildings used as workplaces) shall be provided with sufficient closet accommodation (or privy) according to the intended use of that building and the amount of people using that building. The only exceptions are if the building (in the view of the local authority) has an insufficient water supply and a sewer is not available.

If a building already has a sufficient water supply and sewer available, the local authority have the authority to insist that the owner of the property replaces any other closet (e.g. an earth closet) with a water closet. In these cases the owner is entitled to claim 50% of the expense of doing this off the local authority.

If the local authority completes the work, then they are entitled to claim 50% back from the owner.

 The owner of the property has **no** right of appeal in these cases.

 In the Greater London area, a 'water closet' can **also** be taken to mean a urinal.

Business premises

There may be some additional requirements regarding numbers, types and siting of appliances in business premises. If this applies to you then you will need to look at:

- the Offices, Shops and Railway Premises Act 1963,
- the Factories Act 1961, or
- the Food Hygiene (General) Regulations 1970.

6.17.2 Meeting the requirement

All dwellings (houses, flats or maisonettes) should have at least one closet and one washbasin.	G1
Closets (and/or urinals) should be separated by a door from any space used for food preparation or where washing-up is done.	G1

Washbasins should, ideally, be located in the room containing the closet, or, if not, adjacent to the WC.	G1
The surfaces of a closet, urinal or washbasin should be smooth, non-absorbent and capable of being easily cleaned.	G1
Closets (and/or urinals) should be capable of being flushed effectively.	G1
Closets (and/or urinals) should only be connected to a flush pipe or discharge pipe.	G1
Washbasins should have a supply of hot and cold water.	G1
Closets and/or urinals fitted with flushing apparatus should discharge through a trap and discharge pipe into a discharge stack or a drain.	G1
Closets fitted with macerators and pump can be connected to a discharge pipe discharging to a discharge stack if the macerator and pump system is approved under the current European Technical Approval system.	G1
Washbasins should discharge through a grating, a trap and a branch discharge pipe to a discharge stack or (if it is a ground floor location) into a gully or directly into a drain.	G1
If there is no suitable water supply or means of disposing foul water, closets (and/or urinals) can use chemical treatment.	G1

 Although the above are the minimum requirements for meeting sanitary conveniences and washing facilities, other local authority regulations may apply and it is worth seeking the advice of the local planning officer before proceeding.

Workplace conveniences

If the building is a workplace used by both sexes, then sufficient and satisfactory accommodation is required for persons of each sex.

 This requirement does not apply to premises to which the Offices, Shops and Railway Premises Act of 1963 applies.

Loan of temporary sanitary conveniences

If the local authority is maintaining, improving or repairing drainage systems and this requires the disconnection of existing buildings from these sanitary conveniences, then, on request from the occupier of the building, the local authority are required to supply (on temporary loan and at no charge) sanitary conveniences:

- if the disconnection is caused by a defect in a public sewer;
- if the local authority has order the replacement of earth closets (see above);
- for the first seven days of any disconnection.

Erection of public conveniences

You are not allowed to erect a public sanitary convenience in (or on) any location that is accessible from a street, without the consent of the local authority. Any person who contravenes this requirement is liable to a fine and can be made (at his own expense) to remove or permanently close it.

 This requirement does not apply to sanitary conveniences erected by a railway company within their railway station, yard or approaches and by dock undertakers on land belonging to them.

Protection of openings for pipes

Pipes which pass through a compartment wall or compartment floor (unless the pipe is in a protected shaft), or through a cavity barrier, should conform to one of the following alternatives:

Proprietary seals (any pipe diameter) that maintain the fire resistance of the wall, floor or cavity barrier.	B3 (11.5–11.6)
Pipes with a restricted diameter where fire-stopping is used around the pipe, keeping the opening as small as possible.	B3 (11.5 and 11.7)
Sleeving – a pipe of lead, aluminium, aluminium alloy, fibre-cement or UPVC, with a maximum nominal internal diameter of 160 mm, may be used with a sleeving of non-combustible pipe as shown in Figure 6.164.	B3 (11.5 and 11.8)

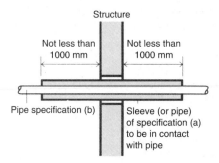

Figure 6.164 Pipes penetrating a structure

Make the opening in the structure as small as possible and provide fire-stopping between pipe and structure.

Ventilation

Buildings (or spaces within buildings) other than those:

- into which people do not normally go; or
- that are used solely for storage; or
- that are garages used solely in connection with a single dwelling;

shall be provided with ventilation to:

- extract water vapour from non-habitable areas where it is produced in significant quantities (e.g. kitchens, utility rooms and bathrooms);   F1 (1.1–1.5)
- extract pollutants (which are a hazard to health) from areas where they are produced in significant quantities (e.g. rooms containing processes that generate harmful contaminants and rest rooms where smoking is permitted);   F1 (2.3–2.5)
- rapidly dilute (when necessary) pollutants and water vapour produced in habitable rooms, occupiable rooms and sanitary accommodation;   F1 (1.1–1.4) F1 (1.6–1.8)
- provide a minimum supply of fresh air for occupants.   F1 (2.6–2.8)

Table 6.46 Ventilation of rooms containing openable windows (i.e. located on an external wall)

Room	Rapid ventilation (e.g. opening windows)	Background ventilation (mm^2)	Extract ventilation fan rates or passive stack (PSV)
Habitable room	1/20th of floor area	8000	
Kitchen	Opening window (no minimum size)	4000	30 litres/second adjacent to a hob or 60 litres/second elsewhere or PSV
Utility room	Opening window (no minimum size)	4000	30 litres/second or PSV
Bathroom (with or without WC)	Opening window (no minimum size)	4000	15 litres/second or PSV
Sanitary accommodation (separate from bathroom)	1/20th of floor area or mechanical extract at 6 litres/second	4000	

Disabled access

The aim of the amended Approved Document is that suitable sanitary accommodation should be available to everybody, including sanitary accommodation specifically designed for wheelchair users, ambulant disabled people, people of either sex with babies and small children and/or people with luggage.

Provision of toilet accommodation

Where sanitary facilities are provided in a building, at least one wheelchair-accessible unisex toilet should be available.   M (5.7b)

If there is only space for one toilet in a building, then it should be a wheelchair-accessible unisex type.   M (5.7a)

In separate-sex toilet accommodation, at least one WC cubicle should be provided for ambulant disabled people.   M (5.7c)

In separate-sex toilet accommodation having four or more WC cubicles, at least one should be an enlarged cubicle for use by people who need extra space.   M (5.7d)

Wheelchair-accessible unisex toilets should always be provided in addition to any wheelchair-accessible accommodation in separate-sex toilet washrooms.   M (5.5)

If there is only space for **one** toilet in a building:
- then it should be a wheelchair-accessible unisex type;   M (5.7a)
- its width should be increased from 1.5 m to 2 m;   M (5.7e)
- it should include a standing height wash basin, in addition to the finger rinse basin associated with the WC.   M (5.7e)

For specific guidance on the provision of sanitary accommodation in sports buildings, refer to 'Access for Disabled People'.

Sanitary accommodation generally

Doors

Doors to WC cubicles and wheelchair-accessible unisex toilets should: M (5.3)

- (ideally) open outwards;
- be operable by people with limited strength or manual dexterity;

- be capable of being opened if a person has collapsed against them while inside the cubicle.

Doors to wheelchair-accessible unisex toilets, changing rooms or shower rooms should:

- be fitted with light action privacy bolts;   M (5.4d)
- be capable of being opened using a force no greater   M (5.4d)
  than 20 N;
- have an emergency release mechanism so that they are   M (5.4e)
  capable of being opened outwards (from the outside) in
  case of emergency.

Door opening furniture should: M (5.4c)

- be easy to operate by people with limited manual dexterity;
- be easy to operate with one hand using a closed fist
  (e.g. a lever handle);
- contrast visually with the surface of the door.

Doors when open should not obstruct emergency   M (5.4f)
escape routes.

Sanitary fittings

The surface finish of sanitary fittings and grab bars should   M (5.4k)
contrast visually with background wall and floor finishes.

Taps should be operable by people with limited strength   M (5.3)
and/or manual dexterity.

Bath and wash basin taps should either be controlled   M (5.4a)
automatically or capable of being operated using a
closed fist, e.g. by lever action.

All terminal fittings should comply with Guidance Note   M (5.4b)
G18.5 of the Guidance Document relating to Schedule 2:
Requirements for Water Fittings, of the Water Supply
(Water Fittings) Regulations 1999, SI 1999/1148.

Outlets, controls and switches

The aim should be to ensure that all controls and switches   M (5.4i)
should be easy to operate, visible and free from obstruction:

- they should be located between 750 mm and 1200 mm
  above the floor;

- they should not require the simultaneous use of both hands (unless necessary for safety reasons) to operate;
- light switches should:
  - have large push pads;
  - align horizontally with door handles;
  - be within 900 to 1100 mm from the entrance door opening;
- switched socket outlets should indicate whether they are 'on';
- mains and circuit isolator switches should clearly indicate whether they are 'on' or 'off';
- individual switches on panels and on multiple socket outlets should be well separated;
- controls that need close vision (e.g. thermostats) should be located between 1200 mm and 1400 mm above the floor;
- emergency alarm pull cords should be:
  - coloured red;
  - located as close to a wall as possible;
  - have two red 50 mm diameter bangles;
- front plates should contrast visually with their backgrounds.

Heat emitters should either be screened or have their exposed surfaces kept at a temperature below 43°C.	M (5.4j)
Where possible, light switches with large push pads should be used in preference to pull cords.	M (5.3)
The colours red and green should not be used in combination as indicators of 'on' and 'off' for switches and controls.	M (5.3)

Wheelchair-accessible unisex toilets

General

Where sanitary facilities are provided in a building, at least one wheelchair-accessible unisex toilet should be available.	M (5.7b)
Wheelchair-accessible unisex toilets should **not** be used for baby changing.	M (5.5)

Wheelchair-accessible unisex toilets should:

- be located as close as possible to the entrance and/or waiting area of the building;	M (5.10a)
- not be located in a way that compromises the privacy of users;	M (5.10b)

- be located in a similar position on each floor of a multi-storey building;   M (5.10c)
- allow for right- and left-hand transfer on alternate floors;   M (5.10c and d)
- be located on accessible routes that are direct and obstruction-free;   M (5.10f)
- always be provided in addition to any wheelchair-accessible accommodation in separate-sex toilet washrooms;   M (5.5)
- not be used for baby changing.   M (5.5)

The minimum overall dimensions of, and arrangement   M (5.10i)
of fittings within, a wheelchair-accessible unisex
toilet should comply with Figure 6.165.

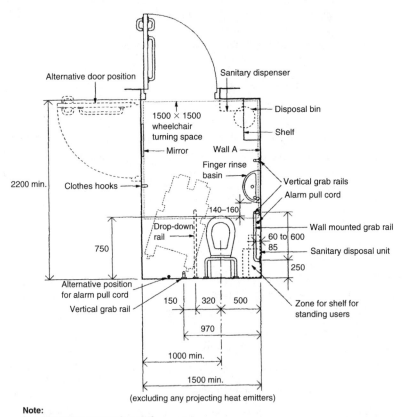

Note:
Layout for right hand transfer to WC

Figure 6.165 Example of a unisex wheelchair-accessible toilet with a corner WC

Accessibility

The approach to a unisex toilet should be separate to other sanitary accommodation.	M (5.9)
Wheelchair users should:	
• not have to travel more than 40 m on the same floor to reach a unisex toilet;	M (5.9 and 5.10h)
• not have to travel more than combined horizontal distance where the unisex toilet accommodation is on another floor of the building (accessible by passenger lift);	M (5.9 and 5.10h)
• be able to approach, transfer to, and use the sanitary facilities provided within a building.	M (5.8)

Heights and arrangements

The heights and arrangement of fittings in a wheelchair-accessible unisex toilet should comply with Figure 6.165 and (as appropriate) Figure 6.166.	M (5.10)
The space provided for manoeuvring should enable wheelchair users to adopt various transfer techniques that allow independent or assisted use.	M (5.8)
The transfer space alongside the WC should be kept clear to the back wall.	M (5.8)
The relationship of the WC to the finger rinse basin and other accessories should allow a person to wash and dry hands while seated on the WC.	M (5.8)
Heat emitters (if located) should not restrict:	
• the minimum clear wheelchair manoeuvring space;	M (5.10p)
• the space beside the WC used for transfer from the wheelchair to the WC.	

Doors

Doors should:	
• preferably open outwards;	M (5.10g)
• be fitted with a horizontal closing bar fixed to the inside face.	

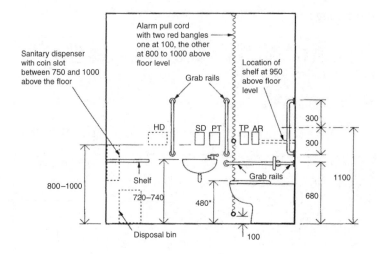

* Height subject to manufacturing tolerance of WC pan

HD: Possible position for automatic hand dryer
SD: Soap dispenser
PT: Paper towel dispenser
AR: Alarm reset button
TP: Toilet paper dispenser

* Height of drop-down rails to be the same as the other horizontal grab rails

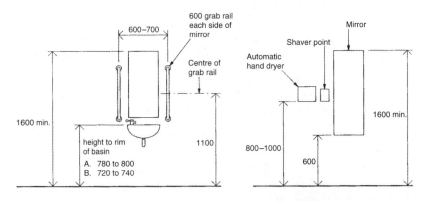

Height of independent wash basin
and location of associated fittings, for
wheelchair users and standing people

A. For people standing
B. For use from WC

Mirror located away from wash basin
suitable for seated and standing people
(Mirror and associated fittings used
within a WC compartment or serving a
range of compartments)

Figure 6.166 Typical heights of arrangement of fittings in a unisex wheelchair-accessible toilet

Support rails

The rail on the open side can be a drop-down rail, but on the wall side, it can either be a wall-mounted grab rail or, alternatively, a second drop-down rail in addition to the wall-mounted grab rail.	M (5.8)
If the horizontal support rail (on the wall adjacent to the WC) is set at the minimum spacing from the wall, an additional drop-down rail should be provided on the wall side, 320 mm from the centre line of the WC.	M (5.10j)
If the horizontal support rail (on the wall adjacent to the WC) is set so that its centre line is 400 mm from the centre line of the WC, there is no additional drop-down rail.	M (5.10k)

Emergency assistance

An emergency assistance alarm system should be provided that has:

- an outside emergency assistance call signal that can be easily seen and heard by those able to give assistance;   M (5.10n)
- visual and audible indicators to confirm that an emergency call has been received;   M (5.10m)
- a reset control reachable from a wheelchair, WC, or from a shower/changing seat;   M (5.10m)
- a signal that is distinguishable visually and audibly from the fire alarms provided.   M (5.10m)

Emergency assistance pull cords should be: M (5.10o)

- easily identifiable;
- reachable from the WC and from the floor close to the WC;
- coloured red;
- located as close to a wall as possible;
- have two red 50 mm diameter bangles.

Alarms

Fire alarms should emit an audio and visual signal to warn occupants with hearing or visual impairments.	M (5.4g)

Emergency assistance alarm systems should have: M (5.4h)

- visual and audible indicators to confirm that an
  emergency call has been received;
- a reset control reachable from a wheelchair, WC, or
  from a shower/changing seat;
- a signal that is distinguishable visually and audibly
  from the fire alarm.

WC pans

WC pans should:

- conform to BS 5503-3 or BS 5504-4;   M (5.10q)
- be able to accept a variable height toilet seat riser;   M (5.10q)
- have a flushing mechanism positioned on the open   M (5.10r)
  or transfer side of the space – irrespective of handing.

See BS 8300 for more detailed guidance on the various techniques used to transfer from a wheelchair to a WC, as well as appropriate sanitary and other fittings.

Toilets in separate-sex washrooms

General

There should be at least the same number   M (5.13)
of WCs (for women) as urinals (for men).

Ambulant disabled people should have the opportunity to use   M (5.11)
a WC compartment within any separate-sex toilet washroom.

A wheelchair-accessible compartment (where provided)   M (5.14f)
shall have the same layout and fittings as the unisex toilet.

Where a separate-sex toilet washroom can be accessed   M (5.13)
by wheelchair users, it should be possible for them
to use both a urinal (where appropriate) and a washbasin
at a lower height than is provided for other users.

Consideration should be given to providing a   M (5.13)
low level urinal for children in male washrooms.

Separate-sex toilet washrooms above a certain size should   M (5.12)
include an enlarged WC cubicle for use by people who

need extra space, e.g. parents with children and babies, people carrying luggage and also ambulant disabled people.

The minimum dimensions of compartments for ambulant   M (5.14b) disabled people should comply with Figure 6.167.

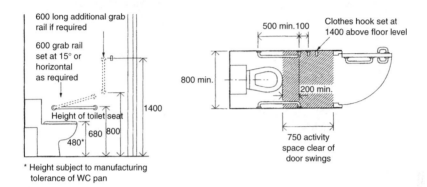

Figure 6.167 Example of a WC cubicle for an ambulant disabled person

Accessibility

The approach to a unisex toilet should be separate   M (5.14e) to other sanitary accommodation.

Wheelchair users should: M (5.14e)

- not have to travel more than 40 m on the same floor to reach a unisex toilet;
- not have to travel more than combined horizontal distance where the unisex toilet accommodation is on another floor of the building (accessible by passenger lift);
- be able to approach, transfer to, and use the sanitary facilities provided within a building.

Height and arrangements

Compartments used for ambulant disabled people should: M (5.14d)

- be 1200 mm wide;
- include a horizontal grab bar adjacent to the WC;
- include a vertical grab bar on the rear wall;
- include space for a shelf and fold-down changing table.

A wheelchair-accessible washroom (where provided)   M (5.14g)
shall have:

- at least one washbasin with its rim set at 720 to
  740 mm above the floor;
- for men, at least one urinal with its rim set at
  380 mm above the floor, with two 600 mm long
  vertical grab bars with their centre lines at 1100 mm
  above the floor, positioned either side of the urinal.

The compartment should: M (5.11)

- be fitted with support rails;
- include a minimum activity space to accommodate
  people who use crutches, or otherwise have impaired
  leg movements.

Doors

Doors to compartments for ambulant disabled people should: M (5.14d)

- preferably open outwards;
- be fitted with a horizontal closing bar fixed to the
  inside face.

The swing of any inward opening doors to standard WC   M (5.14a)
compartments should enable a 450 mm diameter manoeuvring
space to be maintained between the swing of the door,
the WC pan and the side wall of the compartment.

WC pans

WC pans should:

- conform to BS 5503-3 or BS 5504-4;   M (5.14e)
- accommodate the use of a variable height toilet seat riser. M (5.14e)

More detailed guidance on appropriate sanitary and other fittings is given in
BS 8300.

Wheelchair-accessible changing and shower facilities

A choice of shower layout together with correctly located shower controls and
fittings will enable disabled people to independently make use of the facilities –
or be assisted by others where necessary.

General

In large building complexes (e.g. retail parks and large sports centres) there should be one wheelchair-accessible unisex toilet capable of including an adult changing table.	M (5.17)
The dimensions of the self-contained compartment should allow sufficient space for a helper.	M (5.16)
A combined facility should be divided into distinct 'wet' and 'dry' areas.	M (5.16)

 Note: See Section 4.19 for provision of en-suite shower facilities in hotel bedrooms.

For changing and shower facilities

A choice of layouts suitable for left-hand and right-hand transfer should be provided when more than one individual changing compartment or shower compartment is available.	M (5.18a)
Wall mounted drop-down support rails and wall mounted slip-resistant tip-up seats (not spring-loaded) should be provided.	M (5.18b)
Subdivisions (with the same configuration of space and equipment as for self-contained facilities) should be provided for communal shower facilities and changing facilities.	M (5.18c)
In addition to communal separate-sex facilities individual self-contained shower and changing facilities should be available in sports amenities.	M (5.18d)
An emergency assistance alarm system should be provided and which should have: • visual and audible indicators to confirm that an emergency call has been received; • a reset control reachable from a wheelchair, WC, or from a shower/changing seat; • a signal that is distinguishable visually and audibly from the fire alarm.	M (5.18f)
An emergency assistance pull cord should be provided which should: • be easily identifiable and reachable from the wall mounted tip-up seat (or from the floor);	M (5.18e)

- be located as close to a wall as possible;
- be coloured red;
- have two red 50 mm diameter bangles.

Facilities for limb storage should be included for the   M (5.18)
benefit of amputees.

For changing facilities

The floor of a changing area should be level and slip   M (5.18i)
resistant when dry or when wet – particularly when
associated with shower facilities.

There should be a manoeuvring space 1500 mm deep in   M (5.18j)
front of lockers in self-contained and/or communal
changing areas.

The minimum overall dimensions of (and the arrangement   M (5.18h)
of equipment and controls within) individual self-contained
changing facilities should comply with Figure 6.168.

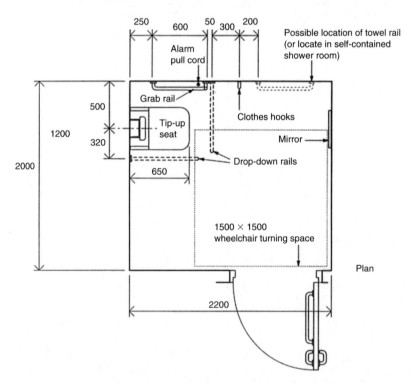

Figure 6.168 Example of a self-contained changing room for individual use

For shower facilities

A shower curtain (enclosing the seat and rails and which can be operated from the shower seat) should be provided.	M (5.18m)
A shelf (that can be reached from the shower seat and/or from the wheelchair) should be provided for toiletries.	M (5.18n)
The floor of the shower and shower area should be slip resistant and self-draining.	M (5.18o)
The shower controls should be positioned between 750 and 1000 mm above the floor in all communal area wheelchair-accessible shower facilities.	M (5.18q)
If showers are provided in commercial developments for the benefit of staff, at least one wheelchair-accessible shower compartment (complying with Figure 6.169) should be made available.	M (5.18l)
Individual self-contained shower facilities should comply with Figure 6.169.	M (5.18k)

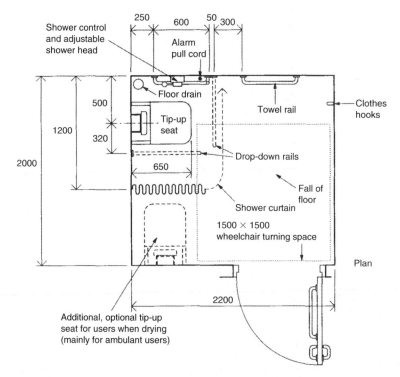

Figure 6.169 Example of a self-contained shower room for individual use

For shower facilities incorporating a WC

 A choice of left-hand and right-hand transfer   M (5.18s)
layouts should be available when more than one
shower area includes a corner WC.

The minimum overall dimensions of (and the arrangement   M (5.18r)
of fittings within) an individual self-contained shower area
incorporating a corner WC – e.g. in a sports building –
should comply with Figure 6.170.

More detailed guidance on appropriate sanitary and other fittings is given in
BS 8300.

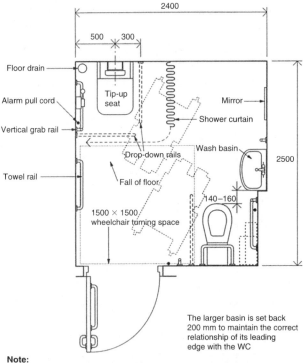

Note:
Layout shown for right-hand transfer to shower seat and WC

Figure 6.170 Example of a shower room incorporating a corner WC for
individual use

Wheelchair-accessible bathrooms

Wheelchair users and ambulant disabled people (in hotels, motels, relatives'
accommodation in hospitals and student accommodation and sports facilities)

should be able to wash or bathe either independently or with assistance from others.

The minimum overall dimensions of (and the arrangement   M (5.21a)
of fittings within) a bathroom for individual use
incorporating a corner WC should comply with Figure 6.171.

A choice of layouts suitable for left-hand and right-hand   M (5.21b)
transfer should be provided when more than one bathroom
is available.

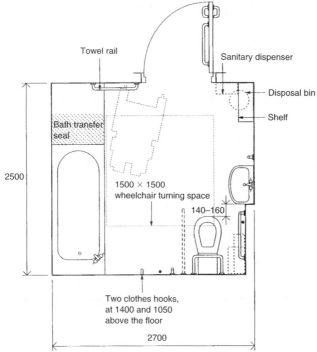

Note:
Layout shown for right-hand transfer to bath and WC.

Figure 6.171 Example of a bathroom containing a WC

The floor of a bathroom should be slip   M (5.21c)
resistant when dry or when wet.

The bath should be provided with a transfer seat that is   M (5.21d)
400 mm deep and equal to the width of the bath.

Outward opening doors, fitted with a horizontal closing   M (5.21e)
bar fixed to the inside face, should be provided.

An emergency assistance pull cord should be provided   M (5.21f)
which should:

• be easily identifiable and reachable from the wall
  mounted tip-up seat (or from the floor);
• be located as close to a wall as possible;
• be coloured red;
• have two red 50 mm diameter bangles.

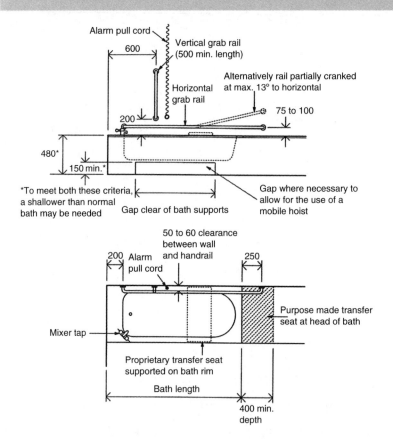

Figure 6.172 Grab rails and fitting associated with a bath

 Note:
(1) For guidance on the provision of en-suite bathrooms associated with
     hotel bedrooms, see section 4.19.
(2) More detailed guidance on appropriate sanitary and other fittings, includ-
     ing facilities for the use of mobile and fixed hoists, is given in BS 8300.
(3) Guidance on the slip resistance of floor surfaces is given in Annex C of
     BS 8300:2001.

WC provision in the entrance storey of the dwelling

Whenever possible, a WC should be provided in the entrance storey of the dwelling so that there is no need to negotiate a stair to reach it from the habitable rooms in that storey.

If there is a bathroom in the principal storey, then a WC may be collocated with it.	M (10.2)
The door to the WC compartment should:	M (10.3b)

- open outwards;
- be positioned so as to allow wheelchair users access to it;
- have a clear opening width in accordance with Table 6.47.

Table 6.47 Minimum widths of corridors and passageways for a range of doorway widths

Doorway clear opening width (mm)	Corridor/passageway width (mm)
750 or wider	900 (when approached head on)
750	1200 (when not approached head on)
775	1050 (when not approached head on)
800	900 (when not approached head on)

The WC compartment should:	M (10.3c)

- provide a clear space for wheelchair users to access the WC;
- position the washbasin so that it does not impede access.

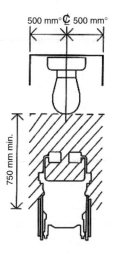

Clear space for frontal access to WC

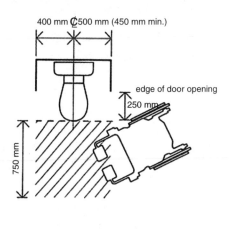

Clear space for oblique access to WC

Note: For further information see the Disability Rights Commission's website at http://www.drc-gb.org.

6.18 Electrical safety

6.18.1 The requirement

Electrical work

Reasonable provision shall be made in the design, installation, inspection and testing of electrical installations in order to protect persons from fire or injury.
(Approved Document P1)

Sufficient information shall be provided so that persons wishing to operate, maintain or alter an electrical installation can do so with reasonable safety.
(Approved Document P2)

 Note: Whilst Part P makes requirements for the safety of fixed electrical installations, this does not cover system functionality (such as electrically powered fire alarm systems, fans and pumps) which are covered in other parts of the Building Regulations and other legislation.

Conservation of fuel and power

Energy efficiency measures shall be provided which:

- *provide lighting systems that utilize energy-efficient lamps with manual switching controls or, in the case of external lighting fixed to the building, automatic switching, or both manual and automatic switching controls as appropriate, such that the lighting systems can be operated effectively as regards the conservation of fuel and power;*
- *provide information, in a suitably concise and understandable form (including results of performance tests carried out during the works), that shows building occupiers how the heating and hot water services can be operated and maintained.*

(Approved Document L1)

 Responsibility for achieving compliance with the requirements of Part L rests with the person carrying out the work. That 'person' may be, for example, a developer, a main (or sub-) contractor, or a specialist firm directly engaged by a private client.

 Note: The person responsible for achieving compliance should either themselves provide a certificate, or obtain a certificate from the sub-contractor, that commissioning has been successfully carried out. The certificate should be made available to the client and the building control body.

Access and facilities for disabled people

In addition to the requirements of the Disability Discrimination Act 1995 precautions need to be taken to ensure that:

- *new non-domestic buildings and/or dwellings (e.g. houses and flats used for student living accommodation, etc.);*
- *extensions to existing non-domestic buildings;*
- *non-domestic buildings that have been subject to a material change of use (e.g. so that they become a hotel, boarding house, institution, public building or shop);*

are capable of allowing people, regardless of their disability, age or gender to:

- *gain access to buildings;*
- *gain access within buildings;*
- *be able to use the facilities of the buildings (both as visitors and as people who live or work in them);*
- *use sanitary conveniences in the principal storey of any new dwelling.*

(Approved Document M)

 Note: See Annex A for guidance on access and facilities for disabled people.

6.18.2 Meeting the requirement

Design

Electrical installations must be designed and installed so that they:

• provide adequate protection against mechanical and thermal damage;	P0.1a
• do not present an electric shock or fire hazard to people.	P0.1a

Note: See Appendix A of Part P to the Building Regulations for details of some of the types of electrical services normally found in dwellings, some of the ways they can be connected and the complexity of wiring and protective systems that can be used to supply them.

Electricity distributors' responsibilities

Distributors are required to:

• provide an earthing facility for new connections;	P3.8
• maintain the supply within defined tolerance limits;	P3.8
• provide certain technical and safety information to consumers to enable them to design their installations.	P3.8

Distributors and meter operators must ensure that their
equipment on consumers' premises:

• is suitable for its purpose;	P3.9
• is safe in its particular environment;	P3.9
• clearly shows the polarity of the conductors.	P3.9

Distributors are prevented by the Regulations from   P3.12
connecting installations to their networks which do
not comply with BS 7671.

Distributors may disconnect consumers' installations   P3.12
which are a source of danger or cause interference
with their networks or other installations.

Note: Detailed guidance on these Regulations is available at
www.dti.gov.uk/electricity-regulations.

The electricity distributor is responsible for:

• installing the cut-out and meter in a safe location;	P1.3
• ensuring that it is mechanically protected and can be safely maintained;	P1.3
• evaluating and agreeing proposals for new installations or significant alterations to existing ones.	P1.4

Extensions, material alterations and material changes of use

Where any electrical installation work is classified as an extension, a
material alteration or a material change of use, the work must consider
and include:

• confirmation that the mains supply equipment is suitable and can carry the additional loads envisaged;	P2.1b–P2.2
• the rating and the condition of existing equipment (belonging to both the consumer and the electricity distributor) is sufficient;	P2.2a
• the amount of additions and alterations that will be required to the existing fixed electrical installation in the building;	P2.1a
• the necessary additions and alterations to the circuits which feed them;	P2.1a
• the protective measures required to meet the requirements;	P2.1a–P2.2b
• the earthing and bonding systems are satisfactory and meet the requirements.	P2.1a–P2.2c

Note: Appendix C to Part P of the Building
Regulations offers guidance on some of the older types
of installations that might be encountered during alteration
work and Appendix D provides guidance on the application
of the now harmonized European cable identification system.

Earthing

Distributors are required to provide an earthing facility for new connections.	P3.8
It is not permitted to use a gas, water or other metal service pipe as a means of earthing for an electrical installation.	P AppA 1
All electrical installations shall be properly earthed.	P AppA 1

Note: The most usual type is an electricity
distributor's earthing terminal, provided for this purpose
near the electricity meter.

New or replacement, non-metallic light fittings, switches or other components must not require earthing (e.g. non-metallic varieties) unless new circuit protective (earthing) conductors are provided.	P AppC
All lighting circuits shall include a circuit protective conductor.	P AppC
Socket outlets that will accept unearthed (2-pin) plugs must not use supply equipment needing to be earthed.	P AppC
Sensitive RCD protection is required for all socket outlets which have a rating of 32 A or less and which may be used to supply portable equipment for use outdoors.	P AppC
Where any electrical installation work is classified as an extension, a material alteration or a material change of use, the work must consider and include the earthing and bonding systems are satisfactory and meet the requirements.	P2.1a–P2.2c

Equipotential bonding conductors

Main equipotential bonding conductors are required for water service pipes, gas installation pipes, oil	P AppC

supply pipes plus certain other 'earthy' metal-work that
may be present on the premises.

The installation of supplementary equipotential bonding conductors is required for installations and locations of increased electric shock risk, such as bathrooms and shower rooms.	P AppC
The minimum size of supplementary equipotential bonding conductors (without mechanical protection) is 4 mm².	P AppC

Types of wiring or wiring system

PVC insulated and sheathed cables are likely to be suitable for much of the wiring in a typical dwelling.	P AppA 2d
Heat-resisting flexible cables are required for the final connections to certain equipment (see makers' instructions).	P AppA 2d
Cables to an outside building (e.g. garage or shed) if run underground, should be routed and positioned so as to give protection against electric shock and fire as a result of mechanical damage to a cable.	P AppA 2d
Cables concealed in floors and walls (in certain circumstances) are required to have an earthed metal covering, be enclosed in steel conduit, or have additional mechanical protection (see BS 7671 for more information).	P AppA 2d

Electrical components

All electrical work should be inspected (during installation
as well as on completion) to verify that the components have:

- been selected and installed in accordance with   P1.8a (ii)
  BS 7671;
- been made in compliance with appropriate   P1.8a (i)
  British Standards or harmonized European Standards;
- been evaluated against external influences (such   P1.8a (ii)
  as the presence of moisture);
- not been visibly damaged (or are defective) so   P1.8a (iii)
  as to be unsafe;

- been tested to check satisfactory performance   P1.8b
  with respect to continuity of conductors, insulation
  resistance, separation of circuits, polarity, earthing
  and bonding arrangements, earth fault loop impedance
  and functionality of all protective devices including
  residual current devices;
- been inspected for conformance with section   P1.9
  712 of BS 7671: 2001;
- been tested as per section 713 of BS 7671: 2001;   P1.10
- been tested using appropriate and accurate   P1.10
  instruments;
- had their test results recorded using the model   P1.10
  in Appendix 6 of BS 7671;
- had their test results compared with the relevant   P1.10
  performance criteria to confirm compliance.

Note: Inspections and testing of DIY work should
also meet the above requirements.

Electrical installation work

All electrical installation work:
- is to be carried out professionally;   P1.1
- is to comply with the Electricity at Work   P1.1
  Regulations 1989 as amended;
- is only to be carried out by persons that are   P3.4a
  competent to prevent danger and injury while
  doing it, or who are appropriately supervised.

Note: Persons installing domestic combined heat and   P3.11
power equipment must advise the local distributor of their
intentions before or at the time of commissioning the source.

Electrical installations

Electrical installations must be inspected and tested:
- during, at the end of installation and before they are   P1.6
  taken into service to verify that they are reasonably
  safe **and** that they comply with BS 7671: 2001;
- to verify that they meet the relevant equipment   P0.1b
  and installation standards.

Electrical installations:

- should be designed and installed (suitably   P1.2
  enclosed and appropriately separated) to provide
  mechanical and thermal protection;
- shall provide adequate protection for persons against   P1.2
  the risks of electric shock, burn or fire injuries;
- must meet the requirements of the Building Regulations.   P3.1

Fittings, switches and other components

- new or replacement, non-metallic light fittings,   P AppC
  switches or other components must not require
  earthing (e.g. non-metallic varieties) unless new
  circuit protective (earthing) conductors are provided;
- all lighting circuits shall include a circuit   P AppC
  protective conductor;
- older types of socket outlets designed for non-fused   P AppC
  plugs must not be connected to a ring circuit;
- socket outlets that will accept unearthed (2-pin) plugs   P AppC
  must not be used to supply equipment needing to be earthed;
- sensitive RCD protection is required for all   P AppC
  socket outlets which have a rating of 32 A or
  less and which may be used to supply portable
  equipment for use outdoors;
- a cable or stud detector shall be used when   P AppC
  attempting to drill into walls, floors or ceilings.

Consumer units

Accessible consumer units should be fitted with a   P1.5
child-proof cover or installed in a lockable cupboard.

Wall-mounted switches and socket outlets

Wall-mounted switches and socket outlets should   P1.5
comply with the requirements of Part M.

Switches and socket outlets for lighting and other   M (8.2)
equipment should be located so that they are easily reachable.

Switches and socket outlets for lighting and other   M (8.3)
equipment in habitable rooms should be located
between 450 mm and 1200 mm from finished floor
level (see Figure 6.173).

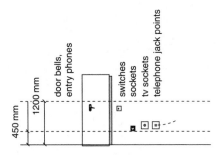

Figure 6.173 Heights of switches and sockets, etc.

 The aim is to help people with limited reach (e.g. seated in a wheelchair) to
access a dwelling's wall-mounted switches and socket outlets.

Internal lighting

The following is an indication of recommended number of locations (excluding
garages, lofts and outhouses) that need to be equipped with efficient lighting.

Number of rooms created (Hall, stairs and landing(s) count as one room as does a conservatory)	Recommended minimum number of locations
1–3	1
4–6	2
7–9	3
10–12	4

In locations, where lighting can be expected to have most   L1 (1.54)
use, fixed lighting (e.g. fluorescent tubes and compact
fluorescent lamps – but not GLS tungsten lamps with
bayonet cap or Edison screw bases – with a luminous efficacy
greater than 40 lumens per circuit watt) should be available.

External lighting fixed to the building

External lighting (including lighting in porches, but not lighting in garages and carports) should:

Automatically extinguish when there is enough daylight, and when not required at night.	L1 (1.57a)
Have sockets that can only be used with lamps having an efficacy greater than 40 lumens per circuit watt (such as fluorescent or compact fluorescent lamp types, and **not** GLS tungsten lamps with bayonet cap or Edison screw bases).	L1 (1.57b)

Portable equipment for use outdoors

Sensitive RCD protection is required for all socket outlets which have a rating of 32 A or less and which may be used to supply portable equipment for use outdoors.	P AppC

Swimming pools and saunas

Swimming pools and saunas are subject to special requirements specified in Part 6 of BS 7671: 2001.	P AppA1

6.19 Combustion appliances

6.19.1 The requirement

Protection of building

Combustion appliances and fluepipes shall be so installed, and fireplaces and chimneys shall be so constructed and installed, as to reduce to a reasonable level the risk of people suffering burns or the building catching fire in consequence of their use.

(Approved Document J3)

Provision of information

Where a hearth, fireplace, flue or chimney is provided or extended, a durable notice containing information on the performance capabilities of the hearth, fireplace, flue or chimney shall be affixed in a suitable place in the building for the purpose of enabling combustion appliances to be safely installed.

(Approved Document J4)

Conservation of fuel and power

Energy efficiency measures shall be provided which:

(a) • *limit the heat loss through the roof, wall, floor, windows and doors, etc. by suitable means of insulation;*
  • *limit unnecessary ventilation heat loss by providing building fabric which is reasonably airtight;*
  • *limit the heat loss from hot water pipes and hot air ducts used* for *space heating;*
  • *limit the heat loss from hot water vessels and their primary and secondary hot water connections by applying suitable thicknesses of insulation (where such heat does not make an efficient contribution to the space heating);*

(b) *provide space heating and hot water systems with reasonably efficient equipment such as heating appliances and hot water vessels where relevant, such that the heating and hot water systems can be operated effectively as regards the conservation of fuel and power.*

(c) *provide lighting systems that utilize energy-efficient lamps with manual switching controls or, in the case of external lighting fixed to the building, automatic switching, or both manual and automatic switching controls as appropriate, such that the lighting systems can be operated effectively as regards the conservation of fuel and power.*

(d) *provide information, in a suitably concise and understandable form (including results of performance tests carried out during the works) that shows building occupiers how the heating and hot water services can be operated and maintained.*

(Approved Document L1)

 Responsibility for achieving compliance with the requirements of Part L rests with the person carrying out the work. That person may be, for example, a developer, a main (or sub-) contractor, or a specialist firm directly engaged by a private client.

The person responsible for achieving compliance should either themselves provide a certificate, or obtain a certificate from the sub-contractor, that commissioning has been successfully carried out. The certificate should be made available to the client and the building control body.

6.19.2 Meeting the requirement

Air supplies for combustion installations

A room containing an open-flued appliance may need permanently open air vents (Figure 6.174(a) and (c)).	J (1.4)

Appliance compartments that enclose open-flued combustion   J (1.5)
appliances should be provided with vents large enough to
admit all of the air required by the appliance for combustion
and proper flue operation, whether the compartment draws its
air from a room directly from outside (Figure 6.174(b) and (c)).

Where appliances require cooling air, appliance compartments   J (1.6)
should be large enough to enable air to circulate and high and
low level vents should be provided.

 In a flueless situation, air for combustion (and to carry away its products) can
be achieved as shown in Figure 6.176.

Where appliances are to be installed within balanced   J (1.7)
compartments, special provisions will be necessary.

If an appliance is room-sealed but takes its combustion air   J (1.8)
from another space in the building (such as the roof void)
or if a flue has a permanent opening to another space in the
building (such as where it feeds a secondary flue in the roof void),
that space should have ventilation openings directly to outside.

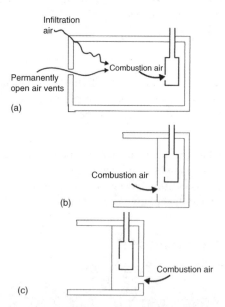

Figure 6.174 Air for combustion and operation of the flue (open flued).
(a) Appliance in room. (b) Appliance in appliance compartment with internal
vent. (c) Appliance in appliance compartment with external vent

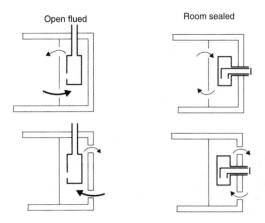

Figure 6.175 Combustion and operation requiring air cooling

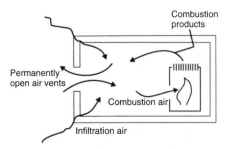

Figure 6.176 Air for combustion and operation of the flue (flueless)

Where flued appliances are supplied with combustion air   J (1.9)
through air vents which open into adjoining rooms or spaces,
the adjoining rooms or spaces should have air vent openings
of at least the same size direct to the outside. Air vents for
flueless appliances however, should open directly to the
outside air.

Air vents

Permanently open air vents should be non-adjustable,   J (1.10)
sized to admit sufficient air for the purpose intended and
positioned where they are unlikely to become blocked.

Air vents should be sufficient for the appliances to be   J (1.11)
installed (taking account where necessary of obstructions
such as grilles and anti-vermin mesh).

> Air vents should be sited outside fireplace recesses and beyond the hearths of open fires so that dust or ash will not be disturbed by draughts.   J (1.11a)
>
> Air vents should be sited in a location unlikely to cause discomfort from cold draughts.   J (1.11b)
>
> Grilles or meshes protecting air vents from the entry of animals or birds should have aperture dimensions no smaller than 5 mm.   J (1.15)
>
> In noisy areas, it may be necessary to install proprietary noise attenuated ventilators to limit the entry of noise into the building.   J (1.16)

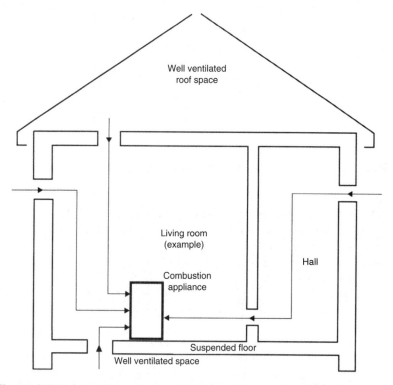

Figure 6.177 Locating permanent air vent openings (examples)

 Discomfort from cold draughts can be avoided by placing vents close to appliances (for example by using floor vents), by drawing air from intermediate spaces such as hallways or by ensuring good mixing of incoming cold air by placing air vents close to ceilings.

In buildings where it is intended to install open-flued   J (1.20)
combustion appliances and extract fans, the combustion
appliances should be able to operate safely whether or
not the fans are running.

For gas appliances where a kitchen contains an   J (1.20a)
open-flued appliance, the extract rate of the kitchen extract
fan should not exceed 20 litres/second (72 m³/hour).

When installing ventilation for solid fuel appliances   J (1.20c)
avoid installing extract fans in the same room.

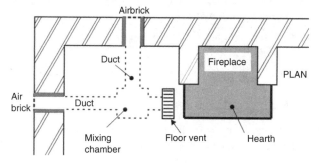

Figure 6.178 Air vent openings in a solid floor

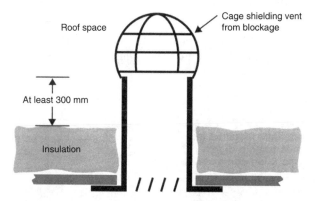

Figure 6.179 Ventilator used in a roof space (e.g. a loft)

Flues

Appliances other than flueless appliances should   J (1.24)
incorporate or be connected to suitable flues which discharge
to the outside air.

Chimneys and flues should provide satisfactory control of water condensation.	J (1.26)
New chimneys should be constructed with flue liners (clay, concrete or pre-manufactured) and masonry (bricks, medium weight concrete blocks or stone) suitable for the intended application.	J (1.27)
Liners should be installed in accordance with their manufacturer's instructions.	J (1.28)
Liners need to be placed with the sockets or rebate ends uppermost to contain moisture and other condensates in the flue.	J (1.28)
Joints should be sealed with fire cement, refractory mortar or installed in accordance with their manufacturer's instructions.	J (1.28)
Spaces between the lining and the surrounding masonry should not be filled with ordinary mortar.	J (1.28)

Space heating system controls

Heating systems other than space heating provided by individual solid fuel, gas and electric fires or room heaters should be provided with:

- zone controls,
- timing controls, and
- boiler control interlocks.

For electric storage heaters the system should have an automatic charge control that detects the internal temperature and adjusts the charging of the heater accordingly.

Zone controls

Hot water central heating systems, fan-controlled electric storage heaters and electric panel heaters should control the temperatures independently in areas (such as separate sleeping and living areas) that have different heating needs.	L1 (1.38)
Room thermostats and/or thermostatic radiator valves (or any other suitable temperature sensing devices) should be used where appropriate.	L1 (1.38)
In most dwellings one timing zone divided into two temperature control sub-zones would be sufficient.	L1 (1.39)
Large dwellings should be divided into zones with floor area no greater than $150 \, m^2$.	L1 (1.39)

Timing controls

For gas-fired and oil-fired systems and for systems with solid-fuel-fired boilers (where forced-draught fans operate when heat is required), timing devices should be provided to control the periods when the heating systems operate.	L1 (1.40)
Separate timing control should be provided for space heating and water heating, except for combination boilers or solid fuel appliances.	L1 (1.40)

Boiler control interlocks

Gas- and oil-fired hot water central heating system controls should switch the boiler off when no heat is required whether control is by room thermostats or by thermostatic radiator valves.	L1 (1.41)
The boiler in systems controlled by thermostats should operate only when a space heating or vessel thermostat is calling for heat.	L1 (1.41a)
A room thermostat or flow switch should also be provided to switch off the boiler when there is no demand for heating or hot water.	L1 (1.41b)

Hot water systems

Systems incorporating integral or separate hot water storage vessels should:

Have an insulating vessel with a 35 mm thick, factory-applied coating of PU-foam having a minimum density of 30 kg/m^3.	L1 (1.43)
Unvented hot water systems need additional insulation to control the heat losses through the safety fittings and pipework.	L1 (1.43)

Operating and maintenance instructions for heating and hot water systems

The building owner and/or occupier should be given information on the operation and maintenance of the heating and hot water systems.	L1 (1.51)

The instructions should be directly related to the system(s)   L1 (1.51)
in the dwelling and should explain to householders how to
operate the systems so that they can perform efficiently and
what routine maintenance is advisable for the purposes
of the conservation of fuel and power.

Insulation of pipes and ducts

Pipes and ducts should be insulated to conserve heat   L1 (1.52)
and hence maintain the temperature of the water or air
heating service.

Space heating pipe work located outside the building   L1 (1.52a)
fabric insulation layer(s) should be wrapped with insulation
material having a thermal conductivity at 40°C not
exceeding 0.035 W/mK and a thickness equal to the outside
diameter of the pipe up to a maximum of 40 mm.

Warm air ducts should be insulated in accordance with   L1 (1.52b)
BS 5422: 2001.

Hot pipes connected to hot water storage vessels   L1 (1.52c)
(including the vent pipe, and the primary flow and return
to the heat exchanger, where fitted) should be insulated
for at least 1 m from their points of connection (or up to
the point where they become concealed).

To protect against freezing, central heating and hot   L1 (1.53)
water pipework in unheated areas may also need
increased insulation thicknesses.

Replacement of controlled services or fittings

Heating boilers: Replacement heating boilers in   L1 (2.3a)
dwellings having a floor area greater than 50 m^2
shall be treated as if it were a new dwelling and
in the case of:

* ordinary oil or gas boilers, shall be by a boiler   L1 (2.3a(1))
  with a SEDBUK not less than the appropriate
  entry in Table 2 of L1;

- back boilers, shall be by a boiler having a SEDBUK   L1 (2.3a(2))
  of not less than three percentage points lower than
  the appropriate entry in Table 2 of L1;
- solid fuel boilers, shall be by a boiler having an   L1 (2.3a(3))
  efficiency not less than that recommended for its
  type in the HETAS certification scheme.

Hot water vessels: Replacements shall all be new   L1 (2.3c)
equipment as if for a new dwelling.

Boiler and hot water storage controls: The work may   L1 (2.3d)
also need to include replacement of the time switch
or programmer, room thermostat, and hot water vessel
thermostat, and provision of a boiler interlock and
fully pumped circulation.

6.20 Hot water storage

6.20.1 The requirement

Hot water storage

A hot water storage system shall:

- *be installed by a competent person*
- *not exceed 100°C*
- *discharge safely*
- *not cause danger to persons in or about the building.*

(Approved Document G3)

 Requirement G3 does not apply to a storage system that:

- has a storage capacity of 15 litres or less;
- is used only for providing space heating only;
- heats or stores water for industrial process.

Conservation of fuel and power

Energy efficiency measures shall be provided which:

(a) • *limit the heat loss through the roof, wall, floor, windows and doors,*
      etc. by suitable means of insulation;
    • *where appropriate permit the benefits of solar heat gains and more*
      efficient heating systems to be taken into account;
    • *limit unnecessary ventilation heat loss by providing building fabric*
      which is reasonably airtight;

- *limit the heat loss from hot water pipes and hot air ducts used for space heating;*
- *limit the heat loss from hot water vessels and their primary and secondary hot water connections by applying suitable thicknesses of insulation (where such heat does not make an efficient contribution to the space heating).*

(b) *Provide space heating and hot water systems with reasonably efficient equipment such as heating appliances and hot water vessels where relevant, such that the heating and hot water systems can be operated effectively as regards the conservation of fuel and power.*

(c) *Provide information, in a suitably concise and understandable form (including results of performance tests carried out during the works) that shows building occupiers how the heating and hot water services can be operated and maintained.*

(Approved Document L1)

 Responsibility for achieving compliance with the requirements of Part L rests with the person carrying out the work. That person may be, for example, a developer, a main (or sub-) contractor, or a specialist firm directly engaged by a private client.

The person responsible for achieving compliance should either themselves provide a certificate, or obtain a certificate from the sub-contractor, that commissioning has been successfully carried out. The certificate should be made available to the client and the building control body.

6.20.2 Meeting the requirement

Insulation of pipes and ducts

Pipes and ducts should be insulated to conserve heat and hence maintain the temperature of the water or air heating service.	L1 (1.52)
Space heating pipework located outside the building fabric insulation layer(s) should be wrapped with insulation material having a thermal conductivity at 40°C not exceeding 0.035 W/mK and a thickness equal to the outside diameter of the pipe up to a maximum of 40 mm.	L1 (1.52a)
Warm air ducts should be insulated in accordance with BS 5422: 2001.	L1 (1.52b)
Hot pipes connected to hot water storage vessels (including the vent pipe, and the primary flow and return to the heat exchanger, where fitted)	L1 (1.52c)

should be insulated for at least 1 m from their points of connection (or up to the point where they become concealed).

To protect against freezing, central heating and hot water pipework in unheated areas may also need increased insulation thicknesses.   L1 (1.53)

Replacement of controlled services or fittings

Heating boilers: Replacement heating boilers in dwellings having a floor area greater than 50 m² shall be treated as if it were a new dwelling and in the case of: L1 (2.3a)

- ordinary oil or gas boilers, shall be by a boiler with a SEDBUK not less than the appropriate entry in Table 2 of L1;   L1 (2.3a(1))
- back boilers, shall be by a boiler having a SEDBUK of not less than three percentage points lower than the appropriate entry in Table 2 of L1;   L1 (2.3a(2))
- solid fuel boilers, shall be by a boiler having an efficiency not less than that recommended for its type in the HETAS certification scheme.   L1 (2.3a(3))

Hot water vessels: Replacements shall all be new equipment as if for a new dwelling.   L1 (2.3c)

Boiler and hot water storage controls: The work may also need to include replacement of the time switch or programmer, room thermostat, and hot water vessel thermostat, and provision of a boiler interlock and fully pumped circulation.   L1 (2.3d)

6.21 Liquid fuel

6.21.1 Meeting the requirement

Storage and supply

Oil and LPG fuel storage installations (including the pipework connecting them to the combustion appliances in the buildings they serve) shall:

- be located and constructed so that they are reasonably protected from fires that may occur in buildings or beyond boundaries;   J (5.1a)

- be reasonably resistant to physical damage and   J (5.1a)
  corrosion;
- be designed and installed so as to minimize the risk   J (5.1bi)
  of oil escaping during the filling or maintenance of
  the tank;
- incorporate secondary containment when there is   J (5.1bii)
  a significant risk of pollution;
- contain labelled information on how to respond to a leak. J (5.1biii)

Oil pollution

The Control of Pollution (Oil Storage) (England) Regulations 2001 (SI 2001/2954) came into force on 1 March 2002. They apply to a wide range of

Table 6.48 Fire protection for oil storage tanks

Location of tank	Protection usually satisfactory
Within a building	Locate tanks in a place of special fire hazard, which should be directly ventilated to outside. Without prejudice to the need for compliance with all the requirements in Schedule 1, the need to comply with Part B should particularly be taken into account.

Less than 1800 mm from any part of a building	(a) Make building walls imperforate:[1]
	   (1) within 1800 mm of tanks with at least 30 minutes' fire resistance;[2]
	   (2) to internal fire and construct eaves within 1800 mm of tanks and extending 300 mm beyond each side of tanks with at least 30 minutes' fire resistance to external fire and with non-combustible cladding;
	or
	(b) Provide a firewall[3] between the tank and any part of the building within 1800 mm of the tank and construct eaves as in (a) above. The firewall should extend at least 300 mm higher and wider than the affected parts of the tank.

Less than 760 mm from a boundary	Provide a firewall between the tank and the boundary or a boundary wall having at least 30 minutes' fire resistance on either side. The firewall or the boundary wall should extend at least 300 mm higher and wider than the top and sides of the tank.
At least 1800 mm from the building and at least 760 mm from a boundary	No further provisions necessary.

Notes:
1   Excluding small openings such as air bricks etc.
2.   Fire resistance in terms of insulation, integrity and stability.
3.   Firewalls are imperforate non-combustible walls or screens, such as masonry walls or steel screens.

oil storage installations in England, but they do not apply to the storage of oil on any premises used wholly or mainly as one or more private dwellings, if the capacity of the tank is 3500 litres or less.

LPG storage

LPG installations are controlled by legislation enforced by the HSE which includes the following requirements applicable to dwellings:

The LPG tank should be installed outdoors and not within an open pit.	J (5.15)
The tank should be adequately separated from buildings, the boundary and any fixed sources of ignition to enable safe dispersal in the event of venting or leaks and in the event of fire to reduce the risk of fire spreading (see Figure 6.180).	J (5.15)
Firewalls may be free-standing built between the tank and the building, boundary and fixed source of ignition (see Figure 6.180(b)) or a part of the building or a fire resistance (insulation, integrity and stability) boundary wall belonging to the property.	J (5.16)
Where a firewall is part of the building or a boundary wall, it should be located in accordance with Figure 6.180(c).	J (5.16)
If the firewall is part of the building then it should be constructed as shown in Figure 6.180(d).	J (5.16)
Firewalls should be imperforate and of solid masonry, concrete or similar construction.	J (5.17)
Firewalls should have a fire resistance (insulation, integrity and stability) of at least 30 minutes.	J (5.17)
If firewalls are part of the building as shown in Figure 6.180(d), they should have a fire resistance (insulation, integrity and stability) of at least 60 minutes.	J (5.17)
To ensure good ventilation, firewalls should not normally be built on more than one side of a tank.	J (5.17)
A firewall should be at least as high as the pressure relief valve.	J (5.18)

 Where an LPG storage installation consists of a set of cylinders, a way of meeting the requirements is as shown in Figure 6.181.

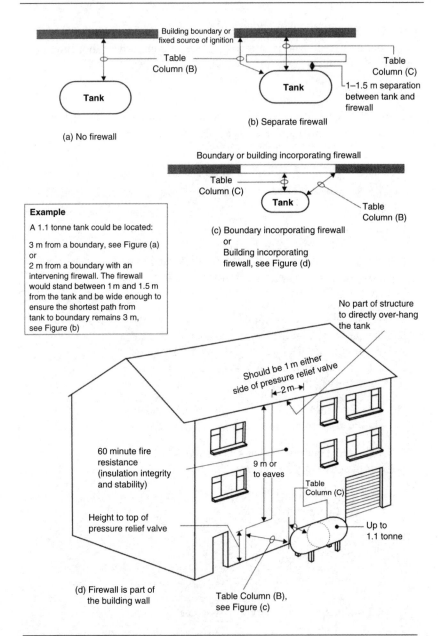

Example

A 1.1 tonne tank could be located:

3 m from a boundary, see Figure (a)
or
2 m from a boundary with an intervening firewall. The firewall would stand between 1 m and 1.5 m from the tank and be wide enough to ensure the shortest path from tank to boundary remains 3 m, see Figure (b)

A Capacity of tank (tonnes)	Minimum separation	
	B tank with no firewall	C tank shielded by firewall
0.25	2.5	0.3
1.1	3.0	1.5

Figure 6.180 Separation or shielding of LPG tanks from building, boundaries and fixed sources of ignition

Cylinders should stand upright and be secured by straps (or chains) against a wall outside the building, in a well ventilated position at ground level.	J (5.20)
Cylinders should be provided with a firm, level base such as concrete at least 50 mm thick or paving slabs bedded on mortar.	J (5.20)

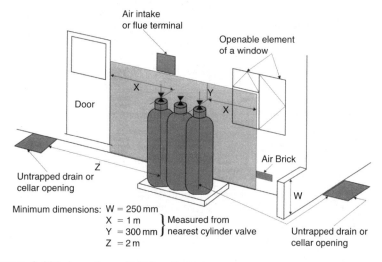

Figure 6.181 Location of LPG cylinders

6.22 Cavities and concealed spaces

6.22.1 The requirement

Internal fire spread (structure)

- *ideally the building should be sub-divided by elements of fire-resisting construction into compartments;*
- *any hidden voids in the construction shall be sealed and sub-divided to inhibit the unseen spread of fire and products of combustion, in order to reduce the risk of structural failure, and the spread of fire.*

(Approved Document B3)

6.22.2 Meeting the requirement

All concealed spaces and/or cavities formed in the construction of a building shall restrict smoke and flame spread.	B3 (10.1–10.4)

Cavities

> Cavities should be constructed to provide at least B3 (10.6)
> 30 minutes' fire resistance.

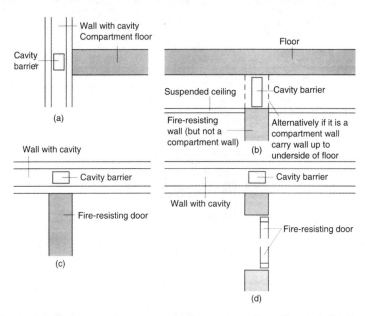

Figure 6.182 Interrupting concealed spaces and cavities. (a), (b) Sections.
(c), (d) Plans

> Cavity barriers should be tightly fitted to and B3 (10.8–10.9)
> mechanically fixed in position so that they are
> unlikely to be affected by building movement
> or failure.
>
> Any openings in a cavity barrier should be limited to B3 (10.14)
> those for:
>
> - doors that have at least 30 minutes' fire resistance;
> - the passage of pipes;
> - the passage of cables or conduits;
> - openings fitted with a suitably mounted automatic
>   fire damper;
> - ducts that (unless they are fire-resisting) are fitted
>   with a suitably mounted automatic fire damper
>   where they pass through the cavity barrier.

Table 6.49 Provision of cavity barriers

Cavity barriers to be provided	Dwelling houses	Flats or maisonettes	Other residential or institutional	Non-residential (e.g. office, shop, storage)
At the junction between an external cavity wall and a compartment wall that separates buildings; and at the top of such an external cavity wall.	X	X	X	X
Above the enclosures to a protected stairway in a house with a floor more than 4.5 m above ground level.	X			
At the junction between an external cavity wall and every compartment floor and compartment wall.		X	X	X
At the junction between a cavity wall and every compartment floor, compartment wall, or other wall or door assembly that forms a fire-resisting barrier.		X	X	X
In a protected escape route, above and below any fire-resisting construction that is not carried full storey height, or (in the case of a top storey) to the underside of the roof covering.		X	X	X
Where the corridor should be sub-divided to prevent fire or smoke affecting two alternative escape routes simultaneously above any such corridor enclosures that are not carried full storey height, or (in the case of the top storey) to the underside of the roof covering.			X	X
Above any bedroom partitions that are not carried full storey height, or (in the case of the top storey) to the underside of the roof covering.			X	
To sub-divide any cavity (including any roof space but excluding any underfloor service void).			X	X
Within the void behind the external face of rainscreen cladding at every floor level, and on the line of compartment walls abutting the external wall, of buildings that have a floor 18 m or more above ground level.		X	X	
At the edges of cavities (including around openings).	X	X	X	X

6.23 Kitchens and utility rooms

6.23.1 The requirement

Ventilation

Ventilation (mechanical and/or air-conditioning systems designed for domestic buildings) shall be capable of restricting the accumulation of moisture and pollutants originating within a building.

(Approved Document F1)

6.23.2 Meeting the requirement

Buildings (or spaces within buildings) other than those:

- into which people do not normally go; or
- that are used solely for storage; or
- that are garages used solely in connection with a single dwelling;

shall be provided with ventilation to:

• extract water vapour from non-habitable areas where it is produced in significant quantities (e.g. kitchens, utility rooms and bathrooms);	F1 (1.1–1.5)
• extract pollutants (which are a hazard to health) from areas where they are produced in significant quantities (e.g. rooms containing processes that generate harmful contaminants and rest rooms where smoking is permitted);	F1 (2.3–2.5)
• rapidly dilute (when necessary) pollutants and water vapour produced in habitable rooms, occupiable rooms and sanitary accommodation;	F1 (1.1–1.4) F1 (1.6–1.8)
• provide a minimum supply of fresh air for occupants.	F1 (2.6–2.8)

6.24 Storage of food

6.24.1 The requirement (Building Act 1984 Sections 28 and 70)

All houses or buildings that have been converted into houses must provide sufficient and suitable accommodation for storing food or '*sufficient and suitable space for the provision of such accommodation by the occupier*'.

 This could prove to be a problem when submitting plans for approval and you would be wise to consider the possibilities.

Table 6.50 Ventilation of rooms containing openable windows (i.e. located on an external wall)

Room	Rapid ventilation (e.g. opening windows)	Background ventilation (mm^2)	Extract ventilation fan rates or passive stack (PSV)
Habitable room	1/20th of floor area	8000	
Kitchen	Opening window (no minimum size)	4000	30 litres/second adjacent to a hob or 60 litres/second elsewhere or PSV
Utility room	Opening window (no minimum size)	4000	30 litres/second or PSV
Bathroom (with or without WC)	Opening window (no minimum size)	4000	15 litres/second or PSV
Sanitary accommodation (separate from bathroom)	1/20th of floor area or mechanical extract at 6 litres/second	4000	

6.25 Refuse facilities

6.25.1 The requirement (Building Act 1984 Section 23)

Probably due to new EU agreements, local authorities have now become far stricter in seeing that the requirements contained in the Building Act for storage and collection of refuse are applied. This means that you have to ensure that the building is equipped with a satisfactory method for storing refuse (with a house this would normally be a simple dustbin; with a block of flats or a factory, however, the system for storage would have to be more sophisticated). The local council will also need to be able to collect and remove this refuse easily and so this should be borne in mind when siting the refuse collection point.

Under the Building Act 1984 it is **unlawful** for any person (except with the consent of the local authority) to close or obstruct the means of access by which refuse or faecal matter is removed from a building.

Airborne and impact sound

Dwellings shall be designed so that the noise from domestic activity in an adjoining dwelling (or other parts of the building) is kept to a level that:

- *does not affect the health of the occupants of the dwelling;*
- *will allow them to sleep, rest and engage in their normal activities in satisfactory conditions.*

(Approved Document E1)

*Dwellings shall be designed so that any domestic noise that is generated inter-
nally does not interfere with the occupants' ability to sleep, rest and engage in
their normal activities in satisfactory conditions.*

(Approved Document E2)

*Domestic buildings shall be designed and constructed so as to restrict the
transmission of echoes.*

(Approved Document E3)

*Schools shall be designed and constructed so as to reduce the level of ambient
noise (particularly echoing in corridors).*

(Approved Document E4)

6.25.2 Meeting the requirement

Refuse chutes

A wall separating a habitable room or kitchen and a refuse chute should have mass (including any finishes) of at least 1320 kg/m².	E (2.28)
A wall separating a non-habitable room, which is in a dwelling, from a refuse chute should have a mass (including any finishes) of at least 220 kg/m².	E (2.28)

6.26 Fire resistance

The overall aim of fire safety precautions is to ensure that:

- a satisfactory means of giving an alarm of fire is available;
- a satisfactory means of escape for persons in the event of fire in a building is available (see Approved Document B1);
- that fire spread over the internal linings of buildings is inhibited (see Approved Document B2);
- the stability of buildings is ensured in the event of fire;
- there is a sufficient degree of fire separation within buildings and between adjoining buildings;
- the unseen spread of fire and smoke in concealed spaces in buildings is inhibited (see Approved Document B3);
- external walls and roofs have adequate resistance to the spread of fire over the external envelope;
- the spread of fire from one building to another is restricted (see Approved Document B4);
- there is satisfactory access for fire appliances to buildings;
- there are facilities in buildings to assist fire fighters in the saving of life of people in and around buildings (see Approved Document B5);

- fire stopping should be flexible to prevent a rigid contact between the pipe and the floor (see Approved Document E3 and E4);
- if a fire stop is required in the cavity between frames, then it should either be flexible or only be fixed to one frame (see Approved Document E2);
- the junction between the separating wall and the roof should be filled with a flexible closer which is also suitable as a fire stop (see Approved Document E2);
- penetrations through a separating floor by ducts and pipes should have fire protection to satisfy Building Regulation Part B – Fire safety (see Approved Document E3 and E4);
- where a staircase performs a separating function it shall conform to Building Regulation Part B – Fire safety (see Approved Document E4);
- all corridor doors shall meet the requirements for fire safety as described in Building Regulations Part B – Fire safety (see Approved Document E2);
- all doors shall satisfy the Requirements of Building Regulation Part B – Fire safety (see Approved Document E4);
- if there is an existing lath and plaster ceiling it should be retained as long as it satisfies Building Regulation Part B – Fire safety (see Approved Document E);
- the ceiling void and roof space detail can only be used where the Requirements of Building Regulation Part B – Fire safety can also be satisfied (see Approved Document E).

Many of the requirements are, of course, closely interlinked. For example, there is a close link between the provisions for means of escape (B1) and those for the control of fire growth (B2), fire containment (B3) and facilities for the fire service (B5). Similarly there are links between B3 and the provisions for controlling external fire spread (B4), and between B3 and B5. Interaction between these different requirements should be recognized where variations in the standard of provision are being considered.

Factors that should be taken into account include:

- the anticipated probability of a fire occurring;
- the anticipated fire severity;
- the ability of a structure to resist the spread of fire and smoke;
- the consequential danger to people in and around the building.

Measures that could be incorporated include:

- the adequacy of means to prevent fire;
- early fire warning by an automatic detection and warning system;
- the standard of means of escape;
- provision of smoke control;
- control of the rate of growth of a fire;
- the adequacy of the structure to resist the effects of a fire;
- the degree of fire containment;
- fire separation between buildings or parts of buildings;
- the standard of active measures for fire extinguishment or control;
- facilities to assist the fire service;

- availability of powers to require staff training in fire safety and fire routines, e.g. under the Fire Precautions Act 1971, the Fire Precautions (Workplace) Regulations 1997, or registration or licensing procedures;
- consideration of the availability of any continuing control under other legislation that could ensure continued maintenance of such systems;
- management.

 The design of fire safety in hospitals is covered by Health Technical Memorandum (HTM) 81 Fire precautions in new hospitals (revised 1996).

 Building Regulations are intended to ensure that a reasonable standard of life safety is provided, in case of fire. The protection of property, including the building itself, may require additional measures, and insurers will in general seek their own higher standards, before accepting the insurance risk. Guidance is given in the LPC *Design guide for the fire protection of buildings*.

 Guidance for assisting protection in Civil and Defence Estates is given in the *Crown Fire Standards* published by the Property Advisers to the Civil Estate (PACE).

6.26.1 The requirement

The building shall be designed and constructed so that there are appropriate provisions for the early warning of fire, and appropriate means of escape in case of fire from the building to a place of safety outside the building capable of being safely and effectively used at all material times.

(Approved Document B1)

 For a typical one- or two-storey dwelling, the requirement is limited to the provision of smoke alarms and to the provision of openable windows for emergency exit (see B1.i).

As a fire precaution, all materials used for internal linings of a building should have a low rate of surface flame spread and (in some cases) a low rate of heat release.

(Approved Document B2)

Internal fire spread (structure)

- *all loadbearing elements of structure of the building shall be capable of withstanding the effects of fire for an appropriate period without loss of stability;*
- *ideally the building should be subdivided by elements of fire-resisting construction into compartments;*
- *all openings in fire-separating elements shall be suitably protected in order to maintain the integrity of the continuity of the fire separation;*

- *any hidden voids in the construction shall be sealed and subdivided to inhibit the unseen spread of fire and products of combustion, in order to reduce the risk of structural failure, and the spread of fire.*

  (Approved Document B3)

External fire spread

- *external walls shall be constructed so that the risk of ignition from an external source, and the spread of fire over their surfaces, is restricted;*
- *the amount of unprotected area in the side of the building shall be restricted so as to limit the amount of thermal radiation that can pass through the wall;*
- *the roof shall be constructed so that the risk of spread of flame and/or fire penetration from an external fire source is restricted;*
- *the risk of a fire spreading from the building to a building beyond the boundary, or vice versa shall be limited.*

  (Approved Document B4)

Access facilities for the fire service

The following should be available:

- *vehicle access for fire appliances;*
- *access for fire-fighting personnel;*
- *the provision of fire mains within the building (for non-domestic buildings);*
- *venting for heat and smoke from basement areas.*

  (Approved Document B5)

6.26.2 Meeting the requirements

For dwellings and other small buildings, it is usually only necessary to ensure that the building is sufficiently close to a point accessible to fire brigade vehicles. In more detail this includes:

There should be vehicle access for a pump appliance to small buildings (those of up to 2000 m² with a top storey up to 11 m above ground level) to either: B4 (17.2)

- 15% of the perimeter; or
- within 45 m of every point on the projected plan area (or 'footprint') of the building;

whichever is the less onerous.

Note: For single family dwelling houses, the 45 m may be measured to a door to the dwelling.

There should be vehicle access for a pump appliance to blocks of flats/maisonettes to within 45 m of every dwelling entrance door.	B4 (17.3)
Every elevation should have a suitable door, not less than 750 mm wide, giving access to the interior of the building.	B4 (17.5)
Fire mains installed in a building should be equipped with valves, etc. so that the fire service may connect hoses for water to fight fires inside the building.	B4 (16.1)
Buildings provided with firefighting shafts should be provided with fire mains in those shafts.	B4 (16.2)
There should be one fire main in every firefighting shaft.	B4 (16.4)

Fire resistance

Table 6.51 sets out the minimum periods of fire resistance for elements of structure for domestic residential buildings (see Approved Document B Table A2).

Table 6.51 Minimum periods of fire resistance (dwellings)

Purpose group of building	Minimum periods for elements for a structure in a:					
	Basement storey (depth of lowest basement)		Ground or upper storey (height of top floor above ground)			
	>10 m	<10 m	<5 m	<18 m	<30 m	>30 m
(a), (b) and (c)						
Flats and maisonettes	90 mins	60 mins	30 mins	60 mins	90 mins	120 mins
Dwelling houses	N/A	30 mins	30 mins	60 mins	N/A	N/A

6.27 Means of escape

6.27.1 The requirement

Subject to Section 30(3) of the Fire Precautions Act 1971, if a building (or proposed building) exceeds two storeys in height and the floor of any upper storey is more than 20 ft above the surface of the street or ground on any side of the building and is:

- *let out as flats or tenement dwellings;*
- *used as an inn, hotel, boarding-house, hospital, nursing home, boarding school, children's home or similar institution; or is*

- *used as a restaurant, shop, store or warehouse and has on an upper floor sleeping accommodation for persons employed on the premises.*

then it must be equipped with adequate means of escape in case of fire, from each storey.

(Building Act 1984 Section 72)

The building shall be designed and constructed so that there are appropriate provisions for the early warning of fire, and appropriate means of escape in case of fire from the building to a place of safety outside the building capable of being safely and effectively used at all material times.

(Approved Document B1)

 For a typical one- or two-storey dwelling, the requirement is limited to the provision of smoke alarms and to the provision of openable windows for emergency exit (see B1.i).

6.27.2 Meeting the requirement

Fire detection and alarms

Dwellings of one or two storeys

Ideally dwellings should:

• be protected with an automatic fire detection and alarm system in accordance with the relevant recommendations of BS 5839: Part 1 (Fire detection and alarm systems for buildings, Code of practice for system design, installation and servicing) to at least an L3 standard, or BS 5839: Part 6 (Code of practice for the design and installation of fire detection and alarm systems in dwellings) to at least a Grade E type LD3 standard;	B1 (1.2–1.3)
• either be provided with a suitable number of mains-operated smoke alarms (conforming to BS 5446 (Components of automatic fire alarm systems for residential premises) Part 1 (Specification for self-contained smoke alarms and point-type smoke detectors) or a battery (either rechargeable or replaceable) device, or capacitor.	B1 (1.4)

Large houses

A large house of more than three storeys (including basement storeys) with any of its storeys exceeding 200 m²:

• should either be fitted with an L2 system or a Grade B type LD3 as described in BS 5839: Part 1: 1988.	B1 (1.6 and 1.7)
Lofts in a one- or two-storey house that have been converted into habitable accommodation, should have an automatic smoke detection and alarm system linked to the main house alarm system.	B1 (1.8)

Flats and maisonettes

The same principles apply within flats and maisonettes as for houses, while noting that:

• a flat with accommodation on more than one level (i.e. a maisonette) should be treated in the same way as a house with more than one storey;	B1 (1.9)
• where groups of students share one flat with its own entrance door, an automatic detection system should be provided within each flat.	B1 (1.9)

Smoke alarms – dwellings

Smoke alarms should normally be positioned in the circulation spaces between sleeping spaces and places where fires are most likely to start (e.g. kitchens and living rooms) to pick up smoke in the early stages, while also being close enough to bedroom doors for the alarm to be effective when occupants are asleep.

Requirement

There should be at least one smoke alarm on every storey of a house (including bungalows).	B1 (1.12)
If more than one smoke alarm is installed in a dwelling then they should be linked so that if a unit detects smoke it will operate the alarm signal of all the smoke detectors.	B1 (1.13)
There should be a smoke alarm in the circulation space within 7.5 m of the door to every habitable room.	B1 (1.14a)

If a kitchen area is not separated from the stairway or circulation space by a door, there should also be an additional heat detector in the kitchen, that is interlinked to the other alarms.

B1 (1.14b)

Installation

Smoke alarms should ideally be ceiling mounted, 25 mm and 600 mm below the ceiling (25–150 mm in the case of heat detectors) and at least 300 mm from walls and light fittings.

B1 (1.14c–d)

Units designed for wall mounting may also be used provided that the units are above the level of doorways opening into the space.

Smoke alarms should not be fixed over a stair shaft or any other opening between floors.

B1 (1.15)

Smoke alarms should not be fixed next to or directly above heaters or air conditioning outlets.

B1 (1.16)

Smoke alarms should not be fixed in bathrooms, showers, cooking areas or garages, or any other place where steam, condensation or fumes could give false alarms.

B1 (1.16)

Smoke alarms should not be fitted in places that get very hot (such as a boiler room), or very cold (such as an unheated porch).

B1 (1.16)

Smoke alarms should not be fixed to surfaces that are normally much warmer or colder than the rest of the space.

B1 (1.16)

Power supplies

The power supply for a smoke alarm system:

- should be derived from the dwelling's mains electricity supply from a single independent circuit at the dwelling's main distribution board (consumer unit);

B1 (1.17)

- should include a stand-by power supply that will operate during mains failure;

B1 (1.17–18)

- should preferably not be protected by any residual current device (rcd).

B1 (1.20)

Smoke alarms may be interconnected using radio-links, provided that this does not reduce the lifetime or duration of any stand-by power supply.

Smoke alarms – other than dwellings

The type of fire alarm/smoke detection system required for a particular building depends on the type of occupancy and the means of escape strategy (e.g. simultaneous, phased or progressive horizontal evacuation) – (B1 (1.23)).

Fire alarms

All fire detection and fire-warning systems shall be properly designed, installed and maintained (B1 (1.32)).

All buildings should have arrangements for detecting fire.	B1 (1.25)
All buildings should have the means of raising an alarm in case of fire (e.g. rotary gongs, handbells or shouting 'fire') or be fitted with a suitable electrically operated fire warning system (in compliance with BS 5839).	B1 (1.26)
The fire warning signal should be distinct from other signals that may be in general use.	B1 (1.28)
In premises used by the general public, e.g. large shops and places of assembly, a staff alarm system (complying with BS 5839) may be used.	B1 (1.29)

Dwellings

In the event of fire, suitable means shall be provided for emergency egress from each storey.	B1 (2.1)
Floors more than 7.5 m above ground level (where the risk that the stairway will become impassable before occupants of the upper parts of the house have escaped is appreciable) will be provided with an alternative route.	B1 (2.1)
Except for kitchens, all habitable rooms in the upper storey(s) of a house served by only one stair should be provided with a window (or external door) that could be used as an emergency exit.	B1 (2.7)
Except for kitchens, all habitable rooms in the ground storey should either: • open directly onto a hall leading to the entrance or other suitable exit; or • be provided with a window (or door).	B1 (2.8)

Where a sleeping gallery is provided: B1 (2.9a–c)

- the gallery should not be more than 4.5 m above ground level;
- the distance between the foot of the access stair to the gallery and the door to the room containing the gallery should not exceed 3 m;
- galleries longer than 7.5 m should be provided with a separate alternative exit.

Any cooking facilities within a room containing   B1 (2.9d)
a gallery should either:

- be enclosed with fire-resisting construction; or
- be positioned so that they do not prejudice escape from the gallery.

Inner rooms

A room whose only escape route is through another room is termed an inner room and is at risk if a fire starts in that other room (access room). Such an arrangement is only acceptable where the inner room is:

- a kitchen;
- a laundry or utility room;
- a dressing room;
- a bathroom, WC, or shower room;
- any other room on a floor not more than 4.5 m above ground level.

Balconies and flat roofs

A flat roof being used as a means of escape should:

- be part of the same building from which escape is being made;
- should lead to a storey exit or external escape route;
- should provide 30 minutes' fire resistance.

 Where a balcony or flat roof is provided for escape purposes, guarding may be needed.

- all external balconies and edges of roofs shall have   K3 (3.2)
  a wall, parapet, balustrade or similar guard at least
  1100 mm high.

Basements

If the basement storey contains a habitable room it should have either:

- an external door or window suitable for emergency egress; or
- a protected stairway leading from the basement to a final exit.

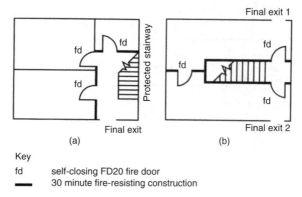

Key

fd   self-closing FD20 fire door
━━   30 minute fire-resisting construction

Figure 6.183 Alternative arrangements for final exits. In loft conversions, existing doors need only be made self-closing

Emergency egress windows and external doors

> Any window provided for emergency egress purposes   B1 (2.11)
> or external door provided for escape should have an
> unobstructed openable area of at least 0.33 m² that is a
> minimum of 450 mm high, 450 mm wide and not more
> than 1100 mm from the floor (see also Figure 6.184).

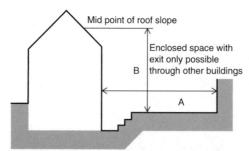

Depth of back garden A should exceed the height B of the house
above ground level for it to be acceptable for an escape route
from the ground or basement storey to open into the garden.

Figure 6.184 Ground or basement storey exit into an enclosed space

Houses with floors more than 4.5 m above ground level

> The top storey should be separated from the lower storeys   B1 (2.13b)
> by fire-resisting construction and be provided with an
> alternative escape route leading to its own final exit.
>
> If a house has two or more storeys with floors   B1 (2.14)
> more than 4.5 m above ground level (typically a house
> of four or more storeys), then an alternative escape
> route should be provided from each storey or level
> situated 7.5 m or more above ground level.

Means of escape

Loft conversions

Where an existing two-storey house is being enlarged by converting the existing roof space into habitable rooms then, provided the new second storey does
not exceed 50 m² in floor area, or contain more than two habitable rooms, then:

> - the stairs in the ground and first storeys should be   B1 (2.18a)
>   enclosed with walls and/or partitions that are
>   fire-resisting;
>
> - the enclosure should either: B1 (2.18b)
>   – extend to a final exit, see Figure 6.183(a); or
>   – give access to at least two escape routes at
>   ground level.

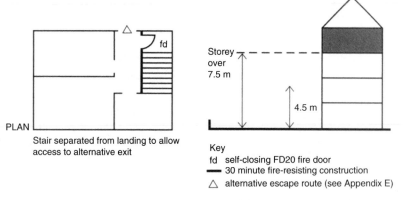

PLAN

Stair separated from landing to allow
access to alternative exit

Storey
over
7.5 m

4.5 m

Key
fd   self-closing FD20 fire door
▬ 30 minute fire-resisting construction
△ alternative escape route (see Appendix E)

Figure 6.185 Fire separation in houses with more than one floor over
4.5 m above ground level

- every doorway within the enclosure to the existing stair   B1 (2.19)
  should be fitted with a door, which (in the case of
  doors to habitable rooms) should be fitted with a
  self-closing device;

- any new door to a habitable room should be a fire door;   B1 (2.19)

- any glazing (whether new or existing) in the   B1 (2.20)
  enclosure to the existing stair, including all doors
  (whether or not they need to be fire doors), but
  excluding glazing to a bathroom or WC, should be
  fire-resisting and retained by a suitable glazing
  system and beads compatible with the type of glass;

- the new storey should be served by a stair that is   B1 (2.21)
  an extension of an existing stairway;

- the new storey should be separated from the rest of   B1 (2.22)
  the house by a fire-resistant door;

- the room (or rooms) in any new storey should each   B1 (2.24)
  have an openable window or rooflight that meets the
  relevant provisions in Figure 6.186. A door to a roof
  terrace is also acceptable.

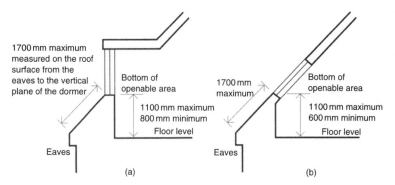

Note 1: The window or rooflight should have a clear opening which complies with paragraph 2.11a of Approved Document B and quoted on page 523.

Note 2: It is not considered necessary for the window in (b) to be provided with safety glazing.

Figure 6.186 Position of dormer window or rooflight that is suitable for emergency exit from a loft conversion of a two-storey dwelling house. (a) Dormer window (the window may be in the end wall of the house, instead of the roof as shown). (b) Rooflight or roof window

Flats and maisonettes

The means of escape from a flat or maisonette 4.5 m above ground level are very similar, in most respects, to those for dwellinghouses. Emphasis is, however, made on the necessity to ensure that means are provided for early warning in the event of fire and that suitable means are available for emergency escape. For complete details of additional requirements, see Approved Document B Sections 3 and 6.

Buildings other than dwellings

The prime differences between the design of buildings other than dwellings, flats and/or maisonettes, is the necessity to address the availability of horizontal as well as vertical escape.

Horizontal escape

The general principle to be followed when designing facilities for means of escape is that any person confronted by an outbreak of fire within a building can turn away from it and make a safe escape. Full details of the requirements for horizontal escape are contained in B2 Section 4. The following are intended to provide an indication of these requirements.

The number of escape routes and exits to be provided depends on the number of occupants in the room, tier or storey in question.	B1 (4.2 and 4.7)
Separate means of escape should be provided from any storeys (or parts of storeys) used for residential or assembly and recreation purposes.	B1 (4.4)
In order to avoid occupants being trapped by fire or smoke, there should be alternative escape routes from all parts of the building.	B1 (4.5)
The occupant capacity of the inner room (i.e. a room from which the only escape route is through another room) should not exceed 60.	B1 (4.9a)
Buildings with more than one exit in a central core should be planned so that storey exits are remote from one another, and no two exits are approached from the same lift hall, common lobby or undivided corridor.	B1 (4.10)
All escape routes should have a clear headroom of not less than 2 m except in doorways.	B1 (4.15)
Whilst the width of escape routes and exits depends on the number of persons needing to use them, they should never be less than 750 mm wide.	B1 (4.16–18)
Every corridor more than 12 m long that connects two or more storey exits, should be sub-divided by self-closing fire doors.	B1 (4.23)

> Unless the escape stairway(s) and corridors are   B1 (4.24)
> protected by a pressurization system complying with
> BS 5588, every dead-end corridor exceeding 4.5 m in
> length should be separated by self-closing fire doors.

> Where an external escape route (other than a stair) is   B1 (4.27)
> beside an external wall of the building, that part of the
> external wall within 1800 mm of the escape route
> should be of fire-resisting construction, up to a height
> of 1100 mm above the paving level of the route.

Vertical escape

The availability of adequately sized and protected escape stairs as a means of escape in multi-storey buildings (other than dwellinghouses, flats and/ or maisonettes) needs careful consideration. The actual number, width and protection of escape stairs needed in a building (or part of a building) will vary according to the constraints imposed by the design of horizontal escape routes, whether a single stair is acceptable and whether independent stairs are required in mixed occupancy buildings, etc. (for full details see B1 Sections 5 and 6).

Uninsulated glazed elements on escape routes

Table 6.52 sets out limitations on the use of uninsulated fire-resisting glazed elements. These limitations do not apply to the use of insulated fire-resisting glazed elements.

6.28 Bathrooms

6.28.1 The requirement

All dwellings (whether they are a house, flat or maisonette) should have at least one bathroom with a fixed bath or shower, and the bath or shower should be equipped with hot **and** cold water. This ruling applies to all plans for:

- new houses;
- new buildings, part of which are going to be used as a dwelling;
- existing buildings that are going to be converted, or partially converted into dwellings.

(Building Act 1984 Section 27)

Table 6.52 Limitations on the use of uninsulated glazed elements on escape routes

| Position of glazed element | Maximum total glazed area in parts of the building with access to | | | |
| | A single stairway | | More than one stairway | |
	Walls	Door leaf	Walls	Door leaf
Single family dwelling houses				
1 (a) Within the enclosures of	Fixed fanlights only	Unlimited	Fixed fanlights only	Unlimited
(i) protected stairway				
(ii) existing stair	Unlimited	Unlimited	Unlimited	Unlimited
(b) Within fire resisting separation	100 mm from floor	100 mm from floor	100 mm from floor	100 mm from floor
(c) Existing window between an attached/ integral garage and the house	Unlimited	N/A	Unlimited	N/A
Flats and maisonettes				
2 Within the enclosures of a protected entrance hall or protected landing	Fixed fanlights only	1100 mm from floor	Fixed fanlights only	1100 mm from floor

Ventilation

Ventilation (mechanical and/or air-conditioning systems designed for domestic buildings) shall be capable of restricting the accumulation of moisture and pollutants originating within a building.

(Approved Document F1)

Sanitary conveniences

All dwellings (houses, flats or maisonettes) should have at least one closet and one washbasin which should:

- *be separated by a door from any space used for food preparation or where washing-up is done in washbasins;*
- *ideally, be located in the room containing the closet;*
- *have smooth, non-absorbent surfaces and be capable of being easily cleaned;*
- *be capable of being flushed effectively;*
- *only be connected to a flush pipe or discharge pipe;*
- *washbasins should have a supply of hot and cold water.*

(Approved Document G1)

All dwellings (house, flat or maisonette) should have at least one bathroom with a fixed bath or shower.

(Approved Document G3)

6.28.2 Meeting the requirement

All dwellings (houses, flats or maisonettes) should have at least one bathroom with a fixed bath or shower and the bath or shower should:

• have a supply of hot and cold water;	G2
• discharge through a grating, a trap and branch discharge pipe to a discharge stack or (if on a ground floor) discharge into a gully or directly to a foul drain;	G2 G2 H1
• be connected to a macerator and pump (of an approved type) if there is no suitable water supply or means of disposing foul water.	G2

 Requirement G2 only applies to dwellings.

Bathrooms shall have either a fixed bath or shower bath that is provided with hot and cold water and connected to a foul water drainage system.

Ventilation

Buildings (or spaces within buildings) other than those:

- into which people do not normally go; or
- that are used solely for storage; or
- that are garages used solely in connection with a single dwelling;

shall be provided with ventilation to:

• extract water vapour from non-habitable areas where it is produced in significant quantities (e.g. kitchens, utility rooms and bathrooms);	F1 (1.1–1.5)
• extract pollutants (which are a hazard to health) from areas where they are produced in significant quantities (e.g. rooms containing processes that generate harmful contaminants and rest rooms where smoking is permitted);	F1 (2.3–2.5)
• rapidly dilute (when necessary) pollutants and water vapour produced in habitable rooms, occupiable rooms and sanitary accommodation;	F1 (1.1–1.4) F1 (1.6–1.8)
• provide a minimum supply of fresh air for occupants.	F1 (2.6–2.8)

Table 6.53 Ventilation of rooms containing openable windows (i.e. located on an external wall)

Room	Rapid ventilation (e.g. opening windows)	Background ventilation (mm²)	Extract ventilation fan rates or passive stack (PSV)
Habitable room	1/20th of floor area	8000	
Kitchen	Opening window (no minimum size)	4000	30 litres/second adjacent to a hob or 60 litres/second elsewhere or PSV
Utility room	Opening window (no minimum size)	4000	30 litres/second or PSV
Bathroom (with or without WC)	Opening window (no minimum size)	4000	15 litres/second or PSV
Sanitary accommodation (separate from bathroom)	1/20th of floor area or mechanical extract at 6 litres/second	4000	

Washbasins

Washbasins should have a supply of hot and cold water.	G1
Washbasins should discharge through a grating, a trap and a branch discharge pipe to a discharge stack or (if it is a ground floor location) into a gully or directly into a drain.	G1

6.29 Loft conversions

6.29.1 The requirements

Stairs, ladders and ramps

All stairs, steps and ladders shall provide reasonable safety between levels in a building.

(Approved Document K1)

Protection from falling

Pedestrian guarding should be provided for any part of a floor (including the edge below an opening window) gallery, balcony, roof (including rooflight and other openings), any other place to which people have access and any light well, basement area or similar sunken area next to a building.

(Approved Document K2)

 Requirement K2 (a) applies only to stairs and ramps that form part of the building.

6.29.2 Meeting the requirements

Loft conversions

Where an existing two-storey house is being enlarged by converting the existing roof space into habitable rooms then, provided the new second storey does not exceed 50 m² in floor area, or contain more than two habitable rooms, then:

- the stairs in the ground and first storeys should be enclosed with walls and/or partitions which are fire-resisting;   B1 (2.18a)

- the enclosure should either: B1 (2.18b)
  − extend to a final exit, see Figure 6.182(a); or
  − give access to at least two escape routes at ground level

- every doorway within the enclosure to the existing stair should be fitted with a door, which (in the case of doors to habitable rooms) should be fitted with a self-closing device;   B1 (2.19)

- any new door to a habitable room should be a fire door;   B1 (2.19)

- any glazing (whether new or existing) in the enclosure to the existing stair, including all doors (whether or not they need to be fire doors), but excluding glazing to a bathroom or WC, should be fire-resisting and retained by a suitable glazing system and beads compatible with the type of glass;   B1 (2.20)

- the new storey should be served by a stair that is an extension of an existing stairway;   B1 (2.21)

- the new storey should be separated from the rest of the house by a fire resistant door;   B1 (2.22)

- the room (or rooms) in any new storey should each have an openable window or rooflight that meets the relevant provisions in Figure 6.187. A door to a roof terrace is also acceptable.   B1 (2.24)

Stairs, ladders and ramps

The rise of a stair shall be between 155 mm and 220 mm with any going between 245 mm and 260 mm and a maximum pitch of 42°.   K1 (1.1–1.4)

☀ The normal relationship between the dimensions of the rise and going is that twice the rise plus the going (2R + G) should be between 550 mm and 700 mm.

Stairs with open risers that are likely to be used by children under five years should be constructed so that a 100 mm diameter sphere cannot pass through the open risers.	K1 (1.9)
Stairs that have more than 36 risers in consecutive flights should make at least one change of direction, between flights, of at least 30°.	K1 (1.14)
If a stair has straight and tapered treads, then the going of the tapered treads should not be less than the going of the straight tread.	K1 (1.20)
The going of tapered treads should measure at least 50 mm at the narrow end.	K1 (1.18)
The going should be uniform for consecutive tapered treads.	K1 (1.19) K1 (1.22–1.24)
Stairs should have a handrail on both sides if they are wider than 1 m and on at least one side if they are less than 1 m wide.	K1 (1.27)
Handrail heights should be between 900 mm and 1000 mm measured to the top of the handrail from the pitch line or floor.	K1 (1.27)
Spiral and helical stairs should be designed in accordance with BS 5395.	K1 (1.21)

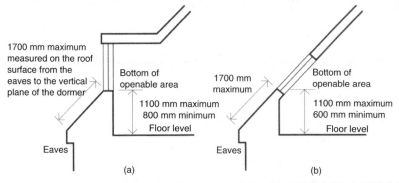

Note 1: The window or rooflight should have a clear opening which complies with paragraph 2.11a of Approved Document B and quoted on page 523.

Note 2: It is not considered necessary for the window in (b) to be provided with safety glazing.

Figure 6.187 Position of dormer window or rooflight that is suitable for emergency purposes from a loft conversion of a two-storey dwelling house. (a) Dormer window (the window may be in the end wall of the house, instead of the roof as shown). (b) Rooflight or roof window

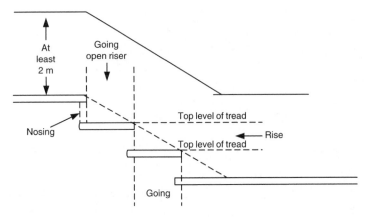

Figure 6.188 Rise and going plus headroom

Steps

• Steps should have level treads.	K1 (1.8)
• Steps may have open risers, but treads should then overlap each other by at least 16 mm.	K1 (1.8)
Steps should be uniform with parallel nosings, the stair should have handrails on both sides and the treads should have slip-resistant surfaces.	
• The headroom on the access between levels should be no less than 2 m.	K1 (1.10)
• Landings should be provided at the top and bottom of every flight.	K1 (1.15)
• The width and length of every landing should be the same (or greater than) the smallest width of the flight.	K1 (1.15)
• Landings should be clear of any permanent obstruction.	K1 (1.16)
• Landings should be level.	K1 (1.17)
• Any door (entrance, cupboard or duct) that swings across a landing at the top or bottom of a flight of stairs must leave a clear space of at least 400 mm across the full width of the flight.	K1 (1.16)
• Flights and landings should be guarded at the sides when there is a drop of more than 600 mm.	K1 (1.28–1.29)

For stairs that are likely to be used by children
under five years the construction of the guarding
shall be such that a 100 mm sphere cannot pass
through any openings in the guarding and
children will not easily climb the guarding.

For loft conversions, a fixed ladder should have fixed   K1 (1.25)
handrails on both sides.

 Whilst there are no recommendations for minimum stair widths, designers
should bear in mind the requirements of Approved Documents B (means of
escape) and M (access for disabled people).

Protection from falling

All stairs, landings, ramps and edges of internal floors shall have a wall, parapet, balustrade or similar guard at least 900 mm high.	K3 (3.2)
All guarding should be capable of resisting at least the horizontal force given in BS 6399: Part 1: 1996.	K3 (3.2)
If glazing is used as (or part of) the pedestrian guarding, see Approved Document N: Glazing – safety in relation to impact, opening and cleaning.	N
If a building is likely to be used by children under five years, the guarding should not have horizontal rails, should stop children from easily climbing it, and the construction should prevent a 100 mm sphere being able to pass through any opening of that guarding.	K3 (3.3)
All external balconies and edges of roofs shall have a wall, parapet, balustrade or similar guard at least 1100 mm high.	K3 (3.2)

Requirement K2 (a) applies only to stairs and ramps that form part of the
building.

6.30 Entrance and access

6.30.1 The requirement (Building Act 1984 Sections 24 and 71)

The Building Act is very specific about exits, passageways and gangways and
local authorities are required to consult with the fire authority to ensure that

proposed methods of ingress and egress are deemed satisfactory (depending on the type of building). The purpose for which the building is going to be used needs to be considered as each case can be different. In particular, is the building going to be:

- a theatre, hall or other public building that is used as a place of public resort;
- a restaurant, shop, store or warehouse to which members of the public are likely to be admitted;
- a club;
- a school;
- a church, chapel or other place of worship (erected or used after the Public Health Acts Amendment Acts 1890 came into force).

At all times, the means of ingress and egress and the passages and gangways, while persons are assembled in the building, are to be kept free and unobstructed.

You are required by the Building Act 1984 to ensure that all courts, yards and passageways giving access to a house, industrial or commercial building (not maintained at the public expense) are capable of allowing satisfactory drainage of its surface or subsoil to a proper outfall.

The local authority can require the owner of any of the buildings to complete such works as may be necessary to remedy the defect.

All entrances to courts and yards must allow the free circulation of air and are not allowed to be closed, narrowed, reduced in height or in any other way altered so that it can impede the free circulation of air through the entrance.

(Building Act 1984 Section 85)

Private houses are not restricted by the actual Building Act of 1984 provided that members of the public are only admitted occasionally or exceptionally.

Vehicle barriers and loading bays

- *Vehicle barriers should be provided that are capable of resisting or deflecting the impact of vehicles.*
- *Loading bays shall be provided with an adequate number of exits (or refuges) to enable people to avoid being crushed by vehicles.*

(Approved Document K3)

Disabled people

During 2002/3, Approved Document M was thoroughly overhauled and restructured in order to meet the changed requirements of the Disability Discrimination Act 1995 (which are enforced by the *Discrimination Act 1995 (Amendment) Regulations 2003* (SI 2003/1673)). Part M now covers:

- the use of a building to disabled people (redefined to include parents with children, elderly people and people with all types of disabilities – such as mobility, sight and hearing etc.) whether as residents, visitors, spectators, customers or employees, or participants in sports events, performances and conferences.

 Note: See Annex A for further guidance on access and facilities for disabled people.

Access and facilities for disabled people

In addition to the requirements of the Disability Discrimination Act 1995 precautions need to be taken to ensure that:

- *new non-domestic buildings and/or dwellings (e.g. houses and flats used for student living accommodation etc.);*
- *extensions to existing non-domestic buildings;*
- *non-domestic buildings that have been subject to a material change of use (e.g. so that they become a hotel, boarding house, institution, public building or shop);*

are capable of allowing people, regardless of their disability, age or gender, to:

- *gain access to buildings;*
- *gain access within buildings;*
- *be able to use the facilities of the buildings (both as visitors and as people who live or work in them);*
- *use sanitary conveniences in the principal storey of any new dwelling.*

(Approved Document M)

 Note: See Annex A for guidance on access and facilities for disabled people.

Access Statements

To assist building control bodies it is recommended that an 'Access Statement' is also provided when plans are deposited, a building notice is given, or details of a project are provided to an approved inspector.

 Note: A building control file should also be prepared for all new buildings, changes of use and where extensive alterations are being made to existing buildings.

In its simplest form, an Access Statement should show where an applicant wishes to deviate from the guidance in Approved Document M, either to:

- make use of new technologies (e.g. infrared activated controls);
- provide a more convenient solution; or
- address the constraints of an existing building.

The Access Statement should include:

- the reasons for departing from the guidance;
- the rationale for the design approach adopted;

- constraints imposed by the existing structure and its immediate environment (why it is not practicable to adjust the existing entrance or provide a suitable new entrance);
- convincing arguments that an alternative solution will achieve the same, a better, or a more convenient outcome (e.g. why a fully compliant independent access is considered impracticable);
- evidence (e.g. current validated research) to support the design approach;
- the identification of buildings (or particular parts of buildings) where access needs to be restricted (e.g. processes that are carried out which might create hazards for children, disabled people or frail, elderly people).

 Note: Further guidance on Access Statements is available on the Disability Rights Commission's website at http://www.drc-gb.org.

6.30.2 Meeting the requirement

Vehicular access

If vehicles have access to a floor or roof edge, barriers of at least 375 mm should be provided to any edges that are level with (or above) the adjoining floor or ground.	K3 (3.7)
If vehicles have access to a ramp edge, barriers of at least 610 mm should be provided to any edges that are level with (or above) the adjoining floor or ground.	K3 (3.7)
Any wall, parapet, balustrade or similar obstruction may serve as a barrier.	K3 (3.8)
All barriers should be capable of resisting forces set out in BS 6399.	K3 (3.8)
Loading bays should be provided with at least one exit point from the lower level (preferably near the centre of the rear wall).	K3 (3.9)

Access and facilities for disabled people

Access (i.e. approach, entry or exit) to a building is frequently a problem for wheelchair users, people who need to use walking aids, people with impaired sight and mothers with prams etc. In designing the approach to a building (and routes between buildings within a complex) the following should, therefore, always be taken into consideration:

- changes in level between the entrance storey and the site entry point should be minimized;

- access routes should be wide enough to let people pass each other;
- potential hazards (e.g. windows from adjacent buildings opening onto access routes) should be avoided.

 Note: See also '*Mobility: A Guide to Best Practice on Access to Pedestrian and Transport Infrastructure*'.

General

The primary aim should be to make it reasonably possible for a disabled person to approach and gain access into the dwelling from the entrance point at the boundary of the site (and from any car parking that is provided on the site) to the building. It is also important that routes between buildings within a complex are also accessible.

Approach to a building

Access from the boundary of the site (and/or from car parking designated for disabled people) to the principal entrance should be level.	M (1.2, 1.4, 1.6 and 1.13) M (6.2)
If a difference in level is unavoidable (i.e. due to site constraints) the approach can have a gentle gradient (provided that it is over a long distance) or can include a number of shorter parts (at steeper gradients) as long as level landings are provided as rest points.	M (1.7)
The principal entrance (entrances, main staff entrance and any lobbies) should be accessible to disabled people and mothers pushing prams etc.	M (2.1)
If this is not possible, an alternative accessible entrance should be provided.	M (2.2)
Risks to people when entering the building should be minimal.	M (2.3)
Access routes should be wide enough to let people pass each other.	M (1.11)
Note: A surface width of 1800 mm is ideal but this can be reduced on restricted sites to 1200 mm, provided that a case is made in the Access Statement.	
The route to the principal entrance (or alternative accessible entrance) should be clearly identified and well lit.	M (1.13g)
A separate pedestrian route should be provided.	M (1.13h)

Uncontrolled vehicular crossing points should be   M (1.13h)
identified by a buff coloured blister surface
(see Figure 6.189).

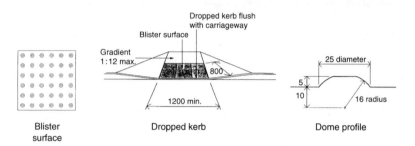

Blister
surface

Dropped kerb

Dome profile

Figure 6.189 An example of tactile paving used at an uncontrolled crossing

Gradients

Approach gradients:

- should ideally be no steeper than 1:60 along their   M (1.13c)
  whole length;
- if steeper than 1:20, should be designed as a   M (1.8)
  ramped access;
- if less than 1:20, should be provided with level   M (1.13c)
  landings for each 500 mm rise of the access.

Cross-fall gradients should be no steeper than 1:40.   M (1.13c)

Surface

The surface of all access routes should:

- allow people to travel along them easily,   M (1.9)
  without excessive effort and without the risk of
  tripping or falling;
- be at least 1.5 m wide;   M (1.13a)
- be firm, durable and slip resistant;   M (1.13d)
- have undulations not exceeding 3 mm   M (1.13d)
  (under a 1 m straight edge);
- be made of the same material and similar   M (1.13d and e)
  frictional characteristics (loose sand or gravel
  should **not** be used).

Building perimeters

The perimeter of the building should be well lit.	M (1.12)

Passing places

Passing places should be:

• free of obstructions to a height of 2.1 m;	M (1.13a)
• at least 1.8 m wide and at least 2 m long.	M (1.13b)

Joints

Joints should be:

• filled flush or (if recessed) no deeper than 5 mm;	M (1.13f)
• no wider than 10 mm or (if unfilled) no wider than 5 mm.	M (1.13f)
The difference in level at joints between paving units should be no greater than 5 mm.	M (1.13f)

On-site car parking and setting down – parking bays

At least one parking bay designated for disabled people should be provided as close as possible to the principal entrance of the building.	M (1.18a)
The dimensions of the designated parking bays should be as per Figure 6.190 (with a 1200 mm accessibility zone between and a 1200 mm safety zone on the vehicular side of the parking bays and with a dropped kerb when there is a pedestrian route at the other side of the parking bay).	M (1.18b)
A clearly signposted setting down point should be located on firm, level ground as close as possible to the principal (or alternative) entrance.	M (1.18e)
The surface of the accessibility zone should be firm, durable and slip resistant, with undulations not exceeding 3 mm (under a 1 m straight edge).	M (1.18c)

The surface of a parking bay designated for disabled   M (1.15)
people (in particular the area surrounding the bay)
should allow the safe transfer of a passenger or driver
to a wheelchair and transfer from the parking bay to
the access route to the building without undue effort.

Ticket machines should: M (1.16
• have their controls between 750 mm and 1200 mm   and 1.18d)
  above ground level;
• be located near parking bays designated for
  disabled people;
• be located so that a person in a wheelchair (or a
  person of short stature) is able to reach the controls.

  The plinth of the ticket machine should not   M (1.18d)
project in front of the face of the machine.

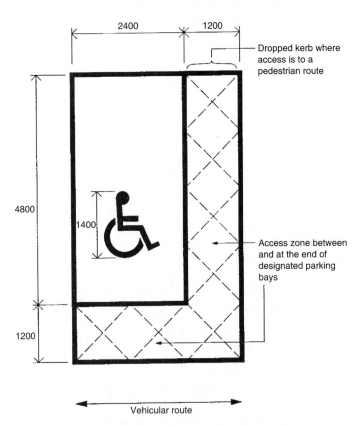

Figure 6.190 Parking bay designated for disabled people

Note: See also BS 8300 for guidance on:

- provision of parking bays designated for disabled people;
- ticket dispensing machines;
- vehicular control barriers; and
- multi-storey car parks.

Ramped access

If the constraints of the site mean that there is an approach gradient of 1 in 20 or steeper, then a ramped access should be provided as they are beneficial not only for wheelchair users but also people pushing prams and bicycles.

Where a ramped access is provided:	
• the approach should be clearly signposted;	M (1.26a)
• the going should be no greater than 10 m;	M (1.26c)
• the rise should be no more than 500 mm;	M (1.26c)
• if the total rise is greater than 2 m then an alternative means of access (e.g. a lift) should be provided for wheelchair users;	M (1.26d)
• the surface width should be at least 1.5 m;	M (1.26e)
• gradients should be as shallow as practicable;	M (1.20)
• the ramp surface should be slip resistant;	M (1.26f)
• the ramp surface should be of a contrasting colour to that of the landings;	M (1.26f)
• frictional characteristics of ramp and landing surfaces should be similar;	M (1.26g)
• landings at the foot and head of a ramp should be at least 1.2 m long and clear of any obstructions;	M (1.26h)
• intermediate landings should be at least 1.5 m long and clear of obstructions;	M (1.26i)
• intermediate landings (at least 1800 mm wide and 1800 mm long) should be provided at passing places;	M (1.26j)
• all landings should: – be level; – have a maximum gradient of 1:60 along their length; – have a maximum cross fall gradient of 1:40;	M (1.26k)
• there should be a handrail on both sides;	M (1.26l)
• in addition to the guarding requirements of Part K, there should be a visually contrasting kerb on the open side of the ramp (or landing) at least 100 mm high;	M (1.26m)
• when the rise of the ramp is greater than two 150 mm steps, signposted steps should be provided;	M (1.26n)

> • the gradient of a ramp flight and its going between   M (1.26b)
>   landings should be in accordance with Table 6.54
>   and Figure 6.191.

 Note: Approved Document K (*Protection from falling, collision and impact*) contains general guidance on stair and ramp design. The guidance in Approved Document M reflects more recent ergonomic research conducted to support BS 8300 and takes precedence over Approved Document K in conflicting areas.

Possible future amendment	Further research on stairs is currently being undertaken and will be reflected in future revisions of Approved Document K.

Table 6.54 Limits for ramp gradients

Going of a flight (m)	Maximum gradient	Maximum rise (mm)
10	1:20	500
5	1:15	333
2	1:12	166

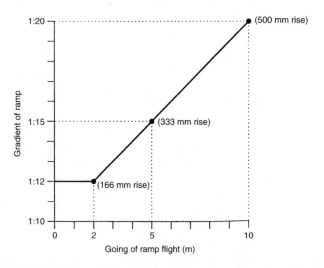

Figure 6.191 Relationship for ramp gradients to the going of a flight

Stepped access

People with impaired sight risk tripping or losing their balance if there is no warning that there are steps that provide a change in level. The risk is most hazardous at the head of a flight of stairs when a person is descending.

A stepped access should:

- have a level landing at the top and bottom of each flight; M (1.33a)
- be 1200 mm long each landing and unobstructed; M (1.33b)
- have a corduroy hazard warning surface at top and M (1.33c)
  bottom landings of a series of flights so as to give
  advance warning of a change in level (see Figure 6.192);

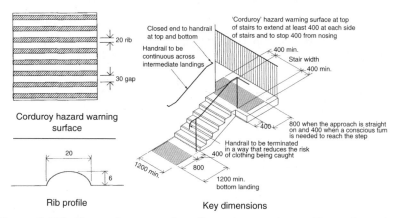

Corduroy hazard warning surface

Rib profile

Key dimensions

Figure 6.192 Stepped access – key dimensions and use of hazard warning surfaces

 Note: Approved Document K (*Protection from falling, collision and impact*) contains general guidance on stair and ramp design. The guidance in Approved Document M reflects more recent ergonomic research conducted to support BS 8300 and takes precedence over Approved Document K.

In addition:

- side accesses onto intermediate landings should M (1.33d)
  have a 400 mm deep corduroy hazard warning surface;
- doors should not swing across landings; M (1.33e)
- the surface width of flights between enclosing M (1.33f)
  walls, strings or upstands, should not be less
  than 1.2 m;
- there should be no single steps; M (1.33g)
- the rise of a flight between landings should contain M (1.33h)
  no more than:
  - 12 risers for a going of **less** than 350 mm
  - 18 risers for a going of 350 mm or **greater**
    (see Figure 6.193).

Figure 6.193 External steps and stairs – key dimensions

📖 **Note:** For school buildings, the preferred dimensions are a rise of 150 mm and a going of 280 mm.

Nosings for the tread and the riser should be 55 mm wide and of a contrasting material.	M (1.33i)
Step nosings should not project over the tread below by more than 25 mm (see Figure 6.194).	M (1.33j)

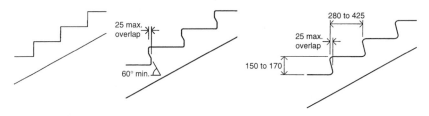

The rise and going dimensions apply to all step profiles

Figure 6.194 Examples of step profiles and key dimensions for external stairs

The rise and going of each step should be consistent throughout a flight.	M (1.33k)
The rise of each step should be between 150 mm and 170 mm.	M (1.33l)
Rises should not be open.	M (1.33n)
The going of each step should be between 280 mm and 425 mm.	M (1.33m)
There should be a continuous handrail on each side of a flight and landings.	M (1.33o)
If additional handrails are used to divide the flight into channels, then they should not be less than 1 m wide or more than 1.8 m wide.	M (1.33p)

Warnings should be placed sufficiently in advance of a hazard to allow time to stop.	M (1.28)
Warnings should not be so narrow that they could be missed in a single stride.	M (1.28)
Materials for treads should not present a slip hazard.	M (1.29)

Handrails to external stepped and for ramped access

Handrails to external stepped or ramped access should be positioned as per Figure 6.195.	M (1.37a)

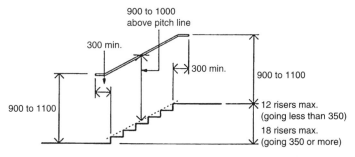

Figure 6.195 Handrails to external stepped and ramped access – key dimensions

Handrails to external stepped or ramped access should:

• be continuous across flights and landings;	M (1.37c)
• extend at least 300 mm horizontally beyond the top and bottom of a ramped access;	M (1.37d)
• not project into an access route;	M (1.37d)
• contrast visually with the background;	M (1.37e)
• have a slip resistant surface which is **not** cold to the touch;	M (1.37f)
• terminate in such a way that reduces the risk of clothing being caught;	M (1.37g)
• either be circular (with a diameter of between 40 and 45 mm) or oval with a width of 50 mm (see Figure 6.196).	M (1.37h)

In addition, handrails should:

• not protrude more than 100 mm into the surface width of the ramped or stepped access	M (1.37i)

where this would impinge on the stair width
requirement of Part B1;

- have a clearance of between 60 and 75 mm   M (1.37j)
  between the handrail and any adjacent wall surface;
- have a clearance of at least 50 mm between a   M (1.37k)
  cranked support and the underside of the handrail;
- ensure that its inner face is located no more than 50 mm   M (1.37l)
  beyond the surface width of the ramped or stepped access;
- be spaced away from the wall and rigidly   M (1.35)
  supported in a way that avoids impeding finger grip;
- be set at heights that are convenient for all   M (1.36)
  users of the building.

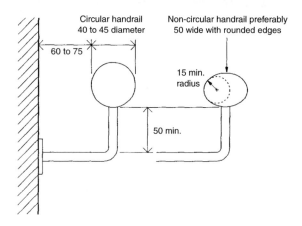

Figure 6.196 Handrail designs

Hazards on access routes

Features of a building (e.g. windows and doors)   M (1.38)
that can occasionally obstruct an access route should
not present a hazard.

Areas below stairs or ramps with a soffit less than   M (1.39b)
2.1 m above ground level should be protected by guarding
and low level cane detection.

Any feature projecting more than 100 mm   M (1.39b)
onto an access route should be protected by guarding that
includes a kerb (or other solid barrier) that can be
detected using a cane (see Figure 6.198).

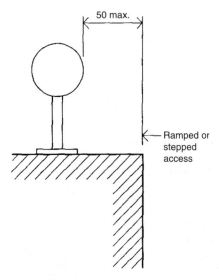

Figure 6.197 Handrail location

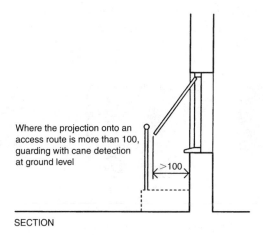

SECTION

Figure 6.198 Avoiding hazards on access routes

Accessible entrances

Accessible entrances should be clearly signposted (e.g. with the International Symbol of Access) and easily recognized.   M (2.5 and 2.7a)

Accessible entrances should also:

- be easily identifiable (e.g. by lighting and/or visual contrast);   M (2.7b)

• have a level landing at least 1500 × 1500 mm, clear of any door swings, immediately in front of the entrance;	M (2.7d)
• avoid raised thresholds (if unavoidable, then the total height should not be more than 15 mm, with a minimum number of upstands and slopes);	M (2.7e)
• ensure that all door entry systems are accessible to deaf and hard of hearing people, plus people who cannot speak;	M (2.7f)
• not have internal floor surface material (e.g. coir matting) adjacent to the threshold that could impede the movement of wheelchairs;	M (2.7h)
• not have changes in floor materials that could create a potential trip hazard;	M (2.7h)
• if mat wells are provided, have the surface of the mat level with the surface of the adjacent floor finish;	M (2.7i)
• have the route from the exterior across the threshold weather protected;	M (2.6 and 2.7a)
• not present a hazard for visually impaired people (e.g. have structural elements such as canopy supports).	M (2.5 and 2.7c)

The premises have no more than three steps

The premises are fully accessible for wheelchair users

Wheelchair assistance required

Figure 6.199 Typical access signs for disabled people

 Note: See BS 8300 for further guidance on signposting.

Doors to accessible entrances

Doors to the principal (or alternative accessible) entrance should be accessible to all, particularly for wheelchair users and people with limited physical dexterity.	M (2.8)
Entrance doors should be capable of being held closed when not in use.	M (2.8)

A power operated door opening and closing system   M (2.13a)
should be used if a force greater than 20 N is required
to open or shut a door.

Once open, all doors to accessible entrances
should be wide enough to allow unrestricted passage for a
variety of users, including wheelchair users, people carrying
luggage, people with assistance dogs, and parents with
pushchairs and small children.

People should be able to see other people approaching   M (2.12)
from the opposite direction.

The effective clear width through a single leaf door   M (2.13b)
(or one leaf of a double leaf door) should be in
accordance with Table 6.55.

Table 6.55 Minimum effective clear widths of doors

Direction and width of approach	New buildings (mm)	Existing buildings (mm)
Straight-on (without a turn or oblique approach)	800	750
At right angles to an access route at least 1500 mm wide	800	750
At right angles to an access route at least 1200 mm wide	825	775
External doors to buildings used by the general public	1000	775

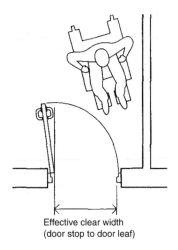

Effective clear width
(door stop to door leaf)

Figure 6.200 Effective clear width and visibility requirements of doors

> Door leaves and side panels wider than 450 mm   M (2.13c)
> should incorporate vision panels:
>
> • towards the leading edge of the door;
> • between 500 mm and 1500 mm from the floor (see
>   Figure 6.200)

Manually operated non-powered entrance doors

Self-closing devices on manually operated (i.e. non-powered) doors can be a great disadvantage to people who have limited upper body strength, or people who are pushing prams or carrying heavy objects. To rectify this matter:

> 💣 The opening force at the leading edge of the   M (2.17a)
> door should be no greater than 20 N.
>
> A space alongside the leading edge of a door should   M (2.15)
> be provided to enable a wheelchair user to reach and
> grip the door handle.
>
> 💡 Door opening furniture should:
>
> • be easy to operate by people with limited   M (2.16)
>   manual dexterity;
> • be capable of being operated with one hand   M (2.17c)
>   using a closed fist (e.g. a lever handle);
> • contrast visually with the surface of the door and   M (2.17d)
>   not be cold to the touch.
>
> 💣 There should be an unobstructed space of at least   M (2.17b)
> 300 mm on the pull side of the door between the
> leading edge of the door and any return wall (unless
> the door is a powered entrance door – see Figure 6.201).

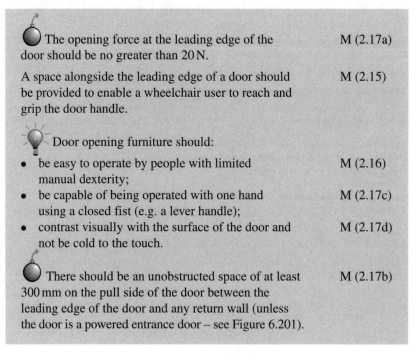

Figure 6.201 Effective clear width and visibility requirements of doors

Powered operated entrance doors

Power operated entrance doors should have a sliding, swinging or folding action controlled manually (by a push pad, card swipe, coded entry, or remote control) or be automatically controlled by a motion sensor or proximity sensor such as a contact mat.

Power operated entrance doors should:

• open towards people approaching the doors;	M (2.21a)
• provide visual and audible warnings that they are operating (or about to operate);	M (2.21c)
• incorporate automatic sensors to ensure that they open early enough (and stay open long enough) to permit safe entry and exit;	M (2.21c)
• incorporate a safety stop that is activated if the doors begin to close when a person is passing through;	M (2.21b)
• revert to manual control (or fail safe) in the open position in the event of a power failure;	M (2.21d)
• when open, not project into any adjacent access route;	M (2.21e)
• ensure that its manual controls:	M (2.21f)
– are located between 750 mm and 1000 mm above floor level;	
– are operable with a closed fist;	
• be set back 1400 mm from the leading edge of the door when fully open;	M (2.21g)
• be clearly distinguishable against the background;	M (2.21g)
• contrast visually with the background.	M (2.19 and 2.21g)

 Note: Revolving doors are not considered 'accessible' as they create particular difficulties (and possible injury) for people who are visually impaired, people with assistance dogs or mobility problems and for parents with children and/or pushchairs.

Glass entrance doors and glazed screens

The presence of the door should be apparent not only when it is shut but also when it is open.	M (2.23)
Glass entrance doors and glazed screens should:	
• be clearly marked (i.e. with a logo or sign) on the glass at two levels, 850 to 1000 mm and 1400 to 1600 mm above the floor;	M (2.24a)

> **Note:** The logo or sign should be at least 150 mm high (repeated if on a glazed screen), or a decorative feature such as broken lines or continuous bands, at least 50 mm high.
>
> - when adjacent to, or forming part of, a glazed screen,   M (2.24c)
>   be provided with a high contrast strip at the top,
>   and on both sides;
> - ensure that glass entrance doors (if capable of being   M (2.24d)
>   held open) are protected by guarding to prevent
>   the leading edge from becoming a possible hazard.

For dwellings:

An external door providing access for disabled   M (6.23)
people should have a minimum clear opening width
(taken from the face of the door stop on the latch side
to the face of the door when open at 90°) of 775 mm.

The door opening width should be sufficient to enable   M (6.22)
a wheelchair user to manoeuvre into the dwelling.

Possible future amendment	Approved Document N (Glazing – safety in relation to impact, opening and cleaning) contains guidance on the use of symbols and markings on glazed doors and screens. The guidance now given in Approved Document M is as a result of more recent experience of 'door manifestation' and takes precedence over the guidance currently provided in Approved Document N in conflicting areas until such time as Approved Document N is revised.

Entrance (and internal) lobbies

Lobbies should be: M (2.27 and 3.15)

- large enough to allow a wheelchair user or a
  person pushing a pram to move clear of one
  door before opening the second door;

- capable of accommodating a companion helping a wheelchair user to open doors and guide the wheelchair through.

The minimum length of the lobby is related to the chosen door size, the swing of each door, the projection of the door into the lobby and the size of an occupied wheelchair with a companion pushing.

Within the lobby:

• glazing should not create distracting reflections;	M (2.29d and 3.16d)
• floor surface materials should do not impede the movement of wheelchairs etc.;	M (2.29e)
• changes in floor materials should not create a potential trip hazard;	M (2.29e and 3.16e)
• the floor surface should assist in removing rainwater from shoes and wheelchairs;	M (2.29f)
• any columns and ducting etc. that project into the lobby by more than 100 mm should be protected by a visually contrasting guard rail.	M (2.29h and 3.16a–c)

The length and width of an entrance and/or an internal lobby should be as per Table 6.56.	M (2.29a, b and c)

Table 6.56 Entrance lobbies – dimensions

	Length	Width
Entrance/internal lobby with single swing door	as per Figure 6.202	at least 1200 mm (or the width of the two doors plus 300 mm whichever is the greater)
Entrance/internal lobby with double swing doors	at least the size (i.e. width) of the two doors plus 1570 mm	at least 1800 mm

Entrance hall and reception area

If there is a reception point:

• it should be easily accessible and convenient to use;	M (3.2)
• it should be located away from the principal entrance;	M (3.6a)
• it should be easily identifiable from the entrance doors or lobby;	M (3.6b)

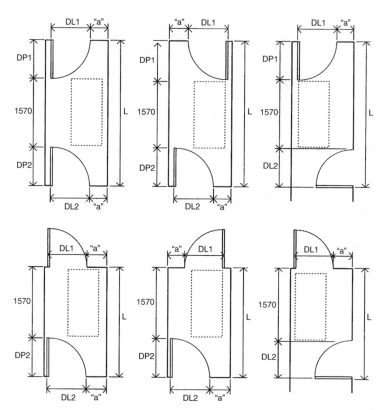

DL1 and DL2 = door leaf dimensions of the doors to the lobby
DP1 and DP2 = door projection into the lobby (normally door leaf size)
L = minimum length of lobby, or length up to door leaf for side entry lobby
"a" = at least 300 mm wheelchair access space (can be increased to reduce L)
1570 = length of occupied wheelchair with a companion pushing (or a large scooter)
NB: For every 100 mm increase above 300 mm in the dimension "a" (which gives a greater overlap of the wheelchair footprint over the door swing), there can be a corresponding reduction of 100 mm in the dimension L, up to a maximum of 600 mm reduction.

Figure 6.202 Key dimensions – lobbies and entrance doors

- relevant information about the building should be clearly available from notice boards and signs; M (3.5)
- the floor surface should be slip resistant; M (3.6)
- the approach to it should be direct and free from obstructions; M (3.6b)
- the design of the approach should allow space for wheelchair users to gain access; M (3.6c)
- there should be a clear manoeuvring space in front of any reception desk at least 1200 mm deep and 1800 mm wide. M (3.6d)

If there is a reception desk or counter:

- it should be designed to accommodate both standing   M (3.6e)
  and seated visitors;
- if there is a knee recess, then this should be at least   M (3.6d)
  500 mm deep (or 1400 mm deep and 2200 mm wide
  if there is no knee recess);
- at least one section of the reception desk should be   M (3.6e)
  no less than 1500 mm wide, its surface no higher than
  760 mm and a knee recess not less than 700 mm,
  above floor level;
- it should be provided with a hearing enhancement   M (3.6f)
  system (e.g. an induction loop).

Note: See BS 8300 for guidance on aids to communication.

Domestic buildings

If there is a reception point:

- it should be easily accessible and convenient to use;   M (3.2 to 3.5)
- information about the building should be clearly
  available from notice boards and signs;
- the floor surface should be slip resistant.

Approach to a dwelling

On plots which are reasonably level, wheelchair users should normally be able to approach the principal entrance.	M (6.2)
Wheelchair users (having approached the entrance) should be able to gain access into the dwelling-house and/or entrance level of flats.	M (6.3)
A suitable approach should be provided from the point of access to the entrance of the dwelling.	M (6.11)
The whole, or part, of the approach may be a driveway.	M (6.12)
The approach should:	
• not have crossfalls greater than 1 in 40;	M (6.11)

- be safe and convenient for disabled people as is   M (6.6)
  reasonable possible;
- ideally be level or ramped.   M (6.8)

Note: On steeply sloping plots, a stepped approach is permissible.

If a stepped approach to the dwelling is unavoidable,   M (6.7)
the aim should be for the steps to be designed to
suit the needs of ambulant disabled people (see
paragraph 6.19).

The surface of the wheelchair user's approach should: M (6.9)

- be firm enough to support the weight of the user and
  his or her wheelchair;
- be smooth enough to permit easy manoeuvre;
- not be made up of loose laid materials (such as
  gravel and shingle);
- take account of the needs of stick and crutch users.

For steeply sloping plots, it would be reasonable to
provide for stick or crutch users.

Level approach

A 'level' approach should: M (6.13)

- be no steeper than 1 in 20;
- have a firm and even surface;
- have a width not less than 900 mm.

Note: The width of the approach, excluding space
for parked vehicle, should take account of the needs
of a wheelchair user, or a stick or crutch user.

Ramped approach

If a plot gradient exceeds 1 in 20, a ramped approach   M (6.14)
may be provided in which case:

- the surface should be firm and even;   M (6.15a)
- the flights should have unobstructed widths of   M (6.15b)
  at least 900 mm;

- individual flights should be no longer than 10 m   M (6.15c)
  (for gradients not steeper than 1 in 15) or 5 m for
  gradients not steeper than 1 in 12;
- it should have top and bottom landings   M (6.15c)
  (and intermediate landings if necessary) not less
  than 1.2 m in length – exclusive of the swing of any
  door or gate which opens onto it.

Stepped approach

A stepped approach should be used if the plot gradient is   M (6.16)
greater than 1 in 15.

A stepped approach should:

- have flights with an unobstructed width of   M (6.17a)
  at least 900 mm;
- have a flight rise of not more than 1.8 m;   M (6.17b)
- have a top and bottom (and if necessary intermediate)   M (6.17c)
  landing not less than 900 mm in length;
- have steps: M (6.17d)
  – with suitable tread nosing profiles (see Figure 6.203);
  – with a uniform rise between 75 mm and 150 mm;   M (6.17e)
- ensure that the going of each step is not less than   M (6.17f)
  280 mm;
- comprise three or more risers;
- have a suitable continuous handrail on one side of the
  flight between 850 mm and 1000 mm above the pitch
  line of the flight; and extend 300 mm beyond the top
  and bottom nosings.

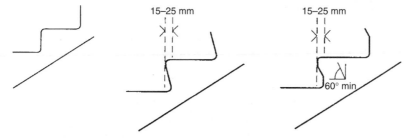

Figure 6.203 External step profiles

Approach using a driveway

Where a driveway provides a means of approach towards the entrance, the approach past any parked cars must be in accordance with requirements M6.11–6.27 above.	M (6.18)

Access into the dwelling

The point of access should be reasonably level.	M (6.11)
Where the approach to the entrance consists of a level or ramped approach, an accessible threshold at the entrance should be provided.	M (6.19)
Note: An accessible threshold into entrance level flats should also be provided.	
If a stepped approach is provided into the dwelling, the rise should be no more than 150 mm.	M (6.20)
An accessible threshold should be provided into the entrance.	M (6.21)

Note: The design of an accessible threshold should also satisfy the requirements of Part C2: 'Dangerous and offensive substances' and Part C4: 'Resistance to weather and ground moisture'.

6.31 Extensions to buildings

6.31.1 The requirement

Conservation of fuel and power

Energy efficiency measures shall be provided that:

(a) • *limit the heat loss through the roof, wall, floor, windows and doors, etc. by suitable means of insulation;*
     • *where appropriate permit the benefits of solar heat gains and more efficient heating systems to be taken into account;*
     • *limit unnecessary ventilation heat loss by providing building fabric that is reasonably airtight;*
     • *limit the heat loss from hot water pipes and hot air ducts used for space heating;*
     • *limit the heat loss from hot water vessels and their primary and secondary hot water connections by applying suitable thicknesses of*

> *insulation (where such heat does not make an efficient contribution to the space heating).*
>
> *(b)* *Provide space heating and hot water systems with reasonably efficient equipment such as heating appliances and hot water vessels where relevant, such that the heating and hot water systems can be operated effectively as regards the conservation of fuel and power.*
>
> *(c)* *Provide lighting systems that utilize energy-efficient lamps with manual switching controls or, in the case of external lighting fixed to the building, automatic switching, or both manual and automatic switching controls as appropriate, such that the lighting systems can be operated effectively as regards the conservation of fuel and power.*
>
> *(d)* *Provide information, in a suitably concise and understandable form (including results of performance tests carried out during the works) that shows building occupiers how the heating and hot water services can be operated and maintained.*
>
> *(Approved Document L1)*

Responsibility for achieving compliance with the requirements of Part L rests with the person carrying out the work. That person may be, for example, a developer, a main (or sub-) contractor, or a specialist firm directly engaged by a private client.

The person responsible for achieving compliance should either themselves provide a certificate, or obtain a certificate from the sub-contractor, that commissioning has been successfully carried out. The certificate should be made available to the client and the building control body.

6.31.2 Meeting the requirement

The fabric U-values given in Table 6.57 can be applied when proposing extensions to dwellings.	L1 (1.11)
The target U-value and carbon index methods can be used only if applied to the whole enlarged dwelling.	L1 (1.11)
The total rate of heat loss is the product of (Area $\times$ U-value) for all exposed elements.	L1 (1.12)
For small extensions to dwellings (e.g. porches where the new heated space created has a floor area of not more than about $6\,m^2$) energy performance should not be less than that in the existing building.	L1 (1.13)
The area-weighted average U-value of windows, doors and rooflights ('openings') in extensions to existing dwellings should not exceed the relevant values in Table 6.57.	L1 (1.14)

Table 6.57 Elemental method – U-values (W/m²K) for construction elements

Exposed element	U-value
Pitched roof with insulation between rafters[1,2]	0.2
Pitched roof with integral insulation	0.25
Pitched roof with insulation between joists	0.16
Flat roof[3]	0.25
Walls, including basement walls	0.35
Floors, including ground floors and basement floors	0.25
Windows, doors and rooflights (area-weighted average), glazing in metal frames[4]	2.2
Windows, doors and (area-weighted average) rooflights[4], glazing in wood or PVC frames[5]	2.0

Notes:
1. Any part of a roof having a pitch of 70° or more can be considered as a wall.
2. For the sloping parts of a room-in-the-roof constructed as a material alteration, a U-value of 0.3 W/m²K would be reasonable.
3. Roof of pitch not exceeding 10°.
4. Rooflights include roof windows.
5. The higher U-value for metal-framed windows allows for additional solar gain due to the greater glazed proportion.

Figure 6.204 summarizes the fabric insulation standards and allowances for windows, doors and rooflights given in the elemental method.

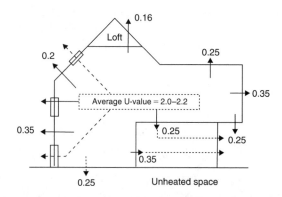

Figure 6.204 Summary of the elemental method

6.32 External balconies

6.32.1 The requirements

Protection from falling

Pedestrian guarding should be provided for any part of a floor (including the edge below an opening window) gallery, balcony, roof (including rooflight and other openings), any other place to which people have access and any light well, basement area or similar sunken area next to a building.

(Approved Document K2)

Requirement K2 (a) applies only to stairs and ramps that form part of the building.

6.32.2 Meeting the requirements

> All external balconies and edges of roofs shall have a wall, parapet, balustrade or similar guard at least 1100 mm high. K3 (3.2)

6.33 Garages

6.33.1 The requirements

Electrical work

Reasonable provision shall be made in the design, installation, inspection and testing of electrical installations in order to protect persons from fire or injury.
(Approved Document P1)

Sufficient information shall be provided so that persons wishing to operate, maintain or alter an electrical installation can do so with reasonable safety.
(Approved Document P2)

Fire precautions

As a fire precaution, all materials used for internal linings of a building should have a low rate of surface flame spread and (in some cases) a low rate of heat release.
(Approved Document B2)

6.33.2 Meeting the requirements

> Cables to an outside building (e.g. garage or shed) if run underground, should be routed and positioned so as to give protection against electric shock and fire as a result of mechanical damage to a cable. P AppA 2d
>
> Cables concealed in floors and walls (in certain circumstances) are required to have an earthed metal covering, be enclosed in steel conduit, or have additional mechanical protection (see BS 7671 for more information). P AppA 2d
>
> If a domestic garage is attached to (or forms an integral part of a house), the garage should be separated from the rest of the house, as shown in Figure 6.205. B3 (9.14)

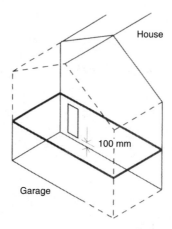

Figure 6.205 Separation between garage and dwellinghouse

The wall and any floor between the garage and the house shall have a 30 minute standard of fire resistance. Any opening in the wall should be at least 100 mm above the garage floor level with an FD30 door.

6.34 Conservatories

6.34.1 Requirements

Conservation of fuel and power

Energy efficiency measures shall be provided that:

(a) • *limit the heat loss through the roof, wall, floor, windows and doors, etc. by suitable means of insulation;*
 • *where appropriate permit the benefits of solar heat gains and more efficient heating systems to be taken into account;*
 • *limit unnecessary ventilation heat loss by providing building fabric that is reasonably airtight;*
 • *limit the heat loss from hot water pipes and hot air ducts used for space heating;*
 • *limit the heat loss from hot water vessels and their primary and secondary hot water connections by applying suitable thicknesses of insulation (where such heat does not make an efficient contribution to the space heating).*
(b) *Provide space heating and hot water systems with reasonably efficient equipment such as heating appliances and hot water vessels where relevant, such that the heating and hot water systems can be operated effectively as regards the conservation of fuel and power.*
(c) *Provide lighting systems that utilize energy-efficient lamps with manual switching controls or, in the case of external lighting fixed to the building, automatic switching, or both manual and automatic switching controls as*

appropriate, such that the lighting systems can be operated effectively as regards the conservation of fuel and power.

(d) Provide information, in a suitably concise and understandable form (including results of performance tests carried out during the works) that shows building occupiers how the heating and hot water services can be operated and maintained.

(Approved Document L1)

Responsibility for achieving compliance with the requirements of Part L rests with the person carrying out the work. That person may be, for example, a developer, a main (or sub-) contractor, or a specialist firm directly engaged by a private client.

The person responsible for achieving compliance should either themselves provide a certificate, or obtain a certificate from the sub-contractor, that commissioning has been successfully carried out. The certificate should be made available to the client and the building control body.

6.34.2 Meeting the requirement

A conservatory that is not separated from the rest of the dwelling shall be treated as an integral part of the dwelling.	L1
Where the conservatory is separated from the rest of the dwelling and has a **fixed** heating installation, the heating in the conservatory shall have its own separate temperature and on/off controls.	L1
Where an opening is created or enlarged, provision shall be made to limit the heat loss from the dwelling such that it is no worse than before the work was undertaken.	L1
Measurements of thermal conductivity should be made according to BS EN 12664, BS EN 12667, or BS EN 12939	L1 (0.9)
Measurements of thermal transmittance should be made according to BS EN ISO 8990 or, in the case of windows and doors, BS EN ISO 12567.	L1 (0.9)

For the purposes of the guidance in Part L, a conservatory is considered to not have less than three-quarters of the area of its roof and not less than one-half of the area of its external walls made of translucent material.

If a conservatory is attached to, and built as part of, a new dwelling it should be treated as an integral part of the building.	L1 (1.59)

If there is no physical separation between the conservatory and the dwelling, then the conservatory should be treated as an integral part of the dwelling.	L1 (1.59a)
If the conservatory is separated from the dwelling, energy savings can be achieved if the conservatory is not heated.	L1 (1.59b)
If fixed heating installations are used in a conservatory, they should have their own separate temperature and on/off controls.	L1 (1.59b)

When a conservatory is attached to an existing dwelling and an opening is enlarged or newly created as a material alteration, the heat loss from the dwelling should be limited by: L1 (1.60)

- retaining the existing separation where the opening is not to be enlarged;
- providing separation with an average U-value less than $2.0\,\text{W/m}^2\,\text{K}$.

Separating walls and floors between a conservatory and a dwelling should be insulated to at least the same degree as the exposed walls and floors.	L1 (1.61)
Separating windows and doors between a conservatory and a dwelling should have the same U-value and draught-stripping provisions as the exposed windows and doors elsewhere in the dwelling.	L1 (1.61b) AD N

6.35 Rooms for residential purposes

6.35.1 The requirement

'Rooms for residential purposes' are defined in Regulation 2 of the Building Regulations 2000 (as amended) and need to conform with the applicable Approved Documents (see Section 5.23) and meet the requirements for airborne and impact sound insulation shown in Table 6.58.

Table 6.58 Dwelling-houses and flats – performance standards for separating walls, separating floors and stairs that have a separating function

	Airborne sound insulation $D_{nT,w} + C_{tr}$ dB (minimum values)	Impact sound insulation $L'_{nT,w}$ dB (maximum values)
Purpose built rooms for residential purposes		
Walls	43	–
Floors and stairs	45	62

6.35.2 Meeting the requirement

Separating walls in new buildings containing rooms for residential purposes

Separating wall types 1and 3 are considered to be the most suitable for use in new buildings containing rooms for residential purposes.

 Note: Wall types 2 and 4 can be used **provided** that care is taken to maintain isolation between the leaves. Specialist advice may be needed.

Wall type 1 (solid masonry)

When using a solid masonry wall, the resistance to airborne sound depends mainly on the mass per unit area of the wall. As shown below, there are three different categories of solid masonry walls:

Table 6.59 Wall type 1 – categories

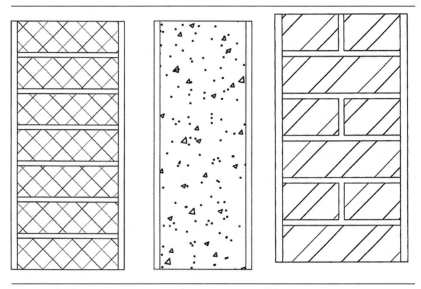

Wall type 1.1 Solid masonry	Wall type 1.2 Dense aggregate concrete	Wall type 1.3 Brick
Dense aggregate concrete block, plaster on both room faces	Dense aggregate concrete cast in-situ, plaster on both room faces	Brick, plaster on both room faces

 Note: Plasterboard may be used as an alternative wall finish, provided a sheet of minimum mass per unit area $10\,\mathrm{kg/m^2}$ is used on each room face.

Wall type 3 (masonry between independent panels)

Wall types 3.1 and 3.2 provide a high resistance to the transmission of both air-borne sound and impact sound on the wall. Their resistance to sound depends partly on the type (and mass) of the core and partly on the isolation and mass of the panels.

Table 6.60 Wall type 3 – categories

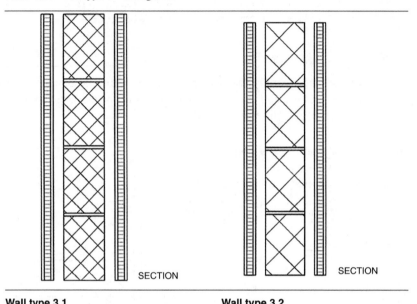

Wall type 3.1	Wall type 3.2
Solid masonry core (dense aggregate concrete block), independent panels on both room faces	Solid masonry core (lightweight concrete block), independent panels on both room faces

Corridor walls and doors in new buildings containing rooms for residential purposes

Separating walls as described in Table 6.60 should be used between rooms for residential purposes and corridors in order to control flanking transmission and to provide the required sound insulation between the dwelling and the corridor. Sound insulation will be reduced by the presence of a door.

All corridor doors shall have a good perimeter sealing (including the threshold where practical).	E6.6
All corridor doors shall have a minimum mass per unit area of 25 kg/m².	E6.6

All corridor doors shall have a minimum sound reduction index of 29 dB Rw (measured according to BS EN ISO 140-3:1995 and rated according to BS EN ISO 717-1:1997).	E6.6
All corridor doors shall meet the requirements for fire safety as described in Building Regulations Part B – Fire safety.	E6.6
Noisy parts of the building should preferably have a lobby, double door or high performance doorset to contain the noise.	E6.7

Separating floors in new buildings containing rooms for residential purposes

Although only one of the separating floor types described in Section 3 of Part E is considered most suitable for use in new buildings containing rooms for residential purposes, provided that floating floors and ceilings are not continuous between rooms, floor types 2 and 3 can also be used.

Table 6.61 Floor type 1 – categories

Floor type 1.1C (with ceiling treatment C) Solid concrete slab (cast in-situ or with permanent shuttering), soft floor covering	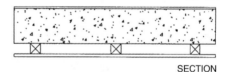 SECTION
Floor type 1.2 (with ceiling treatment B) Concrete planks (solid or hollow), soft floor covering	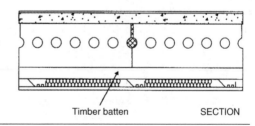 Timber batten SECTION

 Note: Specialist advice may be needed.

6.36 Rooms for residential purposes resulting from a material change of use

6.36.1 The requirement

'Rooms for residential purposes' formed by material change of use need to conform with the applicable Approved Documents (see Section 2.3) and meet the requirements for airborne and impact sound insulation shown in Table 6.62 below.

Table 6.62 Dwelling-houses and flats – performance standards for separating walls, separating floors and stairs that have a separating function

	Airborne sound insulation $D_{nT,w} + C_{tr}$ dB (minimum values)	Impact sound insulation $L'_{nT,w}$ dB (maximum values)
Rooms for residential purposes formed by material change of use		
Walls	43	–
Floors and stairs	43	64

6.36.2 Meeting the requirement

In some cases it may be that an existing wall, floor or stair in a building will achieve these performance standards without the need for remedial work, for example if the existing construction was already compliant. If this is not the case then the building work should be in compliance with the regulations concerning walls and floors as described previously in this book.

Rooms for residential purposes

Junction details

> If there is a junction between a solid masonry separating wall type 1 and the ceiling void and roof space the solid wall need not be continuous to the underside of the structural floor or roof provided that:
>
> - there is a ceiling consisting of two or more layers of plasterboard, of minimum total mass per unit area 20 kg/m²;   E6.14a
> - there is a layer of mineral wool (minimum thickness 200 mm, minimum density 10 kg/m³) in the roof void;   E6.14b
> - the ceiling is not perforated.   E6.14c

As shown in Figure 6.206, the ceiling joists and plasterboard sheets should not be continuous between rooms for residential purposes.

 Note: The ceiling void and roof space detail can only be used where the Requirements of Building Regulation Part B – Fire safety can also be satisfied. The Requirements of Building Regulation Part L – Conservation of fuel and power should also be satisfied.

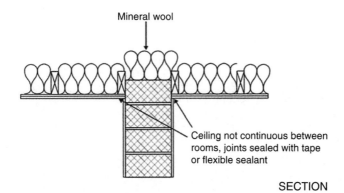

SECTION

Figure 6.206 Ceiling void and roof space (only applicable to rooms for residential purposes)

Room layout and building services – design considerations

As internal noise levels are affected by room layout, building services and sound insulation, the actual layout of rooms should be considered at the design stage particularly to avoid placing noise sensitive rooms next to rooms in which noise is generated.

 Note: See also:

- BS 8233:1999 Sound insulation and noise;
- Reduction for Buildings – Code of practice and sound control for homes.

6.37 Reverberation in the common internal parts of buildings containing flats or rooms for residential purposes

6.37.1 The requirement

Requirement E3 requires that *'Domestic buildings shall be designed and constructed so as to restrict the transmission of echoes.'*

6.37.2 Meeting the requirement

The guidance notes provided in Part E cover two methods. Method A and Method B assist in determining the amount of additional absorption to be used in corridors, hallways, stairwells and entrance halls that give access to flats and rooms for residential purposes.

	Entrance halls	Corridors	Hallways	Stairwells
Method A	Yes	Yes	Yes	Yes
Method B	Yes	Yes	Yes	No

Method A

Cover the ceiling area with the additional absorption.	E7.10
Cover the underside of intermediate landings, the underside of the other landings, and the ceiling area on the top floor.	E7.11
The absorptive material should be equally distributed between all floor levels.	E7.12
For stairwells (or a stair enclosure), calculate the combined area of the stair treads, the upper surface of the intermediate landings, the upper surface of the landings (excluding ground floor) and the ceiling area on the top floor. Either cover an area equal to this calculated area with a Class D absorber, or cover an area equal to at least 50% of this calculated area with a Class C absorber or better.	E7.11

 Note: Method A can generally be satisfied by the use of proprietary acoustic ceilings.

Method B

In comparison to Method A, Method B takes account of the existing absorption provided by all of the surfaces. Section 7 of Part E provides details of how to calculate the total absorption area based on the material's absorption coefficient (α).

This can become a fairly specialist area and it would probably be advisable to seek professional advice and/or peruse BS EN 20354:1993 Acoustics – Measurement of sound absorption in a reverberation room or BS EN ISO 11654:1997 Acoustics – Sound absorbers for use in buildings – Rating of sound absorption for detailed information.

Absorption areas should be calculated for each octave band (in square metres) using the formula

$$A_T = \alpha_1 S_1 + \alpha_2 S_2 + \ldots \alpha_n S_n$$

Where:
A_T = total absorption area in sq. m
$\alpha_1 S_1$ = absorption coefficient for material 1
$\alpha_2 S_2$ = absorption coefficient for material 2
$\alpha_n S_n$ = absorption coefficient for the last type of material n

Table 6.63 Absorption coefficient data for common materials in buildings

Material	Sound absorption coefficient (α) in octave frequency bands (Hz)				
	250	500	1000	2000	4000
Fair-faced concrete or plastered masonry	0.01	0.01	0.02	0.02	0.03
Fair-faced brick	0.02	0.03	0.04	0.05	0.07
Painted concrete brick	0.05	0.06	0.07	0.09	0.08
Windows glass facade	0.08	0.05	0.04	0.03	0.02
Doors (timber)	0.10	0.08	0.08	0.08	0.08
Glazed tile/marble	0.01	0.01	0.01	0.02	0.02
Hard floor coverings (e.g. lino, parquet) on concrete floor	0.03	0.04	0.05	0.05	0.06
Soft floor coverings (e.g. carpet) on concrete floor	0.03	0.06	0.15	0.30	0.40
Suspended plaster or plasterboard ceiling (with large airspace behind)	0.15	0.10	0.05	0.05	0.05

Requirement E3 will be satisfied when:

For entrance halls, provide a minimum of $0.20\,\text{m}^2$ total absorption area per cubic metre of the volume.	E7.17
For corridors or hallways, provide a minimum of $0.25\,\text{m}^2$ total absorption area per cubic metre of the volume.	E7.18

For corridors or hallways, provide a minimum of $0.25\,\text{m}^2$ total absorption area per cubic metre of the volume.

6.38 Internal walls and floors (new buildings)

6.38.1 The requirement

Dwellings shall be designed so that the noise from domestic activity in an adjoining dwelling (or other parts of the building) is kept to a level that:

- *does not affect the health of the occupants of the dwelling;*
- *will allow them to sleep, rest and engage in their normal activities in satisfactory conditions.*

(Approved Document E1)

Dwellings shall be designed so that any domestic noise that is generated internally does not interfere with the occupants' ability to sleep, rest and engage in their normal activities in satisfactory conditions.

(Approved Document E2)

Domestic buildings shall be designed and constructed so as to restrict the transmission of echoes.

(Approved Document E3)

Table 6.64 Dwelling-houses and flats – performance standards for separating walls, separating floors and stairs that have a separating function

	Airborne sound insulation $D_{nT,w} + C_{tr}$ dB (minimum values)	Impact sound insulation $L'_{nT,w}$ dB (maximum values)
Purpose built rooms for residential purposes		
Walls	43	–
Floors and stairs	45	62
Rooms for residential purposes formed by material change of use		
Walls	43	–
Floors and stairs	43	64

6.38.2 Meeting the requirement

 Note: To avoid air paths between rooms, all gaps around internal walls and floors should be filled.

There are four main types of internal wall and three types of internal floor as detailed below.

Internal wall type A or B *Timber or metal frame*	The resistance to airborne sound depends on the mass per unit area of the leaves, the cavity width, frame material and the absorption in the cavity between the leaves.
Internal wall type C or D *Concrete or aircrete block*	The resistance to airborne sound depends mainly on the mass per unit area of the wall.
Internal floor type A or B *Concrete planks or concrete beams with infilling blocks*	The resistance to airborne sound depends on the mass per unit area of the concrete base or concrete beams and infilling blocks. A soft covering will reduce impact sound at source.
Internal floor type C *Timber or metal joist*	The resistance to airborne sound depends on the structural floor base, the ceiling and the absorbent material. A soft covering will reduce impact sound at source.

Internal wall type A

Timber or metal frames with plasterboard linings on each side of frame

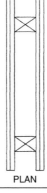

PLAN

- each lining to be two or more layers of plasterboard,
- each sheet of minimum mass per unit area 10 kg/m^2
- linings fixed to timber frame with a minimum distance between linings of 75 mm (or metal frame with a minimum distance between linings of 45 mm)
- all joints are to be sealed

Internal wall type B

Timber or metal frames with plasterboard linings on each side of frame and absorbent material

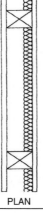

PLAN

- single layer of plasterboard of minimum mass per unit area 10 kg/m^2
- linings fixed to timber frame with a minimum distance between linings of 75 mm (or metal frame with a minimum distance between linings of 45 mm)
- an absorbent layer of unfaced mineral wool batts or quilt (minimum thickness 25 mm, minimum density 10 kg/m^3) suspended in the cavity (may be wire reinforced)
- all joints well sealed

Internal wall type C

Concrete block wall, plaster or plasterboard finish on both sides

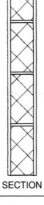

SECTION

- minimum mass per unit area, excluding finish 120 kg/m^2
- all joints well sealed
- plaster or plasterboard finish on both sides

Internal wall type D

Aircrete block wall, plaster or plasterboard finish on both sides

Note: Internal wall type D should only be used with the separating walls described in section Ei (i.e. where there is no minimum mass requirement on the internal masonry walls) and it should not be used as a load-bearing wall or be rigidly connected to the separating floors.

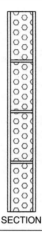

SECTION

- for plaster finish, minimum mass per unit area, including finish 90 kg/m²
- for plasterboard finish, minimum mass per unit area, including finish 75 kg/m²
- all joints well sealed

Internal floor type A

Concrete planks

Note: Insulation against impact sounds can be improved by adding a soft covering (e.g. carpet).

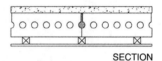

SECTION

- minimum mass per unit area 180 kg/m²
- regulating screed optional
- ceiling finish optional

Internal floor type B

Concrete beams with infilling blocks, bonded screed and ceiling

Note: Insulation against impact sounds can be improved by adding a soft covering (e.g. carpet).

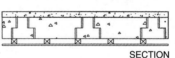

SECTION

- minimum mass per unit area of beams and blocks 220 kg/m²
- bonded sand cement screeds with a minimum thickness of 40 mm
- ceiling finish required

Internal floor type C

Timber or metal joist, with wood-based board and plasterboard ceiling, and absorbent material

Note: Insulation against impact sounds can be improved by adding a soft covering (e.g. carpet).

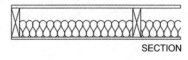

SECTION

- floor of timber or wood-based board, minimum mass per unit area 15 kg/m²
- ceiling treatment of single layer of plasterboard, minimum mass per unit area 10 kg/m², fixed using any normal fixing method

- an absorbent layer of
  mineral wool
  (minimum thickness
  100 mm, minimum
  density 10 kg/m^2) laid
  in the cavity

6.38.3 Other considerations

Doors

Lightweight doors with poor perimeter sealing provide a lower standard of sound insulation than walls.

This will reduce the effective sound insulation of the internal wall. Ways of improving sound insulation include ensuring that there is good perimeter sealing or by using a doorset.

Stairs

If the stair is not enclosed, then the potential sound insulation of the internal floor will not be achieved; nevertheless, the internal floor should still satisfy Requirement E2.

Noise reduction

It is good practice to consider the layout of rooms at the design stage to avoid placing noise sensitive rooms next to rooms in which noise is generated. Guidance on layout is provided in BS 8233:1999 Sound insulation and noise reduction for buildings – Code of Practice.

Electrical cables

Electrical cables give off heat when in use and special precautions may be required when they are covered by thermally insulating materials. See BRE BR 262, Thermal Insulation: avoiding risks, Section 2.3.

6.39 Regulation 7 – Materials and workmanship

6.39.1 The requirement

Building work shall be carried out –

(a)   *with adequate and proper materials which –*
   (i)   *are appropriate for the circumstances in which they are used;*
   (ii)   *are adequately mixed or prepared; and*
   (iii) *are applied, used or fixed so as adequately to perform the functions for which they are designed; and*

(b)   in a workmanlike manner.

(Regulation 7)

Parts A to K and N of Schedule 1 to these regulations shall not require anything to be done except for the purpose of securing reasonable standards of health and safety for persons in or about buildings (and any others who may be affected by buildings or matters connected with buildings).

6.39.2 Meeting the requirements

Materials (including products, components, fittings, naturally occurring materials, e.g. stone, timber and thatch, items of equipment, and backfilling for excavations in connection with building work) shall be:

- of a suitable nature and quality in relation to the purposes and conditions of their use;
- adequately mixed or prepared (where relevant); and
- applied, used or fixed so as to perform adequately the functions for which they are intended.

Environmental impact of building work

The environmental impact of building work can be minimized by careful choice of materials and the use of recycled and recyclable materials.	Reg. 7 (0.2)
The use of such materials must not have any adverse implications for the health and safety standards of the building work.	Reg. 7 (0.2)

Limitations

Reasonable standards of health or safety for persons in or about the building shall be assured.	Reg. 7 (0.3)
Materials and workmanship shall ensure that fuel and power is conserved.	Reg. 7 (0.3)
Access and facilities for disabled people shall be provided.	Reg. 7 (0.3)

There are no provisions under the Building Regulations for continuing control over the use of materials following the completion of building work.

Fitness of materials

The suitability of material used for a specific purpose shall be assessed using:	Reg. 7 (1.2 to 1.7)

- British Standards
- European Standards (ENs)
- other national (and international) technical specifications (from other European Community member states)
- technical approvals (covered by a national or European certificate issued by a European technical approvals issuing body)

CE Marking (which provides a presumption of conformity with the stated minimum legal requirements)

- independent (UK) product certification schemes (such as those accredited by UKAS)
- tests and calculations (for example, those in conformance with UKAS's Accreditation Scheme for Testing Laboratories)
- past experience (that the material can be shown by experience, to be capable of performing the function for which it is intended)
- sampling (of materials by local authorities).

Short-lived materials and materials susceptible to changes in their properties should only be used in works where these changes do not adversely affect their performance.

Resistance to moisture

Any material that is likely to be adversely affected by condensation, moisture from the ground, rain or snow shall either resist the passage of moisture to the material or be treated or otherwise protected from moisture.	Reg. 7 (1.8)

Resistance to substances in the subsoil

Any material in contact with the ground (or in the foundations) shall be capable of resisting attacks by harmful materials in the subsoil such as sulphates.	Reg. 7 (1.9)

Establishing the adequacy of workmanship

The adequacy (and competence) of workmanship will normally be established by using:

- British Standard code of practice;
- workmanship specified and covered by a national or European certificate issued by a European technical approvals issuing body;
- the recommendations of an integrated management system (such as ISO 9001: 2000);
- past experience (of workmanship that is capable of performing the function for which it is intended);
- tests (for example, the local authority has the power to test sewers and drains in or in connection with buildings).

Workmanship on building sites

In the main, local authorities will use the codes of practice as contained and detailed in BS 8000 (Workmanship on Building Sites) to monitor and inspect all building work. These codes of practice consist of:

- Part 1: 1989 Code of practice for excavation and filling
- Part 2: Code of practice for concrete work
    Section 2.1: 1990 Mixing and transporting concrete
    Section 2.2: 1990 Sitework with in situ and precast concrete
- Part 3: 1989 Code of practice for masonry
- Part 4: 1989 Code of practice for waterproofing
- Part 5: 1990 Code of practice for carpentry, joinery and general fixings
- Part 6: 1990 Code of practice for slating and tiling of roofs and claddings
- Part 7: 1990 Code of practice for glazing
- Part 8: 1994 Code of practice for plasterboard partitions and dry linings
- Part 9: 1989 Code of practice for cement/sand floor screeds and concrete floor toppings
- Part 10: 1995 Code of practice for plastering and rendering
- Part 11: Code of practice for wall and floor tiling
    Section 11.1: 1989 Ceramic tiles, Terrazzo tiles and mosaics (confirmed 1995)
    Section 11.2: 1990 Natural stone tiles
- Part 12: 1989 Code of practice for decorative wall coverings and painting
- Part 13: 1989 Code of practice for above ground drainage and sanitary appliances
- Part 14: 1989 Code of practice for below ground drainage
- Part 15: 1990 Code of practice for hot and cold water services (domestic scale)
- Part 16: 1997 Code of practice for sealing joints in buildings using sealants.

6.40 Work on existing constructions

If an existing building is going to be converted into dwelling houses (and/or flats) via a 'material change of use' then a certain amount of remedial work to the existing construction will first have to be completed. In the paragraphs below I have listed the most important areas that should be addressed, but if you are contemplating carrying out this sort of work, then it would be best to have a chat with your local authorities as they might have specific requirements and rulings for your own particular area.

6.40.1 Walls

In any type of building, the amount of sound resistance provided by a wall depends on:

- its construction;
- the type of independent panel(s) it uses (if any);
- the isolation of these panel(s);
- the type of absorbent material that has been used.

If the existing wall is masonry (and has a thickness of at least 100 mm and is plastered on both faces) then the following wall treatment (commonly referred to as *Wall treatment 1: Independent panel(s) with absorbent material*) is recommended.

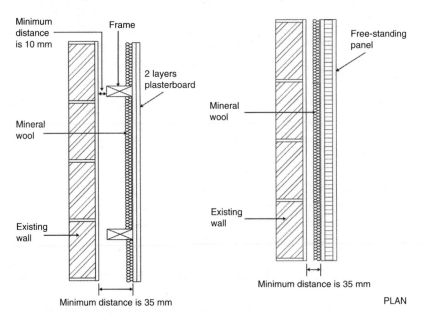

Figure 6.207 Wall treatment 1: Independent panel(s) with absorbent material

In particular:

The independent panel and its supporting frame should not be in contact with the existing wall.	E4.25a
The perimeter of the independent ceiling should be sealed with tape or sealant.	E4.25b
The absorbent material should not be tightly compressed as this may bridge the cavity.	E4.25

In addition:

- the minimum mass per unit area of panel (excluding any supporting framework) should be $20\,kg/m^2$;
- each panel should consist of at least 2 layers of plasterboard with staggered joints;
- if the panels are free-standing they should be at least 25 mm from masonry core;
- if the panels are supported on a frame, there should be a gap of at least 10 mm between the frame and the face of the existing wall;
- mineral wool (minimum density $10\,kg/m^3$ and minimum thickness 35 mm) should be used in the cavity between the panel and the existing wall.

 Note: Wall linings may be required to reduce flanking transmission.

6.40.2 Floors

In buildings, the amount of resistance to airborne and impact sound will depend on:

- the combined mass of the existing floor and the independent ceiling;
- the amount of absorbent material;
- the isolation of the independent ceiling;
- the airtightness of the whole construction.

Two types of floor treatment are recommended dependent on the type of construction and material that has been used for the existing floor. The following requirements are common to both treatments:

- if the existing floor is timber, then gaps in the floor boarding should be sealed by overlaying with hardboard (alternatively, the gap should be filled with sealant);
- if floor boards are going to be replaced, boarding (minimum thickness 12 mm) and mineral wool (minimum thickness 100 mm, minimum density $10\,kg/m^3$) should be laid between the joists in the floor cavity;

- if the existing floor is concrete (and the mass per unit area of the concrete floor is less than $300 \, kg/m^2$) then the mass of the floor should be increased to at least $300 \, kg/m^2$;
- any air gaps through the concrete floor should be sealed.

Floor treatment 1: Independent ceiling with absorbent material

The resistance to airborne and impact sound from Floor treatment 1 depends on:

- the combined mass of the existing floor;
- the independent ceiling;
- the absorbent material;
- the isolation of the independent ceiling;
- the airtightness of the whole construction.

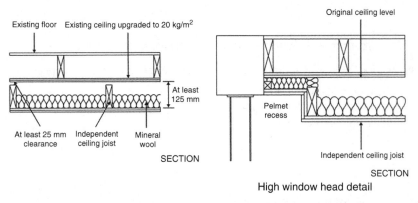

Figure 6.208 Floor treatment 1: Independent ceiling with absorbent material

Specifically:

The ceiling should have: E4.27

- at least 2 layers of plasterboard with staggered joints;
- a minimum total mass per unit area $20 \, kg/m^2$;
- an absorbent layer of mineral wool laid on the ceiling; (minimum thickness $100 \, mm$, minimum density $10 \, kg/m^3$).

The ceiling should be supported by either: E2.47

- independent joists fixed only to the surrounding walls;

📖 **Note:** A clearance of at least $25 \, mm$ should be left between the top of the independent ceiling joists and the underside of the existing floor construction

or

- independent joists fixed to the surrounding walls
  with additional support provided by resilient hangers
  attached directly to the existing floor base.

Where a window head is near to the existing ceiling, the new independent ceiling may be raised to form a pelmet recess.	E4.29
The perimeter of the independent ceiling should be sealed with tape or sealant.	E4.30
A rigid or direct connection should not be created between the independent ceiling and the floor base.	E4.30
The absorbent material should not be tightly compressed as this may bridge the cavity.	E4.30

Floor treatment 2: Platform floor with absorbent material

With Floor treatment 2, the resistance to airborne and impact sound depends on:

- the total mass of the floor;
- the effectiveness of the resilient layer;
- the absorbent material.

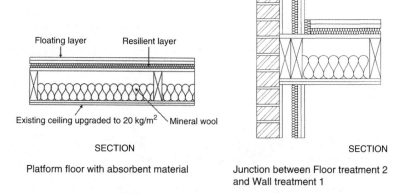

Floating layer   Resilient layer

Existing ceiling upgraded to 20 kg/m² — Mineral wool

SECTION

Platform floor with absorbent material

SECTION

Junction between Floor treatment 2
and Wall treatment 1

Figure 6.209 Floor treatment 2: Platform floor with absorbent material

Specifically:

Where this treatment is used to improve an existing timber floor, a layer of mineral wool (100 mm, minimum density 10 kg/m³) should be laid between the joists in the floor cavity.	E4.32

The floating layer should be a:	E4.32
• minimum of two layers of board material; • minimum total mass per unit area 25 kg/m².	
Each layer should be:	E4.32
• a minimum thickness of 8 mm; • fixed together (e.g. spot bonded or glued/screwed) with joints staggered; • laid loose on a resilient layer.	
The resilient layer should be:	E4.32
• mineral wool (minimum thickness 25 mm, density 60–100 kg/m³); • paper faced on the underside.	
The correct density of resilient layer (to carry the anticipated load) should be assured.	E4.32
The probable movement of materials after laying (e.g. expansion of chipboard) should be taken into account.	E4.32
The resilient layer should be carried up at all room edges to isolate the floating layer from the wall surface.	E4.32
A small 5 mm gap should be left between the skirting and floating layer and filled with a flexible sealant.	E4.32
Resilient materials should be laid in sheets with joints tightly butted and taped.	E4.32
The perimeter of any new ceiling should be sealed with tape or sealant.	E4.32
The floating layer and the base or surrounding walls should not be bridged with services or fixings that penetrate the resilient layer.	E4.32

6.40.3 Ceilings

If there is an existing lath and plaster ceiling it should be retained as long as it satisfies Building Regulation Part B – Fire safety.

If the existing ceiling is not lath and plaster, it should be upgraded to provide at least two layers of plasterboard with joints staggered, total mass per unit area 20 kg/m².

Note: Care should be taken at the design stage to ensure that adequate ceiling height is available in all rooms to be treated.

6.40.4 Corridor walls and doors

A separating wall should be used between the new dwellings (and/or flats) and the adjacent corridor in order to control flanking transmission and provide the required amount of sound insulation.

> **Note:** It is likely that the sound insulation will be reduced by the presence of a door.

In particular, measures should be taken to ensure that any door:

• has good perimeter sealing (including the threshold where practical);	E4.20
• has a minimum mass per unit area of 25 kg/m^2 or a minimum sound reduction index of 29 dB R_W (measured according to BS EN ISO 1403: 1995 and rated according to BS EN ISO 7171: 1997);	
• satisfies the Requirements of Building Regulation Part B – Fire safety.	E4.20
Noisy parts of the building should preferably have a lobby, double door or high performance doorset to contain the noise	E4.21

Note: These facilities should also comply with Building Regulation Part M – Access and facilities for disabled people.

6.40.5 Stairs

If stairs form a separating function then they are subject to the same sound insulation requirements as floors. In all cases, the resistance to airborne sound depends mainly on:

- the mass of the stair;
- the mass and isolation of any independent ceiling;
- the airtightness of any cupboard or enclosure under the stairs;
- the staircovering (which reduces impact sound at source).

The following wall treatment (commonly referred to as *Stair treatment 1: Stair covering and independent ceiling with absorbent material*) is recommended.
    It should be noted:

The soft covering should be:	E4.37
• at least 6 mm thickness;	
• laid over the stair treads;	

- be securely fixed (e.g. glued) so it does not
  become a safety hazard.

If there is a cupboard under all, or part, of the stair: E4.37

- the underside of the stair within the cupboard
  should be lined with plasterboard (minimum mass
  per unit area $10\,kg/m^2$) together with an absorbent
  layer of mineral wool (minimum density $10\,kg/m^3$);
- the cupboard walls should be built from two layers of
  plasterboard (or equivalent), each sheet with a
  minimum mass per unit area $10\,kg/m^2$;
- a small, heavy, well fitted door should be fitted to
  the cupboard.

If there is no cupboard under the stair, an independent   E4.37
ceiling should be constructed below the stair
(see Floor treatment 1).

Where a staircase performs a separating function it shall   E4.38
conform to Building Regulation Part B – Fire safety.

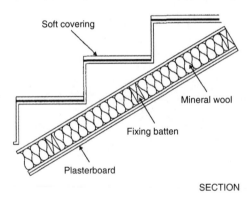

SECTION

Figure 6.210 Stair treatment 1: Stair covering and independent ceiling with absorbent material

6.40.6 Junction requirements for material change of use

There are three recommended types of junctions that can be made:

Junctions with abutting construction

The perimeter of any new ceiling should be sealed   E4.41
with tape or caulked with sealant.

For floating floors:

- the resilient layer shall be carried up at all room   E4.39
  edges to isolate the floating layer from the wall surface;
- a small 5 mm gap, filled with flexible sealant, should   E4.40
  be left between the skirting and floating layer.

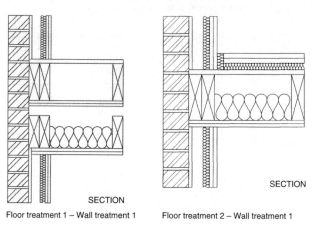

Floor treatment 1 – Wall treatment 1   Floor treatment 2 – Wall treatment 1

Figure 6.211 Junctions with abutting construction

Junctions with external or load bearing walls

If the adjoining masonry wall has a mass per unit area less than
375 kg/m^2, then the walls should be lined with either:

- an independent layer of plasterboard, or   E4.43a
- a laminate of plasterboard and mineral wool.   E4.43b

Note: Specialist advice may be needed on the diagnosis and control of flanking transmission.

Junctions with floor penetrations

Piped services (excluding gas pipes) and ducts which pass   E4.45
through separating floors in conversions should be surrounded
with sound absorbent material for their full height and
enclosed in a duct above and below the floor.

The joint between casings and ceiling should be   E4.46
sealed with tape or

A nominal 5 mm gap, sealed with sealant, should be left between the casing and any floating layer.	E4.46
Pipes and ducts that penetrate a floor separating habitable rooms in different flats should be enclosed for their full height in each flat.	E4.46
The enclosure should be constructed of material having a mass per unit area of at least $15\,kg/m^2$.	E4.47
Either the enclosure should be lined or the duct or pipe within the enclosure should be wrapped with 25 mm unfaced mineral wool.	E4.48
The enclosure may go down to the floor base if Floor treatment 2 is used but ensure isolation from the floating layer.	E4.49
Penetrations through a separating floor by ducts and pipes should have fire protection to satisfy Building Regulation Part B – Fire safety.	E4.50
Fire stopping should be flexible and be capable of preventing a rigid contact between the pipe and floor.	E4.50
Gas pipes may be contained in a separate (ventilated) duct or can remain unenclosed.	E3.120
If a gas service is installed it shall comply with the Gas Safety (Installation and Use) Regulations 1998, SI 1998 No 2451.	E3.120

 Note: All of these facilities should also comply with Building Regulation Part M – Access and facilities for disabled people.

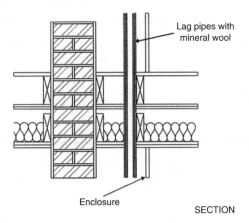

Figure 6.212 Floor penetrations

Appendix A

Access and facilities for disabled people

A.1 Background

During 2002/3, Approved Document M was thoroughly overhauled and restructured in order to meet the changed requirements of the Disability Discrimination Act 1995.

This major rewrite of Part M became effective on 1 May 2004 and was as a result of amendments made to Section 6 of the Disability Discrimination Act 1995 (DDA) which previously stated that '*reasonable adjustments to physical features of premises … shall be made … in certain circumstances*' and '*shall apply to all employers with 15 or more employees*'. As such, an employer with only a few employees (as well as occupations such as police, fire fighters and prison officers) was not required to alter the physical characteristics of a building that met existing requirements at the time the building works were carried out and **still** continued to meet those particular requirements.

 From 1 October 2004, however, under the Disability Discrimination Act 1995 (Amendment) Regulations 2003 (SI 2003/1673), this exemption ceases.

According to the Department of Works and Pensions, it is estimated that the new requirements will affect almost one and a half million private service and over 100 000 public service providers.

 If this situation affects you, then it would be best to seek advice from either the Social Services or the Local Planning Officer.

In a similar manner, up to 30 September 2004, there is no requirement for service providers to make any adjustments to the physical features of their premises.

 From 1 October 2004, however, '*all those who provide services to the public, irrespective of their size*' are required '*to take reasonable steps to remove, alter or provide a reasonable means of avoiding a physical feature of their premises, which makes it unreasonably difficult or impossible for disabled people to make use of their services*'.

In rewriting Part M, the drafting committee also took into consideration BS 8300:2001 (which provides guidance on the design of domestic and non-domestic buildings and their approaches), recent ergonomic research conducted in support of this standard and the need for all people (no matter their disability) to have access and be able to use buildings and their facilities.

 Note: BS 8300:2001 (*Design of buildings and their approaches to meet the needs of disabled people – Code of Practice*) supersedes BS 5619:1978 and BS 5810:1979.

These amendments have naturally resulted in Part M being completely overhauled and it now covers:

- the conversion of a building for use as a shop being redefined as a 'material change of use';
- amendments to omit specific references to (and a definition of) disabled people;
- expansion of the terms to include parents with children, elderly people and people with all types of disabilities (e.g. mobility, sight and hearing etc.);
- the use of a building to disabled people as residents, visitors, spectators, customers or employees, or participants in sports events, performances and conferences (which resulted in amendments being made to M1 (accessibility), M2 (sanitary accommodation) and M3 (audience and/or spectator seating)).

 The current edition, therefore, no longer primarily concentrates on wheelchair users, but includes people using walking aids, people with impaired sight (and other mobility and sensory problems), mothers with prams as well as people with luggage etc.

A.1.1 Benefits resulting from these changes

In producing these changes, the government considers that the following benefits to society will result:

Benefits to disabled people:

- improved legal rights of access to goods, facilities and services;
- better opportunity to play as full a role as possible in the economy and in society;
- reduced financial cost of injuries resulting from negotiating inadequately accessible premises;
- reduced travelling costs as more services closer to home/work become accessible;
- wider range of facilities disabled people can enjoy with carers, friends and family.

Benefits to other people with difficulties such as:

- people with young children;
- elderly people;

- people encumbered with luggage, shopping bags etc.;
- people with temporary impairments (e.g. people with broken limbs).

Benefits to business/service providers:

- better public image;
- easier movement of goods;
- reduced need for home support services;
- reduction in accidents where there are lifts, safer stairs, handrails, better lighting, fewer obstructions and even floor surfaces;
- increased tourism where there is sufficient accommodation for wheelchair users.

A.2 The requirements

M1 – Access and use

Reasonable provision shall be made for people to:

(a)   gain access to and
(b)   use

the buildings and its facilities

(Approved Document M1)

M2 – Access to extensions of buildings other than dwellings

Suitable independent access shall be provided in any building that is to be extended. Reasonable provision shall be made within the extension for sanitary convenience.

(Approved Document M2)

M3 – Sanitary conveniences in extensions to buildings other than dwellings

If sanitary conveniences are provided in any building that is to be extended, reasonable provision shall be made within the extension for sanitary conveniences.

(Approved Document M3)

M4 – Sanitary conveniences in dwellings

Reasonable provision shall be made in the entrance storey for sanitary conveniences, or where the entrance contains no habitable rooms, reasonable provision for sanitary convenience shall be made in either the entrance storey or principal storey.

(Approved Document M4)

A.2.1 Requirement in a nutshell

In addition to the requirements of the *Disability Discrimination Act 1995 (Amendment) Regulations 2003*, precautions need to be taken to ensure that:

- new non-domestic buildings and/or dwellings (e.g. houses and flats used for student living accommodation etc.);
- extensions to existing non-domestic buildings;
- non-domestic buildings that have been subject to a material change of use (e.g. so that they become a hotel, boarding house, institution, public building or shop);

are capable of allowing people, regardless of their disability, age or gender, to:

- gain access to buildings;
- gain access within buildings;
- be able to use the facilities of the buildings (both as visitors and as people who live or work in them).

 Note: The requirements of this part do **not** apply to:

(a)   an extension of, or material alteration of, a (domestic) dwelling; or
(b)   any part of a building which is used solely to enable the building or service (or fitting in the building) to be inspected, repaired or maintained.

  The requirements are also limited if:

- a building is listed as being of architectural or historic interest;
- the cost of providing a fully accessible route is disproportionate to the cost of the intended facility;
- the physical restraints imposed by the building make full compliance impossible or impracticable.

A.2.2 Changes to Approved Document M

Extensions to non-domestic buildings

With the new revision of Part M, an extension to a non-domestic building should now be treated in the same manner as a new building for compliance with Approved Document M which means that:

- there must be '*suitable independent access to the extension where reasonably practicable*';
- if a building is to be extended, '*reasonable provision must be made within the extension for sanitary conveniences*'.

 Note: This requirement does not apply if it is possible for people using the extension to gain access to and be able to use sanitary conveniences in the existing building.

A.2.3 Material alterations of non-domestic buildings

Under Regulation 4, where an alteration of a non-domestic building is a material alteration:

- the work itself must comply, where relevant, with Requirement M1;
- reasonable provision must be made for people to gain access to and use new or altered sanitary conveniences.

A.2.4 Material changes of use

Where there is a material change of use of a whole building to a hotel, boarding house, institution, public building or a shop (restaurant, bar or public house) the building must be upgraded, if necessary, so as to comply with M1 (Access and use).

If an existing building does undergo a change of use so that part of it can be used as a hotel, boarding house, institution, public building or a shop, the work being carried out must ensure that:

- people can gain access from the site boundary and any on-site car parking space;
- sanitary conveniences are provided in that part of the building or it is possible for people (no matter their disability) to use sanitary conveniences elsewhere in the building.

A.2.5 What requirements apply

Buildings other than dwellings

1. Regardless of disability, age or gender it should be possible for people:
    - to reach the principal entrance to the building from the site boundary and from car parking within the site and from other buildings on the same site (such as a university campus, a school or a hospital);
    - to have access into and within any storey of the building;
    - to have access into and use of the building's facilities.
2. The structure and amenities of a building should not constitute a hazard to users (especially people with impaired sight).
3. Suitable accommodation should be made available for people in wheelchairs (or people with other disabilities), in audience or spectator seating.
4. People with a hearing or sight impairment should be provided with some form of aid to communication in auditoria, meeting rooms, reception areas, ticket offices and at information points.
5. Sanitary accommodation should be available for **all** users of the building.

Dwellings

People, including disabled people, should be able:

- to reach the principal, or suitable alternative, entrance to the dwelling from the point of access;

- to gain access into and within the principal storey of the dwelling;
- to gain access to sanitary conveniences at no higher storey than the principal storey.

A.2.6 Educational establishments

From 1 April 2001, maintained schools ceased to have exemption from the Building Regulations and school-specific standards have now been incorporated into the latest editions of Approved Documents.

Purpose-built student living accommodation (including flats) should thus be treated as hotel/motel accommodation in respect of space requirements and internal facilities.

A.2.7 Access Statements

To assist building control bodies it is recommended that an 'Access Statement' is also provided when plans are deposited, a building notice is given, or details of a project are provided to an approved inspector.

 Note: A building control file should also be prepared for all new buildings, changes of use and where extensive alterations are being made to existing buildings.

In its simplest form, an Access Statement should show where an applicant wishes to deviate from the guidance in Approved Document M, either to:

- make use of new technologies (e.g. infrared activated controls);
- provide a more convenient solution; or
- address the constraints of an existing building.

The Access Statement should include:

- the reasons for departing from the guidance;
- the rationale for the design approach adopted;
- constraints imposed by the existing structure and its immediate environment (why it is not practicable to adjust the existing entrance or provide a suitable new entrance);
- convincing arguments that an alternative solution will achieve the same, a better, or a more convenient outcome (e.g. why a fully compliant independent access is considered impracticable);
- evidence (e.g. current validated research) to support the design approach;
- the identification of buildings (or particular parts of buildings) where access needs to be restricted (e.g. processes that are carried out which might create hazards for children, disabled people or frail, elderly people).

 Note: Further guidance on Access Statements is available on the Disability Rights Commission's website at http://www.drc-gb.org.

A.3 Meeting the requirements

A.3.1 Access

Access (i.e. approach, entry or exit) to a building is frequently a problem for wheelchair users, people who need to use walking aids, people with impaired sight and mothers with prams etc. In designing the approach to a building (and routes between buildings within a complex) the following should, therefore, always be taken into consideration:

- changes in level between the entrance storey and the site entry point should be minimized;
- access routes should be wide enough to let people pass each other;
- potential hazards (e.g. windows from adjacent buildings opening onto access routes) should be avoided.

For all new buildings, therefore, the primary aim should be to make it reasonably possible for a disabled person to approach and gain access into the dwelling from the point of alighting from a vehicle (which may be within or outside the plot). In particular the following should be observed.

Access to a building

Approach

- the principal entrance should be accessible to disabled people;
- access from the boundary of the site to the principal entrance should be level;
- access from car parking designated for disabled people to the principal entrance should be level;
- risks to people when entering the building should be minimal;
- access routes should be wide enough to let people pass each other;
- the route to the principal entrance should be clearly identified and well lit;
- the route to the principal entrance should be well lit;
- uncontrolled vehicular crossing points should be identified.

Approach gradients

- should ideally be no steeper than 1:60 along its whole length;
- cross-fall gradients should be no steeper than 1:40.

Building perimeters

- the perimeter of the building should be well lit.

Surface of access routes

The surface of access routes should:

- allow people to travel along them easily, without excessive effort and without the risk of tripping or falling;
- be at least 1.5 m wide;

- be firm, durable and slip resistant;
- have undulations not exceeding 3 mm;
- be made of the same material and similar frictional characteristics.

Joints

Joints should:

- be filled flush or (if recessed) no deeper than 5 mm;
- be no wider than 10 mm or (if unfilled) no wider than 5 mm;
- have a difference in level (at joints between paving units) no greater than 5 mm.

Passing places

Passing places should be:

- free of obstructions to a height of 2.1 m;
- at least 1.8 m wide and at least 2 m long.

On-site car parking and setting down – parking bays

At least one parking bay designated for disabled people should be provided as close as possible to the principal entrance of the building and these should:

- be clearly signposted;
- have the setting down point located on firm, level ground;
- have the surface of the parking bay firm, durable and slip resistant;
- have ticket machines located nearby and positioned so that a person in a wheelchair (or a person of short stature) is able to reach the controls;
- not allow the plinth of the ticket machine to project in front of the face of the machine.

Ramped access

- the approach should be clearly signposted;
- the going should be no greater than 10 m;
- the rise should be no more than 500 mm;
- the surface width should be at least 1.5 m;
- gradients should be as shallow as practicable;
- the ramp surface should be slip resistant and of a contrasting colour to that of the landings;
- frictional characteristics of ramp and landing surfaces should be similar;
- landings at the foot and head of a ramp should be at least 1.2 m long and clear of any obstructions;
- intermediate landings should be at least 1.5 m long and clear of obstructions;
- all landings should be:
  - level;
  - have a maximum gradient of 1:60 along their length;
  - have a maximum cross fall gradient of 1:40;
  - have a handrail on both sides.

Stepped access

A stepped access should:

- have a level landing at the top and bottom of each flight which is at least 1200 mm long and unobstructed;
- have a corduroy hazard warning surface at top and bottom landings of a series of flights;
- ensure that any side accesses (onto intermediate landings) include a deep corduroy hazard warning surface;
- have no doors that swing across landings;
- have the rise of a flight between landings contain no more than 18 risers;
- have nosings (for the tread and the riser) of a contrasting material;
- not allow step nosings to project over the tread below by more than 25 mm;
- ensure that the rise and going of each step is consistent throughout a flight;
- include a continuous handrail on each side of a flight and landings;
- ensure that tread materials do not present a slip hazard.

Handrails to external stepped and ramped access

Handrails to external stepped or ramped access should:

- be continuous across flights and landings;
- extend at least 300 mm horizontally beyond the top and bottom of a ramped access;
- not project into an access route;
- contrast visually with the background;
- have a slip resistant surface which is **not** cold to the touch;
- terminate in such a way that reduces the risk of clothing being caught;
- either be circular (with a diameter of between 40 and 45 mm) or oval with a width of 50 mm;
- not protrude more than 100 mm into the surface width of the ramped or stepped access;
- have a clearance of between 60 and 75 mm between the handrail and any adjacent wall surface;
- have a clearance of at least 50 mm between a cranked support and the underside of the handrail;
- ensure that its inner face is located no more than 50 mm beyond the surface width of the ramped or stepped access;
- be spaced away from the wall and rigidly supported in a way that avoids impeding finger grip;
- be set at heights that are convenient for all users of the building.

Hazards on access routes

- no 'feature' of a building (e.g. windows and doors) should obstruct an access route;

- areas below stairs or ramps with a soffit less than 2.1 m above ground level should be protected by guarding and low level cane detection;
- any feature projecting more than 100 mm onto an access route should be protected by guarding that includes a kerb (or other solid barrier) that can be detected using a cane.

Accessible entrances

Accessible entrances should:

- be clearly signposted;
- be easily identifiable;
- not present a hazard for visually impaired people;
- have a level landing at least 1500 × 1500 mm, clear of any door swings, immediately in front of the entrance;
- avoid raised thresholds;
- ensure that all door entry systems are accessible to deaf and hard of hearing people, plus people who cannot speak;
- not have internal floor surface material (e.g. coir matting) adjacent to the threshold;
- not have changes in floor materials that could create a potential trip hazard;
- have the route from the exterior across the threshold weather protected.

Doors to accessible entrances

Doors to the principal (or alternative accessible) entrance should:

- be easily reached (particularly for wheelchair users and people with limited physical dexterity);
- be capable of being held closed when not in use;
- not require a great deal of force to open or shut a door;
- be wide enough to allow unrestricted passage for a variety of users, including wheelchair users, people carrying luggage, people with assistance dogs, and parents with pushchairs and small children;
- enable people to see other people approaching from the opposite direction;
- incorporate vision panels if the door leaves, and side panels are wider than 450 mm.

Entrance lobbies

Entrance lobbies should:

- be large enough to allow a wheelchair user or a person pushing a pram to move clear of one door before opening the second door;
- be capable of accommodating a companion helping a wheelchair user to open doors and guide the wheelchair through.

Within the lobby:

- glazing should not create distracting reflections;
- floor surface materials should do not impede the movement of wheelchairs etc.;

- changes in floor materials should not create a potential trip hazard;
- the floor surface should assist in removing rainwater from shoes and wheelchairs;
- any columns and ducting etc. that projects into the lobby by more than 100 mm should be protected by a visually contrasting guard rail.

Manually operated non-powered entrance doors

- the opening force at the leading edge of the door should be no greater than 20 N;
- a space alongside the leading edge of a door should be provided to enable a wheelchair user to reach and grip the door handle;
- door opening furniture should:
  - be easy to operate by people with limited manual dexterity;
  - be capable of being operated with one hand using a closed fist (e.g. a lever handle);
  - contrast visually with the surface of the door;
  - not be cold to the touch;
- there should be an unobstructed space of at least 300 mm on the pull side of the door between the leading edge of the door and any return wall.

Powered-operated entrance doors

Power-operated entrance doors should:

- have a sliding, swinging or folding action that can be operated manually, remotely or automatically;
- open towards people approaching the doors;
- provide visual and audible warnings that they are operating (or about to operate);
- incorporate automatic sensors to ensure that they open early enough (and stay open long enough) to permit safe entry and exit;
- incorporate a safety stop that is activated if the doors begin to close when a person is passing through;
- revert to manual control (or fail safe) in the open position in the event of a power failure;
- when open, not project into any adjacent access route;
- ensure that its manual controls:
  - are located between 750 mm and 1000 mm above floor level;
  - are operable with a closed fist;
- be set back 1400 mm from the leading edge of the door when fully open;
- be clearly distinguishable against the background;
- contrast visually with the background.

Glass entrance doors and glazed screens

- the presence of the door should be apparent not only when it is shut but also when it is open;

- glass entrance doors and glazed screens should:
  - be clearly marked (i.e. with a logo or sign) on the glass at two levels;
  - be provided with a high contrast strip at the top and side when adjacent to (or forming part of) a glazed screen;
  - be capable of being held open;
  - be protected by guarding to prevent the leading edge from becoming a possible hazard.

Access to a dwelling

Approach

For dwellings:

- an external door providing access for disabled people should have a minimum clear opening width of 775 mm;
- the door opening width should be sufficient to enable a wheelchair user to manoeuvre into the dwelling;
- the approach to a principal entrance should be reasonably level;
- wheelchair users (having approached the entrance) should be able to gain access into the dwelling-house and/or entrance level of flats;
- a suitable approach should be provided from the point of access to the entrance of the dwelling;
- the whole, or part, of the approach may be a driveway;
- the approach should:
  - not have crossfalls greater than 1 in 40;
  - be safe and convenient for disabled people as is reasonably possible;
  - ideally be level or ramped;
  - be firm enough to support the weight of the user and his or her wheelchair;
  - be smooth enough to permit easy manoeuvre;
  - not be made up of loose laid materials (such as gravel and shingle);
  - take account of the needs of stick and crutch users.

Level approach

A 'level' approach should:

- be no steeper than 1 in 20;
- have a firm and even surface;
- have a width not less than 900 mm.

Ramped approach

If a plot gradient exceeds 1 in 20, a ramped approach may be provided in which case:

- the surface should be firm and even;
- the flights should have unobstructed widths of at least 900 mm;
- individual flights should be no longer than 10 m (for gradients not steeper than 1 in 15) or 5 m for gradients not steeper than 1 in 12;

- it should have top and bottom landings (and intermediate landings if necessary) not less than 1.2 m in length – exclusive of the swing of any door or gate which opens onto it.

Stepped approach

A stepped approach should:

- be used if the plot gradient is greater than 1 in 15;
- have flights with an unobstructed width of at least 900 mm;
- have a flight rise of not more than 1.8 m;
- have a top and bottom (and if necessary intermediate landings) not less than 900 mm in length;
- have steps with suitable tread nosing profiles and a uniform rise between 75 mm and 150 mm;
- ensure that the going of each step is not less than 280 mm.

Access into the dwelling

- an accessible threshold should be provided into the entrance;
- the point of access should be reasonably level;
- where the approach to the entrance consists of a level or ramped approach, an accessible threshold at the entrance should be provided.

Vertical circulation within the building

For all buildings, a passenger lift is considered the most suitable form of access for people moving from one storey of the building to another. The actual size of the lift should be chosen to suit the anticipated density of use of the building and the needs of disabled but in all cases should **not** be less than 1100 mm wide and 1400 mm deep. In some circumstances a lifting platform may be provided but only to transfer wheelchair users and people with impaired mobility (plus their companions) vertically between levels or storeys.

Lifting devices

- wherever possible a passenger lift (or a lifting platform) serving all storeys should be provided;
- the location of lifting devices should be clearly visible from the building entrance;
- signs should be available at each landing to identify the floor reached by the lifting device;
- in addition to the lifting device, internal stairs (designed to suit ambulant disabled people and those with impaired sight) should always be provided.

General requirements for lifting devices

- an unobstructed manoeuvring space should be available in front of each lifting device;

- landing call buttons and lifting device control button symbols:
  - should contrast visually with the surrounding face plate;
  - should be raised to facilitate tactile reading;
  - should be accessible for wheelchair users;
- the floor of the lifting device should be of a light colour;
- a handrail should be provided on at least one wall;
- an emergency communication system should be fitted.

Passenger lifts

Passenger lifts should:

- be accessible from the remainder of the storey;
- have power-operated horizontal sliding doors with an effective clear width of at least 800 mm;
- have doors fitted with timing devices (and reopening activators);
- locate the car controls and landing call buttons within easy reach of wheelchair users;
- be fitted with lift landing and car doors that are visually distinguishable from the adjoining walls;
- include (in the lift car **and** the lift lobby) audible and visual information that a lift has arrived, which floor it has reached and where in a bank of lifts it is located;
- ensure that all glass areas can be easily identified by people with impaired vision;
- conform to BS 5588-8 if the lift is to be used to evacuate disabled people in an emergency;
- not have visually and acoustically reflective wall surfaces.

Lifting platforms

Lifting platforms should:

- restrict the vertical travel distance to no more than 2 m if there is no liftway enclosure and/or floor penetration;
- restrict the rated speed of the platform so that it does not exceed 0.15 m/s;
- locate their controls between 800 mm and 1100 mm from the floor of the lifting platform and at least 400 mm from any return wall;
- locate all landing call buttons within easy reach of wheelchair users;
- have continuous pressure controls (e.g. push buttons);
- have doors with an effective clear width of at least 800 mm;
- be fitted with clear instructions for use;
- have their entrances accessible from the remainder of the storey;
- have doors that are visually distinguishable from adjoining walls;
- have an audible and visual announcement of platform arrival and level reached;
- ensure that all areas of glass can be identified by people with impaired vision.

Wheelchair platform stairlifts

- wheelchair platform stairlifts are only intended for the transportation of wheelchair users and should only be considered for conversions and alterations where it is not practicable to install a conventional passenger lift or a lifting platform;
- the speed of the platform should not exceed 0.15 m/s;
- continuous pressure controls (e.g. joystick) should be provided;
- the platform should have minimum clear dimensions of 800 mm wide and 1250 mm deep;
- wheelchair platform stairlifts should:
  - be fitted with clear instructions for use;
  - provide a clear width of at least 800 mm;
  - not be installed where their operation restricts the safe use of the stair by other people.

Internal steps, stairs and ramps

Stepped access

- a stepped access should have a level landing (1200 mm long) at the top and bottom of each flight;
- doors should not be allowed to swing across landings;
- the surface width of flights between enclosing walls should not be less than 1.2 m;
- there should be no single steps;
- tread and riser nosings should be 55 mm wide and of a contrasting material;
- step nosings should not project over the tread below by more than 25 mm;
- the rise and going of each step should be consistent (e.g. between 150 mm and 170 mm) throughout a flight;
- the rise of each step should be between 150 mm and 170 mm;
- the going of each step should be should be at least 250 mm;
- rises should not be open;
- there should be a continuous handrail on each side of a flight and landings;
- flights between landings should contain no more than 12 risers;
- areas below stairs or ramps with a soffit less than 2.1 m above ground level should be protected by guarding and low level cane detection;
- features projecting more than 100 mm onto an access route should be protected by guarding that includes a kerb.

Internal ramps

- if the change in level is no greater than 300 mm, a ramp should be provided instead of a single step;
- if the change in level is greater than 300 mm, at least two clearly signposted steps should be provided;
- the approach should be clearly signposted;
- the going should be no greater than 10 m;

- the rise should be no more than 500 mm;
- the ramp surface should be slip resistant and of a contrasting colour to that of the landings;
- the frictional characteristics of ramp and landing surfaces should be similar;
- landings at the foot and head of a ramp should be at least 1.2 m long and clear of any obstructions;
- intermediate landings should be at least 1.5 m long and clear of obstructions;
- all landings should be level;
- there should be a handrail on both sides;
- there should be a visually contrasting kerb on the open side of the ramp or landing;
- areas below stairs should be protected by guarding and/or low level cane detection;
- no feature should/project more than 100 mm onto an access route unless protected by a guard;
- gradients should be as shallow as practicable.

Handrails to internal steps, stairs and ramps

Handrails to external stepped or ramped access should:

- be continuous across flights and landings;
- extend at least 300 mm horizontally beyond the top and bottom of a ramped access;
- not project into an access route;
- contrast visually with the background;
- have a slip resistant surface which is not cold to the touch;
- terminate in such a way that reduces the risk of clothing being caught;
- either be circular or oval;
- not protrude more than 100 mm into the surface width of the ramped or stepped access;
- have a clearance of between 60 and 75 mm between the handrail and any adjacent wall surface;
- have a clearance of at least 50 mm between a cranked support and the underside of the handrail;
- ensure that its inner face is located no more than 50 mm beyond the surface width of the ramped or stepped access;
- be spaced away from the wall and rigidly supported in a way that avoids impeding finger grip;
- be set at heights that are convenient for all users of the building.

Entrance hall and reception area

- if there is a reception point it should be:
  - easily accessible and convenient to use;
  - located away from the principal entrance;
  - easily identifiable from the entrance doors or lobby;

- designed to accommodate both standing and seated visitors;
- provided with a hearing enhancement system;
- relevant information about the building should be clearly available from notice boards and signs;
- the floor surface should be slip resistant;
- the approach should:
  - be direct and free from obstructions;
  - allow space for wheelchair users to gain access;
- there should be a clear manoeuvring space in front of the reception desk (at least 1200 mm deep and 1800 mm wide);
- if there is a knee recess it should be at least 500 mm.

Provision of toilet accommodation

General

- wheelchair-accessible unisex toilets should always be provided in addition to any wheelchair-accessible accommodation in separate-sex toilet washrooms;
- where sanitary facilities are provided in a building, at least one wheelchair-accessible unisex toilet should be available;
- at least one WC cubicle should be in separate-sex toilet accommodation;
- if there is only space for **one** toilet in a building:
  - then it should be a wheelchair-accessible unisex type;
  - its width should be increased from 1.5 m to 2 m;
  - it should include a standing height wash basin, in addition to the finger rinse basin associated with the WC.

Sanitary accommodation generally

Doors

- doors to WC cubicles and wheelchair-accessible unisex toilets should:
  - (ideally) open outwards;
  - be operable by people with limited strength or manual dexterity;
  - be capable of being opened if a person has collapsed against them while inside the cubicle;
- doors to wheelchair-accessible unisex toilets/changing rooms or shower rooms should:
  - be fitted with light action privacy bolts;
  - be capable of being opened using a force no greater than 20 N;
  - have an emergency release mechanism so that they are capable of being opened outwards (from the outside) in case of emergency;
- door opening furniture should:
  - be easy to operate by people with limited manual dexterity;
  - be easy to operate with one hand using a closed fist (e.g. a lever handle);
  - contrast visually with the surface of the door;
- doors when open should not obstruct emergency escape routes.

Sanitary fittings

- surface finish of sanitary fixtures/fittings should contrast visually with wall and floor finishes;
- taps should be operable by people with limited strength and/or manual dexterity;
- bath and wash basin taps should either be controlled automatically or capable of being operated using a closed fist, e.g. by lever action.

Alarms

- fire alarms should emit an audio and visual signal to warn occupants with hearing or visual impairments;
- emergency assistance alarm systems should have:
  - visual and audible indicators to confirm that an emergency call has been received;
  - a reset control reachable from a wheelchair, WC, or from a shower/changing seat.

Outlets, controls and switches

- all controls and switches should:
  - be easy to operate, visible and free from obstruction;
  - be located between 750 mm and 1200 mm above the floor;
  - not require the simultaneous use of both hands (unless necessary for safety reasons) to operate;
- light switches should:
  - have large push pads;
  - align horizontally with door handles;
  - be within the 900 to 1100 mm from the entrance door opening;
- mains/circuit isolator switches and switched socket outlets should clearly indicate whether they are 'on';
- individual switches on panels and on multiple socket outlets should be well separated;
- controls that need close vision (e.g. thermostats) should be located between 1200 mm and 1400 mm above the floor;
- emergency alarm pull cords should:
  - be coloured red;
  - be located as close to a wall as possible;
  - have two red 50 mm diameter bangles;
- front plates should contrast visually with their backgrounds;
- heat emitters should either be screened or have their exposed surfaces kept at a temperature below 43°C;
- where possible, light switches with large push pads should be used in preference to pull cords;
- the colours red and green should not be used in combination as indicators of 'on' and 'off' for switches and controls.

Wheelchair-accessible unisex toilets

General

- where sanitary facilities are provided in a building, at least one wheelchair-accessible unisex toilet should be available;
- wheelchair-accessible unisex toilets should not be used for baby changing;
- wheelchair-accessible unisex toilets should:
  - be located as close as possible to the entrance and/or waiting area of the building;
  - not be located in a way that compromises the privacy of users;
  - be located in a similar position on each floor of a multi-storey building;
  - allow for right- and left-hand transfer on alternate floors;
  - be located on accessible routes that are direct and obstruction-free;
  - always be provided in addition to any wheelchair-accessible accommodation in separate-sex toilet washrooms;
  - not be used for baby changing.

Accessibility

- the approach to a unisex toilet should be separate to other sanitary accommodation;
- wheelchair users should:
  - not have to travel more than 40 m on the same floor to reach a unisex toilet;
  - not have to travel more than combined horizontal distance where the unisex toilet accommodation is on another floor of the building (accessible by passenger lift);
  - be able to approach, transfer to, and use the sanitary facilities provided within a building.

Doors

Doors should:

- preferably open outwards;
- be fitted with a horizontal closing bar fixed to the inside face.

Support rails

- a drop-down and/or wall-mounted grab rail should be provided.

Heights and arrangements

- the space provided for manoeuvring should enable wheelchair users to adopt various transfer techniques that allow independent or assisted use;
- the transfer space alongside the WC should be kept clear to the back wall;
- the relationship of the WC to the finger rinse basin and other accessories should allow a person to wash and dry hands while seated on the WC;
- heat emitters (if located) should not restrict:
  - the minimum clear wheelchair manoeuvring space;
  - the transfer space beside the wheelchair.

Emergency assistance

- emergency assistance alarm systems should have:
  - an outside emergency assistance call signal;
  - visual and audible indicators to confirm that an emergency call has been received;
  - a reset control reachable from a wheelchair, WC, or from a shower/changing seat;
  - a signal that is distinguishable visually and audibly from the fire alarms provided;
- emergency assistance pull cords should:
  - be easily identifiable;
  - be reachable from the WC and from the floor close to the WC;
  - be coloured red;
  - be located as close to a wall as possible;
  - have two red 50 mm diameter bangles.

Toilets in separate-sex washrooms

General

- there should be at least the same number of WCs (for women) as urinals (for men) and vice versa;
- ambulant disabled people should have the opportunity to use a WC compartment within any separate-sex toilet washroom;
- wheelchair-accessible compartments shall have the same layout and fittings as the unisex toilet;
- where a separate-sex toilet washroom can be accessed by wheelchair users, it should be possible for them to use both a urinal and a washbasin at a lower height than is provided for other users;
- where possible, a low level urinal for children should be provided in male washrooms;
- separate-sex toilet washrooms above a certain size should include an enlarged WC cubicle for use by people who need extra space, e.g. parents with children and babies, people carrying luggage and also ambulant disabled people.

Doors

- doors to compartments for ambulant disabled people should:
  - preferably open outwards;
  - be fitted with a horizontal closing bar fixed to the inside face.

Accessibility

- the approach to a unisex toilet should be separate to other sanitary accommodation;
- wheelchair users should:
  - not have to travel more than 40 m on the same floor to reach a unisex toilet;

- not have to travel more than combined horizontal distance where the unisex toilet accommodation is on another floor of the building (accessible by passenger lift);
- be able to approach, transfer to, and use the sanitary facilities provided within a building.

Heights and arrangements

- compartments used for ambulant disabled people should:
  - be at least 1200 mm wide;
  - include a horizontal grab bar adjacent to the WC;
  - include a vertical grab bar on the rear wall;
  - include space for a shelf and fold-down changing table;
- a wheelchair-accessible washroom (where provided) shall have:
  - at least one washbasin;
  - and, for men, at least one urinal;
- the compartment should:
  - be fitted with support rails;
  - include a minimum activity space to accommodate people who use crutches, or otherwise have impaired leg movements.

Wheelchair-accessible changing and shower facilities

General

- in large building complexes (e.g. retail parks and large sports centres) there should be at least one wheelchair-accessible unisex toilet;
- a combined facility should be divided into distinct 'wet' and 'dry' areas.

For changing and shower facilities

A choice of layouts suitable for left-hand and right-hand transfer should be provided when more than one individual changing compartment or shower compartment is available:

- wall mounted drop-down support rails and slip-resistant tip-up seats should be provided;
- subdivisions should be provided for communal shower and changing facilities;
- individual self-contained shower and changing facilities should be available in sports amenities;
- emergency assistance alarm systems should be provided which should have:
  - visual and audible indicators to confirm that an emergency call has been received;
  - a reset control reachable from a wheelchair, WC, or from a shower/changing seat;
  - a signal that is distinguishable visually and audibly from the fire alarm;
- an emergency assistance pull cord should be provided which should:
  - be easily identifiable and reachable from the wall mounted tip-up seat (or from the floor);

- be located as close to a wall as possible;
- be coloured red;
- have two red 50 mm diameter bangles;
- facilities for limb storage should be included for the benefit of amputees.

For changing facilities

- the floor should be level and slip resistant when dry or when wet;
- there should be a manoeuvring space 1500 mm deep in front of lockers.

For shower facilities

- if showers are provided in commercial developments for the benefit of staff, at least one wheelchair-accessible shower compartment should be made available;
- a shower curtain should be provided;
- a shelf should be provided for toiletries;
- the floor of the shower and shower area should be slip resistant and self-draining;
- shower controls should be positioned between 750 and 1000 mm above the floor.

For shower facilities incorporating a WC

General

- a choice of left-hand and right-hand transfer layouts should be available when more than one shower area includes a corner WC.

Wheelchair-accessible bathrooms

General

- a choice of layouts suitable for left-hand and right-hand transfer should be provided when more than one bathroom is available;
- the floor should be slip resistant when dry or when wet;
- the bath should be provided with a transfer seat;
- outward opening doors, fitted with a horizontal closing bar fixed to the inside face, should be provided;
- an emergency assistance pull cord should be provided which should:
  - be easily identifiable and reachable from the wall mounted tip-up seat (or from the floor);
  - be located as close to a wall as possible;
  - be coloured red;
  - have two red 50 mm diameter bangles.

WC provision in the entrance storey of the dwelling

- if there is a bathroom in the principal storey, then a WC may be collocated with it;
- the door to the WC compartment should:
  - open outwards;
  - be positioned so as to allow wheelchair users access to it;
  - have a clear opening width.
- the WC compartment should:
  - provide a clear space for wheelchair users to access the WC;
  - position the washbasin so that it does not impede access.

Accessible switches and socket outlets in the dwelling

- switches and socket outlets for lighting and other equipment should be:
  - located so that they are easily reachable;
  - located between 450 mm and 1200 mm from finished floor level.

Facilities in buildings other than dwellings

General

- all floor areas should be accessible;
- in hotels, motels and student accommodation:
  - a proportion of the sleeping accommodation should be designed for wheelchair users;
  - the remainder should include facilities suitable for people with sensory, dexterity or learning difficulties;
- if there is a reception point:
  - it should be easily accessible and convenient to use;
  - information about the building should be clearly available from notice boards and signs;
  - the floor surface should be slip resistant;
- disabled people should be able to have:
  - a choice of seating location at spectator events;
  - a clear view of the activity taking place (whilst not obstructing the view of others);
- bars and counters in refreshment areas should be at a suitable level for wheelchair users.

Audience and spectator facilities

General

- disabled people should be provided with a selection of spaces into which they can manoeuvre easily and which offer them a clear view of an event – taking particular care that these do not become segregated into 'special areas'.

For audience seating generally

- the route to wheelchair spaces should be accessible to users;
- stepped access routes to audience seating should be provided with fixed handrails.

Handrails to external stepped or ramped access should:

- be continuous across flights and landings;
- extend at least 300 mm horizontally beyond the top and bottom of a ramped access;
- not project into an access route;
- contrast visually with the background;
- have a slip resistant surface which is not cold to the touch;
- terminate in such a way that reduces the risk of clothing being caught;
- either be circular or oval with a width of 50 mm;
- not protrude more than 100 mm into the surface width;
- have a clearance of between 60 and 75 mm between:
  - the handrail and any adjacent wall surface;
  - a cranked support and the underside of the handrail;
- ensure that its inner face is located no more than 50 mm beyond the surface width of the ramped or stepped access;
- should be spaced away from the wall and rigidly supported;
- should be set at heights that are convenient for all users of the building.

Seating

- some wheelchair spaces should be provided in pairs, with standard seating on at least one side;
- if more than two wheelchair spaces are provided, they should be located so as to give a range of views of the event;
- the minimum clear space for:
  - access to wheelchair spaces should be 900 mm;
  - wheelchair spaces in a parked position should be 900 mm wide by 1400 mm deep;
- the floor of each wheelchair space should be horizontal;
- seats at the ends of rows and next to wheelchair spaces should have detachable (or lift-up) arms.

Lecture/conference facilities

- all people should be able to use presentation facilities;
- people with hearing impairments should be able to participate fully in conferences, committee meetings and study groups;
- acoustic environment should ensure that audible information can be heard clearly;
- artificial lighting should give good colour rendering of all surfaces;
- glare and reflections from shiny surfaces should be avoided;
- uplighters mounted at low or floor level should be avoided;

- wheelchair users should have access to the podium or stage;
- an audible public address system should be supplemented by visual information;
- the availability of an induction loop or infrared hearing enhancement system should be indicated by the standard symbol;
- hearing enhancement systems should be installed:
  - in rooms and spaces designed for meetings, lectures, classes, performances, spectator sports or films;
  - at service or reception counters (especially when situated in noisy areas or behind glazed screens);
- telephones suitable for hearing aid users should be:
  - clearly indicated by the standard ear symbol;
  - incorporate an inductive coupler and volume control;
- text telephones should be clearly indicated by the standard symbol;
- artificial lighting should be designed to be compatible with other electronic and radio frequency installations.

Entertainment, leisure and social facilities

Theatres and cinemas

- in theatres and cinemas special care should be given to the design and location of wheelchair spaces.

Refreshment facilities

- restaurants and bars should be designed so that they can be reached and used by all people independently or with companions;
- all people should have access to:
  - all parts of the facility;
  - staff areas;
  - public areas (e.g. lavatory accommodation, public telephones and external terraces);
  - self-service facilities (when provided);
- all (changes to) floor levels should be accessible;
- raised thresholds should be avoided;
- part of the working surface of a bar or serving counter should:
  - be permanently accessible to wheelchair users;
  - be at a level of not more than 850 mm above the floor;
  - have worktops of shared refreshment facilities (e.g. tea making) 850 mm above the floor with a clear space beneath at least 700 mm above the floor;
  - have basin taps that can either be controlled automatically or be capable of being operated using a closed fist (e.g. by lever action).

Sleeping accommodation

General

- sleeping accommodation in hotels, motels and student accommodation should be convenient for all types of people;

- a proportion of rooms should be available for wheelchair users;
- wheelchair users should be able to:
  - reach all the facilities available within the building;
  - manoeuvre around and use the facilities in the room;
  - operate switches and controls;
- en-suite sanitary facilities are the preferred option for wheelchair-accessible bedrooms;
- there should be at least as many en-suite shower rooms as en-suite bathrooms;
- built-in wardrobes and shelving should be accessible and convenient to use;
- bedrooms not designed for independent use by a person in a wheelchair should have an outer door wide enough to be accessible to a wheelchair user.

For all bedrooms

- built-in wardrobe swing doors should open through 180°;
- handles on hinged and sliding doors should:
  - be easy to grip and operate;
  - contrast visually with the surface of the door;
- windows and window controls should be:
  - located between 800 and 1000 mm above the floor;
  - easy to operate without using both hands simultaneously;
- a visual fire alarm signal should be provided;
- room numbers should be embossed characters.

Wheelchair-accessible bedrooms should:

- be located on accessible routes;
- be designed to provide a choice of location;
- have a standard of amenity equivalent to that of other bedrooms;
- be large enough to enable a wheelchair user to manoeuvre with ease;
- if a balcony is provided:
  - have a level threshold;
  - have no horizontal transoms between 900 mm and 1200 mm above the floor;
- have no permanent obstructions in a zone 1500 mm back from any balcony doors;
- be provided with emergency assistance alarms;
- have the door from the access corridor to a wheelchair-accessible bedroom that should:
  - not require more than 20 N opening force;
  - have an effective clear width through a single leaf door;
  - have an unobstructed space of at least 300 mm on the pull side of the door.

Switches, outlets and controls

- light switches should:
  - have large push pads;
  - align horizontally with door handles;
  - be located close to the entrance door opening;

- the operation of all switches, outlets and controls should not require the simultaneous use of both hands;
- switched socket outlets should indicate whether they are 'on' or 'off';
- individual switches on panels should be well separated;
- socket outlets should be wall mounted;
- telephone points and TV sockets should be located 400 mm to 1000 mm above the floor;
- socket outlets should be located no nearer than 350 mm from room corners;
- controls that need close vision should be located between 1200 mm and 1400 mm above the floor;
- emergency alarm pull cords should be:
  - coloured red;
  - located as close to a wall as possible;
- red and green should not be used in combination as indicators of 'on' and 'off' for switches and controls.

Aids to communication

- hearing enhancement systems should be installed:
  - in all rooms and spaces designed for meetings, lectures, classes, performances, spectator sport or films;
  - at service or reception counters (especially when they are situated in noisy areas or they are behind glazed screens);
- the availability of an induction loop or infrared hearing enhancement system should be indicated by the standard symbol;
- telephones suitable for hearing aid users should be clearly indicated and incorporate an inductive coupler and volume control;
- artificial lighting should be designed to:
  - be compatible with other electronic and radio frequency installations;
  - give good colour rendering of all surfaces;
- uplighters mounted at low or floor level should be avoided as they can disorientate some visually impaired people and should be avoided.

The problem disabled people face accessing society

There are currently 10 million people with physical and sensory impairments in the UK. Only a fraction are out and about, working, shopping, relaxing and taking part in the day-to-day life of wider society. A key reason is the great friction people encounter when attempting to access the shops, services and leisure facilities that the rest of the community takes for granted.

Trying to go out for a meal, visit a dentist, take a train, find a solicitor or even pop to the shops all have to be researched before they are possible. There is no easy way to do it. Should you phone a venue to enquire, the person who answers is unlikely to be able to answer your questions; in fact, people often become the subject of searching personal enquiries such as 'What is wrong with you?' or 'What can you do?' As a result people lose confidence, withdraw and become excluded from family and community life.

In this way society disables people with physical and sensory impairments and limits their opportunities. Businesses miss out on a market of millions of people who, according to government figures, have £80 billion to spend each year.

Disabled*Go*.info

Bringing disabled people and business together

Accurate and reliable access information is the key to tackling this problem.

Founded by Dr Gregory Burke, a wheelchair user, and launched in January 2003 to mark the European Year of Disabled People, DisabledGo provides detailed and accurate data for people with hearing, vision or mobility impairments about access to shops, pubs, restaurants and offices – tens of thousands of goods and service providers across the UK.

For the first time business and service providers have a means to communicate directly with disabled people and explain their access to them.

People can judge for themselves whether or not a venue will suit their needs. You can check, for example: whether a cinema offers wheelchair access, has an accessible toilet, or a hearing loop; whether a hotel has a vibrating fire alarm; what side of a museum's steps the handrails are located on; if a solicitor will make home visits; if a cafe offers its menu in large print or in Braille or even which shops will welcome an assistance dog!

Service providers often do not realize that simple changes to the way they deliver their service could open up their business to many more customers. DisabledGo can offer advice and consultancy on issues surrounding access audits, the training of staff and the Disability Discrimination Act.

A free service developed by disabled people for disabled people

DisabledGo's innovative service was developed in consultation with over 100 disability groups across the UK. It is free for people to access and is paid for by a partnership of private sector sponsorship, from Marks & Spencer, and sponsorship from local authorities who are committed to opening up their localities to everybody. The guide can be accessed on www.DisabledGo.info.

For more information about DisabledGo please contact Chris Sherwood on 01727 739 700 or visit www.DisabledGo.info.

Key

The following signs have been used in the access guides to assist disabled people in determining the suitability of a venue they may wish to visit.

 The service provider has pledged to give a little more time to customers with any additional needs, if the customers so wish. A business will only be listed in the guide if it claims to be positive about disabled people.

 The premises/services are fully accessible to a wheelchair user who is travelling unaccompanied.

The premises/services are accessible to a wheelchair user so long as there is someone to assist them.

The premises have no more than three steps. If there is more than one step, there must be a handrail.

A seat is available on the premises for customers, otherwise one may be requested.

There are toilets on the premises that have been adapted for people with mobility impairments.

There are level access toilets (excluding adapted ones) on the premises.

Depending on the nature of the service this means either:
1. there are baby changing facilities within the adapted toilets; or
2. there are adapted changing rooms for people with mobility impairments.

Bills, menus, programmes or any other written information are available in a larger print size if requested.

Bills, menus, programmes or any other written information are available in Braille if requested.

The service provider is able to welcome assistance dogs onto the premises.

A hearing system is available at certain locations within the premises.

Either the service provider has a mini-com/text phone facility, or they are prepared to conduct any business via fax or email. Please note the contact details of the listing to know which of these facilities are available.

A home delivery or home visit service is available if requested.

Staff have received disability-awareness training.

A personal shopper service is available.

Appendix B

Conservation of fuel and power

B.1 Main changes in the 2002 edition

The main changes made by the Building Regulations (Amendment) Regulations 2001 primarily concerned the legal requirements for:

- the conservation of fuel and power in buildings (Domestic L1 and non-domestic L2);
- heating and hot water systems (including boiler efficiency);
- efficient internal and external lighting systems fixed to the dwelling;
- replacement work on windows (and other glazed elements), boilers and hot water vessels.

These changes provoked an improved method for calculating the U-values by bringing them in line with European standards and by setting lower (i.e. better) values. For example:

- The nominal average U-value for windows and other glazed elements has been improved and is now based on sealed double units with low-emissivity inner panes and 16 mm air gaps.
- Metal windows are given some credit for their thinner framing members and hence the increased useful solar gain.
- The energy rating method that used SAP ratings has been replaced by the new carbon index method.
- The guidance on reducing thermal bridging has been improved and there is new guidance on reducing unwanted air leakage.
- New guidance is given on complying with the requirements for commissioning heating and hot water systems, for lighting and for the provision of information to householders.
- The guidance on conservatories has been extended to cover conservatories attached to existing dwellings.
- Guidance on material alterations and changes of use remains unchanged but has been placed in a new section along with new guidance work on controlled services in existing dwellings and the conservation and restoration work in historic buildings.

B.2 The requirement

B.2.1 Dwellings

Reasonable provision shall be made for the conservation of fuel and power in
dwellings by:

(a) *limiting the heat loss:*
    (i) through the fabric of the building;
    (ii) from hot water pipes and hot air ducts used for space heating;
    (iii) from hot water vessels;
(b) *providing space heating and hot water systems which are energy efficient;*
(c) *providing lighting systems with appropriate lamps and sufficient controls
so that energy can be used efficiently;*
(d) *providing sufficient information with the heating and hot water services so
that building occupiers can operate and maintain the services in such a
manner as to use no more energy than is reasonable in the circumstances.*

 The requirement for sufficient controls in requirement L1 (c) applies only to
external lighting systems fixed to the building.

 In the context of Approved Document L1, 'building work' consists of the provision of a window, rooflight, roof window, door (being a door which together
with its frame has more than 50% of its internal face area glazed), a space heating
or hot water service boiler, or a hot water vessel.

B.3 Meeting the requirement

Energy efficiency measures shall be provided that:

(a) • limit the heat loss through the roof, wall, floor, windows and doors, etc.
by suitable means of insulation;
• where appropriate permit the benefits of solar heat gains and more
efficient heating systems to be taken into account;
• limit unnecessary ventilation heat loss by providing building fabric that
is reasonably airtight;
• limit the heat loss from hot water pipes and hot air ducts used for space
heating;
• limit the heat loss from hot water vessels and their primary and
secondary hot water connections by applying suitable thicknesses of
insulation (where such heat does not make an efficient contribution to the
space heating).

(b) Provide space heating and hot water systems with reasonably efficient
equipment such as heating appliances and hot water vessels where relevant, such that the heating and hot water systems can be operated effectively as regards the conservation of fuel and power.

(c)   Provide lighting systems that utilize energy-efficient lamps with manual switching controls or, in the case of external lighting fixed to the building, automatic switching, or both manual and automatic switching controls as appropriate, such that the lighting systems can be operated effectively as regards the conservation of fuel and power.

(d)   Provide information, in a suitably concise and understandable form (including results of performance tests carried out during the works) that shows building occupiers how the heating and hot water services can be operated and maintained.

 Responsibility for achieving compliance with the requirements of Part L rests with the person carrying out the work. That person may be, for example, a developer, a main (or sub-) contractor, or a specialist firm directly engaged by a private client.

The person responsible for achieving compliance should either themselves provide a certificate, or obtain a certificate from the sub-contractor, that commissioning has been successfully carried out. The certificate should be made available to the client and the building control body.

B.3.1 General

A conservatory that is not separated from the rest of the dwelling shall be treated as an integral part of the dwelling.	L1
Where the conservatory is separated from the rest of the dwelling and has a **fixed** heating installation, the heating in the conservatory shall have its own separate temperature and on/off controls.	L1
Where an opening is created or enlarged, provision shall be made to limit the heat loss from the dwelling such that it is no worse than before the work was undertaken.	L1
Measurements of thermal conductivity should be made according to BS EN 12664, BS EN 12667, or BS EN 12939.	L1 (0.9)
Measurements of thermal transmittance should be made according to BS EN ISO 8990 or, in the case of windows and doors, BS EN ISO 12567.	L1 (0.9)

 Thermal conductivity (i.e. the λ-value) of a material is a measure of the rate at which that material will pass heat and is expressed in units of watts per metre per degree of temperature difference (W/mK).

Thermal transmittance (i.e. the U-value) is a measure of how much heat will pass through 1 m^2 of a structure when the air temperatures on either side differ by 1°C.

B.3.2 Calculation of U-values

U-values should be calculated using the methods given in: L1 (0.11)

- for walls and roofs: BS EN ISO 6946;
- for ground floors: BS EN ISO 13370;
- for windows and doors: BS EN ISO 10077–1 or prEN (SO 10077–2);
- for basements: BS EN ISO 13370 or the BCA/NHBC Approved Document.

 Appendix A to Approved Document L1 contains tables of U-values and examples of their use, which provide a simple way to establish the amount of insulation needed to achieve a given U-value for some typical forms of construction.

When calculating U-values the thermal bridging effects of   L1 (0.13)
timber joists, structural and other framing, normal mortar
bedding and window frames should generally be taken
into account.

Thermal bridging can be disregarded where the difference   L1 (0.13)
in thermal resistance between the bridging material and
the bridged material is less than $0.1\,\mathrm{m^2\,K/W}$.

Normal mortar joints need not be taken into account in   L1 (0.13)
calculations for brickwork.

Walls containing in-built meter cupboards, and ceilings   L1 (0.13)
containing loft hatches, recessed light fittings, etc. should
be calculated using area-weighted average U-values.

B.3.3 Design and construction

Under separate provisions in the Building Regulations a new dwelling created by building work, or by a material change of use in connection with which building work is carried out, must be given an energy rating, using the SAP edition having the Secretary of State's approval at the relevant time in the particular case; and the rating must be displayed in the form of a notice. Administrative guidance on producing and displaying SAP ratings is given in DETR Circular No 07/2000 dated 13 October 2000.

Three methods are currently used for demonstrating reasonable provision for limiting heat loss through the building fabric:

1. An elemental method;
2. A target U-value method;
3. A carbon index method.

SAP means the Government's Standard Assessment Procedure for Energy Rating of Dwellings. The SAP provides the methodology for the calculation of the carbon index, which can be used to demonstrate that dwellings comply with Part L.

B.3.4 Elemental method

The Elemental method can be used only when the heating system is either a gas boiler, oil boiler, heat pump, community heating with CHP, biogas or biomass fuel, but **not** for direct electric heating or other systems.

The elemental method is suitable for alterations and extension work, and for newbuild work when it is desired to minimize calculations.	L1 (1.3)
When using the elemental method, the requirement will be met for new dwellings by selecting construction elements that provide the U-value thermal performances given in Table B1.	L1 (1.3)

 Exposed element means an element exposed to the outside air (including a suspended floor over a ventilated or unventilated void, and elements so exposed indirectly via an unheated space), or an element in the floor or basement in contact with the ground. In the case of an element exposed to the outside air via an unheated space (previously known as a 'semi-exposed element') the U-value should be determined using the method given in the SAP 1998 (replaced by

Table B1 Elemental method – U-values (W/m²K) for construction elements

Exposed element	U-value
Pitched roof with insulation between rafters[1,2]	0.2
Pitched roof with integral insulation	0.25
Pitched roof with insulation between joists	0.16
Flat roof[3]	0.25
Walls, including basement walls	0.35
Floors, including ground floors and basement floors	0.25
Windows, doors and rooflights (area-weighted average), glazing in metal frames[4]	2.2
Windows, doors and (area-weighted average) rooflights[4], glazing in wood or PVC frames[5]	2.0

Notes:
1. Any part of a roof having a pitch of 70° or more can be considered as a wall.
2. For the sloping parts of a room-in-the-roof constructed as a material alteration, a U-value of 0.3 W/m²K would be reasonable.
3. Roof of pitch not exceeding 10°.
4. Rooflights include roof windows. (A roof window is a window in the plane of a pitched roof and may be considered as a rooflight for the purposes of this Approved Document.)
5. The higher U-value for metal-framed windows allows for additional solar gain due to the greater glazed proportion.

SAP 2001 in 2001). Party walls, separating two dwellings or other premises that can reasonably be assumed to be heated to the same temperature, are assumed not to need thermal insulation.

> Single-glazed panels can be acceptable in external doors   L1 (1.5)
> provided that the heat loss through all the windows, doors
> and rooflights matches the relevant figure in Table B1 and
> the area of the windows, doors and rooflights together does
> not exceed 25% of the total floor area.
>
> Boiler efficiency would be demonstrated by using a boiler   L1 (1.7)
> with SEDBUK not less than the appropriate entry in Table B2.

 SEDBUK is the Seasonal Efficiency of a Domestic Boiler in the UK, defined in the Government's Standard Assessment Procedure for the Energy Rating of Dwellings.

Table B2 Minimum boiler SEDBUK corresponding to Table B1 for use with the target U-value

Central heating system fuel	SEDBUK percentage
Mains natural gas	78
LPG	80
Oil	85

> The average U-value of windows, doors and rooflights   L1 (1.7)
> should match the relevant figure in Table B1 and the area
> of the windows, doors and rooflights together does not
> exceed 25% of the total floor area.
>
> Examples of how the average U-value is calculated are
> given in Appendix D to Approved Document L1.

Extensions to dwellings

> The fabric U-values given in Table B1 can be applied   L1 (1.11)
> when proposing extensions to dwellings.
>
> The target U-value and carbon index methods can be   L1 (1.11)
> used only if applied to the whole enlarged dwelling.
>
> The total rate of heat loss is the product of (area × U-value) L1 (1.12)
> for all exposed elements.

For small extensions to dwellings (e.g. porches where the   L1 (1.13)
new heated space created has a floor area of not more than
about 6 m²) energy performance should not be less than
those in the existing building.

The area-weighted average U-value of windows, doors and   L1 (1.14)
rooflights ('openings') in extensions to existing dwellings
should not exceed the relevant values in Table B1.

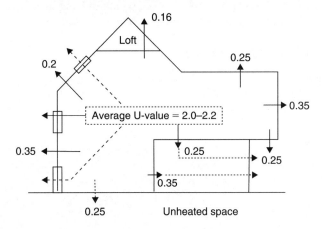

Figure B1 Summary of the elemental method

Figure B1 summarizes the fabric insulation standards and allowances for windows, doors and rooflights given in the elemental method.

 Examples of the procedures used in this method are given in Appendices A to C of Approved Document L1.

B.3.5 Target U-value

This method allows greater flexibility than the elemental method in selecting the areas of windows, doors and rooflights, and the insulation levels of individual elements in the building envelope, taking into account the efficiency of the heating system and enabling solar gain to be addressed. It can be used for any heating system.

The requirement would be met if the calculated average U-value of the dwelling does not exceed the target U-value, corrected for the proposed method of heating, as determined from the following equation:

$$U_T = [0.35 - 0.19(A_R/A_T) - 0.10(A_{GF}/A_T) + 0.413(A_F/A_T)]$$ (B.1)

where:

U_T is the target U-value prior to any adjustment for heating system performance or solar gain

A_R is the exposed roof area
A_T is the total area of exposed elements of the dwelling (including the ground floor)
A_{GF} is the ground floor area
A_F is the total floor area (all storeys).

 The target U-value equation assumes equal distribution of glazed openings on north and south elevations.

 For dwellings whose windows have metal frames (including thermally broken frames) the target U-value can be increased by multiplying by a factor of 1.03, to take account of the additional solar gain due to the greater glazed proportion.

B.3.6 Carbon index method

The aim in this method is to provide more flexibility in the design of new dwellings whilst achieving similar overall performance to that obtained by following the elemental method. The carbon index method can be used with any heating system.

> The carbon index for a dwelling (or each dwelling in a   L1 (1.27)
> block of flats or converted building) should not be less than
> 8.0. (Examples of dwellings with a carbon index of 8.0 or
> more are given in Appendix F to Approved Document L1.)

B.3.7 Limiting thermal bridging at junctions and around openings

> The building fabric should be constructed so that there   L1 (1.30)
> are no significant thermal bridges or gaps in the insulation
> layer(s) within the various elements of the fabric, at the
> joints between elements, and at the edges of elements such
> as those around window and door openings.

B.3.8 Limiting air leakage

> Reasonable provision should be made to reduce   L1 (1.33)
> unwanted air leakage.

> A continuous barrier to air movement around the habitable   L1 (1.34)
> space (including separating walls and the edges of intermediate
> floors) that is in contact with the inside of the thermal
> insulation layer should be maintained wherever possible.

B.3.9 Space heating system controls

Heating systems other than space heating provided by individual solid fuel, gas and electric fires or room heaters should be provided with:

- zone controls,
- timing controls, and
- boiler control interlocks.

For electric storage heaters the system should have an automatic charge control that detects the internal temperature and adjusts the charging of the heater accordingly.

B.3.10 Zone controls

Hot water central heating systems, fan controlled electric storage heaters and electric panel heaters should control the temperatures independently in areas (such as separate sleeping and living areas that have different heating needs)	L1 (1.38)
Room thermostats and/or thermostatic radiator valves (or any other suitable temperature sensing devices) should be used where appropriate.	L1 (1.38)
In most dwellings one timing zone divided into two temperature control sub-zones would be sufficient.	L1 (1.39)
Large dwellings should be divided into zones with floor area no greater than $150\,m^2$.	L1 (1.39)

B.3.11 Timing controls

For gas-fired and oil-fired systems and for systems with solid-fuel-fired boilers (where forced-draught fans operate when heat is required), timing devices should be provided to control the periods when the heating systems operate.	L1 (1.40)
Separate timing control should be provided for space heating and water heating, except for combination boilers or solid fuel appliances.	L1 (1.40)

B.3.12 Boiler control interlocks

Gas- and oil-fired hot water central heating system controls should switch the boiler off when no heat is	L1 (1.41)

required whether control is by room thermostats or by thermostatic radiator valves.

The boiler in systems controlled by thermostats should operate only when a space heating or vessel thermostat is calling for heat.   L1 (1.41a)

A room thermostat or flow switch should also be provided to switch off the boiler when there is no demand for heating or hot water.   L1 (1.41b)

B.3.13 Hot water systems

Systems incorporating integral or separate hot water storage vessels should:

- have an insulating vessel with a 35 mm thick, factory-applied coating of PU-foam having a minimum density of $30 \, kg/m^3$.   L1 (1.43)

Unvented hot water systems need additional insulation to control the heat losses through the safety fittings and pipework.   L1 (1.43)

B.3.14 Operating and maintenance instructions for heating and hot water systems

The building owner and/or occupier should be given information on the operation and maintenance of the heating and hot water systems.   L1 (1.51)

The instructions should be directly related to the system(s) in the dwelling and should explain to householders how to operate the systems so that they can perform efficiently, and what routine maintenance is advisable for the purposes of the conservation of fuel and power.   L1 (1.51)

B.3.15 Insulation of pipes and ducts

Pipes and ducts should be insulated to conserve heat and hence maintain the temperature of the water or air heating service.   L1 (1.52)

Space heating pipework located outside the building   L1 (1.52a)
fabric insulation layer(s) should be wrapped with
insulation material having a thermal conductivity at 40°C
not exceeding 0.035 W/m²K and a thickness equal to the
outside diameter of the pipe up to a maximum of 40 mm.

Warm air ducts should be insulated in accordance with   L1 (1.52b)
BS 5422: 2001.

Hot pipes connected to hot water storage vessels   L1 (1.52c)
(including the vent pipe, and the primary flow and return
to the heat exchanger, where fitted) should be insulated for
at least 1 m from their points of connection (or up to
the point where they become concealed).

To protect against freezing, central heating and hot water   L1 (1.53)
pipework in unheated areas may also need increased
insulation thicknesses.

B.3.16 Internal lighting

Table B3 gives an indication of recommended number of locations (excluding
garages, lofts and outhouses) that need to be equipped with efficient lighting
(L1 (1.55–1.56)).

In locations where lighting can be expected to have most   L1 (1.54)
use, fixed lighting (e.g. fluorescent tubes and compact
fluorescent lamps – but not GLS tungsten lamps with
bayonet cap or Edison screw bases) with a luminous
efficacy greater than 40 lumens per circuit-watt should
be available.

📖 **Circuit-watts** means the power consumed in lighting
circuits by lamps and their associated control gear and
power factor correction equipment.

Table B3

Number of rooms created (see Note)	Recommended minimum number of locations
1–3	1
4–6	2
7–9	3
10–12	4

Note: Hall, stairs and landing(s) count as one room, as does a conservatory.

B.3.17 External lighting fixed to the building

External lighting (including lighting in porches, but not lighting in garages and carports) should:

• automatically extinguish when there is enough daylight, and when not required at night;	L1 (1.57a)
• have sockets that can only be used with lamps having an efficacy greater than 40 lumens per circuit-watt (such as fluorescent or compact fluorescent lamp types, and **not** GLS tungsten lamps with bayonet cap or Edison screw bases).	L1 (1.57b)

B.3.18 Conservatories

For the purposes of the guidance in Part L, a conservatory is considered to not have less than three-quarters of the area of its roof and not less than one-half of the area of its external walls made of translucent material.

If a conservatory is attached to and built as part of a new dwelling it should be treated as an integral part of the building.	L1 (1.59)
If there is no physical separation between the conservatory and the dwelling, then the conservatory should be treated as an integral part of the dwelling.	L1 (1.59a)
If the conservatory is separated from the dwelling, energy savings can be achieved if the conservatory is not heated.	L1 (1.59b)
If fixed heating installations are used in a conservatory, they should have their own separate temperature and on/off controls.	L1 (1.59b)
When a conservatory is attached to an existing dwelling and an opening is enlarged or newly created as a material alteration, the heat loss from the dwelling should be limited by: • retaining the existing separation where the opening is not to be enlarged; • providing separation with an average U-value less than $2.0\,\mathrm{W/m^2\,K}$.	L1 (1.60)
Separating walls and floors between a conservatory and a dwelling should be insulated to at least the same degree as the exposed walls and floors.	L1 (1.61)

> Separating windows and doors between a conservatory   L1 (1.61b)
> and a dwelling should have the same U-value and   AD N
> draught-stripping provisions as the exposed windows
> and doors elsewhere in the dwelling.

B.3.19 Material alterations

Depending on the circumstances, material alterations with respect to the requirements for conserving fuel and power as shown in Approved Document L could be considered to be:

- **Roof insulation**: when substantially replacing any of the major elements of a roof structure in a material alteration, providing insulation to achieve the U-value for new dwellings.
- **Floor insulation**: where the structure of a ground floor or exposed floor is to be substantially replaced, or re-boarded, providing insulation in heated rooms to the standard for new dwellings.
- **Wall insulation**: when substantially replacing complete exposed walls or their external renderings or cladding or internal surface finishes, or the internal surfaces of separating walls to unheated spaces, providing a reasonable thickness of insulation.
- **Sealing measures**: when carrying out any of the above work, including reasonable sealing measures to improve airtightness.
- **Controlled services and fittings**: when replacing controlled services and fittings.

B.3.20 Replacement of controlled services or fittings

For the purpose of L1, 'building work' consists of the provision of a window, rooflight, roof window, door (being a door which together with its frame has more than 50% of its internal face area glazed), a space heating or hot water service boiler, or a hot water vessel.

Replacement work on controlled services or fittings (whether replacing with new but identical equipment or with different equipment) depends on the circumstances in the particular case and would also need to take account of historic value (L1 (2.3)).

> **Windows, doors and rooflights**: when windows, etc.   L1 (2.3a)
> are being replaced with new draught-proofed ones either
> with an average U-value not exceeding the appropriate
> entry in Table B1, or with a centre-pane U-value not
> exceeding $1.2 \, \text{W/m}^2\text{K}$.

> This requirement does not apply to repair work on parts of these elements, such as replacing broken glass or sealed double-glazing units or replacing rotten framing members.

Heating boilers: replacement heating boilers in dwellings having a floor area greater than 50 m² shall be treated as if it were a new dwelling and in the case of: L1 (2.3a)

- ordinary oil or gas boilers, shall be by a boiler with a SEDBUK not less than the appropriate entry in Table B2 L1 (2.3a(1))
- back boilers, shall be by a boiler having a SEDBUK of not less than three percentage points lower than the appropriate entry in Table B2 L1 (2.3a(2))
- solid fuel boilers, shall be by a boiler having an efficiency not less than that recommended for its type in the HETAS certification scheme. L1 (2.3a(3))

Hot water vessels: replacements shall all be new equipment as if for a new dwelling. L1 (2.3c)

Boiler and hot water storage controls: the work may also need to include replacement of the time switch or programmer, room thermostat, and hot water vessel thermostat, and provision of a boiler interlock and fully pumped circulation. L1 (2.3d)

Commissioning and providing operating and maintenance instructions: where heating and hot water systems are to be altered as in paragraphs (a) to (e), reasonable provision would also include appropriate commissioning and the provision of operating and maintenance instructions. L1 (2.3f)

B.3.21 Material changes of use

Depending on the circumstances, material changes of use for the purpose of conserving fuel and power as required by Approved Document L, could be satisfied by:

- **Accessible lofts**: when upgrading insulation in accessible lofts, providing additional insulation to achieve a U-value not exceeding 0.25 W/m²K where the existing insulation provides a U-value worse than 0.35 W/m²K.
- **Roof insulation**: when substantially replacing any of the major elements of a roof structure, providing insulation to achieve the U-value considered reasonable for new dwellings.

- **Floor insulation**: when substantially replacing the structure of a ground floor, providing insulation in heated rooms to the standard for new dwellings.
- **Wall insulation**: when substantially replacing complete exposed walls or their internal or external renderings or plaster finishes or the internal renderings and plaster of separating walls to an unheated space, providing a reasonable thickness of insulation.
- **Sealing measures**: when carrying out any of the above work, including reasonable sealing measures to improve airtightness.

B.3.22 Historic buildings

Historic buildings include:

- listed buildings,
- buildings situated in Conservation Areas,
- buildings that are of architectural and historical interest and that are referred to as a material consideration in a local authority's development plan,
- buildings of architectural and historical interest within national parks, areas of outstanding natural beauty, and World Heritage sites.

The need to conserve the special characteristics of such historic buildings needs to be recognized. In such work, the aim should be to improve energy efficiency where practically possible, always provided that the work does not prejudice the character of the historic building, or increase the risk of longterm deterioration to the building fabric or fittings. In arriving at an appropriate balance between historic building conservation and energy conservation, it would be appropriate to take into account the advice of the local planning authority's conservation officer.

B.3.23 Additional guidance

Attached to Approved Document L1 are a number of appendices that provide the reader with additional guidance and worked examples concerning fuel and power conservation and methods for obtaining U-values. These consist of:

- Appendix A Tables of U-values
- Appendix B Calculating U-values
- Appendix C U-values of ground floors
- Appendix D Determining U-values for glazing
- Appendix E Target U-value examples
- Appendix F SAP ratings and carbon indexes
- Appendix G Carbon Index.

Appendix C _____
Sound insulation

C.1 Requirement E1

C.1.1 The requirement

Dwellings shall be designed so that the noise from domestic activity in an adjoining dwelling (or other parts of the building) is kept to a level that:

- *does not affect the health of the occupants of the dwelling*
- *will allow them to sleep, rest and engage in their normal activities in satisfactory conditions*

(Approved Document E1)

C.1.2 Meeting the requirement

Walls, floors and stairs that have a separating function should achieve the sound insulation values for dwelling-houses and flats as set out in Table C1.	E0.1
Walls, floors and stairs that have a separating function should achieve the sound insulation values for rooms for residential purposes as set out in Table C1.	E0.1
For walls that separate rooms for residential purposes from adjoining dwelling-houses and flats should achieve the sound insulation values for dwelling-houses and flats as set out in Table C1.	E0.1

 Note: The sound insulation values in these tables include a built-in allowance for 'measurement uncertainty' and so if any these test values are not met, then that particular test will be considered as failed.

 Note: Occasionally a higher standard of sound insulation may be required between spaces used for normal domestic purposes and noise generated in and to an adjoining communal or non-domestic space. In these cases it would be best to seek specialist advice before committing yourself.

Table C1 Dwelling-houses and flats – performance standards for separating walls, separating floors and stairs that have a separating function

	Airborne sound insulation $D_{nT,w} + C_{tr}$ dB (minimum values)	Impact sound insulation $L'_{nT,w}$ dB (maximum values)
Purpose built dwelling-houses and flats		
Walls	45	–
Floors and stairs	45	62
Dwelling-houses and flats formed by material change of use		
Walls	43	–
Floors and stairs	43	64

Figure C1 illustrates the relevant parts of the building that should be protected from airborne and impact sound in order to satisfy Requirement E1.

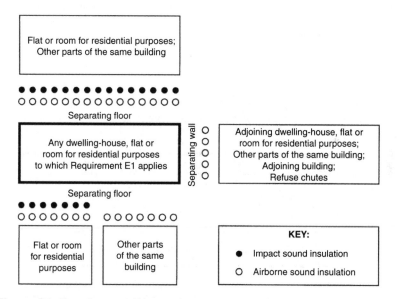

Figure C1 Requirement E1 – resistance to sound

Material change of use

In some circumstances (for example, when a historic building is undergoing a material change of use) it may not be practical to improve the sound insulation to the standards set out in Approved Document E1 particularly if the special characteristics of such a buildings need to be recognized. In these circumstances the aim should be to improve sound insulation to the *'extent that it is practically possible'*.

Note: BS 7913:1998 *The principles of the conservation of historic buildings* provides guidance on the principles that should be applied when proposing work on historic buildings.

C.2 Requirement E2

> Constructions for new walls and floors within a dwelling-house   E0.9
> (flat or room for residential purposes) – whether purpose built
> or formed by a material change of use – shall meet the
> laboratory sound insulation values set out in Table C2.

Table C2 Laboratory values for new internal walls within dwelling-houses, flats and rooms for residential purposes – whether purpose built or formed by a material change of use

	Airborne sound insulation R_W dB (minimum values)
Purpose built dwelling houses and flats	
Walls	40
Floors	40

Figures C2 and C3 illustrate the relevant parts of the building that should be protected from airborne and impact sound in order to satisfy Requirement E2.

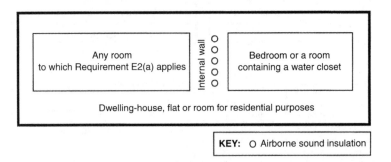

Figure C2 Requirement E2a – internal walls

C.3 Requirement E3

> Sound absorption measures described in Section 7 of   E0.11
> Approved Document N shall be applied.

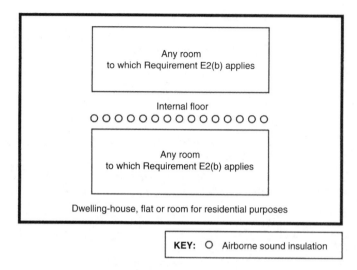

Figure C3 Requirement E2(b) – internal floors

C.4 Requirement E4

The values for sound insulation, reverberation time and   E0.12
indoor ambient noise as described in Section 1 of Building
Bulletin 93 'The Acoustic Design of Schools' (produced
by DFES and published by the Stationery Office
(ISBN 0 11 271105 7)) shall be satisfied.

C.5 Sound insulation testing

Sound insulation testing has to be completed for:

(a)   purpose built dwelling-houses and flats;
(b)   dwelling-houses and flats formed by material change of use;
(c)   purpose built rooms for residential purposes;
(d)   rooms for residential purposes formed by material change of use.

The person carrying out the building work is responsible for ensuring that sound insulation testing is carried out by a test body with appropriate third party accreditation (preferably UKAS accredited) for completing field measurements. The person is also responsible for the cost of the testing.

 Note: The procedures for sound insulation testing are described in Annex B to Part E of the Regulations.

Sound insulation testing should be carried out in accordance with the procedure described in Annex B of this Approved Document E.	E0.3
The person carrying out the building work should arrange for sound insulation testing to be carried out by a test body with appropriate third party accreditation.	E0.4
Test bodies conducting testing should preferably have UKAS accreditation (or a European equivalent) for field measurements.	E0.4
Sound insulation testing (to demonstrate compliance with Requirement E1) should be carried out on site as part of the construction process (i.e. pre-completion testing).	E1.2
Testing should not be carried out between living spaces, corridors, stairwells or hallways.	E1.8
Tests should be carried out between rooms or spaces that share a common area of separating wall or separating floor.	E1.9
Tests should be carried out once the dwelling-houses, flats or rooms for residential purposes either side of a separating element are essentially complete, except for decoration.	E1.10
Impact sound insulation tests should be carried out without a soft covering (e.g. carpet, foam backed vinyl etc.) on the floor.	E1.10

 Note: Some properties, for example loft apartments, may be sold before being fitted out with internal walls and other fixtures and fittings. In these cases sound insulation measurements should be made between the available spaces.

Table C3 shows the types of tests that have to be carried out on dwelling-houses and flats.

Table C3 Sets of tests

	A test of insulation against airborne sound between one pair of rooms (where possible suitable for use as living rooms) on opposite sides of the separating wall	A test of insulation against airborne sound between another pair of rooms (where possible suitable for use as bedrooms) on opposite sides of the separating wall	Tests of insulation against both airborne and impact sound between one pair of rooms (where possible suitable for use as living rooms) on opposite sides of the separating floor	Tests of insulation against both airborne and impact sound between another pair of rooms (where possible suitable for use as bedrooms) on opposite sides of the separating floor

Sets of tests in dwelling-houses (including bungalows)	Yes	Yes		
Sets of tests in flats with separating floors but without separating walls			Yes	Yes
Sets of tests in flats with a separating floor and a separating wall	Yes	Yes	Yes	Yes

 Note: To conduct a full set of tests, access to at least three flats will be required.

C.5.1 Failed tests

In the event of a failed set of tests, '*appropriate remedial treatment*' should be applied to the rooms that failed the test.	E1.33
After a failed set of tests, the rate of testing should be increased until the building control body is satisfied that the problem has been solved.	E1.36

C.5.2 Remedial treatment

Appropriate remedial treatment should be applied following a failed set of tests.	
Note: Guidance is available in BRE information paper IP 14/02.	E1.37
Where remedial treatment has been completed, the building control body should be satisfied with its efficacy – normally this will be assessed through additional sound insulation testing.	E1.39
Building control bodies should be satisfied that everything reasonable has been done to improve the sound insulation.	E1.40

Appendix D

Guidance to the requirements of Part P – Electrical safety

Background

For many years, the UK has managed to maintain relatively high electrical safety standards with the support of voluntary controls based on BS 7671, but with a growing number of electrical accidents occurring in the 'home', the government have been forced to consider legal requirements for safety in electrical installation work in dwellings.

As from 1 January 2005, therefore, all new electrical wiring or electrical components for domestic premises (or small commercial premises linked to domestic accommodation) will have to be designed and installed in accordance with the Building Regulations, Part P, which is based on the fundamental principles set out in Chapter 13 of the BS 7671: 2001 (i.e. '*The IEE Wiring Regulations*'). In addition, all fixed electrical installations (i.e. wiring and appliances fixed to the building fabric such as socket outlets, switches, consumer units and ceiling fittings) shall be designed, installed, inspected, tested and certified to BS 7671.

Part P also introduces a requirement for new cable core colours for AC power circuits as shown in Table D1 below.

 For single-phase installations in domestic premises, the new colours are the same as those for flexible cables to appliances (namely green-and-yellow, blue and brown for the protective, neutral and phase conductors respectively).

- These new (harmonized) colour cables may be used on site from 31 March 2004.

Table D1 Identification of conductors in AC power and lighting circuits

Conductor	Colour
Protective conductor	Green-and-yellow
Neutral	Blue
Phase of single phase circuit	Brown
Phase 1 of 3-phase circuit	Brown
Phase 2 of 3-phase circuit	Black
Phase 3 of 3-phase circuit	Grey

- New installations or alterations to existing installations may use either new or old colours (but **not** both!) from 31 March 2004 until 31 March 2006.
- Only the new colours may be used after 31 March 2006.
- Further information, concerning cable identification colours for extra-low voltage and DC power circuits, is available from the IEE website at www.iee.org/cablecolours.

 Note: Part P applies only to fixed electrical installations that are intended to operate at low voltage or extra-low voltage which are not controlled by the Electricity Supply Regulations 1988 as amended, or the Electricity at Work Regulations 1989 as amended.

What is the aim of Approved Document P?

To increase the safety of householders by improving the design, installation, inspection and testing of electrical installations in dwellings when they (i.e. the installations) are being newly built, extended or altered.

 It is understood that the government is also intending to introduce a scheme whereby domestic installations are checked at regular intervals (as well as when they are sold and/or purchased) to make sure that they comply. This would mean, of course, that if you had an installation which was not correctly certified, then your house insurance might well **not** be valid.

Who is responsible for what?

The owner – needs to ensure the works carried out are either minor or notifiable work. If the work is notifiable, then the owner needs to make sure that the person(s) carrying out the work is either registered under one of the self-certified schemes (see Figure D1) or is able to certify their work under the local authority building control approval route.

The designer – needs to ensure that all electrical work is designed, constructed, inspected and tested in accordance with BS 7671 and either falls under a competent persons scheme or the local authority building control approval route.

The builder/developer – needs to ensure that they have electricians who can self-certify their work or who are qualified/experienced enough to enable them to sign off under the Electrical Installation Certification form.

What are the statutory requirements?

In future all electrical installations will need to:

- be designed and installed to protect against mechanical and thermal damage;
- be designed and installed so that they will not present an electrical shock and/or fire hazard;
- be tested and inspected to meet relevant equipment/installation standards;

- provide sufficient information so that persons wishing to operate, maintain or alter an electrical installation can do so with reasonable safety;
- comply with such requirements placed by:
  - Part A (Structure): depth of chases in walls, and size of holes and notches in floor and roof joists;
  - Part B (Fire safety): fire safety of certain electrical installations; provision of fire alarm and fire detection systems; fire resistance of penetrations through floors and walls;
  - Part C (Site preparation and resistance to moisture): moisture resistance of cable penetrations through external walls;
  - Part E (Resistance to the passage of sound): penetrations through floors and walls;
  - Part F (Ventilation): ventilation rates for dwellings;
  - Part L (Conservation of fuel and power): energy-efficient lighting; reduced current carrying capacity of cables in insulation;
  - Part M (Access to and use of buildings): heights of switches and socket outlets.

What does all this mean?

With a few exceptions any electrical work undertaken in your home which includes the addition of a new electrical circuit, or involves work in your:

- kitchen
- bathroom
- garden area

must from 1 January 2005 be reported to the local authority Building Control for inspection. This includes any work undertaken professionally, by you or another family member or by a friend.

The **ONLY** exception is when the installer has been approved by a Competent Persons organization such as ELECSA (see below).

What types of building does Approved Document P cover?

Part P applies to electrical installations in buildings or parts of buildings comprising:

- dwelling-houses and flats;
- dwellings and business premises that have a common supply;
- shops and public houses with a flat above;
- common access areas in blocks of flats such as corridors and stairways;
- shared amenities of blocks of flats such as laundries and gymnasiums;
- in or on land associated with domestic buildings;
- fixed lighting and pond pumps in gardens.

Table D2 provides the details of works that are notifiable to a local authority and/or must be completed by a company registered as a 'competent firm'.

Table D2 Notifiable work

	Extensions and modifications to circuits	New circuits
Bathrooms (work within 3 m of the bath)	Yes	Yes
Bathrooms (work greater than 3 m from the bath)		Yes
Bedrooms		Yes
Bedrooms containing a shower or basin	Yes	Yes
Ceiling (overhead) heating	Yes	Yes
Communal area of flats	Yes	Yes
Computer cabling		
Conservatories		Yes
Dining rooms		Yes
Extra low voltage (ELV) non-pre-assembled CE marked lights	Yes	Yes
Garden – lighting	Yes	Yes
Garden – power	Yes	Yes
Greenhouses	Yes	Yes
Halls		Yes
Integral garages		Yes
Kitchen	Yes	Yes
Kitchen diners	Yes	Yes
Landings		Yes
Lounge		Yes
Remote buildings	Yes	Yes
Saunas	Yes	Yes
Sheds	Yes	Yes
Shower rooms	Yes	Yes
Small-scale generators	Yes	Yes
Solar power systems	Yes	Yes
Stairways		Yes
Studies		Yes
Swimming pools	Yes	Yes
Telephone cabling		Yes
TV rooms		Yes
Underfloor heating	Yes	Yes
Workshops (remote)	Yes	Yes

What is a competent firm?

For the purposes of Part P, the government has defined 'competent firms' as electrical contractors:

- who work in conformance with the requirements to BS 7671;
- whose standard of electrical work has been assessed by a third party;
- who are registered under the NICEIC Approved Contractor scheme and the Electrotechnical Assessment Scheme.

What is a competent person responsible for?

When a competent person undertakes installation work, that person is responsible for:

- ensuring compliance with BS 7671: 2001 and all relevant Building Regulations;

- providing the person ordering the work with a signed Building Regulations self-certification certificate;
- providing the relevant building control body with an information copy of the certificate;
- providing the person ordering the work with a completed Electrical Installation Certificate.

Who is entitled to self-certify an installation?

Part P affects every electrical contractor carrying out fixed installation and alteration work in homes. **Only** registered installers are entitled to self-certify the electrical work, however, and they must be registered as a competent person under one of the following schemes.

 Note: Registered Installers are in the process of becoming members of the new building trades Quality Schemes which will be launched in 2005. Quality Schemes members will offer consumers more protection than the minimum protection registered installers must offer.

What are the consequences of not obtaining approval?

Failure to comply with the requirements of Part P is a criminal offence and local authorities have the power to require the removal or alteration of work that does not comply with the Building Regulations. In addition, a Completion Certificate for works will not be issued – which could cause severe problems in the future as:

- the electrical installation may not be safe;
- you will have no record of the work done;
- you may have difficulty in selling your home without records of the installation and the relevant safety certificates.

Do I have to inform the local authority building control body?

All proposals to carry out electrical installation work **must** be notified to the local authority's building control body before work begins, **unless** the proposed installation work is undertaken by a person who is a competent person registered with an electrical self-certification scheme and does not include the provision of a new circuit.

Non-notifiable work such as:

- replacing any electrical fitting including socket outlets, control switches and ceiling roses;

Authorized competent person self-certification schemes for installers who can do all electrical installation work	Authorized competent person self-certification schemes for installers who can do electrical work only if it is necessary when they are carrying out other work

BRE Certification Ltd
Phone: 0870 609 6093
www.partp.co.uk

CORGI Services Limited
Phone: 01256 372200
www.corgi-gas-safety.com

British Standards Institution
Phone: 01442 230442
www.bsiglobal.com/kitemark

ELECSA Limited
Phone: 0870 749 0080
www.elecsa.org.uk

ELECSA Limited
Phone: 0870 749 0080
www.elecsa.org.uk

NAPIT Certification Limited
Phone: 0870 444 1392
www.napit.org.uk

NAPIT Certification Limited
Phone: 0870 444 1392
www.napit.org.uk

NICEIC Certification Services Ltd
Phone: 0800 013 0900
www.niceic.org.uk

OFTEC (Oil Firing Technical Association)
Phone: 0845 658 5080
www.oftec.co.uk

Figure D1 Authorized competent person self-certification schemes for installers

- replacing the cable for a single circuit cable (where damaged, for example, by fire, rodent or impact) (provided that the replacement cable has the same current carrying capacity, follows the same route and does not serve more than one sub-circuit through a distribution board);
- re-fixing or replacing the enclosures of existing installation components (provided that the circuit's protective measures are unaffected);
- providing mechanical protection to existing fixed installations (provided that the circuit's protective measures and current-carrying capacity of conductors are unaffected by increased thermal insulation);
- work that is not in a kitchen (bathroom or garden) and does not involve special installation and only consists of:
  - adding lighting points (light fittings and switches) to an existing circuit;
  - adding socket outlets and fused spurs to an existing ring or radial circuit (provided that the existing circuit protective device is suitable and supplies adequate protection for the modified circuit);
  - installing or upgrading main or supplementary equipotential bonding (provided that the work complies with other applicable legislation, such as the Gas Safety (Installation and Use) Regulations);
- installing or replacing computer cabling;

can be completed by a DIY enthusiast (family member or friends) but needs to be installed in accordance with manufacturer's instructions and done in such a way that it does not present a safety hazard. This work does **not** need to be notified to a local authority building control body (unless it is installed in an area of high risk such as a kitchen or a bathroom, etc.) **but** all DIY electrical work (unless completed by a qualified professional – who is responsible for issuing a Minor Electrical Installation Certificate) will still need to be checked, certified and tested by a competent electrician.

Any work that involves adding a new circuit to a dwelling will need to be either notified to the building control body (who will then inspect the work) or needs to be carried out by a competent person who is registered under a Government Approved Part P Self-Certification Scheme.

Work involving any of the following will also have to be notified:

- locations containing a bath tub or shower basin;
- swimming pools or paddling pools;
- hot air saunas;
- electric floor or ceiling heating systems;
- garden lighting or power installations;
- solar photovoltaic (PV) power supply systems;
- small-scale generators such as microCHP units;
- extra-low voltage lighting installations, other than pre-assembled, CE-marked lighting sets.

 Note: Where a person who is **not** registered to self-certify intends to carry out the electrical installation, then a Building Regulation (i.e. a building notice or full plans) application will need to be submitted together with the appropriate fee, based on the estimated cost of the electrical installation. The building control body will then arrange to have the electrical installation inspected at first fix stage and tested upon completion.

In any event the electrical work will still need to be certified under BS 7671 by a suitably competent person who will be responsible for the design, installation, inspection and testing of the system (on completion) and have the confidence of completing a certificate to say that the work is satisfactory and complies with current codes of practice.

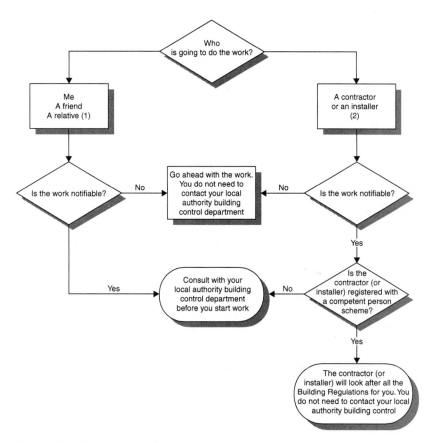

Figure D2 How to meet the new rules

The main things to remember are:

- is the work notifiable or non-notifiable?
- does the person undertaking the work need to be registered as a competent person?
- what records (if any) need to be kept of the installation?

Figure D2 is a quick guide to the requirements.

(1) Work completed by a friend, a relative or myself

You do **not** need to tell your local authority's building control department about non-notifiable work such as:

- repairs, replacements and maintenance work;
- extra power points or lighting points or other alterations to existing circuits (**unless** they are in a kitchen, bathroom, or are outdoors)

You **do** need to tell them about most other work.

 Note: If you are not sure about this, or you have any questions, ask your local authority's building control department.

(2) Work completed by a contractor or an installer

If the work is of a notifiable nature then the installer(s) must be registered with one of the schemes shown in Figure D2.

What inspections and tests will have to be completed and recorded?

As shown in Table D3, there are four types of installation:

Table D3 Types of installation

Type of inspection	When is it used?	What should it contain?	Remarks
Minor Electrical Installation Works Certificate	For a new electrical installation or for new work associated with an alteration or addition to an existing installation	Relevant provisions of Part 7 of BS 7671	
Full Electrical Installation	For the design, construction, inspection and testing of an installation	A schedule of inspections and test results as required by Part 7 of BS 7671. A certificate, including guidance for recipients (standard form from Appendix 6 of BS 7671)	For safety reasons, the electrical installation will need to be inspected at appropriate intervals by a competent person

Type of inspection	When is it used?	What should it contain?	Remarks
Electrical Installation Certificate (short form)	For use when a person is responsible for the design, construction, inspection and testing of an installation	A schedule of inspections and a schedule of test results as required by Part 7 of BS 7671	For safety reasons, the electrical installation will need to be inspected at appropriate intervals by a competent person
Periodic Inspection Report	For the inspection of an existing electrical installation	A schedule of inspections and a schedule of test results as required by Part 7 of BS 7671	For safety reasons, the electrical installation will need to be inspected at appropriate intervals by a competent person

Figure D3 (p. 656) indicates how to choose what type of inspection is required.

What should be included in the records of the installation?

All 'original' certificates should be retained in a safe place and be shown to any person inspecting or undertaking further work on the electrical installation in the future. If you later vacate the property, this certificate will demonstrate to the new owner that the electrical installation complied with the requirements of British Standard 7671 at the time the certificate was issued. The Construction (Design and Management) Regulations require that for a project covered by those Regulations, a copy of this certificate, together with schedules, is included in the project health and safety documentation.

Where can I get more information?

Further guidance concerning the requirements of Part P (Electrical safety) is available from the

- IEE (Institution of Electrical Engineers) at www.iee.org/Publish/WireFiegs/IEE-Building-Res.pdf

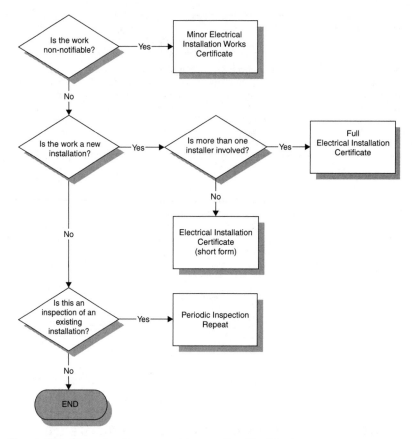

Figure D3 Choosing the correct Inspection Certificate

- the NICEIC (National Inspection Council for Electrical Installation Contracting) at
  www.niceic.org.uk
- the ECA (Electrical Contractors' Association) at
  www.niceic.org.uk or www.eca.co.uk

To download .pdf copies, go to the following:

- for Part P
  http://www.odpm.gov.uk/stellent/groups/odpm_buildreg/
  documents/page/odpm_breg_029960.pdf
- for Part P (modified Dec 04)
  http://www.odpm.gov.uk/stellent/groups/odpm_buildreg/documents/page/
  odpm_breg_033693.pdf
- for details of fixed wire colour changes
  http://www.niceic.org.uk/downloads/WiringSupp.pdf.

Appendix 1

Form 2   Form No.   /2

ELECTRICAL INSTALLATION CERTIFICATE (notes 1 and 2)

(REQUIREMENTS FOR ELECTRICAL INSTALLATIONS - BS 7671 [IEE WIRING REGULATIONS])

DETAILS OF THE CLIENT (note 1) ....
INSTALLATION ADDRESS ....Postcode...

DESCRIPTION AND EXTENT OF THE INSTALLATION Tick boxes as appropriate	
(note 1) Description of installation: .... Extent of installation covered by this Certificate:	New installation ☐ Addition to an existing installation ☐ Alteration to an existing installation ☐

FOR DESIGN

I/We being the person(s) responsible for the design of the electrical installation (as indicated by my/our signatures below), particulars of which are described above, having exercised reasonable skill and care when carrying out the design, hereby CERTIFY that the design work for which I/we have been responsible is to the best of my/our knowledge and belief in accordance with BS 7671: ...., amended to.......... (date) except for the departures, if any, detailed as follows:

> Details of departures from BS 7671 (Regulations 120-01-03, 120-02):

The extent of liability of the signatory or the signatories is limited to the work described above as the subject of this Certificate. For the DESIGN of the installation.   **(Where there is mutual responsibility for the design)

Signature: .... Date ....   Name (BLOCK LETTERS): .... Designer No. 1

Signature: .... Date ....   Name (BLOCK LETTERS): .... Designer No. 2**

FOR CONSTRUCTION

I/We being the person(s) responsible for the construction of the electrical installation (as indicated by my/our signatures below), particulars of which are described above, having exercised reasonable skill and care when carrying out the construction, hereby CERTIFY that the construction work for which I/we have been responsible is to the best of my/our knowledge and belief in accordance with BS 7671: ....amended to ....(date) except for the departures, If any, detailed as follows:

> Details of departures from BS 7671 (Regulations 120-01-03, 120-02):

The extent of liability of the signatory is limited to the work described above as the subject of this Certificate. For CONSTRUCTION of the installation:

Signature: .... Date ....

Name (BLOCK LETTERS): .... Constructor

FOR INSPECTION & TESTING

I/We being the person(s) responsible for the inspection & testing of the electrical installation (as indicated by my/our signatures below), particulars of which are described above, having exercised reasonable skill and care when carrying out the inspection & testing, hereby CERTIFY that the work for which I/we have been responsible is to the best of my knowledge and belief in accordance with BS 7671:........., amended to .... (date) except for the departures, if any, detailed as follows:

> Details of departures from BS 7671 (Regulations 120-01-03, 120-02):

The extent of liability of the signatory is limited to the work described above as the subject of this Certificate. For INSPECTION & TEST of the installation: **(Where there is mutual responsibility for the design)

Signature: .... Date ....

Name (BLOCK LETTERS): .... Inspector

NEXT INSPECTION (notes 4 and 7)

I/We the designer(s) recommend that this installation is further inspected and tested after an interval of not more than............ years/months

Figure D4 Electrical Installation Certificate (*Continued over page*)

Appendix 1 (*Continued*)

PARTICULARS OF THE SIGNATORIES TO THE ELECTRICAL INSTALLATION CERTIFICATE (note 3)
Designer (No. 1) Name: .... Company: .... Address: .... Postcode: .... Tel No.: ....
Designer (No. 2) (if applicable) Name: .... Company: .... Address: .... Postcode: .... Tel No.: ....
Constructor Name: .... Company: .... Address: .... Postcode: .... Tel No.: ....
Inspector Name: .... Company: .... Address: .... Postcode: .... Tel No.: ....

SUPPLY CHARACTERISTICS AND EARTHING ARRANGEMENTS Tick boxes and enter details, as appropriate

Earthing arrangements	Number and Type of Live Conductors		Nature of Supply Parameters	Supply Protective Device Characteristics
TN-C ☐	a.c. ☐	d.c. ☐	Nominal voltage, $U/U_o^{(1)}$ V	
TN-S ☐	1-phase, 2-wire ☐	2-pole ☐	Nominal frequency, $f^{(1)}$ .... Hz	Type: ....
TN-C-S ☐	1-phase, 3-wire ☐	3-pole ☐	Prospective fault current, $Ipf^{(2)}$ (note 6)....... kA	
TT ☐	2-phase, 3-wire ☐	other ☐	External loop impedance, $Ze^{(2)}$....................Ω	
IT ☐	3-phase, 3-wire ☐		(Note: (1) by enquiry, (2) by enquiry or by measurement)	Nominal current rating
Alternative source ☐ of supply (to be detailed on attached schedules)	3-phase, 4-wire ☐			..............................A

PARTICULARS OF INSTALLATION REFERRED TO IN THE CERTIFICATE Tick boxes and enter details, as appropriate

Means of Earthing	Maximum Demand	
Distributor's facility ☐	Maximum demand (load) ....Amps per phase	
	Details of Installation Earth Electrode (where applicable)	
Installation ☐ earth electrode	Type (e.g. rod(s), tape etc.)	Location .... Electrode resistance to earth ....Ω

Main Protective Conductors

Earthing conductor: material .... csa ....mm^2 connection verified ☐

Main equipotential bonding conductors: material .... csa ....mm^2 connection verified ☐

To incoming water and/or gas service ☐   To other elements ....

Main Switch or Circuit-breaker

BS. Type .... No. of poles .... Current rating ....A Voltage rating V

Location .... Fuse rating or setting ....A

Rated residual operating current $I_{\Delta n}$ = ... mA, and operating time of ....ms (at $I_{\Delta n}$)
(applicable only where an RCD is suitable and is used as a main circuit-breaker)

COMMENTS ON EXISTING INSTALLATION: (In the case of an alternation or addition see Section 743)
....
....
....
....

SCHEDULES (note 2)
The attached Schedules are part of this document and this Certificate is valid only when they are attached to it.
.... Schedules of Inspections and .... Schedules of Test Results are attached. (Enter quantities of schedules attached)

Figure D5 Electrical Installation Certificate (*Continued*)

Appendix 2

MINOR ELECTRICAL INSTALLATION WORKS CERTIFICATE
(REQUIREMENTS FOR ELECTRICAL INSTALLATIONS - BS 7671 [IEE WIRING REGULATIONS])

To be used only for minor electrical work which does not include the provision of a new circuit

PART 1: Description of minor works

1. Description of the minor works: ....

2. Location/Address: ....

3. Date minor works completed: ....

4. Details of departures, if any, from BS 7671

....

....

....

PART 2: Installation details

1. System earthing arrangement: TN-C-S ☐   TN-S ☐   TT ☐

2. Method of protection against indirect contact: ....

3. Protective device for the modified circuit: Type BS .... Rating ....A

4. Comments on existing installation, including adequacy of earthing and bonding arrangements:
   (see Regulation 130-07) ....

   ....

   ....

   ....

PART 3: Essential Tests

1. Earth continuity: satisfactory ☐

2. Insulation resistance:

   Phase/neutral ....MΩ

   Phase/earth ....MΩ

   Neutral/earth ....MΩ

3. Earth fault loop impedance: ....Ω

4. Polarity: satisfactory ☐

5. RCD operation (if applicable): Rated residual operating current I∆n ....mA and operating time of ....ms (at I∆n)

PART 4: Declaration

1. I/We CERTIFY that the said works do not impair the safety of the existing installation, that the said works have been designed, constructed, inspected and tested in accordance with BS 7671: .... (IEE Wiring Regulations), amended to .... and that the said works, to the best of my/our knowledge and belief, at the time of my/our inspection, complied with BS 7671 except as detailed in Part 1.

2. Name: ....   3.   Signature: ....

   For and on behalf of: ....   Position: ....

   Address: ....

   ....   Date: ....

   ....

   ....Postcode: ....

Figure D6 Minor Electrical Installation Certificate

Appendix 3

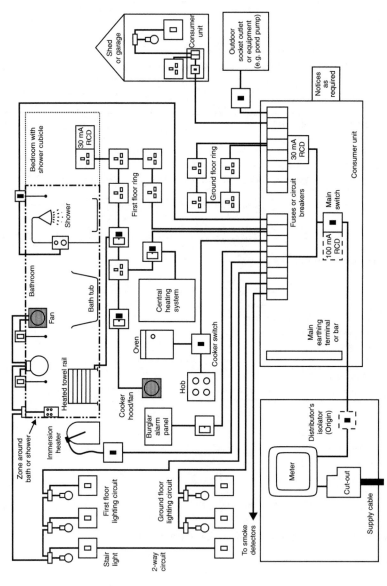

Figure D7 Typical fixed installations that might be encountered in new (or upgraded) existing dwellings

Appendix 4

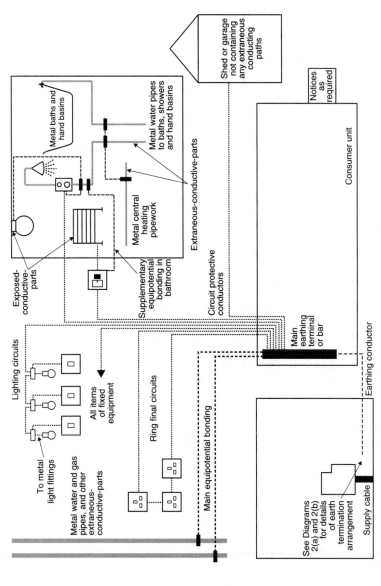

Figure D8 Typical earth and bonding conductors that might be part of the electrical installation shown in Figure D7

Bibliography

Standards referred to

Title	Standard
Acoustics – Measurement of sound absorption in a reverberation room	BS EN 20354: 1993
Acoustics – Measurement of sound insulation in buildings and of building elements Part 3: Laboratory measurement of airborne sound insulation of building elements	BS EN ISO 140-3: 1995
Acoustics – Measurement of sound insulation in buildings and of building elements Part 4: Field measurements of airborne sound insulation between rooms	BS EN ISO 140-4: 1998
Acoustics – Measurement of sound insulation in buildings and of building elements Part 8: Laboratory measurements of the reduction of transmitted impact noise by floor coverings on a heavyweight standard floor	BS EN ISO 140-8: 1998
Acoustics – Measurement of sound insulation in buildings and of building elements Part 6: Laboratory measurements of impact sound insulation of floors	EN ISO 140-6: 1998
Acoustics – Measurement sound insulation in buildings and of building elements Part 7: Field measurements of impact sound insulation of floors	BS EN ISO 140-7: 1998
Acoustics – Method for the determination of dynamic stiffness Part 1: Materials used under floating floors in dwellings	BS EN 29052-1: 1992
Acoustics – Rating of sound insulation in buildings and of building elements Part 1: Airborne sound insulation	BS EN ISO 717-1: 1997
Acoustics – Rating of sound insulation in buildings and of building elements Part 2: Impact sound insulation	BS EN ISO 717-2: 1997
Acoustics – Sound/absorbers for use in buildings – Rating of sound absorption	BS EN ISO 11654: 1997

Aggregates for concrete	BS EN 12620: 2002
Building components and building elements – Thermal resistance and thermal transmittance – Calculation method	BS EN ISO 6946: 1997
Building materials and products – Hydrothermal properties – Tabulated design values	BS EN 12524: 2000
Capillary and compression tube fittings of copper and copper alloy	BS 864
Part 2: 1983 Specification for capillary and compression fittings for copper tubes	
Cast iron pipes and fittings, their joints and accessories for the evacuation of water from buildings. Requirements, test methods and quality assurance	BS EN 877: 1999
Cement	BS EN 197
Part 1: 2000: Composition, specifications and conformity criteria for common elements	
Part 2: 2000: Conformity evaluation	
Chimneys. Clay/ceramic flue blocks for single wall chimneys. Requirements and test methods	BS EN 1806: 2000
Chimneys. Clay/ceramic flue liners. Requirements and test methods	BS EN 1457: 1999
Chimneys. General requirements	BS EN 1443: 1999
Chimneys. Metal chimneys. Test methods	BS EN 1859: 2000
Code of practice for accommodation of building services in ducts	BS 8313: 1989
Code of practice for assessing exposure of walls to wind-driven rain	BS 8104: 1992
Code of practice for building drainage	BS 8301: 1985
Code of practice for design and installation of damp-proof courses in masonry construction	BS 8215: 1991
Code of practice for design and installation of natural stone cladding and lining	BS 8298: 1994
Code of practice for design and installation of non-loadbearing precast concrete cladding	BS 8297: 2000
Code of practice for design and installation of small sewage treatment works and cesspools	BS 6297: 1983
Code of practice for design of non-loadbearing external vertical enclosures of buildings	BS 8200: 1985
Code of practice for drainage of roofs and paved areas	BS 6367: 1983
Code of practice for earth retaining structures	BS 8002: 1994
Code of practice for external renderings	BS 5262: 1991
Code of practice for fire door	BS 8214: 1990
Code of practice for flues and flue structures in buildings	BS 5854: 1980 (1996)

Title	Standard
Code of practice for foundations	BS 8004: 1986
Code of practice for mechanical ventilation and air-conditioning in buildings	BS 5720: 1979
Code of practice for oil firing	BS 5410
Part 1: 1977 Installations up to 44 kW output capacity for space heating and hot water supply purposes	
Part 2: 1978 Installations of 44 kW or above output capacity for space heating, hot water and steam supply purposes	
Code of practice for powered lifting platforms for use by disabled persons (amendment due 1999)	BS 6440: 1983
Code of practice for protection of structures against water from the ground	BS 8102: 1990
Code of practice for protective barriers in and about buildings	BS 6180: 1995
Code of practice for sanitary pipework	BS 5572: 1978
Code of practice for sheet roof and wall coverings	BS CP 143
Code of practice for sheet roof and wall coverings:	BS 5247
Part 14: 1975 Corrugated asbestos cement	
Code of practice for site investigations	BS 5930: 1999
Code of practice for stone masonry	BS 5390: 1976 (1984)
Code of practice for the audio-frequency induction-loop systems (AFILS)	BS 7594: 1993
Code of practice for the control of condensation in buildings	BS 5250: 2002
Code of practice for the storage and on-site treatment of solid waste from buildings	BS 5906: 1980 (1987)
Code of practice for thermal insulation of cavity walls (with masonry or concrete inner and outer leaves) by filling with urea-formaldehyde (UF) foam systems	BS 5618: 1985
Code of practice for use of masonry	BS 5628
Part 1: 1992 Structural use of unreinforced masonry	
Part 2: 2000 Structural use of reinforced and prestressed masonry	
Part 3: 2001 Materials and components, design and workmanship	
Code of practice for ventilation principles and designing for natural ventilation	BS 5925: 1991

Components for smoke and heat control systems Part 2: 1990 Specification for powered smoke and heat exhaust ventilators	BS 7346
Components of automatic fire alarm systems for residential premises Part 1: 1990 Specification for self-contained smoke alarms and point-type smoke detectors	BS 5446

Concrete
Part 1: 2002 Method of specifying and guidance
   for the specifier
Part 2: 2002 Specification for constituent
   materials and concrete
	BS 8500

Concrete
Part 1: 1990 Guide to specifying concrete
Part 2: 1990 Method for specifying concrete
   mixes
Part 3: 1990 Specification for the procedures to
   be used in producing and transporting concrete
Part 4: 1990 Specification for the procedures to
   be used in sampling, testing and assessing
   compliance of concrete
	BS 5328

Construction and testing of drains and sewers	BS EN 1610: 1998

Copper and copper alloys. Plumbing fittings
Part 1: Fittings with ends for capillary
   soldering or capillary brazing to copper tubes
Part 2: Fittings with compression ends for use
   with copper tubes
Part 3: Fittings with compression ends for use
   with plastic pipes
Part 4: Fittings combining other end connections
   with capillary or compression ends
Part 5: Fittings with short ends for capillary
   brazing to copper tubes
	BS EN 1254: 1998

Copper and copper alloys. Seamless, round copper tubes for water and gas in sanitary and heating applications	BS EN 1057: 1996
Copper indirect cylinders for domestic purposes. Specification for double feed indirect cylinders	BS 1566-1: 1984
Design of buildings and their approaches to meet the needs of disabled people – Code of Practice	BS 8300: 2001
Discharge and ventilating pipes and fittings, sand-cast or spun in cast iron Part 1: 1990 Specification for spigot and socket systems Part 2: 1990 Specification for socketless systems	BS 416
Drain and sewer systems outside buildings	BS EN 752

Title	Standard
Part 1: 1996 Generalities and definitions	
Part 2: 1997 Performance requirements	
Part 3: 1997 Planning	
Part 4: 1997 Hydraulic design and environmental aspects	
Part 5: 1997 Rehabilitation	
Part 6: 1998 Pumping installations	
Part 7: 1998 Maintenance and operations	
Ductile iron pipes, fittings, accessories and their joints for sewerage applications. Requirements and test methods	BS EN 598: 1995
Emergency lighting	BS 5266
Part 1: 1988 Code of practice for the emergency lighting of premises other than cinemas and certain other specified premises used for entertainment	
Factory-made insulated chimneys	BS 4543
Part 1: 1990 (1996) Methods of test, AMD 8379	
Part 2: 1990 (1996) Specification for chimneys with stainless steel flue linings for use with solid fuel fired appliances	
Part 3: 1990 (1996) Specification for chimneys with stainless steel flue linings for use with oil fired appliances	
Fibre cement flue pipes, fittings and terminals	BS 7435
Part 1: 1991 (1998) Specification for light quality fibre cement flue pipes, fittings and terminals	
Specifications for heavy quality cement flue pipes, fittings and terminals Part 2: 1991	
Fire detection and alarm systems for buildings	BS 5839
Part 1: 1988 Code of practice for system design, installation and servicing	
Part 2: 1983 Specification for manual call points	
Part 6: 1995 Code of practice for the design and installation of fire detection and alarm systems in dwellings	
Part 8: 1998 Code of practice for the design, installation and servicing of voice alarm systems	
Fire extinguishing installations and equipment on premises	BS 5306
Part 1: 1976 (1988) Hydrant systems, hose reels and foam inlets	
Part 2: 1990 Specification for sprinkler systems	
Fire precautions in the design, construction and use of buildings	BS 5588

Part 0: 1996 Guide to fire safety codes of
   practice for particular premises
Part 1: 1990 Code of practice for residential
   buildings
Part 4: 1998 Code of practice for smoke control
   using pressure differentials
Part 5: 1991 Code of practice for firefighting
   stairs and lifts
Part 6: 1991 Code of practice for places
   of assembly
Part 7: 1997 Code of practice for the incorporation
   of atria in buildings
Part 8: 1999 Code of practice for means of escape
   for disabled people
Part 9: 1989 Code of practice for ventilation
   and air conditioning ductwork
Part 10: 1991 Code of practice for shopping
   complexes
Part 11: 1997 Code of practice for shops,
   offices, industrial, storage and other similar
   buildings

Fire safety signs, notices and graphic symbols   BS 5449
Part 1: 1990 Specification for fire safety signs

Fire safety signs, notices and graphic symbols   BS 5499
Part 1: 1990 Specification for fire safety signs

Fire tests on building materials and structures   BS 476
Part 3: 1958 External fire exposure roof tests
Part 4: 1970 (1984) Non-combustibility test for
   materials
Part 6: 1981 Method of test for fire propagation
   for products
Part 6: 1989 Method of test for fire propagation
   for products
Part 7: 1971 Surface spread of flame tests for
   materials
Part 7: 1987 Method for classification of the surface
   spread of flame of products
Part 7: 1997 Method of test to determine the
   classification of the surface spread of flame of
   products
Part 8: 1972 Test methods and criteria for the
   fire resistance of elements of building
   construction
Part 11: 1982 (1988) Method for assessing the
   heat emission from building materials
Part 20: 1987 Method for determination of the
   fire resistance of elements of construction
   (general principles)

Title	Standard
Part 21: 1987 Methods for determination of the fire resistance of loadbearing elements of construction	
Part 22: 1987 Methods for determination of the fire resistance of non-loadbearing elements of construction	
Part 23: 1987 Methods for determination of the contribution of components to the fire resistance of a structure	
Part 24: 1987 Method for determination of the fire resistance of ventilation ducts	
Part 31: Methods for measuring smoke penetration through doorsets and shutter assemblies	
Section 31.1: 1983 Measurement under ambient temperature conditions	
Flue blocks and masonry terminals for gas appliances	BS 1289-1: 1986
Part 1: 1986 Specification for precast concrete flue blocks and terminals	
Part 2: 1989 Specification for clay flue blocks and terminals	
Fuel oils for non-marine use	BS 2869: 1998
Part 2: 1988 Specification for fuel oil for agricultural and industrial engines and burners (classes A2, C1, C2, D, E, F, G and H)	
Glossary of terms relating to solid fuel burning equipment	BS 1846
Part 1: 1994 Domestic appliances	
Gravity drainage systems inside buildings	BS EN 12056: 2000
Part 1: Scope, definitions, general and performance requirements	
Part 2: Wastewater systems, layout and calculation	
Part 3: Roof drainage layout and calculation	
Part 4: Effluent lifting plants, layout and calculation	
Part 5: Installation, maintenance and user instructions	
Guide for design, construction and maintenance of single-skin air supported structures	BS 6661: 1986
Guide to assessment of suitability of external cavity walls for filling with thermal insulants	BS 8208
Part 1: 1985 Existing traditional cavity construction	
Guide to development and presentation of fire tests and their use in hazard assessment	BS 6336: 1998

Guide to the principles of the conservation of historic buildings	BS 7913: 1998
Heating boilers. Heating boilers with forced draught burners. Terminology, general requirements, testing and marketing	BS EN 303-1: 1999
Hydrothermal performance of building components and building elements. Internal surface temperature to avoid critical surface humidity and interstitial condensation. Calculation methods	BS EN ISO 13788: 2001

Installation and maintenance of flues and
ventilation for gas appliances of rated input
not exceeding 70 kW net   BS 5440
Part 1: 2000 Specification for installation and
    maintenance of flues
Part 2: 2000 Specification for installation   BS 6461
    of chimneys and flues for domestic
    appliances burning solid fuel (including
    wood and peat)
Part 1: 1984 (1998) Code of practice for
    masonry, chimneys and flue pipes

Installation of domestic heating and cooking   BS 8303
appliances burning solid mineral fuels
Part 1: 1994 Specification for the design
    of installations
Part 2: 1994 Specification for installing and
    commissioning on site
Part 3: 1994 Recommendations for design
    and on site installation

Installation of factory-made chimneys to   BS 7566
BS 4543 for domestic appliances
Part 1: 1992 (1998) Method of specifying
    installation design information
Part 2: 1992 (1998) Specification for
    installation design
Part 3: 1992 Specification for site installation
Part 4: 1992 (1998) Recommendations for
    installation design and installation

Installations for separation of light liquids   BS EN 858: 2001
(e.g. petrol or oil)
Part 1: Principles of design, performance and
    testing, marking and quality control

Internal and external wood doorsets,   BS 4787
door leaves and frames
Part 1: 1980 (1985) Specification for
    dimensional requirements

Title	Standard
Investigation of potentially contaminated land. Code of practice	BS 10175: 2001
Investigation of potentially contaminated land. Code of practice	BS 10175: 2001
Lifts and service lifts: Part 1: 1986 Safety rules for the construction and installation of electric lifts (Part 1 to be replaced by BS EN 81-1, when published) Part 2: 1988 Safety rules for the construction and installation of hydraulic lifts (Part 2 to be replaced by BS EN 81-2, when published) Part 5: 1989 Specifications for dimensions for standard lift arrangements Part 7: 1983 Specification for manual control devices, indicators and additional fittings amendment slip	BS 5655
Loading for buildings Part 1: 1996 Code of practice for dead and imposed loads Part 2: 1997 Code of practice for wind loads Part 3: 1988 Code of practice for imposed roof loads	BS 6399
Measurement of sound insulation in buildings and of building elements Part 1: 1980 Recommendations for laboratories Part 3: 1980 Laboratory measurement of airborne sound insulation of building elements Part 4: 1980 Field measurement of airborne sound insulation between rooms Part 6: 1980 Laboratory measurement of impact sound insulation of floors Part 7: 1980 Field measurements of impact sound insulation of floors	BS 2750
Method for specifying thermal insulating materials for pipes, tanks, vessels, ductwork and equipment operating within the temperature range $-40°C$ to $+70°C$	BS 5422: 2001
Method of test for ignitability of fabrics used in the construction of large tented structures	BS 7157: 1989
Methods for rating the sound insulation in building elements Part 1: 1984 Method for rating the airborne sound insulation in buildings and interior building elements	BS 5821

Part 2: 1984 Method for rating the impact
   sound insulation

Methods of test for flammability of textile   BS 5438: 1989
fabrics when subjected to a small igniting
flame applied to the face or bottom edge of
vertically oriented specimens, Test 2

Methods of testing plastics   BS 2782
Part 1: Thermal properties: Methods 120A to
   120E: 1990 Determination of the Vicat
   softening temperature of thermoplastics

Oil burning equipment   BS 799
Part 5: 1987 Specification for oil storage tanks

Particleboards. Specifications. Requirements   BS EN 312-5: 1997
for load-bearing boards for use in humid
conditions

Plastics piping systems for non-pressure   BS EN 1401-1: 1998
underground drainage and sewerage.
Unplasticized poly(vinylchloride) (PVC-U).
Specifications for pipes, fittings and the system

Plastics piping systems for soil and waste   BS EN 1455-1: 2000
(low and high temperature) within the
building structure. Acryionitrilebutadiene-styrene
(ABS). Specifications for pipes, fittings and
the system

Plastics piping systems for soil and waste   BS EN 1329-1: 2000
discharge (low and high temperature) within
the building structure. Unplasticized polyvinyl
chloride (PVC-U). Specifications for pipes,
fittings and the system

Plastics piping systems for soil and waste   BS EN 1451-1: 2000
discharge (low and high temperature) within
the building structure. Polypropyiene (PP).
Specifications for pipes, fittings and the system

Plastics piping systems for soil and waste   BS EN 1519-1: 2000
discharge (low and high temperature) within
the building structure. Polyethylene (PE).
Specifications for pipes, fittings and the system

Plastics piping systems for soil and waste   BS EN 1565-1: 2000
discharge (low and high temperature) within
the building structure

Plastics piping systems for soil and waste   BS EN 1566-1: 2000
discharge (low and high temperature) within
the building structure. Chlorinated polyvinyl
chloride (PVC-C). Specification for pipes,
fittings and the system

Title	Standard
Precast concrete masonry units Part 1: 1981 Specification for precast concrete masonry units	BS 6073

Precast concrete pipes fittings and ancillary
products
Part 2: 1982 Specification for inspection
   chambers and street gullies
Part 100: 1988 Specification for unreinforced
   and reinforced pipes and fittings with
   flexible joints
Part 101: 1988 Specification for glass
   composite concrete (GCC) pipes and fittings
   with flexible joints
Part 120: 1989 Specification for reinforced
   jacking pipes with flexible joints
Part 200: 1989 Specification for unreinforced
   and reinforced manholes and soakaways
   of circular cross-section
	BS 5911

Pressure sewerage systems outside buildings	BS EN 1671: 1997
Profiled fibre cement. Code of practice	BS 8219: 2001
Protection of buildings against water from the ground	BS CP 102: 1973
Requirements for electrical installations (IEE Wiring Regulations 16th Edition)	BS 7671: 2001 (incorporating Amendments No. 1: 2002 and No. 2: 2004)

Safety and control devices for use in
hot water systems
Part 2: 1991 Specification for temperature
   relief valves for pressures from 1 bar to 10 bar
Part 3: 1991 Specification for combined
   temperature and pressure relief valves for
   pressures from 1 bar to 10 bar
	BS 6283

Safety rules for the construction and installation of lifts – Particular applications for passenger and good passenger lifts – Accessibility to lifts for persons including persons with disability	BS EN 81-70: 2003
Sanitary installations Part 1: 1984 Code of practice for scale of provision, selection and installation of sanitary appliances	BS 6465
Sanitary tapware. Waste fittings for basins, bidets and baths. General technical specifications	BS EN 274: 1993
Small wastewater treatment plants less than 50 PE	BS EN 12566-1: 2000

Sound insulation and noise reduction for buildings – Code of practice	BS 8233: 1999
Specification for aggregates from natural sources for concrete	BS 882: 1983
Specification for ancillary components for masonry Part 1: 2001 Ties, tension straps, hangers and brackets Part 2: 2001 Lintels Part 3: 2001 Bed joint reinforcement of steel meshwork	BS EN 845
Specification for asbestos-cement pipes, joints and fittings for sewerage and drainage	BS 3656: 1981 (1990)
Specification for calcium silicate (sandlime and flintlime) bricks	BS 187: 1978
Specification for cast iron spigot and socket drain pipes and fittings	BS 437: 1978
Specification for cast iron spigot and socket flue or smoke pipes and fittings	BS 41: 1973 (1981)
Specification for clay and calcium silicate modular bricks	BS 6649: 1985
Specification for clay bricks	BS 3921: 1985
Specification for clay flue linings and flue terminals	BS 1181: 1999
Specification for copper and copper alloys. Tubes Part 1: 1971 Copper tubes for water, gas and sanitation	BS 2871
Specification for copper direct cylinders for domestic purposes	BS 699: 1984
Specification for copper hot water storage combination units for domestic purposes	BS 3198: 1981
Specification for dedicated liquefied petroleum gas appliances. Domestic flueless space heaters (including diffusive catalytic combustion heaters)	BS EN 449: 1997
Specification for design and construction of fully supported lead sheet roof and wall coverings	BS 6915: 2001
Specification for design, installation, testing and maintenance of services supplying water for domestic use within buildings and their curtilages	BS 6700: 1987
Specification for direct surfaced wood chipboard based on thermosetting resins	BS 7331: 1990

Title	Standard
Specification for electrical controls for household and similar general purposes	BS 3955: 1986
Specification for fabrics for curtains and drapes Part 2: 1980 Flammability requirements	BS 5867
Specification for fibre boards	BS 1142: 1989
Specification for flexible joints for grey or ductile cast iron drain pipes and fittings (BS 437) and for discharge and ventilating pipes and fittings (BS 416)	BS 6087: 1990
Specification for impact performance requirements for flat safety glass and safety plastics for use in buildings	BS 6206: 1981
Specification for installation in domestic premises of gas-fired ducted-air heaters of rated input not exceeding 60 kW	BS 5864: 1989
Specification for installation of domestic gas cooking appliances (1st, 2nd and 3rd family gases)	BS 6172: 1990
Specification for installation of gas-fired catering appliances for use in all types of catering establishments (1st, 2nd and 3rd family gases)	BS 6173: 2001
Specification for installation of gas fires, convector heaters, fire/back boilers and decorative fuel effect gas appliances Part 1: 2001 Gas fires, convector heaters and fire/back boilers and heating stoves (1st, 2nd and 3rd family gases) Part 2: 2001 Inset live fuel effect gas fires of heat input not exceeding 15 kW (2nd and 3rd family gases) Part 3: 2001 Decorative fuel effect gas appliances of heat input not exceeding 20 kW (2nd and 3rd family gases), AMD 7033	BS 5871
Specification for installation of gas-fired hot water boilers of rated input not exceeding 60 kW	BS 6798: 2000
Specification for installation of hot water supplies for domestic purposes, using gas-fired appliances of rated input not exceeding 70 kW	BS 5546: 2000
Specification for ladders for permanent access to chimneys, other high structures, silos and bins	BS 4211: 1987
Specification for masonry units Part 1: 2003 Clay masonry units	BS EN 771

Part 2: 2001 Calcium silicate masonry units
Part 3: Aggregate concrete masonry units
Part 4: 2001 Autoclaved aerated concrete
   masonry units
Part 5: Manufactured stone masonry units
Part 6: 2001 Natural stone masonry units

Specification for metal flue pipes, fittings, terminals and accessories for gas-fired appliances with a rated input not exceeding 60 kW, AMD 8413	BS 715: 1993
Specification for metal ties for cavity wall construction	BS 1243: 1978
Specification for modular co-ordination in building	BS 6750: 1986
Specification for mortar for masonry	**BS EN 998**

Part 2: 2002 Masonry mortar

Specification for open fireplace components	BS 1251: 1987
Specification for performance requirements for cables required to maintain circuit integrity under fire conditions	BS 6387: 1994
Specification for performance requirements for domestic flued oil burning appliances (including test procedures)	BS 4876: 1984
Specification for plastics inspection chambers for drains	BS 7158: 2001
Specification for plastics waste traps	BS 3943: 1979 (1988)
Specification for Portland cements	BS 12: 1989
Specification for powered stairlifts	BS 5776: 1996
Specification for prefabricated drainage stack units in galvanized steel	BS 3868: 1995
Specification for quality of vitreous china sanitary appliances	BS 3402: 1969
Stairs, ladders and walkways. Code of practice for the design, construction and maintenance of straight stairs and winders	BS 5395-1: 2000
Specification for safety aspects in the design, construction and installation of refrigerating appliances and systems	BS 4434: 1989
Specification for sizes of sawn and processed softwood	BS 4471: 1987
Specification for softwood grades for structural use	BS 4978: 1988
Specification for the use of structural steel in building	BS 449

Title	Standard
Part 2: 1969 – Metric units	
Specification for thermoplastics waste pipe and fittings	BS 5255: 1989
Specification for thermostats for gas-burning appliances	BS 4201: 1979 (1984)
Specification for tongued and grooved softwood flooring	BS 1297: 1987
Specification for topsoil	BS 3882: 1994
Specification for unplasticized PVC soil and ventilating pipes, fittings and accessories	BS 4514: 1983
Specification for unplasticized polyvinyl chloride (PVC-U) pipes and plastics fittings of nominal sizes 110 and 160 for below ground drainage and sewerage	BS 4660: 1989
Specification for unplasticized PVC pipe and fittings for gravity sewers	BS 5481: 1977 (1989)
Specification for unvented hot water storage units and packages	BS 7206: 1990
Specification for urea-formaldehyde (UF) foam systems suitable for thermal insulation of cavity walls with masonry or concrete inner and outer leaves	BS 5617: 1985
Specification for vitreous-enamelled low-carbon-steel fluepipes, other components and accessories for solid-fuel-burning appliances with a maximum rated output of 45 kW	BS 6999: 1989 (1996)
Specification for vitrified clay pipes, fittings and ducts, also flexible mechanical joints for use solely with surface water pipes and fittings	BS 65: 1991
Stainless steels. List of stainless steels	BS EN 10 088-1: 1995
Stairs, ladders and walkways Part 1: 1977 Code of practice for stairs Part 2: 1984 Code of practice for the design of helical and spiral stairs Part 3: 1985 Code of practice for the design of industrial type stairs, permanent ladders and walkways	BS 5395
Steel plate, sheet and strip Part 2: 1983 Specification for stainless and heat-resisting steel plate, sheet and strip	BS 1449
Steel plate, sheet and strip. Carbon and carbon manganese plate, sheet and strip. General specifications	BS 1449-1: 1991

Structural design of buried pipelines under   BS EN 1295-1: 1998
various conditions of loading. General requirements

Structural design of low-rise buildings   BS 8103
Part 1: 1995 Code of practice for stability,
    site investigation, foundations and ground
    floor slabs for housing
Part 2: 1996 Code of practice for masonry
    walls for housing
Part 3: 1996 Code of practice for timber
    floors and roofs for housing
Part 4: 1995 Code of practice for suspended
    concrete floors for housing

Structural fixings in concrete and masonry   BS 5080: 1993
Part 1: 1993 Method of test for tensile loading

Structural use of aluminium   BS 8118
Part 1: 1991 Code of practice for design
    amendment slip
Part 2: 1991 Specification for materials,
    workmanship and protection

Structural use of concrete   BS 8110
Part 1: 1997 Code of practice for design
    and construction
Part 2: 1985 Code of practice for special
    circumstances
Part 3: 1995 Design charts for single reinforced beams,
    doubly reinforced beams and rectangular columns

Structural use of steelwork in building   BS 5950
Part 1: 2000 Code of practice for design
Part 2: 2001 Specification for materials,
    fabrication and erection
Part 3: 1990 Design in composite construction
Part 4: 1994 Code of practice for design of
    composite slabs with profiled steel sheeting
Part 5: 1998 Code of practice for design of
    cold formed thin gauge sections

Structural use of timber   BS 5268
Part 2: 2002 Code of practice for permissible
    stress design, materials and workmanship
Part 3: 1998 Code of practice for trussed rafter roofs
Part 6: Code of practice for timber framed walls
Part 6.1: 1988 Dwellings not exceeding
    three storeys

Thermal bridges in building construction –   BS EN ISO 10211-1: 1996
Calculation of heat flows and surface
temperatures
Part 1: General methods

Thermal bridges in building construction –   BS EN ISO 10211-2: 2001
Calculation of heat flows and surface temperatures

Title	Standard
Part 2: Linear thermal bridges	
Thermal insulation – Determination of steady-state thermal transmission properties – Calibrated and guarded hot box	BS EN ISO 8990: 1996
Thermal insulation of cavity walls by filling with blown man-made mineral fibre	BS 6232
Part 1: 1982 Specification for the performance of installation systems	
Part 2: 1982 Code of practice for installation of blown man-made mineral fibre in cavity walls with masonry and/or concrete leaves	
Thermal performance of building materials and products – Determination of thermal resistance by means of guarded hot plate and heat flow meter methods – Dry and moist products of low and medium thermal resistance	BS EN 12664: 2001
Thermal performance of building materials and products – Determination of thermal resistance by means of guarded hot plate and heat flow meter methods – Products of high and medium thermal resistance	BS EN 12667: 2000
Thermal performance of building materials and products – Determination of thermal resistance by means of guarded hot plate and heat flow meter methods – Thick products of high and medium thermal resistance	BS EN 12939: 2001
Thermal performance of buildings – Heat transfer via the ground – Calculation methods	BS EN ISO 13370: 1998
Thermal performance of windows and doors – Determination of thermal transmittance by hot box method	BS EN ISO 12567-1: 2000
Part 1: Complete windows and doors	
Thermal performance of windows, doors and shutters – Calculation of thermal transmittance	BS EN ISO 10077-1: 2000
Part 1: Simplified methods	
Thermoplastics piping systems for non-pressure underground drainage and sewerage – Structure walled piping systems of unplasticized poly(vinyl chloride) (PVC-U), polypropylene (PP) and polyethylene (PE) –	BS EN 13476-1: 2001
Part 1: Specification for pipes, fittings and the system	
Part 1: 1997 Guide to specifying concrete	
Part 2: 1997 Methods for specifying concrete mixes	
Part 3: 1990 Specification for the procedures to be used in producing and transporting concrete	
Part 4: 1990 Specification for the procedures to be used in sampling, testing and assessing compliance of concrete	

Vacuum drainage systems inside buildings	BS EN 12109: 1999
Vacuum sewerage systems outside buildings	BS EN 1091: 1997

Vitreous china washdown WC pans with
horizontal outlet. Specification for WC pans
with horizontal outlet for use with 7.5 L
maximum flush capacity cisterns — BS 5503-3: 1990

Vitrified clay pipes and fittings and pipe
joints for drains and sewers — BS 295
Part 1: 1991 Test requirements
Part 2: 1991 Quality control and sampling
Part 3: 1991 Test methods

Vitrified clay pipes and fittings and pipe
joints for drains and sewers — BS EN 295
Part 1: 1991 Test requirements
Part 2: 1991 Quality control and sampling
Part 3: 1991 Test methods
Part 6: 1996 Requirements for vitrified
    clay manholes

Wall hung WC pan, specification for
WC pans with horizontal outlet for use with
7.5 l maximum flush capacity cisterns — BS 5504-4: 1990

Wastewater lifting plants for buildings
and sites – Principles of construction and testing — BS EN 12050: 2001
Part 1: Lifting plants for wastewater containing
    faecal matter
Part 2: Lifting plants for faecal-free wastewater
Part 3: Lifting plants for wastewater containing
    faecal matter for limited application

Windows, doors and rooflights — BS 8213: Part 1: 1991
Part 1: Code of practice for safety in use and
    during cleaning of windows and doors (including
    guidance on cleaning materials and methods)

Wood preservatives. Guidance on choice,
use and application — BS 1282: 1999

Wood stairs — BS 585
Part 1: 1989 Specification for stairs with closed
    risers for domestic use, including straight and
    winder flights and quarter and half landings

Workmanship on building sites — BS 8000
Part 6: 1990 Code of practice for slating
    and tiling of roofs and claddings
Part 13: 1989 Code of practice for above
    ground drainage and sanitary appliances
Part 14: 1989 Code of practice for below
    ground drainage.

 Note: Copies of British Standards available from: BSI, PO Box 16206, Chiswick,
London W4 4ZL. Website: www.bsonline.techindex.co.uk.

Books by the same author

Title	Details	Publisher
ISO 9001:2000 for Small Businesses (2nd edn)	A guide to cost-effective compliance with the requirements of ISO 9001:2000	Butterworth-Heinemann ISBN: 0 7506 6617 X
ISO 9001:2000 in Brief (3rd edn)	A 'hands-on' book providing practical information on how to cost-effectively set up an ISO 9001:2000 Quality Management System	Butterworth-Heinemann ISBN: 0 7506 6616 1
ISO 9001:2000 Audit Procedures (2nd edn)	A complete set of audit check sheets and explanations to assist authors in completing internal, external and third part audits of *newly implemented, existing* and *transitional* QMSs	Butterworth-Heinemann ISBN: 0 7506 6615 3
Optoelectronic and Fiber Optic Technology	An introduction to the fascinating technology of fiber optics	Butterworth-Heinemann ISBN: 0 7506 5370 1

CE Conformity Marking 	Essential information for any manufacturer or distributor wishing to trade in the European Union. Practical and easy to understand	Butterworth-Heinemann ISBN: 0 7506 4813 9
Environmental Requirements for Electromechanical and Electronic Equipment 	Definitive reference containing all the background guidance, ranges, test specifications, case studies and regulations worldwide	Butterworth-Heinemann ISBN: 0 7506 3902 4
MDD Compliance using Quality Management Techniques 	Easy to follow guide to MDD, enabling the purchaser to customize the Quality Management System to suit his own business	Butterworth-Heinemann ISBN: 0 7506 4441 9
Quality and Standards in Electronics 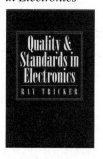	Ensures that manufacturers are aware of all the UK, European and international necessities, and know the current status of these regulations and standards, and where to obtain them	Butterworth-Heinemann ISBN: 0 7506 2531 7

CDs by the same author

Title	Details	ISBN
ISO 9001:2000 Quality Manual & Audit Checksheets	A CD containing a soft copy of the generic Quality Management System featured in *ISO 9001:2000 for Small Businesses* (3rd edn) plus a soft copy of all the checksheets and example audit forms contained in *ISO 9001:2000 Audit Procedures* (2nd edn)	0 9548647 2 7
ISO 9001:2000 Audit Checklists	A CD containing all of the major audit checksheets and forms required to conduct either: • a simple internal review of a particular process; • an external (e.g. third party) assessment of an organization against the formal requirements of ISO 9001:2000; • an assessment of an organization against the formal requirements of other associated management standards such as **ISO 140001** and **OHSAS etc.**	0 9548647 1 9
Aide Mémoire for ISO 9001:2000	An electronic book (in CD format) that cuts out all the lengthy explanations, hypothesis and discussions about ISO 9001:2000 and just provides the reader with the bare details of the standard (e.g. What does the standard actually say? What needs to be checked? Where is a particular requirement covered in the standard? What will an auditor be looking for?, etc.). Also includes additional auditors checksheets and example forms etc. from *ISO 9001:2000 Audit Procedures* (1st edn) and *ISO 9001:2000 for Small Businesses* (2nd edn)	0 9548647 0 0

Useful contact names and addresses

The following professional body is willing to provide general and informal advice about the Act. **However**, any advice given should **not** be seen as being endorsed by the Office of the Deputy Prime Minister!

The Royal Institution of Chartered Surveyors (RICS)
Technical Services Unit
12 Great George Street
London, SW1P 3AD
Tel: 020 7222 7000 (extension 492)
Fax: 020 7222 9430

The following bodies hold lists of their members who may be willing to provide professional advice or act as a 'surveyor' under the Act – again with the proviso that any advice given should **not** be seen as being endorsed by the Office of the Deputy Prime Minister!

Architecture and Surveying Institute
Register of Party Wall Surveyors
St Mary House
15 St Mary Street
Chippenham
Wiltshire, SN15 3WD
Tel: 01249 444505
Fax: 01249 443602

The Association of Building Engineers (ABE)
Private Practice Register
Lutyens House
Billing Brook Road
Weston Favell
Northampton, NN3 8NW
Tel: 01604 404121
Fax: 01604 784220

The Pyramus & Thisbe Club
Florence House
53 Acton Lane
London, NW10 8UX
Tel: 020 8961 3311
Fax: 020 8963 1689

The Royal Institute of British Architects (RIBA)
Clients Advisory Service
66 Portland Place
London, W1N 4AD
Tel: 020 7307 3700
Fax: 020 7436 9112

The Royal Institution of Chartered Surveyors (RICS)
Information Centre
12 Great George Street
London, SW1P 3AD
Tel: 020 7222 7000
Fax: 020 7222 9430

Professional contacts

Asbestos specialists

Asbestos Information Centre Ltd
PO Box 69
Widnes
Cheshire

WA8 9GW
0151 420 5866

Concrete specialist

British Ready Mixed Concrete Association
The Bury
Church Street
Chesham
Buckinghamshire
HP5 1JE
01494 791050

Damp, rot, infestation

British Wood Preserving & Damp Proofing Association
Building No. 6
The Office Village
4 Romford Road
Stratford
London
E15 4EA
020 8519 2588

English Nature
Northminster House
Northminster Road
Peterborough
PE1 1VA
01733 340345

Countryside Council for Wales
Plaspenrhos
Penrhos Road
Bangor
Gwynedd
LL57 2LG
01248 370444

Scottish Natural Heritage
12 Hope Terrace
Edinburgh
EH9 2AS
0131 447 4784

Local Department of Environmental Health
Refer to your local directory

Decorators

British Decorators Association
32 Coton Road
Nuneaton
Warwickshire
CV11 5TW
01203 353776

Scottish Decorators Federation
41A York Place
Edinburgh
EH1 3HT
0131 557 9345

Electricians

National Inspection Council for Electrical Installation Contracting
37 Albert Embankment
London
SE1 7UJ
020 7582 7746

Fencing erectors

Fencing Contractors Association
Warren Rd
Trellech
Monmouthshire
NP25 4PQ
07000 560722

Glazing specialist

Glass and Glazing Federation
44–48 Borough High Street
London
SE1 1XB
020 7403 7177

Heating installers

British Gas Regional Office
Refer to your local directory

Electricity Supply Company
Refer to your local directory

British Coal Corporation
Hobart House
Grosvenor Place

London
SW1X 7AE
020 7235 2020

Heating and Ventilating Contractors Association
Esca House
34 Palace Court
London
W2 4JG
020 7229 2488

National Association of Plumbing, Heating and Mechanical Services Contractors
Ensign House
Ensign Business Centre
Westwood Way
Coventry
CV4 8JA
01203 470626

Home security

Local Crime Prevention Officer
Refer to your local directory

Local Fire Prevention Officer
Refer to your local directory

National Approval Council for Security Systems
Queensgate House
14 Cookham Road
Maidenhead
S16 8AJ
01628 37512

Master Locksmiths Association
Units 4–5
Woodford Halse Business Park
Great Central Way
Woodford Halse
Daventry
NN1 6PZ
01327 62255

British Security Industry Association
Security House
Barbourne Road

Worcester
WR1 1RS
01905 21464

Insulation installers

Draught Proofing Advisory Association Ltd,
External Wall Insulation Association,
National Cavity Insulation Association,
National Association of Loft Insulation Contractors
PO Box 12
Haslemere
Surrey
GU27 3AH
01428 654011

Plasterers

Federation of Master Builders
14 Great James Street
London
WC1N 3DP
020 7242 7583

Plumbers

National Association of Plumbing, Heating and Mechanical Services Contractors
Ensign House
Ensign Business Centre
Westwood Way
Coventry
CV4 8JA
01203 470626

Roofers

Builders' Merchants Federation
15 Soho Square
London
W1V 5FB
020 7439 1753

National Federation of Roofing Contractors
24 Weymouth Street

London
W1G 7LX
020 7436 0387

Ventilation

Heating and Ventilating Contractors Association
Esca House
34 Palace Court
London
W2 4JG
020 7229 2488

Other useful contacts

British Board of Agrément (BBA)
PO Box 195
Bucknalls Lane
Garston
Watford
WD2 7NG
Tel: 01923 665300
Fax: 01923 665301
E-mail: bba@btinternet.com
Internet: Http://www.bbacerts.co.uk

BSI
British Standards Institution
389 Chiswick High Road
London
W4 4AL
Tel: 0181 996 9001
Fax: 0181 996 7001
E-mail: info@bsi.org.uk
Internet: www.bsi.org.uk

Fensa Ltd
Fenestration Self-Assessment Scheme
44–48 Borough High Street
London SE1 1XB
Tel: 020 7207 5874
Fax: 020 7357 7458

HETAS Ltd
HETAS Ltd
12 Kestrel Walk
Letchworth

Hertfordshire
SG6 2TB
Tel: 01462 634721
Fax: 01462 674329

Note: Registration scheme for companies and engineers involved in the installation and maintenance of domestic solid fuel fired equipment.

HMSO
The Stationery Office
The Publications Centre
PO Box 29
Norwich
NR3 1GN
Telephone orders/General enquiries
0870 600 5522
Fax orders 0870 600 5533
www.thestationeryoffice.com

Institute of Plumbing
64 Station Lane
Hornchurch
Essex
RM12 6NH
Tel: 01708 472791
Fax: 01708 448987

Note: Approved Contractor Person Scheme (Building Regulations).

OFTEC
Oil Firing Registration Scheme
Century House
100 High Street
Banstead
Surrey
SM7 2NN
Tel: 01737 373311
Fax: 01737 373553

Robust Details Limited
PO Box 7289
Milton Keynes
MK14 6ZQ
Business Line: 0870 240 8210
Technical Support Line: 0870 240 8209
Fax: 0870 240 8203

UKAS
United Kingdom Accreditation Service
21–47 High Street
Feltham, Middlesex
TW3 4UN
Tel: 0208 917 8400
Fax: 0208 917 8500

WIMLAS
WIMLAS Limited
St Peter's House
6–8 High Street
Aver, Buckinghamshire
SL0 9NG
Tel: 01753 737744
Fax: 01753 792321
E-mail: wimlas@compuserve.com

Useful websites

The Building Act and Building Regulations http://www.odpm.gov.uk/stellent/
groups/odpm_control/documents/contentservertemplate/odpm_index.hcst?n =
235&l = 1

Approved Documents http://www.odpm.gov.uk/stellent/groups/odpm_control/
documents/contentservertemplate/odpm_index.hcst?n = 240&l = 2

Assessable thresholds	www.tso.co.uk
British Standards	www.bsonline.techindex.co.uk
Building near trees	www.nhbc.co.uk
Building Research	www.bre.co.uk
Carbon dioxide from natural sources	www.bgs.ac.uk
and mining areas	www.tso.co.uk
	www.defra.gov.uk
	www.ciria.org.uk
Cladding	www.mcrma.co.uk
Concrete in aggressive ground	www.bre.co.uk
Contaminated land	www.defra.gov.uk
	www.ciria.org.uk
	www.hse.co.uk
Contamination in disused coal mines	www.tso.co.uk
Demolition	www.ciria.org.uk
Environmental aspects	www.arup.com
	www.environment-agency.gov.uk
Excavation and disposal	www.defra.gov.uk
	www.ciria.org.uk
Flood protection	www.ciria.org.uk
	www.safety.odpm.gov.uk
Flooding from sewers	www.defra.gov.uk
	www.ciria.org.uk
Foundations	www.bre.co.uk
Gas contaminated land	www.defra.gov.uk
	www.bre.co.uk
	www.ciria.org.uk

Geoenvironmental and geotechnical investigations	www.ags.org.uk
Glass and glazing	www.ggf.org.uk
Hardcore	www.bre.co.uk
Health and safety	www.hse.co.uk
	www.defra.gov.uk
Land quality	www.environment-agency.gov.uk
Landfill gas	www.ciwm.co.uk
	www.defra.gov.uk
	www.gassim.co.uk
Laying water pipelines in contaminated ground	www.fwr.org
Low-rise buildings	www.bre.co.uk
Materials and workmanship	www.tso.co.uk
Methane	www.bgs.ac.uk
	www.tso.co.uk
	www.defra.gov.uk
	www.ciria.org.uk
Oil seeps from natural sources and mining areas	www.bgs.ac.uk
	www.tso.co.uk
	www.defra.gov.uk
	www.ciria.org.uk
Petroleum retail sites	www.petroleum.co.uk
Pollution control	www.odpm.gov.uk
Protection of ancient buildings	www.spab.org.uk
Radon	www.bre.co.uk
Robust construction details	www.tso.co.uk
Roofing design	www.mcrma.co.uk
Shrinkable clay soils	www.bre.co.uk
Soil sampling	www.defra.gov.uk
Soils, sludge and sediment	www.ciria.org.uk
Subsidence	www.bre.co.uk
Thermal bridging	www.bre.co.uk
	www.tso.co.uk
Thermal insulation	www.bre.co.uk
Timbers	www.bre.co.uk

Index

89/106/EEC, 170, 171
Access
   and facilities for the fire service,
      78
   and use of buildings, 97
   buildings
      see requirements – access
   disabled people, 37
Access Statement, 543, 601
   content, 543, 601
   further guidance, 544
Acoustic environment
   design, 461, 468
Acoustics, 578, 662
Advertising
   planning permission, 146
   regulations, 114
Aerials, 146
Aids to communication, 468
   acoustic environment, 468
   artificial lighting, 469
   hearing enhancement systems, 469
   induction loop, 469
   text telephones, 469
   uplighters, 469
Air paths, 257
   and airborne sound, 257
Air supply, 91
Airborne and impact sound, 239
Airborne sound, 327
   and air paths, 327
   floors and stairs, 258
   insulation, 258, 260
   walls, 326
Alterations, internal, 137
Ambulant disabled people, 447, 455,
      475
   access, 602
   access to a dwelling, 607
   doors, 484
   stairs, 418
Appeals, against local authority
      ruling, 23
Appendix A, Building Regulation
      24–27
Applications
   neighbours, 54

not requiring submission of
      plans, 48–49
   submitting full plans, 50
Approved Documents
   2004, 169
   aim, 31
   amendments, 30
   compliance, 170
   consultation draft proposals, 31
   electrical safety *see* Approved
      Document P
   further information, 61
   future amendments, 39
   non-compliance, 133
   the 2000 list, 169
   types, 31
Approved Document A, 32
Approved Document C, 32
Approved Documents C1, 174–75
Approved Document E
   E1
      flanking transmission, 257
      in a nutshell, 81
      performance standards, 641
      requirements, 640
      resistance to sound, 641
      sound insulation, 641
      sound insulation values,
         AE1, 641
   E2
      in a nutshell, 81, 88
      internal floors, 643
      laboratory values, 642
      requirements, 642
      sound insulation, 642
      sound insulation values, 642
   E3
      in a nutshell, 82
      requirements, 82, 642
      sound absorption, 642
   E4
      in a nutshell, 82
      requirements, 643
      sound insulation, 643
Approved Document K
   future amendments, 36
Approved Document M, 596

Approved Document M (*Continued*)
  access, 596
    see requirements – access
  and Disability Discrimination Act
  Access Statement, 601
  benefits to business/service providers,
    598
  benefits to disabled people, 597
  benefits to other people, 597
  changes, 597, 598
  the requirements, 598
    buildings other than dwellings, 598
Approved Document N
  future amendments, 38
Approved Document P
  aim, 647
  design and installation, 38
  fixed wiring, 492
  further information, 47
  inspection and testing, 38
Approved Document Q, 39
  aim, 40
  broadband, 40
  distribution of services, 41
  existing ducts, 40
  outline requirements, 39
  Q1 – limits of application, 40
  Q1 – requirements, 40
  Q2 – limits of application, 40
  Q2 – requirements, 40
  Q3 – limits of application, 41
  Q3 – requirements, 41
  supply, 41
    around a building, 6, 28, 41
    into a dwelling, 31
    to a dwelling, 40
  terminal chambers, 39, 40
Approved inspector, 15, 46
  assistance, 44
  building notice, 47
  disagreements with, 47
  duties, 59
  final certificate, 17
  full plans, 44, 47
  further information, 61
  initial notice, 16
  plans certificate, 16
Archaeology, 122
Area of Outstanding Natural Beauty, 49
  and planning permission, 28
Artificial lighting, 7, 96, 462
Asbestos works, 177
  contamination, 176
Audience and spectator facilities
  disabled people, 37, 457

Automatic sensors, 559
  doors, 559

BAA, 4
  building regulations, 3
Baby changing, 477
  see Requirements – sanitary
    conveniences, 477–478
Barn owls, 130
Basic requirements
  Planning Permission, Building
    Regulations, 108–111
Bathrooms, 84
  disabled people, 464, 467, 488
  see Requirements – bathrooms, 469
  see Requirements – sanitary
    conveniences, 470
Bats, 129
  vampire, 130
Bedrooms, disabled people, 465
Blister surface, 546
  see Requirements – access
Block plan, 112
Boats, 145
Bonded junction, 339, 347
BRE BR, 264
British Airports Authority, *see* BAA
British Board of Agrément
  Certificate, 170
  preparation of certificates, 172
British Standards Institution, 72
BS 1243: 1978, 294, 675
BS 1289-1: 1986, 668
BS 5588-8, 666
BS 5619: 1978, 597
BS 5628-3: 2001, 664
BS 5810: 1979, 597
BS 7671, 146
BS 7913:1998, 669
BS 8000, 586
BS 8233: 1999, 673
BS 8300: 2001, 665
BS EN 20354: 1993, 578, 662
BS EN ISO 29052-1: 1992, 272, 662
BS EN 81-70: 2003, 448, 672
BS EN ISO 11654: 1997, 578
BS EN ISO 140-3: 1995, 335
BS EN ISO 140-8:1998, 264, 272
BS EN ISO 717-1:1997, 335
BS EN ISO 717-2:1997, 272
Building
  a new house, 111
  classification, 13
  demolition, 111
  design, 78

extensions, 111
  on a site containing offensive
    material, 20
  over an existing sewer, 24
  over sewers, 89
  protection, 91
  standards, 28
Building Act
  aim, 1
  amendments on other Acts, 11
  and its effect on other Acts, 11
  approved methods of
    construction, 6
  availability, 32
  Building Act: 1984, 103
  Building Amendment Regulations
    2001 (SI 2001/3335), 169
  certificates, 10
  contents, 24
  Inner London, 7, 8
  public bodies – notices and
    repeal of other Acts, 12
  schedule 1, 5
  schedule 2, 7
  schedule 3, 7
  schedule 4, 10
  schedule 5, 11
  schedule 6, 11
  schedule 7, 12
  sections inapplicable to London, 8
  supplementary regulations, 5
Building control, application for, 48
Building control approval, 28
Building control bodies, 46
Building control file, 543, 601
Building control officer, 28, 44, 46
  consultation, 50
  responsibilities, 56
Building extensions, see Extensions
Building inspectors
  duties, 56
  initial inspection, 56
  responsibility, 56
Building notice, 45, 47
  advantages, 48
  and unauthorized work, 59
  application, 44
  approved inspector, 47
  building extensions, 51
  cavity wall, 52
  commencement notice, 53
  Conservation Areas, 49
  content, 24
  difference to full plans, 44, 45
  hot water storage systems, 52

listed buildings, 51
  minor works, 51
  neighbours, 54
  procedure, 47, 48
  rejection, 53
  using a local authority, 54, 55
  validity, 53
Building plan, 113
  time to completion, 19
Building Regulations, 3
  Advisory Committee, 12
  aim, 29
  amendments, 18
  and the Fire Authority, 13
  application, 44
  approval – assistance in
    understanding them, 44, 46
  approval – basic
    requirement, 107, 108
  BAA, 4
  CAA, 4
  carrying out work, 59
  completion certificate, 44
  compliance, 5, 6
  contravention of, 5, 59
  Crown buildings, 4
  educational establishments, 601
  exemptions, 3, 41, 43
  guidance, 38
  how to obtain approval, 46
  informal advice, 4
  Inner London, 4
  mandatory requirements, 6
  mothers with prams, 544
  overall responsibility, 15
  people with impaired sight, 544
  people with luggage, 557
  Planning Officers responsibility, 56
  prior notice, 56
  professional advice, 578
  proposed amendments, 36–37
  prosecution for non-compliance,
    42
  purpose, 28
  relaxation, 18
  requirements, 30
  statutory undertakers, 4
  UKAEA, 4
  ventilation, 33–34
  what building works needs it?, 108
  what is covered, 29
  wheelchair users, 38
  where to find technical
    guidance, 30
  where to find the requirements, 30

Building Regulations 1991
    (SI 1985 No. 1065), 169
Building work
    commencement of work, 55
    completion, 56
    control, 58
    excavations, 56
    foundations, 56
    initial inspection, 56
    local authority tests, 57
    notice to backfill, 57
    notice of commencement, 56
    possible causes for rejection, 18, 53
    quality, 56
    requirement for formal approval, 42
    requirements, 58
    without approval, 59
Building work, *see also* Work
    requiring Building Regulations
        approval, 42
    requiring planning permission,
        28, 42
Building works
    meaning, 113
    minor, 137
    without permission, 130
Buildings
    classification of purpose groups,
        198
    classification of size, 200
    dangerous, 20, 167
    defective, 22
    exempt from control, 43
    exemptions from control – other, 43
    size, 198
        annexes, 201
        residential buildings, 200
    suitable for conversion, 163
Buildings other than dwellings
    access, 598
    disabled people, 598
    requirements, 598
    sanitary conveniences, 598
    see Requirements – access, 598
    structure, 600
By-laws, Inner London, 10
By-product works
    contamination, 176, 177

CAA
    building regulations, 4
Car parking, disabled people, 38
Caravans, 146
Carbon dioxide, 177, 181
Cavity insulation, 43

Cavity wall, *see* Wall
    Building Notice, 44, 47, 50
    insulation, 139
Cavity widths
    see Requirements – walls, 337–338
CE Marking Directive (93/68/EEC),
    170
Ceiling treatment A, 255
Ceiling treatment B, 255
Ceiling treatment C, 255
Ceilings
    see Requirements – floors and
        ceilings, 365
Cellars, *see* Requirements – cellars
Central heating, 109, 142
Certificate of lawfulness, 115
Certificate of regularization, 59
Certification schemes, 172
    independent, 172
    product, 172
Cesspools, *see* Requirements – drainage
Changing facilities, 486
    disabled people, 488
Chemical works, contamination, 176
CIC, 61
Civil Aviation Authority, *see* CAA
Cladding, 139
Classification of buildings, 5, 13
Combustion, discharge of products
    of, 91
Commencement notice, 55
    building notice, 55
    full plans, 55
Commercial parking, 145
Completion certificate, 60
    for full plans, 44
Completion notice, 60
Concrete, 253
    base, 254
    beams, 254
    condensed surface resistance, 371
    floor, 253
    floor base, 254
    floor slab, 254
    hollow-core plank floor, 254
    internal concrete floor slab,
        254, 341
    plank floor, 254, 341
    plank, 254
    separating floor, 259
    slab, 253
    slab floor, 253
    solid slab, 253
    suspended floor, 80
Concrete floors, internal, 254

Condensation, in roofs, 33
Conservation, 129
  issues, 129
  of fuel – dwellings, 95
  of fuel – other buildings, 96
  of fuel and power, 96
Conservation Area, 113
  and planning permission, 114
  consent, 113
  designation, 126
  grants, 126
  planning application, 114
  Tree Preservation Order, 114
  trees, 114
  what are they, 113
Conservatory, 110
  building, 152
Construction industry, 61
Construction Industry Council, *see* CIC
Construction Products Directive 1994
    (SI 1994/3051), 170
Contaminants
  gaseous, 178
  resistance to, 32
  types and remedies, 178
Contamination
  asbestos works, 177
  by-product works, 177
  chemical works, 177
  gas works, 177
  hazardous, 176
  industrial, 177
  landfill gas, 177
Control of Pollution (Oil Storage)
    (England) Regulations 2001
    (SI 2001/2954), 512
Control of Pollution Act 1974, 24
Conversion
  of a building, 158
  old building, 158
  residential, 164
  structural condition, 163
  to a shop, 158
  workshop, 163
Conversions, 158
Corridor doors
  fire safety, 335, 521, 575
Corridors and passageways
  disabled people, 455
  entrance storey, 455, 544
  see Requirements – corridors and
    passageways, 453
Covenants, and planning permission,
    125
Crime prevention, 124

Crown buildings
  and building regulations, 4

Damp-proof course, 56, 242, 286
Damp-proof membrane, 240, 241
Dangerous buildings, 20, 167
  emergency measures, 21, 167
Decoration
  external, 137
  internal, 137
Defective buildings, 22
Demolished building replacement, 22
Demolition, 22, 166
  of dangerous buildings, 167
Density of the materials
  walls, 333
Department of the Environment,
    Transport and the Regions,
    see DETR, 54
Department of Works and Pensions, 596
DETR
  further information, 61
  publications, 61
Diesel, storage tanks, 150
Disability Discrimination Act 1995,
    417, 442, 446, 453, 470, 493
  (Amendment) regulations 2003, 596
  new requirements, 596
Disabled
  audience or spectator seating, 597
  sanitary conveniences, 598
Disabled people, *see* Aids to
    communication, 468
  access
    buildings other than dwellings, 600
    statements, 601
    use of buildings, 597
  access and facilities, 417, 429
  Approved Document M, 599
  building control file, 543, 601
  buildings other than dwellings, 600
  material changes of use, 543, 575
  non-domestic buildings, 599
    extensions, 598
    material alterations, 600
  parking bays, 547
  requirements, 598
    access, 577
    dwellings, 600
    in a nutshell, 599
  sanitary conveniences, 598
    buildings other than dwellings, 598
    dwellings, 598
  see Requirements – access, 598
  typical access signs, 556

Disproportionate collapse, 191
  Class 1, 191
  Class 2A, 191–2
  Class 2B, 192
  Class 3, 193–194
Document control, 14
  local authority, 13–16
Door, 441
  replacement, 443
Door opening furniture, 444, 606
  see Requirements – doors, 441
  see Requirements – sanitary
    conveniences, 442
Doors, 592
  ambulant disabled people, 615
  corridor, 592
    sound insulation, 583
    sound reduction index, 592
  disabled people, 442
  emergency escape routes, 612
  entrance, 605
    see Requirements – access, 605
  low energy, 446
    see Requirements – doors
  opening clear width, 452
  reopening activators, 609
  sanitary accommodation, 475
    see Requirements – sanitary
      conveniences
  see Requirements – corridors and
    passageways, 453
  see Requirements – lifting platforms,
    609
  see Requirements – sanitary
    conveniences
  sound insulation, 572
  timed closing devices, 446
Drainage, 203, 204
  and waste water, 205
  foul water – exemptions, 205
  Inner London, 7, 9
  rainwater – exemptions, 45, 206
  separate systems, 89
Driveway, 144
  see Requirements – access
Dwellings, 37
  disabled people, 37
  electrical safety, 38
    see Approved Document P, 38
  requirements, 38
  sanitary conveniences, 37

Educational establishments, 601
Electrical
  cables, 264

installations, 52–53
  safety, 38, 38–39, 100, 140
  see Approved Document P, 38
  supply company, 140
  work, 140
Electricity at Work Regulations 1989,
  140, 497, 647
Electricity Supply Regulations 1988, 647
Electronic communication services, 31
  see Approved Document Q, 31
Emergency alarm pull cords, 613, 622
Emergency assistance
  alarm system, 613
    see Requirements – sanitary
      conveniences
  alarm systems, 482
  disabled people, 482
  see Requirements – sanitary
    conveniences
Emergency communication system
  see Requirements – lifting devices, 608
Emergency escape routes doors, 612
Emergency measures, 21
Energy
  rating, 58
  rating – calculation, 58, 59
Enforcement Notices, 115
Entrance
  disabled people, 605
  see Requirements – access
Entrance doors
  see Requirements – doors
Entrance halls
  requirements for disabled
    people, 611
Entrance lobbies
  disabled people, 77
ENV Eurocodes, 172
Environmentally Sensitive Areas, 164
Escape route, 375
  see Requirements – corridors and
    passageways, 453
European Economic Area, *see* EEA
European pre-standards, 172
European standard, 30, 172
Extensions, 51
  building notice, 51
  limitations, 157
External fire spread, 77
External lighting, 124, 500

Facilities, 457, 617
  audience and spectator, 457
  disabled people, 417
  self service, 620

showers, 487
  disabled people, 417
Factories Act 1961, 471
Fenestration Self-Assessment Scheme,
  139
Fensa Ltd, 139
Final certificate, 10
Fire alarms, 481
  disabled people, 481
  see Requirements – sanitary
    conveniences
Fire Authority
  and the Building Regulations, 13
Fire Brigade
  and dangerous buildings, 168
  full plans responsibilities, 44
Fire doors, 446
  see Requirements – doors
Fire Precautions Act 1971, 11
  and Full Plans, 44
Fire safety, 32
  corridor doors, 82
  see Approved Document P
Five separating element
  lateral support, 248
    ceiling joists, 251–252
    interruption, 251
    small building, 250
    tension strap, 250
Fire, 208
  protection
    pipes and ducts, 208
  safety, 239, 256
    floor penetrations
  stopping, 208
Fire stop, 389
Flagpoles, 146
Flanking transmission, 257, 327
  corridor walls, 288
  wall type, 268
Floor
  airborne sound, 256
  floating layer, 278
  floating, 261
  hangers, 250
  impact sound, 256
  resilient layer, 278
Floor joists, see Requirements, floors
  and roofs
Floor materials, 556, 561
Floor surface, see Requirements –
  reception area, 563
  see Requirements – corridors and
    passageways
Floor type 1, 261

Floor type 1, see Requirements – floors
  and ceilings
  see Requirements – floors and
    ceilings, 261
Floor type 2, 261
Floor type 3, 264
  see Requirements – floors and ceilings
Floorboards, see Requirements, floors
  and ceilings, 238
  ceramic tiles, 379
  floors, 238
  general, 255
  internal, 259
    concrete, 244
  see Requirements – floors and
    ceilings
  wood block, 264
Flue blocks, 337, 346
Fly posting, 147
Food Hygiene (General) Regulations
  1970, 471
Foul water, drainage, 205
Foundation
  concrete, 173
  contaminated ground, 180
  gaseous contaminants, 177, 180
    containment, 184
    corrective measures, 183
    landfill gas, 181
    radon, 180
    remedial measures, 183
    removal, 184
    risk assessment, 181–2
    site preparation, 184–5
    treatment, 184
  general, 179
  hazards, 179
    identification, assessment, 179
  meeting the requirement, 179
  plain concrete, 187–8
  requirements, 173
  risks, 180
  solid, 173
  strip, 189
  timber framed houses, 173
  types of subsoil, 186
Fuel conservation
Full plans, 47
  advantages, 50
  and unauthorized work, 59
  appeal, 52
  application, 49
  approval, 50
  approved inspector, 47
  commencement notice, 55

Full plans (*Continued*)
  completion certificate, 49
  completion notice, 60
  consultation with sewerage
      undertaker, 50
  content, 49
  determination, 56
  difference to building notice, 48
  Fire Brigade involvement, 48
  neighbours, 54
  rejection, 53
  subject to the Fire Precautions Act, 60
  submission, 46
  using the local authority, 55

Garage, 151
  building a, 151
Gas Act 1972, 22, 168
Gas appliances, exemption, 44
Gas Safety (installation and use)
      Regulations 1984 (S1.198411358),
      142
Gas Act 1972 (S1.197211178), 22,
      168
Gas works
  contamination, 176
  landfill gas, 177
General Rate Act 1967, 21
Glazed screens
  see Requirements – corridors and
      passageways, 455
Glazing, 99
Great Fire of London, 28
Greater London Area, water and earth
      closets, 469
Green belt, 120
Ground contamination, 122, 176–7
Ground moisture, 240
Ground movement, 75, 186
Guidance Note G18.5, 476

Handrails, 425, 426, 427
  common stairs, 433
  dimensions, 425
  disabled people, 482
  external stairs, 552
  key requirements, 427
  lifting devices, 447
  location, 432
  ramped access, 432
  see Requirements – access, 438, 451
  see Requirements – handrails, 431
  see Requirements – internal ramps,
      431, 610
  see Requirements – stairs, 416

Hardstanding, 144
  access, 144
  boat, 145
  car, 145
  caravan, 145
  commercial parking, 145
  covenants, 145
Hazardous conditions, contamination, 176
Hazardous materials, 122
Hazards
  access routes, 430
  identification, 179
  see Requirements – stairs
  disabled people, 543
  ground associated, 176
  see Requirements
  access, 430
Hearing enhancement systems, 462, 622
Heat emitters
  see Requirements – sanitary
      conveniences
Hedges, 142
Highway safety, 122
Highways Authority, 123
Historic buildings, 639
  material change of use, 160, 641
HMSO
  availability of Approved
      Documents, 61
Hot water
  storage, 85, 141
Hot water vessel
  exemptions, 45
House
  brick, 173, 174
    typical components, 174
  timber, 173
    typical construction, 175
House-construction, 173
  brick, 173
  timber, 173
House numbers
  approval, 123
Hygiene, 35

IEE Regulations, 140
  see Approved Document
Impact sound
  floors and walls, 262
  insulation, 258
  see Requirements – stairs, 416
Induction loop, 462, 469
Industrial contamination, 176
Industrial Revolution, 29
Infilling, 165

Information
    access to, 115
    approved inspector, 61
    provision of, 91
Infrared system, 462, 468, 469, 620,
        622
Initial notice, 16
    cancellation of, 16
Inner London
    and building regulations, 4
    and the Building Act, 7
    Building Act – sections inapplicable, 7
    buildings, 7
    by-laws, 10
    drainage, 7
Inspection of work
    by local authorities, 5, 13
Insulation
    cavity wall, 52
Internal doors
    see Requirements – doors, 443
Internal ramp, 431, 610
    see Requirements – internal ramps,
        431, 610
International Symbol of Access, 555
Interstitial condensed damaged
Isolated buildings
    suitable for conversion, 163

Junction, bonded, 265
Junctions
    between separating walls, 259
    floor penetrations, 275
    internal framed walls, 275
    with an external solid masonry
        wall, 266, 275
    with an internal concrete floor, 254
    with an internal framed wall, 266
    with an internal masonry floor, 259
    with an internal masonry wall, 266
    with an internal timber floor, 341
    with ceiling and roof space, 350
    with ceilings, 343
    with concrete ground floors, 359
    with an external cavity wall, 265
    with external cavity wall, 361
    with external solid masonry wall,
        362
    with floor penetrations, 267
    with gas pipes, 424
    with inner leaf of wall type 1, 265
    with separating floors, 259
    with wall type 1, 268
    with wall type 2, 269
    with wall type 3, 269

with wall type 3.1 and 3.2, 269
with wall type 3.3, 270
with wall type 4, 270

Kerb
    dropped
        see Requirements – access, 546,
            547

Land Registry, 121
Landfill gas, 177
Landing call buttons
    lifting devices, 447, 608
    see Requirements – lifting devices,
        447
Landings
    see Requirements – internal ramps,
        431
    landscaping, 122
    lath and plaster, 256
    lecture/conference facilities
        disabled people, 454
    lateral support, 251
    roof, floor, 248
Level approach, 564
    see Requirements – access, 564
Leylandii, 143
Lift controls
    see Requirements – passenger lifts
Lift Regulations 1997 SI 1997/831,
        448
Lifting devices, 447
    disabled people, 447
Lifting platform, 450
    see Requirements – wheelchair
        platform
    see Requirements – lifting devices,
        447
    see Requirements – lifting platform,
        450
    stairlifts, see Requirements –
        wheelchair platform stairlifts
Lighting
    external, 124
    security, 124
Light pollution, 122
Light switches, 467
Linings, internal fire spread, 76
Liquid fuel, protection, 91
Listed building
    see Requirements – lifting devices
    and selling one, 126
    enforcement action, 125
    owners responsibility, 125
    planning application, 114

Listed Building Consent, 66, 114
Loading, 75
Loading bays, 94
Lobbies
    glazing, 561, 605
    see Requirements – access, 566
Local authorities, and dangerous
        buildings, 20
Local authority, 13
    and serving notices, 14
    appeals against their ruling, 23
    compensation, 23
    document control, 14
    duties, 13
    inspection of work, 19
    power to enter buildings, 15
    power to relax building regulations, 18
    powers of, 15
    record, 17
Local Government Act 1972, 117
Local Government, Planning and Land
        Act 1980, 14
Local Land Charges Register, 128
Local plan, 101, 103, 136
    and Conservation Areas, 126
    publication, 103
    written statement, 104
Local planning officer, 19, 107, 109,
        110, 111
Local plans, 103
    district, 103
    preparation, 103
    proposals map, 104
    waste and minerals, 103
Loft conversions, 154
Loft space, 41, 255, 343, 351
London, changes to the Building Act, 7
London Building (Amendment) Act
        1939, 167

Maintenance, 137
Marking, 170
Material alteration, 42
    non-domestic buildings, 160
Material changes of use, 600
    disabled access, 600
    existing buildings, 600
    historic building, 162
    meaning, 159
    requirements, 161
    rooms for residential purposes, 572
    sanitary conveniences, 600
Materials
    approved, 171
    CE marked, 171

related change, 161–2
    short lived, 171
Materials and workmanship, 170
Methane, 181
Minor improvements, 137
Minor works, 136
Mixed use developments, 162
Mobility impaired people, 430, 447
Moisture
    resistance to, 32
        see Approved Document C, 32
Mothers with prams
    access, 544

National park, 120
Nature conservation, 122, 129
Nature Conservancy Council, 123
Neighbours, 123
    and their rights, 123
    building applications, 122
Noise
    and disturbance, 122
Non-domestic buildings
    extensions, 157
    material alterations, 159
Notices
    local authority and how served, 14

Offensive material
    building on a site with, 20
Offices, Shops and Railway Premises
        Act 1963, 471
Oil fired appliances
    exemptions, 45
Oil storage, contamination, 177
Oil storage tanks
    exemptions, 45
Ordnance Survey, availability of map
        extracts, 117
Outbuildings, 149
Owls, 130
Owner, rights of, 23

Painting your house, 137
Party Wall etc. Act 1996, 55
    background, 119
    explanatory booklet, 55, 105
    in a nutshell, 54
    useful contacts, 54
Passenger lift, 447, 448, 449, 452, 609
    disabled people, 457, 475, 542, 596
    see Requirements – lifting devices, 448
    see Requirements – passenger lifts, 448
    see Requirements – wheelchair
        platform stairlifts, 451

Passing places, disabled people, 453, 547
Passive infrared detectors, 124
Pathways, 144
Pedestrian route
    see Requirements – access, 598
People accessing facility
People with impaired sight
    access, 544, 550
People with limited physical dexterity,
    access, 556
People with walking aids, access,
    544, 597
Permitted development, 155
    conservatory, 155
    garage, 155
    porch, 155
    right, 101
Petrol, storage tanks, 150
Piped services, 422
Pipistrelle, 129
Plan of work, 19
Planning
    applications – list of, 114
    policies, 103
    responsibility, 103
Planning application
    checklist, 116
    Conservation Area, 114
    decision, 119
    early stages, 112
    erecting an advertisement, 114
    for future development, 120
    important areas, 113
    infringements, 121
    listed building, 114
    objections, 106
    obtaining application forms, 116
    publicity, 115
    register, 114
    submission, 107
    types of plan, 106
    when property not yours, 102
Planning applications, Listed
    Building Consent, 120
Planning authority, 136
Planning consent, 136, 143
Planning control
    administration, 105
    why required, 107
Planning officers
    responsibility, 56
Planning permission, 116–9
    and covenants, 121, 124
    and neighbours, 123
    application, 103

application fee, 116
application form, 117
basic requirements, 107
before you start work, 123
building without permission, 115,
    130, 131
building work requirements, 152
Conservation Area, 114, 138
control, 101, 102
enforcement, 130
examples of where needed, 105
existing, 121
fees, 116
for extensions, 105
listed buildings, 114, 138
objection areas, 120
other approvals, 110
outline, 106
permitted development right, 101
process, 116
protected species, 129
purpose, 107
reason for refusal, 120
refusal – next steps, 120
reserved matters, 106
Tree Preservation Order, 127
types of, 106
validity, 119
what building works needs it?, 108, 137
what it is, 105
when required?, 106
who needs it, 102
Planning Policy Guidance
    Notes, 103
Planning register, 115, 117
Planning system, 101
Plans certificate, 10, 16
Plasterboard
    see Requirements – walls
Plumbing, 141
Podium, 461
Pollution, 122
    protection against, 92
Porch, 155
    building, 148
Portable building, 152
Possible contamination signs, 178
Potential problems, 176
Principal entrance
    see Requirements – access, 545
Privacy, loss of, 122
Professional advice
    before work commences, 46
Propane gas, 150
Prosecution, for non compliance, 59

Protection
  against impact, 99
  from collision with open windows
    etc., 94
  from falling, 93
  impact from doors, 94
  of building, 91
  of liquid fuel, 91
Public address system, 462
Public bodies
  and the Building Act, 10
  final certificate, 10
Public building meaning, 160
Public Health (Control of Disease) Act
    1984, 14
Public Health Act 1875, 29
Public Health Act 1936, 13, 24
Publicity, planning application, 114

Radioactive decay, 177
Radium, 177
Radon, 177
  West Country, 177
Railway land contamination, 177
Rainwater drainage, 88
Ramped access
  disabled people, 433, 457
  see Requirements – access, 432, 458
Ramped approach, 564, 607
Reception area, see Requirements –
    reception area
Refuse chute, 335
Regional Planning Guidance, 103
Regularization certificate, 59
Regulation, 169
  compliance, 170
  guidance, 170
  materials and workmanship, 170
Renovation, 164
Repairs, 138
Replacing, 140
Requirements – access
  accessible entrances, 556, 605
  alternative accessible
  approach gradients, 546, 601
  entrance, 546, 600
  XE Blister surface: see Requirements –
    access
  XE "Surface: blister: see Requirements –
    access blister surface
  and use, 570
  approach, 602, 607
  audience facilities, 457, 618
  audience handrails, 619
  building perimeters, 547, 602

building perimeter, 547
changes in level, 550, 610
cross-fall gradients, 546, 602
parking bays, 547, 603
  disabled people, 548
  key dimensions, 551
design, 542, 602
dome profile, 546
door entry systems, 556, 605
doors to accessible entrances, 30, 35,
    555, 605
driveway, 566
dropped kerb, 547
dwellings, 560
entrance door, 556, 606
entrance lobbies, 605
wheelchair users, 542, 549
external stairs, 552
key dimension
entrance storey, 98
external step profiles, 565
floor materials, 556, 606
from the car parking, 545, 603
from the site boundary, 600
general requirements, 544
glass entrance doors, 559
  marking, 560
handrails, 604
  design, 604
  external ramped access, 604
  external stepped access, 604
  key dimensions, 552
  key requirements, 554
hazards on access routes, 554
hazards, 554, 604
hotels, 457, 620
internal threshold, 566
into a dwelling, 566
level approach, 564, 607
level landing, 556, 610
lobbies, 560, 605
  glazing, 560, 605
  key dimensions, 562
manually operated entrance doors, 606
  force requirements, 558, 605
parking bays, 547, 603
  disabled people, 548
pedestrian route, 545
potential hazards, 545, 604
power operated entrance doors, 559,
    606
principal entrance, 545
ramped access, 549
  dimensions, 547
  key dimensions, 551, 602

ramped approach, 564, 607
reception point, 561
revolving doors, 559, 609
risks, 545, 602
route from the exterior, 556
route width, 605
routes, 612
sign posts, 547, 555
signs for disabled people, 556
step profiles, 552
stepped access, 550
stepped approach, 565
surface, 546
  characteristics, 546
  joints, 547, 603
  material, 547, 605
  width, 545, 604
tactile paving, 546
thresholds, 556
ticket machines for parking bays, 548
to extensions other than dwellings,
  598
upstands and slopes, 556
vehicle crossing points, 545, 602
wheelchair users, 431, 457, 463, 472
Requirements – bathrooms
aim, 536
imposed loads, 195
maximum floor area, 194
sanitary conveniences, 535
size limitations, 200
ventilation, 535
washbasins, 537
Requirements – Buildings maximum
  height, 194
Requirements – cavities and concealed
  spaces
cavity barriers, 516
cavity openings, 516
interrupting, 516
Requirements – ceilings, 365
suspended, 365
Requirements – cellars, 236
existing, 236
new constructions, 237
smoke outlets, 237
venting, 237
Requirements – chimneys, 386
bases for back boilers, 413
bends in flues, 396
connecting fluepipes, 393, 401
construction, 390
construction of fireplace gathers, 403
construction of hearths, 405
down draughts, 390

dry lining around fireplace openings,
  396
end restraints, 387
factory made metal chimneys, 394
fire stopping, 389
fireplace lining components, 408
fireplace recesses, 407
fireplaces, 391
fireplaces for gas fires, 413
flue systems, 395
flueblock chimneys, 392, 413
flues etc., 388
flues in chimneys, 399
functions of a hearth, 391
gas burning devices, additional
  provisions, 409
hearths, 392
hearths of oil appliances, 414
height of flues, 399
kerosene and gas oil burning devices,
  413
large and unusual chimneys, 400
masonry and flueblock chimneys, 402
masonry chimneys, 388
  change of use, 392
  proportions, 388
notice plates (for hearths and flues),
  397
open fire with throat and gather, 398
performance capability, 387
protection of building, 386
relining, 394
relining chimney flues (for oil
  appliances), 414
repairs of flues, 393
re-use of existing flues, 394
shingled roof, flue outlet clearances,
  400
size of flues, 398
  boilers, 398
  cookers, 398
  stoves, 398
solid flue appliances, flue outlet
  positions, 401
solid fuel, additional provisions, 397
thatched roof, flue outlet clearances,
  400
walls adjacent to hearths, 409
Requirements – combustion appliances,
  500
air supplies, 501
air vents, 503
boiler control interlocks, 507
conservation of fuel and energy, 509
flues, 505

Requirements – combustion appliances
        (*Continued*)
    hot water systems, 507
    insulation of pipes and ducts, 508
    operating and maintenance
            instructions, 507
    protection of buildings, 500
    provision of information, 500
    replacement of controlled services
        heating boilers, 508
        hot water vessels, 509
        storage controls, 509
    space heating system controls, 506
    timing controls, 507
    zone controls, 506
Requirements – conservatories, 570
Requirements – corridors and
        passageways, 453
    design, 455
    doors, 455
    escape routes, 455
    floor surface, 455
    glazed screens, 455
    ramp gradients, 454
    school buildings, 454
    width, 454, 455
Requirements – disabled people
    access and facilities for disabled
            people, 442
    access for disabled people, 442
    automatic sensors, 559, 606
    clear width requirements, 558, 605
    conservation of fuel and power, 441
    door opening furniture, 558, 606, 612
    door vision panel, 605
    effective clear width, 557, 609
    entrance, 557, 605
        power operated, 559, 606
    fire doors, 445
    frames, 444
    glass, 559, 606
        logos and decoration, 559
    glazed screens, 559, 606
    internal doors
        key dimension, 443
        opening force, 443
    low energy, 446
    minimum clear width, 557
    power operated, 559, 606
        force requirements, 558, 606
    power operated doors, 442–3
    replacement of controlled service,
            doors, 443
    revolving, 559
    safety stop, 559, 606

    self closing, 446
    single glazed panels, 443
    sliding doors, 442
    surface, 445, 558, 606
    upward opening doors, 442
    visibility, 557
    visibility requirements, 557
    visibility zones, 445
    vision panels, 445, 558, 605
Requirements – drainage
    access points, 222
    air admittance valves, 215
    anti flooding valve, 216
    backfilling trenches, 222
    bedding and backfill, 221
    branch discharge pipes, 209, 210, 211
    branch ventilation stacks, 213
    branched connections, 212
    building over existing drains, 206
    building over existing sewers, 232
    cesspools, 205, 224, 225
    commercial kitchens, 215
    condensate, 212
    connection of sanitary pipework, 215
    connections, 217
    connection to public sewers, 217
    depth of cover, 218
    differential settlement pipes, 221
    discharge stacks, 215
    disconnection of existing drain, 204
    drainage fields, 206
    drain trenches, 218
    earth closet, 203
    emergency repair, 204
    existing rules, 203
    filter drains, 230
    fire protection, 205
    foul water, 205, 208
        and rainwater, 233
    greywater, 224
    gutters, 228
    infiltration basins, 231
    inspection chambers, 223
    manholes, 223
    materials for pipes and jointing, 221
    multi connection, 203
    packaged treatment works, 227
    paved areas, 230
    pipe access points, 217
    pipe supports, 215
    pipes and sizes, 218
    pipes penetrating a structure, 209
    pipes serving a single appliance,
            212
    propriety seals for pipes, 208

protection from settlement, 218
protection of openings, 208
pumping installations, 219
rainwater, 228
rainwater gutters and pipes, 228
rainwater tanks, 228
refuse water, 203
repairs to existing drains, 204
repairs to existing WCs, 204
rodding points, 212, 215
rodent barriers, 217–8
separate systems, 206–7, 233
septic tanks, 205, 224
sewers, 224
soakaways, 231
solid waste, 233
solid waste storage, 207
structural fire resistance, 207
stub stack, 215, 216
support of pipes, 215
surface water, 230–1
swales, 231
system layout, 217
systems, 205
traps, 84, 209
trenched flexible pipes, 222
trenched rigid pipes, 221
ventilating pipes, 214
ventilating stack, 213
ventilation, 215
ventilation of soil pipes, 204
waste containers, 233
wastewater treatment systems, 224
WCs, 202
Requirements – entrances and
    courtyards, 541
Requirements – extensions to buildings,
    566
conservation of fuel and power, 566
Requirements – external balconies, 568
Requirements – facilities
audience, 457
    wheel chair users
handrails – key dimensions, 432, 458
    wheel chair spaces, 431, 456, 459
lecture/conference, 457, 460
    acoustic environment, 461
    artificial lighting, 461
    hearing enhancement systems, 462
    induction loop, 462
    infrared hearing, 462
    public address system, 462
    telephones, 462
    text telephones, 462
    uplighters, 461

leisure, 462
refreshments, 463
restaurants, 463
Requirements – fire resistance
access facilities for the fire service,
    523
aim, 520
factors to be taken into account, 521
fire resistance, 523
fire safety precautions, 520
measures to be incorporated, 521
Requirements – floors
concrete, intermediate floor, 253
Requirements – floors and ceilings,
    238
airborne and impact sound, 239, 255,
    287
alterations and fire safety, 259
ceiling height, 256
ceiling joists, 251, 256
ceiling junctions, 256
ceiling treatment A, 262
ceiling treatment B, 262
ceiling treatment C, 262
ceiling void, 256
ceilings
    and roof junctions, 256
    ceiling treatment A, 263
    ceiling treatment B, 263
    ceiling treatment C, 263
    mass per unit area, 263
    lath and plaster, 256
    junctions with external cavity
        walls, 256
    junctions with independent panels,
        256
    height, 256
    support, 256
    junctions with separating wall, 256
    with wall type 1, 268
    with wall type 2, 269
    with wall type 3, 269
    with wall type 4, 270
ceilings – general, 255
compartmentation, 247, 252
concrete, 238, 253
concrete floors, 291
concrete slab floor, 253
construction, 239
damp proof membrane, 238
design, 239
direct and flanking transmission of
    sound, 257
early warning of fire, 239
enclosures for drainage, 254

Requirements – floors and ceilings
  (*Continued*)
enclosures for water supplies, 254
external cavity walls, floor, 265
fire precautions, 239
fire resistance, 252
fire safety – penetrations, 276
fire stopping – penetrations, 267,
  323
floating layers, type 2 floors, 261
  floor bases, 254
floorboarding, 243
floorboards, 238
floor joists, 238, 259
floor junctions and masonry walls
  between isolated panels, 264
floor penetrations, floor type 1, 267
floor penetrations, floor type 2, 275
floor penetrations, floor type 3, 281
floors
  ceiling treatment – A, 366
  ceiling treatment – B, 366, 575
  ceiling treatment – C, 366, 575
  ceiling treatments, 255, 262, 263
  general requirements, 264, 278
  type 1 – floor penetration, 267, 423
  type 1 – general, 255
  type 1 – junctions, 267
  type 1 – junctions with gas pipes,
    267
  type 1 – junctions with wall type 1,
    276
  type 1 – junctions with wall type 2,
    277
  type 1 – junctions with wall type
    3.1 and 3.2, 277
  type 1 – junctions with wall type
    3.3, 278
  type 1 – with ceiling B, 255
  type 1 – with ceiling C, 255
  type 2 – floating floor, 261
  type 2 – junctions with floor
    penetrations, 275
  type 2 – junctions with gas pipes,
    267
  type 3 – floating layer, 262
  type 3 – general requirements, 278
  type 3 – junctions with an external
    cavity wall, 280
  type 3 – junctions with external
    solid masonry wall, 280
  type 3 – junctions with floor
    penetrations, 281
  type 3 – junctions with internal
    framed wall, 355

type 3 – junctions with internal
    masonry wall, 281
type 3 – junctions with wall type 1,
    282
type 3 – junctions with wall type 2,
    283
type 3 – with an external, 280
floors, 256
  type 1, 261
ground floor, 238
ground moisture, 240
ground supported floor, 240
suspended concrete ground floor, 244
suspended timber ground floor, 242
health and safety, 183
houses with floors above 4.5 m, 246
imposed loads, 196
interruption of lateral support, 251
junctions for concrete based type 1
    floors, 264
junctions for concrete based type 2
    floors, 270
junctions for timber based type 3
    floors, 278
junctions with floor penetrations, 275
junctions with penetrating gas pipes,
    275
lateral support, 247
lateral support by floors, 247
loading on walls, 292
materials, 286
maximum floor area, 194
maximum floor span, 248
maximum height of building, 194,
    288, 371
means of escape, 245
new building, 261
  doors, 583
  floor type A or B, 273
  floor type C, 271
  internal, 579
  internal floor type B, 580
  internal wall type C, 580
  meeting the requirements, 265
  noise reduction, 583
  performance standards, 262
  requirements, 262
  stairs, 583
remedial work and conversions, 587
  floating layer, floor treatment, 590
remedial work and conversions, floor
    treatment, 284
remedial work and conversions,
    pugging, floor treatment, 284
resilient layers, type 2 floors, 261

restraint, concrete, 247
restraint on internal walls, 247
smoke alarms, 239
solid floor, 238
suspended ceilings, 255
suspended concrete floor, 253
suspended concrete ground floor,
    244
suspended floor, 242
suspended timber ground floors, 242
timber, 241
timber fillets, 241
timber floor junctions, 242, 258, 261
timber framed walls, floor type 3,
    278
types of floor bases, floor type 3, 262
types of floor bases, type 2, floors,
    270
types of separating floors, 268
Requirements – foundations
    buttresses, 189
    chemically aggressive soil, 188
    chimneys, 189
    concrete mix, 188
    dead and imposed load, 173
    dimensions, 188
    draining, 186
    imposed loads, 194
    maximum floor level, 194
    maximum height, 194
    minimum thickness, 188
    minimum width of strip, 190
    non-aggressive soils, 187
    on landfill, 181
    piers, 189
    piers and chimneys, 189
    removal of turf, 176, 184
    stepped elevation, 188
    subsoil, 177, 185
    tree roots, 176
    vertical loading, 192
    vertical ties, 192
    walls, 306
Requirements – garages, 569
    separation between garage and
        dwelling houses, 569
Requirements – handrails, 432, 611
    design, 428, 433, 612
    key dimensions, 432, 612
    positioning, 432, 612
    to external stepped access, 432, 611
    to ramped access, 432, 611
Requirements – hot water storage, 509
    conservation of fuel and power, 509
    insulation of pipes and ducts, 510

replacement of controlled services,
    boiler and hot water storage
    controls, 511
Requirements – internal ramps
    frictional characteristics, 431, 611
    general, 431, 611
    gradients, 431, 611
    handrails, 431, 611
    landings, 431, 611
    surface, 431, 611
Requirements – kitchens and utility
    rooms
    extraction of pollutants, 518
    extraction of water vapour, 518
    fresh air, 518
    ventilation, 518
    ventilation of rooms, 519
Requirements – lifting devices
    ambulant disabled people, 447, 608
    floor, 447, 608
    general requirements, 447, 608
    landing call buttons, 609
    lifting platforms, 608
    listed buildings, 608
    location, 33, 447, 608
    passenger lifts, 609
    signs, 447, 608
Requirements – lifting platforms,
    450, 609
    controls and buttons, 450, 608
    doors, 450
    instructions for use, 450, 609
    landing call buttons, 450, 609
    location of controls, 450, 609
    minimum dimensions, 450, 610
    permitted speed, 450, 609
    vertical travel, 450, 609
    wall surfaces, 451, 611
    wheelchair users, 446, 449, 610
Requirements – electrical safety, 492
    access and facilities for disabled
        people, 493
    accessible switches and sockets, 618
    competent firm, 649
    competent person, 649
    component fitting, 498
    conductor, 495–6
    conservation of fuel and energy, 492
    consumer unit, 498
    contractor or an installer, 654
    design, 493
    distribution responsibility, 494
    earthing, 495
    electrical components, 496–7
    electrical fittings, 498

Requirements – electrical safety
(*Continued*)
  equi-potential bonding, 496
  external lighting, fixed to buildings,
    500
  friend, relative or myself, 654
  heights of switches and sockets, 621
  installation, 654
    inclusion in the records, 655
    types, 654
  installation work, 497–8
  internal lighting, 499
  lighting, 499
    external, 500
    internal, 499
  low authority building control body,
    650
  material alteration, changes, 494–5
    electrical installation, 495
  statutory requirements, 647
  wall mount, socket outlet, 498–9
  wiring types/wiring system, 496
  work completion, 654
Requirements – liquid fuel, 511
  location of LPG cylinders, 513
  location of tank, 513
  LPG storage, 513
  oil pollution, 512
  storage and supply, 511
Requirements – loft conversions
  access and facilities for disabled
    people, 543
  protection from falling, 541
  stairs ladders and ramps, 538
  steps, 540
Requirements – materials and
    workmanship, 583
  access for disabled people
  choice of materials, 585
  codes of practice, 586
  conservation of fuel and power, 584
  continued control, 584
  environmental impact, 584
  fitness for use, 584
  health and safety, 584
  limitations, 584
  properties, 585
  recyclable materials, 584
  resistance to moisture, 585
  resistance to other harmful materials,
    585
  schedule 7, 12
  short lived materials, 585
Requirements – means of escape, 524
  buildings other than dwellings, 533

fire alarms, 528
  above 4.5 m, 531
  balconies, 529
  basements, 530
  dwellings, 528
  emergency egress, 530
  flat roofs, 529
  inner rooms, 529
fire detection and alarms,
    dwellings, 525
flats, 532–3
horizontal escape, 533
loft conversions, 531
maisonettes, 532–3
smoke alarms, 526
  dwellings, 526
  installation, 527
  other than dwellings, 528
  power supplies, 527
uninsulated glazed elements on
    escape routes, 534
vertical escape, 534
Requirements – passenger lifts, 449, 609
  car controls, 449, 609
  closing devices – timed, 449, 609
  key dimensions, 449, 609
  landing call buttons, 449, 609
  minimum dimensions, 610
  reopening activators, 449, 609
  size, 449, 609
  sliding doors, 449, 609
  wall surfaces, 449, 609
Requirements – passing places, 453, 547
Requirements – reception area
  availability of information, 562, 612
  floor surface, 562, 612
  location and accessibility, 561, 611,
    612
  manoeuverability, 562, 612
  reception desk, 562, 612
Requirements – refuse facilities, 519
  refuse chutes, 520
Requirements – residential buildings
  maximum height – annexes, 201
  maximum height – single storey, 201
Requirements – roofs, 368
  compartmentation, 376
  fire precautions, 368
  flat roofs with bitumen felt, 378
  imposed loads on roofs, 376
  interruption of lateral support, 375
  jack rafters, 372
  joists for flat roofs, 372
  materials for roof covering, 378
  maximum height of buildings, 371

means of escape, 375
moisture resistance, 369
precipitation, 369
pitched roofs, 377
pitched roofs, covered with self
    supporting sheet, 378
plastic roof lights, 377
purlins, 321, 368, 372
rafters, 368
rating of materials and products, 379
roof covering, 376
roof joists, 374
roof with a pitch >15°, 380
slates, 377
thatch, 377
tiles, 377
trussed rafter roofs, 375
ventilation, 380
ventilating roof void, ceiling
    following pitch of roof, 381
with a pitch >15°, 380
wood shingles, 377
Requirements – rooms for residential
    purposes, 576
corridor doors, 574
corridor walls, 574
doors, 574
fire safety, 575
flanking transmission, 574
floor type, 575
material change of use, 580
    ceiling void, 577
    design considerations, 577
    junction details, 576
    meeting the requirements, 576
    performance standards, 580
    requirement, 577
    reverberation – method A, 577–8
    reverberation – method B, 577–8
    reverberation – sound absorbers, 578
    reverberation – stair enclosure, 578
    reverberation – stair treads, 578
    reverberation – stairwells, 578
    reverberation, 577
    roof space, 576
    sound insulation, 577
meeting the requirement, 576
Requirements – sanitary conveniences
accessibility, 479
accommodation, 479
baby changing, 477
bathrooms, 491
buildings other than dwellings, 456–7
changing facilities, 485
children's urinal, 482

colour combinations, 603
dimensions, 486
    wheelchair accessible unisex
        toilets, 477
door opening furniture, 558, 606, 612
doors, 479, 484
    wheelchair accessible unisex
        toilets, 482
dwellings, 517
emergency assistance alarm system,
    481, 615
emergency assistance alarm, 481, 615
entrance storey, 491
fire alarms, 481, 525
fittings and fixtures, 613
heat emitters, 479, 613
separate sex toilet accommodation, 612
shower facilities, 484
sports amenities, 485
support rails, 481
switches, outlets and controls, 613
toilet accommodation, 612
typical heights and arrangements, 480
washrooms, 482
WC cubicles, 483
WC pans, 482
wheelchair accessible unisex toilets,
    477
Requirements – sleeping
    accommodation, 411
bedrooms, 413, 465
circuit isolator switches, 467
colours, 469
controls, 467
emergency alarm pull cords, 468
outlets, 467
sanitary facilities, 467, 475
telephone points, 622
TV sockets, 622
wheelchair accessible, 477
Requirements – sound insulation
airborne sound, 239
failed tests, 645
impact sound, 239
internal floors, 261
internal walls, 247
laboratory values – new internal
    walls, 260
meeting the requirements, 260
new walls and floors, 253
remedial treatment, 284
resistance to sound, 258, 330
sets of tests, 259
sound absorption, 260
testing, 285

Requirements – stairs, 416
  access and facilities for disabled
    people, 417
  access for disabled people, 417
  airborne and impact sound
    requirements, 416
  airborne and impact sound, 416
  common stairs, 433
  dimensions, 425
  disabled people, 427
  dwellings, 417, 437
  early warning of fire, 521
  entrance storey, 453
  fire safety, 422
  flanking transmission, 418
  for loft conversions, 427
  handrails, 425
  impact sound, 419
  in flats, 419
  internal, 429
    dimensions, 425
    doors, 429
    handrails, 432
    hazards, 430
    key dimensions, 432
    school buildings, 430
    stepped access, 429
    width, 426
  pedestrian guarding, 417
  performance standards, 580
  piped services, 422
  protection from falling, 428
  provisions, 433
  ramps, 425
  remedial work and conversions,
    419
  reverberation – method A, 578
  rise and going, 425
  safety, 424
  smoke alarms, 522
  soft covering, 580
  sound insulation, 572
    testing, 643
  stair treads, 578
  stair treatment, 592–3
  stairwell, 578
  steps, 426
  with cupboard under, 593
Requirements – storage of food, 518
Requirements – ventilation, 518, 535
  fresh air, 518
  pollutant extraction, 535
  requirements of openable windows,
    522
  vapour extraction, 202

Requirements – walls, 286–364
  absorbent material, type 4 walls, 359
  adjacent to hearths, 364
  airborne and impact sound, 287
  and roof, 351
  basic requirements, 288
  building height, 288
  buttressing, 299
  buttressing walls, openings, 299, 300
  buttressing walls, piers, 297, 298
  cavity between isolating panels, 352
  cavity insulation, 287, 311
  cavity masonry, 344
  cavity masonry, type 2, 344
  cavity masonry walls, 334, 344
  cavity separating walls, concrete floor
    based junctions, 349–350
  cavity walls, 294
  cavity wall ties, maximum spacing, 290
  ceiling junctions, 357, 362
  cavity walls, 294
  cavity widths, 334, 344
  chases, 301
  chimneys, 300
  cladding, 313–4
  classification of linings, 319
  compartmentation, 319–320
  compartment wall, 319, 320
  concrete floor based junctions, 349
  construction, brick and block, 293
  construction, cavity masonry,
    type 2, 344
  construction, solid wall, 293, 346
  corridor doors
    fire safety, 335
    mass per unit area, 335
    sound insulation, 335
    sound reduction index, 335
  corridor walls, 335
    flanking transmission, 335
  density of walls, 333
  direct and flanking transmission, of
    sound, 327
  doors, 335
    fire safety, 286
  end restraints, 387
  external wall junctions, 333, 338
    type 1, 339
    type 2, 344
    type 3, 352
  fire precautions, 239
  fire resistance, 323
  fire resisting elements, 320
  flanking transmission, 335
    wall type 1, 268

flats, 319
floor junctions, 342
  timber framed, 359
flow of sound energy, 327
flue blocks, 337
  internal wall type A or B, 580
  internal wall type A, 580
  internal wall type B, 581
  internal wall type C or D, 582
  internal wall type C, 581
  internal wall type D, 582
flue penetrations, concrete floor based
    junctions, 323
flue walls, 322
foundations, 306
harmful effects, 187
insulating material, 325
internal loading of bearing walls, 292
joist hanger, restraint type, 247
junctions between walls, 339
junctions with internal concrete
    floors, 341
  laboratory values, 330
  mass per unit area, 333
  meeting the requirement, 288
lateral support, 290
lining, 318
loadbearing elements of a structure,
    319
loading, 292
maisonette, 319
masonry cores, type 3, 352
masonry walls between isolated
    panels, 332
masonry wall, construction, 325
maximum difference in permitted
    level, 293
maximum floor span, 248, 292
mortar, 289
new building, 328
noise reduction, 328
openings and recesses, 300
overhangs, 301
panels, type 3, 353
parapet walls, 295
partitions, timber framed walls, 340
partitions, type 1, 339
partitions, type 2, 344
partitions, type 3, 352
performance standards, 328
plasterboard, 333
piers, 298
protection of openings for pipes, 322
refuse chutes, 335
resistance to airborne sound, 332

resistance to passage of moisture, 306
resistance to sound, 330
restraint, by concrete floor or roof, 291
roof junctions, 350, 362
  masonry walls, 356
separation between garage and
    dwelling, 321
sockets and chases, 337
solid masonry, type 1, 336
solid walls, 286, 293
solid wall, construction, 346
sound absorption, 331
sound energy, 327
sound insulation, 328
  sound reduction index, 335
  stairs, 332
surface linings, 319
thickness of walls, 289
tension straps, 290
thickness of external walls, 293
thickness of masonry walls, 294
timber framed, 286
timber framed, type 4, 359
timber frames with absorbent
    material, 359
timber frames with masonry
    cores, 358
timber frames, types of, 340
type 1 – bonded junction, 339
type 1 – general requirements, 337
type 1 – junction requirements, 338
type 1 – junctions with ceiling and
    roof, 350, 357
type 1 – junctions with concrete
    ground floors, 342
type 1 – junctions with external cavity
    wall, 366
type 1 – junctions with external solid
    masonry wall, 343
type 1 – junctions with inner leaf,
    338, 340
type 1 – junctions with internal
    framed wall, 343
type 1 – junctions with timber ground
    floors, 343
type 1 – categories, 336
type 1 – fire stop, 343
type 1 – other types, 343
type 1 – position of openings, 339
type 1.1, 336
type 1.2, 336
type 1.3, 337
type 2 – bonded junction, 347
type 2 – categories, 334
type 2 – fire stop, 351

Requirements – walls (*Continued*)
  type 2 – flanking transmission, 337
  type 2 – flue blocks, 337
  type 2 – general requirements, 337–8
  type 2 – junctions with a solid
      external masonry wall, 351
  type 2 – junctions with an external
      cavity wall, 346
  type 2 – junctions with ceiling roof
      space, 350
  type 2 – junctions with concrete
      ground floors, 340
  type 2 – junctions with internal
      concrete floors, 349
  type 2 – junctions with internal
      framed walls, 363
  type 2 – junctions with internal
      masonry walls, 351
  type 2 – junctions with internal
      timber floors, 349
  type 2 – junctions with timber ground
      floors, 351
  type 2 – roof space, 321
  type 2 – tied junction, 339
  type 2 – wall ties, 331
  type 2 concrete floor based junctions,
      internal wall, 341
  type 2 concrete floor based junctions,
      separating wall, 342
  type 2 concrete floor penetrations, 343
  type 2.1, 344
  type 2.2, 344
  type 2.3, 345
  type 2.4, 345
  type 3 – flanking transmission, 360
  type 3 – flue blocks, 352
  type 3 – general requirements, 352
  type 3 – junctions with an external
      cavity wall, 354
  type 3 – junctions with an external
      solid masonry wall, 359
  type 3 – junctions with an internal
      concrete floor, 356
  type 3 – junctions with an internal
      framed wall, 355
  type 3 – junctions with an internal
      masonry floor, 358
  type 3 – junctions with an internal
      masonry wall, 359
  type 3 – junctions with an internal
      timber floor, 356
  type 3 – junctions with ceiling and
      roof space, 357
  type 3 – junctions with concrete
      ground floors, 359
  type 3 – junctions with timber ground
      floors, 358
  type 3 – sound insulation, 352
  type 3 timber floor based junctions,
      external wall, 355
  type 3 timber floor based junctions,
      heavy masonry leaf, 358
  type 3 timber floor based junctions,
      light masonry leaf, 358
  type 3.1, 353
  type 3.2, 353
  type 3.3, 353
  type 4 – junctions with concrete
      ground floors, 363
  type 4 – junctions with internal
      concrete floors, 363
  type 4 – junctions with internal
      framed wall, 275
  type 4 – junctions with internal
      masonry wall, 363
  type 4 – junctions with internal
      timber floor, 341
  type 4 – junctions with separating
      floors, 351
  type 4 – junctions with solid masonry
      wall, 363
  type 4 – junctions with timber ground
      floors, 363
  types, 336
  types
      butterfly, 334
      double triangle, 334
  wall ties, 294
Requirements – water (and earth)
    closets, 469
  access and facilities for disabled
      people, 470
  business premises, 471
  disabled access, 475
  erection of public conveniences, 473
  Greater London Area, 471
  loan of temporary sanitary
      conveniences, 472
  protection of openings for pipes,
      473
  provision, 475
  sanitary conveniences, 470
  ventilation, 470, 474
  workplace conveniences, 472
Requirements – water supplies, 235
  fire precautions (structural), 237
  from the Local Authority, 235
  occupied house, 235
  proprietary seals, 236
  protection of pipe openings, 236

supply of wholesome water
    to more than one property, 235
Requirements – wheelchair platform
    stairlifts, 451
    controls, 452
    dimensions, 452
    speed, 452
Requirements – windows, 434
    conservation of fuel and energy,
        436
    glazed walls and partitions, 445
    glazing, 437
    height of controls, 440
    protection against impact, 435, 438
    protection from falling, 435
    replacement of controlled services
        and fittings, 437
    safe access for cleaning, 441
    safe opening and closing, 439
    transparent glazing, 439
Residential building, 371
Resistance
    to ground moisture, 79
    to weather, 79
Resonance
    in parts of the structure, 327
Restaurants and bars, disabled people,
    463
Reverberation, 422, 577
Right to light, 123
Risk assessments, 178
Road access, 122
Robust Standards, 284
Rodent control, see Requirements –
    drainage, 217
Roof, 371–2
    existing structure, 376
    lateral restrains, 374
    lateral support, 373
    openings, 375
    roof re-covering, 376
    timber, 374
    vertical strapping, 372
Roof percipitations, 369
Roof space, 208
    alterations, 224
    fire safety, 256, 259
Room, for residential purposes, 33

SI 2000 No. 2531, 30
SI 2001 No 3335, 74
SI 1998 No 2451, 267, 276, 282
Sanitary accommodation, 475
    accessibility, 479
    disabled people, 475

Sanitary conveniences, 535, 598
    buildings other than dwellings, 600
    disabled, 601
    dwellings, 600
    material changes of use, 638
Sanitary facilities, 612
    disabled people, 615
Sanitary fittings, 476, 613
        Requirements – sanitary
            conveniences
Satellite dishes, 146
Schedule 1, 6, 30, 169, 170
    Building Act, 4
    materials and workmanship, 170
Schedule 2
    Building Act, 7
Schedule 3
    Building Act, 7
Schedule 4
    Building Act, 10
Schedule 5
    Building Act, 11
Schedule 6
    Building Act, 11
Schedule 7
    Building Act, 12
School buildings, 430, 454
    see Requirements – corridors and
        passageways, 454
    see Requirements – stairs, 416
    sound insulation, 418
Scrap yards, 177
    contamination, 179
Seating
    audience – disabled, 457
    disabled people, 460
Security lights, 124
SEDBUK, 508
Separate-sex toilet accommodation
        see Requirements – sanitary
            conveniences
Separating concrete floor, 264
Septic tanks, see Requirements –
    drainage, 88
Sewerage undertaker, 50
    full plans applications, 50
Sewers
    building over, 24, 206
Shower facilities, 487
    disabled people, 488
    see Requirements – sanitary
        conveniences
SI 2001 No 3335, 74
SI 2002 No 2871, 74
SI 2002 No 440, 74

SI 2003/1673, 538
Site
    contaminant moisturing resistance,
        79–80
    preparation, 30, 178
        Approved Document C, 30
Site plan, 119
Sleeping accommodation, 464
    disabled people, 464
Sliding doors, 442
    see Requirements – passenger lifts,
        448
Sizes
    see openings, 249
Smoke outlets, 237
Social Services, 596
Socket outlets, 495
Solar panels, 137
Solid fuel appliances
    exemptions, 402
Solid waste storage, 90
Sound
    absorbers, 578
    absorption, 331
    airborne, 416
    airborne – resistance to, 421
    energy, 259
    impact, 416
    impact – resistance to, 420
    insulation testing, 419
    resistance to, 420
    sound absorption, 421
    sound energy, 418
Sound insulation, 82, 162, 240
    corridor walls, 335
    higher standard for domestic
        purposes, 259
    historic buildings, 135
    impact, 32
    improvement, 21
    method of improvement, 21, 31, 36, 37
    new buildings, 28
    rooms for residential purposes, 33
Sound insulation testing, 419
Sound insulation values, 258
    values – E1, 81
    values – E2, 81
see Requirements – sound insulation, 162
Sound reduction index, 592
Specifications
    technical, 171
Sports amenities
    see Requirements – sanitary
        conveniences
    stable requirements, 197

Stairs, 418
Stair enclosure, 422
Stairlifts
    ambulant disabled people, 429
    disabled people, 451
    people with impaired sight, 433
    see Requirements – passenger lifts,
        448
    see Requirements – stairs
Stair treads, 421
Stairwells, 420, 422
    see Requirements – stairs, 416
Standards approvals, 13, 18, 19
Statutory undertakers and building
        regulations, 168
Stepped access
    disabled people, 429
Stepped approach, 565
    see Requirements – access, 566
Steps
    external step profiles, 565
    see Requirements – access, 566
Stop notices, 115
Storage
    diesel, 150
    oil, 150
    petrol, 150
    tank, 150
Street names, approval, 123
Strip footing, 22
Structural alterations, 5
Structural safety, 197
Structural stability, 7
Structure, 32
    internal fire spread, 76
Structure plan, 102
    buildings other than dwellings, 161
Subsoil
    drainage, 79, 177, 185–6
    types of, 186
Substances, dangerous and offensive, 6
Sun lounge, building a, 152
Supplementary Regulations
    Building Act, 7
Supply of Machinery (Safety) Regulations
        1992, S.I. 1992/3073, 452
Support rail, 481
see Requirements – sanitary
        conveniences
Surface
    blister, 546
        see Requirements – access, 544
Swimming pools, 146
Switched socket outlets
    disabled people, 467, 477

Switches and controls, 467
    see Requirements – sanitary
        conveniences
Symbols
    control button, 448
    disabled access, 475
    facilities (e.g. hearing), 460
    facility signs, 455

Tanks
    external, 150
    rainwater, 150
Technical
    approvals, 170
    specifications, 170
Technical Services Department, 28, 62,
    70
Telephone points, 468
Telephones
    disabled people, 462
    lecture/conference, 461
Tests, local authority, 20
Text telephones, 462
The Airports Authority Act 1975, 12
The Association of Building Engineers
    (ABE), 683
The Atomic Energy Authority Act 1954,
    12
The Building Regulations 2000, 28
The City of London (Various Powers)
    Act 1977, 12
The Civil Aviation Act 1982, 12
The Clean Air Act 1956, 11
The Control of Pollution Act 1974, 12
The Criminal Law Act 1977, 12
The Development of Rural Wales Act
    1976, 11
The Education Act 1944, 12
The Education Act 1980, 12
The Faculty Jurisdiction Measure 1964,
    11
The Fire Precautions Act 1971, 11, 12
The Gas Safety Regulations 1998, 424
The Health and Safety at Work etc. Act
    1974, 12
The Highways Act 1980, 11, 12
The Housing Act 1957, 11
The Interpretation Act 1978, 11
The Local Government (Miscellaneous
    Provisions) 1976, 11
The Local Government (Miscellaneous
    Provisions) 1982, 11
The Local Government Act 1972,
    11, 12
The Local Land Charges Act 1975, 11

The London Government Act 1963, 11,
    12
The Offices, Shops and Railway
    Premises Act 1963, 11
The Public Health (Control of Disease)
    Act 1984, 11
The Public Health Act 1936, 11
The Public Health Act 1961, 11, 12
The Radioactive Substances Act 1960, 12
The Restriction of Ribbon Development
    Act 1935, 11
The Royal Institution of Chartered
    Surveyors (RICS), 70
The Safety of Sports Grounds act 1975,
    11
The Town and Country Planning Act
    1947, 12
The Water Act 1945, 12
The Water Act 1973, 12
The Water Act 1981, 12
The Workplace (Health, Safety and
    Welfare) Regulations 1992, 172
Thermal insulation, 264
Ticket machines (parking)
    see Requirements – access, 548, 603
Tied junction, 339
Toilets
    in separate-sex washrooms, 482, 615
    see Requirements – sanitary
        conveniences, 475
    WC access, 462, 615
Town and Country Planning Act 1947,
    12
Traffic, generation, 122
Tree Preservation Order, 101, 114
    Conservation Area, 102
    meaning, 102
    owner's responsibilities, 125
Trees, 143
    leylandii, 143

UKAEA, 4
    building regulations, 4
Unauthorized work, 59
Understanding assistance, 44
Unisex toilets, 477
    disabled people, 477
    see Requirements – sanitary
        conveniences, 475
Uplighters, 461, 469
Uranium, 177
Urea formaldehyde, 80, 287

Vampire bats, 130
Vehicle barriers, 543

Ventilation
    means of, 203
    roofs, 195
    rooms, 199
Visibility zones, 445
    see Requirements – zones, 445
Visually impaired people
    hazards, 604

Wall
    boundary, 287
    cavity, 271
    garden, 223
    party, 386
Wall cladding 317–8
Wall surfaces, 325
    see Requirements – lifting platforms,
        447
Wall ties, 294
Wall type 1, 268
    see Requirements – floors and ceilings,
        268
Wall type 2, 269
    see Requirements – floors and ceilings,
        270
    see Requirements – wall
Wall type 3, 269
    see Requirements – floors and ceilings,
        280
Wall type 4, 270
Walls
    airborne sound, 239
    impact sound, 239
    see Requirements – rooms for
        residential purposes, 258
    see Requirements – walls
Warning and escape, 161
Washroom, 482, 484
    wheelchair accessible, 484
    see Requirements – sanitary
        conveniences

Wastewater treatment systems, 205
Water prevention of contamination, 171
Water Act 1945, 12
WC cubicles,
    see Requirements – sanitary
        conveniences
WC pans, 482
    see Requirements – sanitary
        conveniences
    disabled people, 482
Wheelchair platform stairlifts, 451
    see Requirements – wheelchair
        platform stairlifts, 451
Wheelchair users, 431, 444, 450, 451
    access to a dwelling, 470
    see Requirements – corridors and
        passageways, 453
    see Requirements – internal ramps,
        431, 454
    see Requirements – lifting platforms,
        447
    see Requirements – sanitary
        conveniences
Wheelchair-accessible unisex toilets,
    475
    see Requirements – sanitary
        conveniences – unisex toilets, 475
Wildlife and Countryside Act 1981, 123,
    129
Window
    bay, 137
    exemptions, 133
    from collision with, 94
    protection, 142
    replacement, 139
    safe access for cleaning, 99
    safe opening and closing of, 99
Work, 58
    needing approval, 42–3
    unauthorized, 59
World Heritage site, 639